T0350492

Principles and Applications of Distributed Event–Based Systems

Annika M. Hinze
University of Waikato, New Zealand

Alejandro Buchmann
University of Waikato, New Zealand

A volume in the Advances in Systems
Analysis, Software Engineering, and High
Performance Computing (ASASEHPC)
Book Series

Information Science
REFERENCE
An Imprint of IGI Global

Director of Editorial Content:	Kristin Klinger
Director of Book Publications:	Julia Mosemann
Acquisitions Editor:	Lindsay Johnston
Development Editor:	Julia Mosemann
Publishing Assistant:	Keith Glazewski
Typesetter:	Michael Brehm
Production Editor:	Jamie Snavely
Cover Design:	Lisa Tosheff

Published in the United States of America by
Information Science Reference (an imprint of IGI Global)
701 E. Chocolate Avenue
Hershey PA 17033
Tel: 717-533-8845
Fax: 717-533-8661
E-mail: cust@igi-global.com
Web site: http://www.igi-global.com

Library of Congress Cataloging-in-Publication Data

Principles and applications of distributed event-based systems / Annika M. Hinze and Alejandro Buchmann, editors.
    p. cm.
 Includes bibliographical references and index.
 Summary: "This book showcases event-based systems in real-world applications, providing professionals, researchers, and students in systems design with a rich compendium of latest applications in the field"--Provided by publisher.
 ISBN 978-1-60566-697-6 (hardcover) -- ISBN 978-1-60566-698-3 (ebook) 1. Electronic data processing--Distributed processing. 2. Application software--Development. 3. System design. 4. Sensor networks. I. Hinze, Annika. II. Buchmann, Alejandro P.
 QA76.9.D5P755 2010
 004'.36--dc22
                        2010024436

This book is published in the IGI Global book series Advances in Systems Analysis, Software Engineering, and High Performance Computing (ASASEHPC) Book Series (ISSN: 2327-3453; eISSN: 2327-3461)

British Cataloguing in Publication Data
A Cataloguing in Publication record for this book is available from the British Library.

All work contributed to this book is new, previously-unpublished material. The views expressed in this book are those of the authors, but not necessarily of the publisher.

Advances in Systems Analysis, Software Engineering, and High Performance Computing (ASASEHPC) Book Series

Vijayan Sugumaran
Oakland University, USA

ISSN: 2327-3453
EISSN: 2327-3461

MISSION

The theory and practice of computing applications and distributed systems has emerged as one of the key areas of research driving innovations in business, engineering, and science. The fields of software engineering, systems analysis, and high performance computing offer a wide range of applications and solutions in solving computational problems for any modern organization.

The **Advances in Systems Analysis, Software Engineering, and High Performance Computing (ASASEHPC) Book Series** brings together research in the areas of distributed computing, systems and software engineering, high performance computing, and service science. This collection of publications is useful for academics, researchers, and practitioners seeking the latest practices and knowledge in this field.

COVERAGE

- Computer Graphics
- Computer Networking
- Computer System Analysis
- Distributed Cloud Computing
- Enterprise Information Systems
- Metadata and Semantic Web
- Parallel Architectures
- Performance Modeling
- Software Engineering
- Virtual Data Systems

IGI Global is currently accepting manuscripts for publication within this series. To submit a proposal for a volume in this series, please contact our Acquisition Editors at Acquisitions@igi-global.com or visit: http://www.igi-global.com/publish/.

Titles in this Series

Service-Driven Approaches to Architecture and Enterprise Integration
Raja Ramanathan (Independent Researcher, USA) and Kirtana Raja (Independent Researcher, USA)
Information Science Reference • copyright 2013 • 367pp • H/C (ISBN: 9781466641938) • US $195.00 (our price)

Progressions and Innovations in Model-Driven Software Engineering
Vicente García Díaz (Universidad de Oviedo, Spain) Juan Manuel Cueva Lovelle (University of Oviedo, Spain) B. Cristina Pelayo García-Bustelo (University of Oviedo, Spain) and Oscar Sanjuan Martinez (University of Oviedo, Spain)
Engineering Science Reference • copyright 2013 • 352pp • H/C (ISBN: 9781466642171) • US $195.00 (our price)

Knowledge-Based Processes in Software Development
Saqib Saeed (Bahria University Islamabad, Pakistan) and Izzat Alsmadi (Yarmouk University, Jordan)
Information Science Reference • copyright 2013 • 318pp • H/C (ISBN: 9781466642294) • US $195.00 (our price)

Distributed Computing Innovations for Business, Engineering, and Science
Alfred Waising Loo (Lingnan University, Hong Kong)
Information Science Reference • copyright 2013 • 369pp • H/C (ISBN: 9781466625334) • US $195.00 (our price)

Data Intensive Distributed Computing Challenges and Solutions for Large-scale Information Management
Tevfik Kosar (University at Buffalo, USA)
Information Science Reference • copyright 2012 • 352pp • H/C (ISBN: 9781615209712) • US $180.00 (our price)

Achieving Real-Time in Distributed Computing From Grids to Clouds
Dimosthenis Kyriazis (National Technical University of Athens, Greece) Theodora Varvarigou (National Technical University of Athens, Greece) and Kleopatra G. Konstanteli (National Technical University of Athens, Greece)
Information Science Reference • copyright 2012 • 330pp • H/C (ISBN: 9781609608279) • US $195.00 (our price)

Principles and Applications of Distributed Event-Based Systems
Annika M. Hinze (University of Waikato, New Zealand) and Alejandro Buchmann (University of Waikato, New Zealand)
Information Science Reference • copyright 2010 • 538pp • H/C (ISBN: 9781605666976) • US $180.00 (our price)

www.igi-global.com

701 E. Chocolate Ave., Hershey, PA 17033
Order online at www.igi-global.com or call 717-533-8845 x100
To place a standing order for titles released in this series, contact: cust@igi-global.com
Mon-Fri 8:00 am - 5:00 pm (est) or fax 24 hours a day 717-533-8661

Table of Contents

Detailed Table of Contents

Avigdor Gal, Israel Institute of Technology, Israel
Ethan Hadar, CA Labs, Israel

The authors introduce an architecture for a generic system for performing complex event processing. They use a grid approach to achieve scalability and modularity: A system is represented by a grid of complex event processing components amongst which the processing of an event is divided. The fundamental components include event collection, event purification, event storage, event inference, and event situation management. The architecture is presented in two case studies on early warning systems in disaster management and on service availability management.

Rolando Blanco, University of Waterloo, Canada
Paulo Alencar, University of Waterloo, Canada

This chapter compares event models in various distributed event-based systems and in other implicit invocation systems. The event model is identified as a central part of each system as it determines each system's event concept and how events are generated and notified to consumers. The chapter's main focus lies in identifying common features between all systems that may need to be supported in the engineering of any event-based application. The chapter proposes a categorization for event models.

Eiko Yoneki, University of Cambridge, UK

The author suggests methods for temporal and spatial correlation between events as needed, for example, in ubiquitous computing. The chapter introduces an event model that uses interval-semantics for event correlation considering spatio-temporal event information. The model supports stateful subscriptions, parameterization, and aggregation. The authors show how controlled event consumption mechanisms (e.g., time restriction, subset policies, and use of durative events) can reduce the event buffer for composite event detection.

Chapter 4

Opher Etzion, IBM Haifa Research Lab, Israel

The chapter explores the temporal aspects of event processing. It introduces the notion of an event processing network as the underlying model behind event processing. This event processing network is being executed on a collection of processors that span different time zones. The author identifies and discusses temporal aspects of the execution and within the underlying event model. Temporal aspects that are discussed in the chapter are: temporal dimensions of events, time granularities, temporal context, temporal patterns, event order, and retrospective and proactive operations.

Chapter 5

Daniel Cutting, University of Sydney, Australia
Aaron Quigley, University College Dublin, Ireland

The authors address the problem of load balancing for event-based messaging. The authors propose the use of a structured p2p overlay network to ensure that the participating peers perform only limited work and store limited state information. The overlay network ICE uses hierarchical tesseral addressing (using regions of space instead of Carthesian coordinates) and an amortized multicast routing algorithm. The suggested load balancing is shown to be independent of the number of peers, popularity of subscriptions, or frequency of messages.

Chapter 6

Jean Bacon, University of Cambridge, UK
David Eyers, University of Cambridge, UK
Jatinder Singh, University of Cambridge, UK

The authors examine security techniques for different types of event-based systems. The chapter first discusses methods for application-level security. Then security measures within a large-scale distributed broker network are introduced, including distributed cryptographic key management. Thirdly, broker-specific policy enforcement mechanisms are evaluated, where context-sensitive event processing allows more than binary access control. The authors identify a number of open issues in security for event-based systems.

Chapter 7

This chapter discusses testing of distributed publish/subscribe systems. In addition to testing separate components, the chapter proposes two techniques for testing the component integration. One is based on large numbers of random event sequences to identify malfunctioning system states. The second one uses predefined data-flows to be tested together with unit-testing. The authors also discuss runtime testing of component integration. The suggested techniques are exemplified in a small case study on testing in Maritime Safety and Security (MSS) systems.

Chapter 8

The chapter gives a detailed overview of the Padres publish/subscribe system. The authors describe Padres' content-based routing mechanism and discuss techniques to ensure system robustness. The Padres event model supports filtering, aggregation, and correlation of historic and future event data. The system's architecture consists of distributed and replicated data repositories that can be configured according to criteria such as availability, storage and query overhead, query delay, load distribution, parallelism, redundancy and locality. The chapter describes Padres' techniques to ensure robustness, in particular the use of alternate message routing paths, load balancing, and fault resilience techniques in case of broker failures.

The authors introduce a publish/subscribe middleware for wireless sensor networks (WSN). The authors address two issues in traditional WSN: lack of interoperability with access networks and lack of flexibility to customize non-functional properties. For example, an event detection application may require in-network event correlation and filtering as its non-functional properties in order to eliminate false positive sensor data in the network. The chapter gives a detailed description of the implementation of the proposed middleware TinyDDS. TinyDDS allows applications to interoperate regardless of programming languages and protocols across the boundary of WSNs and access networks. It also gives

WSN applications control over application-level and middleware-level non-functional properties such as event filtering, data aggregation and routing.

Chapter 10

John Keeney, Trinity College Dublin, Ireland
Dominic Jones, Trinity College Dublin, Ireland
Song Guo, Trinity College Dublin, Ireland
David Lewis, Trinity College Dublin, Ireland
Declan O'Sullivan, Trinity College Dublin, Ireland

The chapter discusses Knowledge-Based Networking: the routing of events based on filters applied to some semantics of the event data and associated metadata. The chapter introduces techniques of filtered dissemination of semantically enriched knowledge over a loosely coupled network of distributed heterogeneous agents. Ontological semantics is incorporated into event messages to allow subscribers to define semantic filters. The ontological concepts are encoded as additional message attribute type, onto which subsumption relationships, equivalence, type queries and arbitrary ontological relationships can be applied. The chapter authors explore these ontological operators and their application to a publish/subscribe application.

Chapter 11

Christof Fetzer, Dresden University of Technology, Germany
Andrey Brito, Dresden University of Technology, Germany
Robert Fach, Dresden University of Technology, Germany
Zbigniew Jerzak, Dresden University of Technology, Germany

The authors propose a distributed event processing system for large clusters. Instead of traditional event stream pipes, a content-based publish/subscribe middleware is used for the communication between components. To parallelize processing, the system supports speculative execution of events. The execution of CPU intensive components is spread across multiple cores. Resulting ordering conflicts are detected dynamically and rolled back using a Software Transactional Memory. The authors suggest an extension of the pub/sub mechanism to support load balancing across multiple machines.

Chapter 12

Anton Michlmayr, Vienna University of Technology, Austria
Philipp Leitner, Vienna University of Technology, Austria
Florian Rosenberg, Vienna University of Technology, Austria
Schahram Dustdar, Vienna University of Technology, Austria

The chapter proposes an event-based approach to Service-oriented Architectures (SOA) and Web services. The services in an SOA are changing regularly, which creates a challenge to loosely coupled service providers and consumers. Using the publish/subscribe style of communication service consumers can be notified when such changes occur. This chapter describes an approach of an event-based Web service runtime environment based on a rich event model and different event visibilities. The authors describe the full service lifecycle, including runtime information concerning service discovery and service invocation, as well as Quality of Service attributes.

Chapter 13

André Appel, Clausthal University of Technology, Germany
Holger Klus, Clausthal University of Technology, Germany
Dirk Niebuhr, Clausthal University of Technology, Germany
Andreas Rausch, Clausthal University of Technology, Germany

The authors describe how event-based interaction could be used in Dynamic Adaptive Systems. These systems consist of interacting components from different vendors; they have to be automatically composed at runtime to deal with components joining or leaving the system. Traditional event-based interaction is lacking explicit directed dependencies that are needed to derive a system configuration. This chapter proposes an approach for realizing Dynamic Adaptive Systems using event-based interaction among the components while maintaining automatic system composition. The system use is demonstrated for a smart city application.

Chapter 14

Tom Van Cutsem, Vrije Universiteit Brussel, Belgium
Wolfgang De Meuter, Vrije Universiteit Brussel, Belgium

Cutsem and Meuter argue that event-driven approaches are particularly suitable to address the challenges of mobile and ubiquitous computing, and mobile ad hoc networks. The authors observe that since most programming languages do not provide built-in support for event-driven programming, these concepts are often implemented using different paradigms. This chapter studies the difficulties of combining events with the object-oriented paradigm. Various issues at the software engineering level are highlighted and as a solution, the new object-oriented programming language AmbientTalk is proposed. The language is intended for composing software components across a mobile ad hoc network and supports first-class abstractions to represent event-driven communication.

Chapter 15

Guanhong Pei, Virginia Tech, USA
Binoy Ravindran, Virginia Tech, USA

This chapter also addresses publish/subscribe communication in mobile ad-hoc networks. In particular, this chapter focuses on ensuring quality-of-service characteristics such as end-to-end timeliness, reliability properties, and limited message overhead. The chapter provides an overview of interconnection topologies, event routing schemes and architectural support of publish/subscribe systems in MANETs. The authors propose the use of event routing in a two-layer scheme using hierarchical tree-based connections in the Manet. They additionally introduce a dynamic system self-reconfiguring ability for event causal dependencies.

Chapter 16

Ling Liu, Georgia Institute of Technology, USA
Bhuvan Bamba, Georgia Institute of Technology, USA
Myungcheol Doo, Georgia Institute of Technology, USA
Peter Pesti, Georgia Institute of Technology, USA
Matt Weber, Georgia Institute of Technology, USA

The authors introduce is a standing spatial trigger with reference to a spatial region, actions to be taken, and a list location-based triggers to support location-aware mobile applications. A location-based trigger of recipients. The author describes the mTrigger framework in three alternative architectures for location triggers: a client-server architecture, a client-centric architecture for management within mobile environments, and a decentralized peer-to-peer architecture for collaboration between mobile users. Performance optimizations are proposed for each of the architectures, including a number of energy-efficient spatial trigger grouping techniques for optimizing both wake-up times and check times of location trigger evaluations.

Chapter 17

Sasu Tarkoma, Helsinki University of Technology, Finland & Nokia NRC, Finland
Jani Heikkinen, Helsinki University of Technology, Finland
Jilles van Gurp, Nokia NRC, Finland

This chapter discusses an event-based infrastructure that combines the Session Initiation Protocol (SIP) and Web technologies to realize secure converged mobile services. The chapter uses the scenario of proactive information delivery to airline customers, which has been developed in collaboration with an airline company. The distributed system architecture is outlined and the use of the two key technologies is evaluated for the given scenario. The chapter reports about the authors' experience in combining the various asynchronous communication mechanisms that are commonly used in today's Internet in a realistic mobile Internet context.

Chapter 18

Katrine Stemland Skjelsvik, University of Oslo, Norway
Vera Goebel, University of Oslo, Norway
Thomas Plagemann, University of Oslo, Norway

This chapter is motivated by the need to provide event-based notifications (ENS) in rescue and emergency applications. The authors focused on sparse Mobile Ad-hoc Networks (MANETs), since emergency applications have to run, for example, in places lacking infrastructure where the density of nodes may be low, or where physical obstacles may limit the transmission range. The chapter analyzes characteristics of rescue operations to discuss ENS design choices such as subscription language, architecture and routing. The authors also present their own Distributed Event Notification Service (DENS), which is specifically tailored for this application domain.

Chapter 19

    Annika Hinze, University of Waikato, New Zealand
    Jean Bacon, University of Cambridge, UK
    Alejandro Buchmann, Technische Universität Darmstadt, Germany
    Sharma Chakravarthy, The University of Texas at Arlington, USA
    Mani Chandi, California Institute of Technology, USA
    Avigdor Gal, Israel Institute of Technology, Israel
    Dieter Gawlick, Oracle Incorporated, USA
    Richard Tibbetts, StreamBase Systems, USA

This chapter is a panel discussion; it aims to capture the opinions of some of the leading researchers from academia and industry about the future of event-based systems. Seven panelists were interviewed; each of them is introduced with a short biography highlighting their professional background from which they approach the field of event-based systems. This allows one to see how different panelists set their individual focus, mirroring their community of origin. The chapter presents their statements and a discussion of the opinions voiced.

Foreword

Many technologists foresee a future in which trillions of sensors will be deployed around the world, forming a "central nervous system for the Earth" that monitors the vital signs of people, of processes and systems, and of the planet itself. Efforts are under way to create Smart Grids, Smart Cars, Smart Traffic Systems, Smart Factories, Smart Cities, a Smart Planet, all relying on the ability to monitor and capture events as they occur in many different sources, process and analyze the events to extract valuable insights, and use these insights to perform actions that improve the systems or processes. This is exactly the functionality that distributed event-based systems – the subject of this timely book – aim to provide.

Over the past two decades, a number of sub-disciplines of Computer Science have created technologies for monitoring and processing events. Each of these areas has created its own paradigms for modeling events and actions; techniques for monitoring, processing, correlating, and analyzing events; and techniques for reacting to events. These range from active database systems, centered around the event-condition-action rule paradigm, and their recent successors, data stream processing systems; complex event processing (CEP); real-time and reactive control systems; publish-subscribe and event-based messaging from the distributed middleware community; situation-action or production rule-based systems; and business activity monitoring and business process intelligence technologies. Much has been written in these communities about models, techniques, systems, and applications for distributed event processing.

This excellent book by Professors Alex Buchmann and Annika Hinze is unique in two respects. First, it provides a common framework to unify the disparate paradigms and approaches from the different communities. This is reflected in the first half of the book, which includes representative contributions about different aspects of distributed event processing, ranging from architectures to temporal models to security issues to techniques for mining event streams. The second half of the book is a selection of papers on many different applications of distributed event processing, including mobile and ubiquitous computing, sensor networks, web services, and emergency response.

I believe this book is an invaluable resource for readers like me, who want both to understand the models, architectures, and techniques underlying distributed event processing systems, and to see these models and techniques in action in real applications.

Umeshwar Dayal
Palo Alto, California
February 2010

Umeshwar Dayal *is an HP Fellow in Intelligent Information Management at Hewlett-Packard Laboratories, Palo Alto, California, where he leads research programs in data warehousing, business intelligence, analytics, and information visualization. Umesh has 30 years of research experience in data management, and has contributed to research and practice in data integration, federated databases, active rule-based systems, business process management, data mining, and visual analytics. He has published over 170 research papers and holds over 30 patents. In 2001, he received (with two co-authors) the VLDB 10-year best paper award for his paper on a transactional model for long-running activities. Prior to joining HP Labs, Umesh was a senior researcher at DEC's Cambridge Research Lab, Chief Scientist at Xerox Advanced Information Technology and Computer Corporation of America, and on the faculty at the University of Texas-Austin. He received his PhD from Harvard University, and Bachelor and Master degrees from the Indian Institute of Science.*

Preface

THE COMMUNITY

Companies are cooperating in many different ways, from supply chains with tight inventory controls to dynamic partnerships and virtual enterprises. This leads to an emerging class of enterprise wide applications in which monitoring of relevant situations and the timely response play an ever increasing role. Parallel to this trend, we are seeing the massive deployment of sensors. The resulting cyber-physical systems can monitor from low-level sensor signals to high-level abstract events, and react to them. Taking the event processing paradigm a step further, novel applications combine sensor data with other sources to infer threats and opportunities.

In (Hinze, Sachs, Buchmann, 2009) we surveyed 20 event-based applications and analyzed their requirements and main features. In this book additional applications are addressed. The combined applications cover a broad spectrum:

- Financial applications, risk management and fraud detection (see chapters 2 and 9)
- Supply chain management, logistics and real time enterprises (see chapter 9)
- Environmental monitoring and warning systems (see chapters 2 and 7)
- Marine security and rescue systems (see chapters 8 and 19)
- Integrated health care systems and emergency response (see chapters 5 and 7)
- Smart environments ranging from smart cars and homes to smart cities (chapters 4, 6 and 17)
- Traffic control ranging from cars to air traffic control (see chapters 16 and 12)
- Information dissemination, alerter mechanisms and information dashboards (see chapter 18)
- Network management and service collaboration (see chapters 10, 11, 13 and 15)

The list of reactive systems, often called sense-and-respond systems, is continuously growing. The reasons for this explosive growth is a combination of increasing demand for timely response to relevant events (even in the face of a torrential flood of information), the ensuing need for scalability, and the requirements to react fast to new threats and opportunities. At the same time, (complex) event processing is maturing as a discipline and gaining widespread attention.

How vibrant the event processing community is can best be seen by its multiple activities: various web sites and blogs serve as portals to relevant information and timely and informed commentary[1] start ups are gaining traction and are opening up new business domains[2], open source projects are making event processing accessible to a broader academic clientele[3], early workshops, such as DEBS (Distributed

Event-based Systems) have come a long way from their early existence at the fringes of major conferences to a full-fledged conference with ACM sponsorship.

Last but not least, there are books that address a broad public such as the pioneering work by Luckham (Luckham, 2002), or the newest book by Chandy and Schulte (Chandy, Schulte, 2009) that gives practical advice to industrial users, as well as books directed more at the academic community and give a snapshot of the state of the art. Among the latter the book by Mühl, Fiege and Pietzuch (Mühl, Fiege, & Pietzuch, 2006) has a strong background in publish/subscribe technologies while the book by Chakravarthy and Jiang (Chakravarthy, Jiang, 2009) focuses more on quality of service in stream processing. The Event Processing Technical Society[4], which was founded in 2008, is making headway in standardization efforts.

Standardization and a common understanding are major issues in an emerging discipline that has its roots in so many areas of computer science, ranging from databases and data mining to artificial intelligence, from distributed and real-time systems to sensor fusion. Even in a book like this, the authors of the various chapters may have slightly different understanding of the terms we all use. In an attempt to further a common understanding we review here a few basic concepts and issues that will be addressed throughout the chapters of this book before addressing the individual contributions.

THE FOUNDATION

An event was defined by Chandy (Chandy, 2006) as a significant change in the state of the universe. Since time is an inherent dimension of the universe, two observations of the universe at different time constitute two distinct events, even if no other properties have changed. Chandy's definition, however, refers to *significant* changes in the state of the universe, thereby limiting the infinite number of events to those that are relevant to an application. For an application it may be relevant that an object changed its position (change event), or that two different observations of a temperature sensor at different times yielded the same temperature (status event). By considering time as an integral part of the state of the universe, both change and status events can be modeled in a uniform manner.

Events must be observed, interpreted, reported and processed. An observation is a discrete instance of a (possibly continuous) signal. An observation of an event encapsulated in an event object carries a timestamp and descriptive parameters and is typically represented as a tuple of values. Parameters may be absolute values or deltas relative to older reference values.

The type of event and the application domain determine the granularity of the timestamp. In distributed systems the timestamp may carry an imprecision due to clock synchronization problems. Depending on the type of event and application system, the timestamp may be just one point (point semantics of time) or an interval (interval semantics of time). Events may also be characterized by a validity interval.

Events may be simple events or compositions of simple and/or other composite events. Simple events could be individual sensor observations, method invocations or absolute temporal events. Composite events are aggregations of events that are produced from event representations and the operators of an event algebra.

Event contextualization implies the interpretation of observed events with respect to a mental model of the application. Event contextualization leads to derived events. Derived events are caused by other events and often are at a different level of abstraction. For example, five failed logins with the wrong password may cause an intrusion-attempt event to be signaled. Derived events involve semantic knowledge and may be raised explicitly. Event contextualization often relies on external data and derived events are often enriched with data from external sources.

An event model consists of the types of events that are supported and the operators used for their composition and manipulation. Because of the wide variety of application domains and the different emphasis placed on the correlation criteria, such as time, location or other notion of context, no canonic event model is in sight. However, a tendency towards domain specific event models can be observed. The state of event description and manipulation languages parallels the state of the event models. (see chapters 3, 4, 5, 9, 15, 16, 17)

The reference architecture of an event-based system consists of event producers, event detectors, event composers, an event notification mechanism, and event consumers that trigger the application logic. The principles of an Event-Driven Architecture are postulated by Chandy and Schulte as: reporting current events; pushing notifications; responding immediately; communicating one-way; and being free of commands or requests from the consumer.

Event producers may be sensors, a clock or any component that produces events. Event composers can be both event consumers and event producers: they consume simple or composite events, operate on them and produce new events. Event detectors are a type of event producers; they may be software components that observe the environment and generate the corresponding events. For example, a software component in an RFID reader that recognizes the reading of a passive RFID tag, associates a position and timestamp with it and produces an event. Event detectors may implement a polling cycle, but they push the detected events to the consumers.

Events, or more precisely, their representation, must be reported to event consumers. An essential aspect of the event processing paradigm is that events are pushed from the event producers to the event consumers.[5] One or more event representations are packaged into notifications. The header or envelope of a notification may contain routing information. Event notifications are routed from event producers to event consumers by a notification service. The notification service decouples producers and consumers and provides the routing from source to sink. Publish/subscribe middleware is the most flexible (but not the only) notification mechanism: event producers publish events and event consumers subscribe to events that are of interest to them. In the simplest form this may be a low level channel into which event notifications are placed by the publisher and from where they are retrieved by the subscriber (channel based pub/sub). In this case the envelope of the notification is minimal and streams of tuples are delivered over a fixed channel. However, the notification service may be a more sophisticated network of brokers routing the notifications based on type or content. Notifications consist of one or more event representations packaged in an envelope. Routing may occur on the envelope data (topic based pub/sub) or on the content of the notification (content based pub/sub). Content based pub/sub assumes a common name space used by publishers and subscribers. To deal with heterogeneous name spaces concept-based pub/sub introduced the use of ontologies to mediate and resolve ambiguities.

Event consumers are the reactive component in an event-driven architecture. They either implement the application logic in form of rules or provide a triggering mechanism to invoke other application code, e.g. services. Rules may be combined into policies. Rules may have different formats that result in different execution models. Procedural Event-Condition-Action rules are fired whenever the corresponding event (simple, composite or derived) is raised. The condition acts as a guard that can be used to express more complex application logic. Only if the condition is met, the action is executed. A missing condition is always considered to be true and results in an event-action rule. More powerful event expressions decrease the need for explicit conditions but require more powerful event algebras.

This also makes the event detection mechanism heavier and more difficult for users to use properly. The decision on the trade-off between expressiveness of the event language and the lightweight nature of the event system is domain dependent.

The functionality of event-based systems may be partitioned in different ways. In particular the event composition may be implemented at the monitoring component, in the notification service, or as part of the reactive component. The decision of where to realize event composition and contextualization depends on many application and environment specific factors, such as capabilities of the sensing devices, bandwidth of the communication channels, complexity of the composite events, source of the events that are to be composed, and additional information required for event enrichment and interpretation. (see chapters 4, 5, 9)

An additional dimension is introduced when the source and/or the consumer of events is mobile. Mobility may cause the loss or delay of events and requires special provisions for reliable event delivery and forwarding, staging of events, replay of events and makes the ordering of events more difficult. (see chapters 10, 15, 16, 17, 18, 19)

The reference architecture must be mapped to a specific platform. The platform is dictated by the requirements of the application. The capabilities of the hardware, the communication infrastructure, the source of events, and the application requirements will determine how the reference architecture is mapped to a specific platform. Four families of platforms are emerging for applications that are implemented according to the event processing paradigm:

Stream processing engines are typically used for processing very large volumes of homogeneous events that are provided in the form of continuous and high-volume streams. Filters and continuous queries are expressed in a SQL-based language. Scalability and non-blocking behavior are of paramount importance. Stream processing engines exhibit a high degree of centralization and often run on large mainframes or clusters. These platforms are often identified with Complex Event Processing.

Wireless sensor networks are typically used in small, well-contained applications in which homogeneous sensors are connected in the form of multi-hop networks to a single sink. These platforms at present do not scale and are typically used for low volumes of events. Processing is done mostly in the network with relatively simple filtering and event aggregation and some composition across heterogeneous sensors. Communication is wireless, low bandwidth, unreliable, and often the limiting factor.

Messaging systems are based on reliable and scalable message delivery systems that can connect stationary and/or mobile event publishers and subscribers. Event filtering, composition and routing occur in the broker network and the main goal is to decouple event sources and event consumers. Most information dissemination applications use messaging systems of different kinds as platform.

Mixed-mode platforms include a wide variety of nodes, ranging from simple tag readers and sensors to high end servers. Mixed-mode systems are typical for environments in which multiple smaller applications are integrated. Among the applications reviewed in (Hinze, Sachs, Buchmann, 2009), the infrastructure for smart cities, large scale health monitoring and care systems, integrated traffic monitoring and management systems all require mixed mode platforms. The event streams are heterogeneous, both in nature and volume. To scale, the communication must be based on messaging systems, enrichment and event derivation may occur at different nodes in the network, and stream processing engines may be required at selected nodes to detect complex patterns on many event streams. The high volume in mixed mode systems is often the result of many converging low volume streams rather than a few high volume streams. The combination of mobility, heterogeneity, and an extreme distribution compounds the problems these platforms must deal with.

The event based processing paradigm provides the necessary primitives and abstractions to support other architectures. For example, in an SOA, the changes to services are events that are used for notification of the users of these services. In general, event monitoring and processing ideally supports system management, from simple load balancing to control of redundancy and placement of replicas. In the case of self-configuring or adaptive systems, the detection of new components or the disappearance of a component are events that cause system reconfiguration. (see chapters 10, 13, 14)

Quality of service is a major issue in event-based systems. Quality of service (QoS) is a broad term that is often used to mean different things; here we name four aspects relevant to event-based systems. Firstly, quality of service often refers to the guarantees associated with the delivery of events, distinguishing whether events are delivered in order, at least once, at most once, exactly once or only on a best effort base. These delivery guarantees have an impact on the quality of service of the event composition and derivation. Secondly, QoS also refers to dependability, availability and scalability of the event system. A third dimension of QoS refers to performance (throughput) of the event system and the timely delivery of events (real-time behavior). Last but not least QoS often subsumes security. Because of the loose coupling that is inherent to event-based systems, quality of service aspects must be specified and enforced on consumer side based on the expectations of the event consumer. (see chapters 6, 7, 9, 12)

THIS BOOK

In its fledgling stages, the event-based research focused almost exclusively on performance of event delivery but in recent years, a number of new challenges have been identified. They are typically introduced by new application areas entering the scene: for example, health-care monitoring introduced security and privacy, traffic monitoring and gaming introduced mobility, avalanche warning sensors introduced memory limitations, e-commerce introduced transactions.

As described above, industry has given event-based systems increasing attention, which is creating and gaining momentum in the commercial world supported by industry-centered summits[6] and working groups. A substantial number of jargon is newly created and used to such an extent that it has become hard to see the wood for the trees[7]. Also the lack of a common knowledge base of real world applications has been repeatedly identified (Rosenblum, 2005) (Hinze, Sachs, Buchmann, 2009, July).

Different to other works recently published, this book does not present a single uniform view but presents a multidisciplinary approach. This book takes a practical approach to the subject, showcasing real-world applications. Including this introduction, the book contains 20 chapters that can be grouped into four parts. They move from the conceptual view to quality-of-service considerations, from novel architectures to mobile and ubiquitous applications of event-based concepts. The chapters bring together concepts and application examples from a wide basis of contributors.

The first four chapters discuss event models and generic system architectures for distributed event-based systems.

- Chapter 1 introduces an architecture for a generic system for performing complex event processing. The authors use a grid approach to achieve scalability and modularity: A system is represented by a grid of complex event processing components amongst which the processing of an event is divided. The fundamental components include event collection, event purification, event storage, event inference, and event situation management. The architecture is presented in two case studies on early warning systems in disaster management and on service availability management.

- Chapter 2 compares event models in various distributed event-based systems and in other implicit invocation systems. The event model is identified as a central part of each system as it determines each system's event concept and how events are generated and notified to consumers. The chapter's main focus lies in identifying common features between all systems that may need to be supported in the engineering of any event-based application. The chapter proposes a categorization for event models.

- Chapter 3 suggests methods for temporal and spatial correlation between events as needed, for example, in ubiquitous computing. The chapter introduces an event model that uses interval-semantics for event correlation considering spatio-temporal event information. The model supports stateful subscriptions, parameterization, and aggregation. The authors show how controlled event consumption mechanisms (e.g., time restriction, subset policies, and use of durative events) can reduce the event buffer for composite event detection.

- Chapter 4 explores the temporal aspects of event processing. It introduces the notion of an event processing network as the underlying model behind event processing. This event processing network is being executed on a collection of processors that span different time zones. The author identifies and discusses temporal aspects of the execution and within the underlying event model. Temporal aspects that are discussed in the chapter are: temporal dimensions of events, time granularities, temporal context, temporal patterns, event order, and retrospective and proactive operations.

In the second part, five chapters explore detailed quality-of-service aspects of publish-subscribe middleware and event-based messaging systems.

- Chapter 5 addresses the problem of **load balancing** for event-based messaging. The authors propose the use of a structured p2p overlay network to ensure that the participating peers perform only limited work and store limited state information. The overlay network ICE uses hierarchical tesseral addressing (using regions of space instead of Carthesian coordinates) and an amortized multicast routing algorithm. The suggested load balancing is shown to be independent of the number of peers, popularity of subscriptions, or frequency of messages.

- Chapter 6 examines several **security** techniques for various types of event-based systems. The chapter first discusses methods for application-level security. Then security measures within a large-scale distributed broker network are introduced, including distributed cryptographic key management. Thirdly, broker-specific policy enforcement mechanisms are evaluated, where context-sensitive event processing allows more than binary access control. The authors identify a number of open issues in security for event-based systems.

- Chapter 7 discusses **testing** of distributed publish/subscribe systems. In addition to testing separate components, the chapter proposes two techniques for testing the component integration. One is based on large numbers of random event sequences to identify malfunctioning system states. The second one uses predefined data-flows to be tested together with unit-testing. The authors also discuss runtime testing of component integration. The suggested techniques are exemplified in a small case study on testing in Maritime Safety and Security (MSS) systems.

- Chapter 8 gives a detailed overview of the Padres publish/subscribe system. The authors describe Padres' content-based routing mechanism and discuss techniques to ensure system **robustness**. The Padres event model supports filtering, aggregation, and correlation of historic and future event data. The system's architecture consists of distributed and replicated data repositories that can be configured according to criteria such as availability, storage and query overhead, query delay, load distribution, parallelism, redundancy and locality. The chapter describes Padres' techniques to ensure robustness, in particular the use of alternate message routing paths, load balancing, and fault resilience techniques in case of broker failures.

- Chapter 9 introduces a publish/subscribe middleware for wireless sensor networks (WSN). The authors address two issues in traditional WSN: lack of **interoperability** with access networks and lack of flexibility to customize **non-functional properties**. For example, an event detection application may require in-network event correlation and filtering as its non-functional properties in order to eliminate false positive sensor data in the network. The chapter gives a detailed description of the implementation of the proposed middleware TinyDDS. TinyDDS allows applications to interoperate regardless of programming languages and protocols across the boundary of WSNs and access networks. It also gives WSN applications control over application-level and middleware-level non-functional properties such as event filtering, data aggregation and routing.

The next four chapters in the third part of the book describe novel system architectures that can be realized with event-based communication: knowledge-based networking, processing in parallel clusters, service-oriented architectures and dynamic adaptive systems.

- Chapter 10 discusses **Knowledge-Based Networking**: the routing of events based on filters applied to some semantics of the event data and associated metadata. The chapter introduces techniques of filtered dissemination of semantically enriched knowledge over a loosely coupled network of distributed heterogeneous agents. Ontological semantics is incorporated into event messages to allow subscribers to define semantic filters. The ontological concepts are encoded as additional message attribute type, onto which subsumption relationships, equivalence, type queries and arbitrary ontological relationships can be applied. The chapter authors explore these ontological operators and their application to a publish/subscribe application.
- Chapter 11 introduces a distributed event processing system for **large clusters**. Instead of traditional event stream pipes, a content-based publish/subscribe middleware is used for the communication between components. To parallelize processing, the system supports speculative execution of events. The execution of CPU intensive components is spread across multiple cores. Resulting ordering conflicts are detected dynamically and rolled back using a Software Transactional Memory. The authors suggest an extension of the pub/sub mechanism to support load balancing across multiple machines.
- Chapter 12 proposed an event-based approach to **Service-oriented Architectures** (SOA) and Web services. The services in an SOA are changing regularly, which creates a challenge to loosely coupled service providers and consumers. Using the publish/subscribe style of communication service consumers can be notified when such changes occur. This chapter describes an approach of an event-based Web service runtime environment based on a rich event model and different event visibilities. The authors describe the full service lifecycle, including runtime information concerning service discovery and service invocation, as well as Quality of Service attributes.
- Chapter 13 describes how event-based interaction could be used in **Dynamic Adaptive Systems**. These systems consist of interacting components from different vendors; they have to be automatically composed at runtime to deal with components joining or leaving the system. Traditional

event-based interaction is lacking explicit directed dependencies that are needed to derive a system configuration. This chapter proposes an approach for realizing Dynamic Adaptive Systems using event-based interaction among the components while maintaining automatic system composition. The system use is demonstrated for a smart city application.

Part four introduces five approaches to mobile and ubiquitous systems. Chapter 15 and 16 propose solutions for mobile ad-hoc networks.

- Chapter 14 argues that event-driven approaches are particularly suitable to address the challenges of mobile and ubiquitous computing, and mobile *ad hoc* networks. The authors observe that since most programming languages do not provide built-in support for event-driven programming, these concepts are often implemented using different paradigms. This chapter studies the difficulties of **combining events with the object-oriented paradigm**. Various issues at the software engineering level are highlighted and as a solution, the new object-oriented programming language Ambient-Talk is proposed. The language is intended for composing software components across a mobile ad hoc network and supports first-class abstractions to represent event-driven communication.

- Chapter 15 also addresses publish/subscribe communication in mobile ad-hoc networks. In particular, this chapter focuses on ensuring **quality-of-service** characteristics such as end-to-end timeliness, reliability properties, and limited message overhead. The chapter provides an overview of interconnection topologies, event routing schemes and architectural support of publish/subscribe systems in MANETs. The authors propose the use of event routing in a two-layer scheme using hierarchical tree-based connections in the Manet. They additionally introduce a dynamic system self-reconfiguring ability for event causal dependencies.

- Chapter 16 introduces **location-based triggers** to support location-aware mobile applications. A location-based trigger is a standing spatial trigger with reference to a spatial region, actions to be taken, and a list of recipients. The author describes the mTrigger framework in three alternative architectures for location triggers: a client-server architecture, a client-centric architecture for management within mobile environments, and a decentralized peer-to-peer architecture for collaboration between mobile users. Performance optimizations are proposed for each of the architectures, including a number of energy-efficient spatial trigger grouping techniques for optimizing both wake-up times and check times of location trigger evaluations.

- Chapter 17 discusses an event-based infrastructure that combines the Session Initiation Protocol (SIP) and Web technologies to realize **secure converged mobile services**. The chapter uses the scenario of proactive information delivery to airline customers, which has been developed in collaboration with an airline company. The distributed system architecture is outlined and the use of the two key technologies is evaluated for the given scenario. The chapter reports about the authors' experience in combining the various asynchronous communication mechanisms that are commonly used in today's Internet in a realistic mobile Internet context.

- Chapter 18 is motivated by the need to provide event-based notifications (ENS) in rescue and emergency applications. The authors focused on sparse Mobile Ad-hoc Networks (MANETs), since emergency applications have to run, for example, in places lacking infrastructure where the density of nodes may be low, or where physical obstacles may limit the transmission range. The chapter analyzes characteristics of rescue operations to discuss ENS design choices such as subscription language, architecture and routing. The authors also present their own Distributed Event Notification Service (DENS), which is specifically tailored for this application domain.

- Chapter 19 is a panel discussion; the chapter tries to capture the opinions of some of the leading researchers from academia and industry about the future of event-based systems. Seven panelists were interviewed; each of them is introduced with a short biography highlighting their professional background from which they approach the field of event-based systems. This allows one to see how different panelists set their individual focus, mirroring their community of origin. The chapter presents their statements and a discussion of the opinions voiced.

REFERENCES

Chandy, M. (2006, June). Event-driven applications: Costs, benefits and design approaches. In *Proceedings of the Gartner Application Integration and Web Services Summit 2006*, San Diego, CA.

Chakravarthy, S., & , Q. (2009). *Stream Data Processing: A Quality Of Service Perspective: Modeling, Scheduling, Load Shedding, And Complex Event Processing*. New York: Springer.

Chandy, M., & Schulte, W. R. (2009). *Event Processing: Designing IT Systems for Agile Companies*. New York: McGraw-Hill Osborne.

Hinze, A., Sachs, K., Buchmann, A. (2009, July). Event-Based Applications and Enabling Technologies. In *Proceedings of the International Conference on Distributed Event-Based Systems* (DEBS 2009), Nashville, TN.

Luckham, D. (2002). *The Power of Events: An Introduction to Complex Event Processing in Distributed Enterprise Systems*. Reading, MA: Addison Wesley Professional.

Mühl, G., Fiege, L., & Pietzuch, P. (2006). *Distributed Event-Based Systems*. New York: Springer.

Rosenblum, D. (2005). *Content-Based Publish/Subscribe: A Re-Assessment, Keynote Presentation at DOA 2005*. Retrieved from http://www.cs.ucl.ac.uk/staff/D.Rosenblum/Presentations/doa2005-keynote.html

ENDNOTES

[1] A sample of event processing blogs and pages online in 2009 are Complex Event Processing at http://complexevents.com, and Event-based.org at http://event-based.org Apama Event Processing Blog at http://apama.typepad.com/my_weblog, Streambase Event Processing Blog at http://streambase.typepad.com/streambase_stream_process, Cyberstrategics Complex Event Processing (CEP) blog at http://www.thecepblog.com Tibco Complex Event Processing blog at http://tibcoblogs.com/cep Oracle CEp blog at http://blogs.oracle.com/CEP Event Processing Thinking blog, online at http://epthinking.blogspot.com

[2] Examples are Aleri (http://www.aleri.com/), Coral8 (http://www.coral8.com/), Steambase (http://www.streambase.com/), Tibco (http://www.tibco.com/)

[3]   Software is openly available for Esper (http://esper.codehaus.org), Padres (http://padres.msrg.utoronto.ca), Siena (http://www.inf.usi.ch/carzaniga/siena/software/index.html)

[4]   See http://www.ep-ts.com/

[5]   The pushing of events from producers to consumers is a conceptual view. On implementation level, either push, pull, or combinations may be used at different abstraction levels.

[6]   The EPTS Event Processing Symposia are now in their 5th instalment: 5th EPTS Event Processing Symposium September 2009, Trento, Italy, and the Gartner Event Processing Summit is a regular annual event since 2007.

[7]   Several industry presentations at the Dagstuhl seminar on event processing 2007 focused solely on the comparison of acronyms and concepts (e.g, EDA, ESP, BAM, BPM, SOA, CEP).

Chapter 1
Generic Architecture of Complex Event Processing Systems

Avigdor Gal
Israel Institute of Technology, Israel

Ethan Hadar
CA Labs, Israel

ABSTRACT

In recent years, there has been a growing need for the use of active systems, systems that are required to respond automatically to events. Examples include applications such as Business Activity Monitoring (BAM) and Business Process Management (BPM). Complex event processing is emerging as a discipline in its own right, with roots in existing research disciplines such as Databases and Software Engineering. This chapter aims at introducing a generic architecture of complex event processing systems that promotes modularity and flexibility. We start with a brief introduction of the primitive elements of complex event processing systems, namely events and rules. We discuss a grid approach to complex event processing systems. We detail the layers of the proposed architecture, as well as the architecture main components within the context of the major data flow in an event management system, namely: event collection, event purification, event storage, event inferencing, and event situation management. We discuss each of these elements in detail. Our tool of choice is the CA Agile Architecture (C3A) reference description approach (Hadar & Silberman, 2008). Throughout the chapter, we illustrate our discussion with two case studies. The first is that of service availability management, providing safeguards to critical business processes. The second involves disaster management, managing early warning systems.

INTRODUCTION

In recent years, there has been a growing need for the use of active systems. Active systems perform automatic actions that may be reactive, such as responding to provided stimuli. They may also perform automatic actions that are proactive, such as predicting possible phenomena. While earliest active systems were directed to databases (Dayal U. et al. 1988, Widom & Ceri, 1996), a current major need for such active functionality covers other areas such as Business Activity Monitoring

DOI: 10.4018/978-1-60566-697-6.ch001

(BAM) (Balazinska et. al, 2008) and Business Process Management (BPM). Concurrent with the proliferation of such applications, the events to which active systems must respond have expanded from Information Technology (IT) systems and application-level events to business-level events. Active systems may encompass many levels of IT, from IT infrastructure events to business level objectives, and are used in a wide spectrum of sectors including financial and e-Trading, BPM, system management, client relationship management (CRM), workflow management, and military applications.

The main primitive in active systems is an event, a mechanism for delivering information. Events may be generated by some mechanism, external to the active system, and deliver information across distributed systems. Alternatively, events and their related data need to be inferred by the active system itself. Complex event processing (CEP) supports the inference of events based on event notification history and the data delivered by events. From an industrial point of view, CEP covers distributed a-synchronic message-based systems, such as enterprise applications that can immediately act upon business critical events.

Complex event processing is emerging as a discipline in its own right, with roots in existing research disciplines such as Artificial Intelligence, Business Process Management, Databases, Distributed Computing, Programming Languages, Semantics of Specification and Formal Verification methods, Sensor networks, Simulation, and Software Engineering. For example, in database research, events were suggested with the intention of introducing triggers in database management systems for monitoring application state and to automate applications by reducing or eliminating user intervention. This discipline still goes by various names such as Event Processing, Real-time Information Systems, Reactive Systems, Proactive Systems, and Active Technologies.

CEP requires intense computation efforts. Each complex event depends on its underlying events or previously calculated complex events. In addition, each of these complex events might be generated by domain specific business rules using this or that language. Therefore, CEP solutions must be scalable, flexible in configuration, easily managed, and efficient in terms of computation calculation. In this chapter we present a generic architecture for CEP systems adhering to the following principles:

1. A separation of concerns between different CEP instances that can interact in a form of a network of CEP instance components;
2. A simple modular and dynamic way to integrate CEP instances;
3. A means to reduce computation efforts within each of the CEP instances;
4. A centralized governance and compliance supportive arena, enabling CEP service virtualization and management capabilities;
5. A scalable and highly available solution, supporting real-time processing as well as analytics of historical events.

The rest of the chapter is organized as follows: We start with a presentation of two case studies to illustrate the architecture. Then, we present events and rules, and introduce the reader to the essentials of the C3A approach. Details of the architecture follow and we conclude with a discussion and future work.

CASE STUDIES

We begin with the introduction of two case studies to illustrate the architecture. The first is that of service availability management, providing safeguards to critical business processes. The second involves disaster management, and more specifically managing early warning systems.

Service Availability Management

In the domain of Enterprise IT Management (EITM), the IT managers and their Chief Information Officer (CIO) are required to maintain qualitative operation levels of the managed IT resources. It is required to do so while aligning the production IT infrastructure to the business needs of the organization, and measure its service quality levels in several terms, known as Key Performance Indicators (KPI). The managed resources, known as Configuration Items (CI), may be configured and should be able to change their behavior, or be subjected to architectural changes that may require some action, manually or automatically. Moreover, there is a structural dependency between the CIs that can affect the need for an action, or compose an aggregated KPI.

As an example, consider the case where a CI is a logical IT service, such as a Credit Card validation, comprised of a business process that uses mail servers, a Web server and database servers for fulfilling the process. The overall performance metric of the Credit Card IT Service is the accumulative sum of the separate transactions on each server. The overall malfunctioning metric is the number of reported requests for repair (service request). It is aggregated as well and defines the IT Credit Card Service quality metric.

The operational needs of IT managers involve the reduction of the number of malfunctions and the increase of the IT service performance according to a preset threshold. Thus, the KPIs will be the number of aggregated support calls from these CIs, as well as aggregated performance duration for each service call, combined with associated thresholds.

In our example, a complex event is the aggregated sum of the individual metrics accumulated from the separate, service related, servers. Naturally, there are other complex rules such as financial implications of downtime, in which the threshold will be defined in currency measurements, and should be translated into the financial

value of a server, as well as to the potential loss of transactions. Thus, a complex event will need to detect a downtime server event, relate it to the transactional business service it supports, aggregate the overall value, and compare it to the defined financial threshold.

Disaster Management

Disaster management is the discipline of dealing with and avoiding risks (Haddow & Bullock 2004). According to Wikipedia, it is a discipline that involves preparing for disaster before it happens, disaster response, and rebuilding society after natural or human-made disasters have occurred.

A concrete example exists in syndromic surveillance systems, which are designed to detect outbreaks of epidemics or bio-terrorist attacks. Such systems use data from external data sources (*e.g.*, transactional databases) and expert knowledge (see (Shmueli & Fienberg 2006)) to identify outbreaks and to quantify the outbreak attributes, such as the severity of an attack. Responding quickly to such attacks requires recognizing an attack has occurred—a difficult task as no direct indications of it exists. Therefore, hazard events must be deduced based on available data sources, which often provide insufficient information to determine hazard occurrence.

Shmueli & Fienberg (2006) demonstrated that over-the-counter and pharmacy medication sales, calls to nurse hotlines, school absence records, Web hits of medical Web sites, and complaints of individuals entering hospital emergency departments can all serve as indicators of disease outbreak. These measures can be collected by querying local stores, physician databases, or data warehouses. Figure 1 (based on (Shmueli & Fienberg 2006)) provides an illustration of the daily sales of over-the-counter cough medication from a large grocery chain in Pittsburgh, Pennsylvania. The data was gathered by intermittent querying of the chain's transactional database.

Figure 1. Over-the-counter cough medication sales

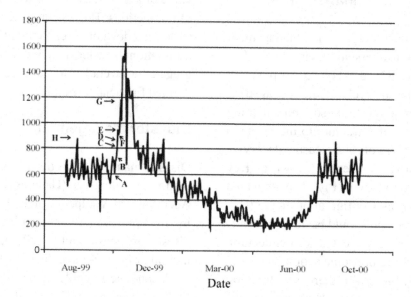

Figure 1 displays the changes in sales during the months of November and December 1999, where a gradual, yet constant and significant increase of sales indicates the outbreak of the flu.

PRELIMINARIES

We start the section with a discussion of events, rules, and their representation. We then discuss related work that deal with complex event processing architectures. Finally, we briefly discuss the CA Agile Architecture framework.

Events

Complex Event Processing is supported by systems from various domains. These include ODE (Gehani et. al 1992) for active databases and the Situation Manager Rule Language (Adi & Etzion 2004), a general purpose event language. For an excellent introduction to Complex Event Processing, see (Luckham 2002).

The main primitive in active systems is an event. An event refers to any mechanism for delivering data. An *event* is an actual occurrence or happening that is **significant** (falls within a domain of interest to the system), **instantaneous** (takes place at a specific point in time), and **atomic** (it either occurs or not). This definition, while limited, suits our specific needs. Examples of events may include a database notification, a termination of workflow activity, a report on daily sales, a person entering or leaving a certain geographical area, a server joining the network, and a CPU reaching a certain utilization threshold.

An event may be generated by several sources. We differentiate between *explicit events* that are signaled by some mechanism external to the active system and *inferred events* that are inferred by the active system based on other events.

Data can be associated with the occurrence of an event. Some data are common to all events (*e.g.*, occurrence time), while others are specific only to some events (*e.g.*, amount of sales). The data items associated with an event are termed *attributes*. An *event history* is the set of all events (of interest to the system), as well as their associated data, whose occurrence time falls in a given time interval.

An event is captured in the system using an Event Instance Data (EID) relation. For example, the structure of an EID for the disaster management case study may be *DS(EID, Date, dailySales)*, representing the date and the daily sales of each event. Events that share the same set of attributes are of the same type, *e.g.*, the *DS* event type in the example above.

A *complex event* refers to any inference that may be drawn from the events. For example, a complex event may be the simple aggregated sum of the individual metrics accumulated from separate, service related, servers. Complex Event Processing supports the inference of events based on event notification history and the data delivered by events. From an industrial point of view, CEP may be performed by distributed, asynchronous, message-based systems, such as enterprise applications that can immediately act upon business critical events.

In most existing work on event inferencing, the possible combinations that define a complex event are defined using a set of operators that constitute event algebra. Numerous studies have been carried out in order to define the semantics of composite events, and some prototypes defining specific event algebras have been implemented. Some of these prototypes are designed specifically for active databases. These include ODE (Gehani et. al 1992), an active object oriented database that supports the specification and detection of composite events and Snoop (Chakravarthy 1994), an expressive event specification language for object oriented databases, implemented in the Snoop object oriented database. There are also general-purpose event composition languages (*e.g.*, the Situation Manager Rule Language (Adi & Etzion 2004)). Each complex event may depend on underlying events and/or previously calculated complex events. In addition, each of these complex events may be generated by domain-specific business rules and languages.

Rules

Complex event processing languages support event inferencing based on a set of rules. A rule defines the necessary conditions for inferencing new, complex events. It defines how many new events should be generated and helps calculate their attributes. Zimmer and Unland suggest a meta-model for the specification of complex events in active databases (Zimmer D., Unland R. 1999). This meta-model is based upon three independent dimensions: *Event Instance Pattern*, *Event Instance Selection*, and *Event Instance Consumption*. These dimensions are further refined into sub-dimensions. The following describes these three main dimensions in further detail:

1. *Event instance pattern* of a complex event type E_i describes at an abstract level the event instance sequences that will trigger event instances of E_i. It considers the following aspects:

   a. The *event types* whose instances must (or must not) occur in an event instance sequence and the restrictions concerning their order.

   b. A *delimiter* that restricts the number of event instances of a composite event type, which must occur to specify the event instance pattern.

   c. *Operator modes*, which are used to define coupling and concurrency. *Coupling modes* define whether event instance patterns may be interrupted by event instances that are not relevant to the event detection. *Concurrency mode* defines whether the time interval associated with the triggering event instances may overlap.

   d. *Context conditions* define whether the values of a parameter of different instances must be the same, different, or without any restrictions.

2. *Event instance selection* defines which event instances are bound to a complex event. This selection is performed individually for each component event and selects the first, last or every instance of the component that satisfies the event operator.

3. *Event instance consumption* determines the points in time in which events cannot be considered for the detection of further complex events.

As a derivative of this meta-model, a rule can be defined to contain a *selection* function that filters events, relevant to the rule. Such a selection function receives an event history as input, and returns a filtered, subset of history, containing *selectable events*. It is the role of the selection expression to filter out events that are irrelevant for inferencing, according to the rule.

A rule also includes a predicate, defined over a filtered event history, determining when events become candidates for the rule. The selection function eliminates events that are, by themselves, irrelevant, while the predicate determines the relevance of sets of events.

Example 1. *As an example, consider a rule that detects a flu outbreak whenever there is an increase in over-the-counter cough medication sales for four sequential days with a minimum increase of 350 units. In this case, both the selection operator and the predicate will keep over-the-counter cough medication sales as the relevant events. The association function looks for consecutive sales with an increase of 350 units. Finally, a mapping function will generate a new event, determining a flu outbreak and assigning a date value to it.*

To the best of our knowledge, the most semantically expressive and general purpose event specification model is the Situation Manager Rule Language (Adi & Etzion 2004). This specification has the following features:

1. *Semantic Event Model*: The model is based on three fundamental generic kinds of semantic relationship between objects. The *has-attribute* relationship connects an event type to one or more objects, which describe some aspects of the event. These objects are called *event attributes*, and describe the schema of the event types. The *has-instance* relationship links an event instance to an event type. Finally, the *has-subtype* relationship logically links an event type with another event type.

2. *Lifespan management*: A *lifespan* defines the time interval during which event composition is of interest. It is bounded by two events termed *initiator* and *terminator*, which respectively initiate and terminate the lifespan. Lifespans are defined using lifespan types, which describe common properties of a similar set of lifespans. Examples of such properties are the set of events that are possible initiators and terminators. Composite event detection is relevant during the time window defined by a lifespan. Composite events and lifespans are related through a many-to-many relationship.

3. *Partitioning*: A mechanism by which semantically related events are grouped together.

4. *Situations*: Composite events are defined using situations. For each composite event a situation defining its detection conditions is specified. A situation definition includes the relevant event instances (types and filtering conditions) to the composite event. It also contains the lifespan during which composite event detection is relevant. Finally, it includes a *situation expression*, with an operator, a predicate, a detection mode, and event consumption, as follows.

An *operator* defines an event pattern. The situation manager rule language has four operator types: Joining operators (sequence, strict sequence, conjunction, disjunction, and aggregation); Temporal operators (at, after, and every); Selection operators (first, until, since, and range); and Assertion operators (never, sometimes, last, min, max, and unless).

A *predicate* must hold for the event instances. A *detection mode* determines when the composite event must be detected. This can be either *immediate, i.e.*, as soon as all conditions (specified both by operators and predicates) are satisfied, or *deferred, i.e.*, at the termination of the lifespan, if all conditions for the situation were satisfied during the lifespan.

Finally, *event Consumption* determines whether one event instance can be used in the inference of more than one composite event.

Event Architectures

Event-driven architecture has become popular with many industrial products such as Progress[1]. Several industrial perspective presentations of event-driven architecture exist, *e.g.*, (Michelson 2006)[2]. The main focus of these works is on the distribution of the event sources while we focus on the distribution of the inferencing as well.

The architecture presented in (Dayal U. et al. 1988, Adi & Etzion 2004) involves a central mechanism for rule management. Other architectures in related areas such as data fusion (Hall & Llina 2001) also assume a central event processor. The CoopWARE project (Gal 1999) supports a central Complex Event Processing system to which different clients contribute events and rules in a collaborative manner. We extend such architectures to a network of interacting, fully-functional complex event processors.

In a tutorial at DEBS'2008[3] Etzion has provided his point of view of the architectural side of event processing networks. In his talk, Etzion separates event producers from event consumers and introduced the event processing network, which is a graph representing the inter-relationships among events. Within this setting, Etzion has discussed issues of layering, stratification, hierarchy, parallelism and partition. Our work is different in that we discuss the distribution of the event processing mechanisms, as well as the distribution of event generation.

CA Agile Architecture (C3A)

CA Agile Architecture (C3A) is a reference description approach proposed by Hadar & Silberman (2008). It supports a common strategic vision of products as well as an agile structure implementation of loosely-coupled components with accurately described interfaces. In addition, it provides a perspective of both major and minor releases. C3A addresses the current dichotomy between strategic thinkers and tactical implementers, which may be traced to three main factors. The first involves the control over abstraction barriers, limiting architectural granularity details, organized in four architectural perspectives. The second involves minimal documentations and artifacts for capturing component's functional and non-functional structure and requirements. The third and final factor involves synchronization constraints, determining the time horizon of each architectural task. These factors led to the main C3A dimensions, namely: (1) levels of abstractions; (2) number and nature of the documented artifacts; and (3) the time horizon of each artifact.

The reference architecture is intended to provide both a conceptual framework for the system and functional components that form a structure that supports customer needs. The implementation details, such as the exact algorithms, can vary from one solution to another; however, the purpose of the reference architecture is to unify the main concerns under a coherent and well-defined blueprint. Using the definitions of C3A reference architecture, we provide several Level 1 architectural components that are contained within Level 0 components, supporting Complex Event Processing.

The Level 0 components describe the system's logically-separated and self-maintained modules,

Figure 2. Scalable and modular grid of Complex Event Processing (CEP) units

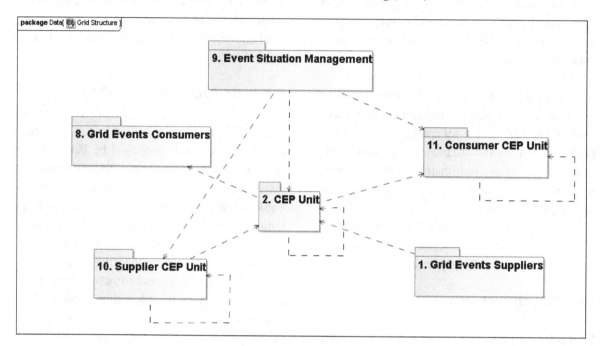

with distinct and disjoint functional responsibilities. Each Level 0 component is composed of one or more Level 1 physical components, which are deployable units that provide a collection of cohesive services, using well-defined interfaces. These interfaces should support customer needs and enable reuse. Level 1 components (marked as the UML component symbol) provide higher granularity and more transparent abstraction barriers. These barriers are represented by the detailed interfaces, including parameters, protocols and messages, channels and ports. The spotlight of this level is on a good interface design and a clear separation of functional concerns. The abstraction barriers for this level are deployment capabilities. Each Level 1 component is self-deployed, and therefore may be replaced locally without the need to re-install a full Level 0 component. The separable Level 1 components enable a local replacement of such a unit, without the need to reinstall the full CEP application. The lower abstraction barriers for the Level 1 components are internal components that do not require separable deployment capabilities

in terms of customer needs. Such components can be changed and reused together in terms of maintenance.

THE GRID CEP

Figure 2 is a block diagram illustrating a generic system for performing complex event processing. The figure presents reference architecture of CEP as a grid of suppliers and consumers of events. A CEP system generally includes one or more event suppliers, CEP units, event consumers, and event situation management services. CEP units receive and/or retrieve events from event suppliers and provide complex events to event consumers. CEP units are managed by an event situation management service.

A more detailed explanation is required to clarify our use of the term "grid." We consider a grid to be a topological network of components, where a component in our case is a complex event processor. A situation manager allocates

Figure 3. Integration points of a single CEP unit and the central event Situation management system

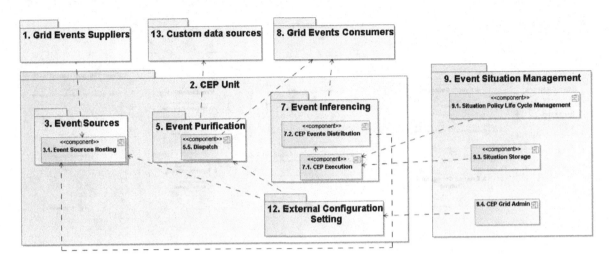

tasks to the components based on load and needs. Therefore, the task of event inferencing becomes a distributed task, where pieces of the task (in terms of event and complex event processing) are divided among several processors, sometimes up to many thousands. We prefer the use of the term "grid" over cluster computing since grids tend to be more loosely coupled, heterogeneous, and geographically dispersed.

From the point of view of a single CEP system, the main CEP unit (2) receives events from a supplier CEP unit (10) as well as from a grid event supplier (1) (representing any event source). Each single CEP unit provides aggregated complex events to the grid event consumers (8), or another Consumer CEP unit (11) (serving as a supplier CEP unit to that consumer). All CEP units are fed with events in real-time, and are managed centrally by the Event Situation Management (9) service as detailed below.

This grid CEP provides a modular representation of the interaction between CEP units, where each CEP can generate or consume events and centrally manage redundancy and configuration by the situation management system. This approach enables **service transparency** in which the exact implementation instance of the CEP is hidden from the consumer of events. It enables the weaving

of different CEP unit technologies, although, as detailed below, the same CEP unit structure is preferred. Also, overall grid performance characteristics can be optimized globally amongst all the CEP units providing load-balancing solutions, high-availability, and systems redundancy. The events pass from one unit to another, based on the dynamic configuration and weaving of the CEP units as defined by (9).

Figure 3 represents the Level 1 integration points between our implementation of specific components of a CEP Unit (2), the Event Situation Management (9) and external data sources. These integration points, as will be detailed below, are aimed to enable grid control and external enrichment capabilities, as well as to improve the overall grid performance.

Example 2. *As an example, consider the disaster management case study. Events reporting on over-the-counter cough medication may be explicit events, generated by each store. Alternatively, they may be events that were "massaged," e.g., by rounding sale numbers to the nearest tenth. Either way, the CEP unit (2) receives the events from a Supplier CEP Unit (10), hiding the specific details of implementation.*

Figure 4. **Level 1** *CEP unit internals linked to the external Event Situation Management System*

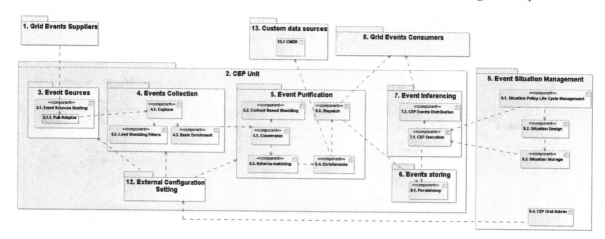

CEP SYSTEM STRUCTURE

The CEP unit within Figure 3 represents a single CEP system, showing its interaction with the outside world, using a grid structure. We now expose the internals of such a unit in Figure 4 and elaborate on its internal components.

The CEP Unit contains a component for each event source (3) for monitoring incoming events. Event Collection (4) captures events and filters redundant ones. Then, Event Purification (5) adds information to events, as well as adapting their structure and format. Next, Event Storing (6) supports persistency as well as historical analytics of events. Finally, Event Inferencing (7) provides the actual near real-time engine to calculate a situation if detected and dispatching the events.

This flow illustrates the main steps in CEP as supported by the modular grid architecture for CEP. From sampling of events through collection, purifications, storing, inferencing, design and management, this model supports modular combinations of deployment and solutions.

As we show below, the internal recursive mechanism enables the weaving of a grid network of CEP units. It also minimizes the required computation efforts in detecting situations. It supports proactive triggering of additional sampling of events based on need, thus improving consider-

ably real-time root-cause capabilities, in an ad-hoc manner without overusing system resources.

Moreover, a centralized control unit in the Event Situation Management enables remote configuration of a distributed CEP unit behavior in terms of shedding of information, according to a global grid need. As an example, one can stop sampling an explicit event at the entry to the grid, in order to prevent its consumption by several CEP units further ahead, effectively eliminating processing of that event. Thus, overall sampling frequency of events can be balanced. The remote admin enables the replacement of rules between CEP units, event source links, as well as overall system load balancing and redundancy tuning.

The remote centralized management provides CEP service virtualization, seamlessly to the users. It governs compliance and regulations in defining rules for situations detections, thus enable monitoring business process and underlying activities in the operation domain, as well as detect system behavior.

The centralized storage, integrated as federated distributed storage, enables analysis and trends detection, thus serves as a fundamental block in automatic discovery of business processes, based on analyzing the measured qualitative events. It may enable problem detection and awareness

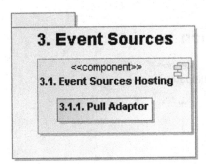

Figure 5. Event Sources internals

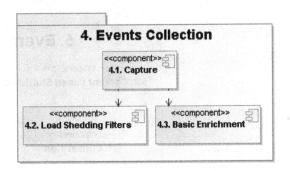

Figure 6. Event Collection internals

based on previous patterns, thus support problem prevention capabilities in IT.

In the next sections we describe, in further details, each component of the CEP unit.

Event Sources

Event sources and their hosting capabilities (see Figure 5) are in charge of extracting real world events from external sources. Typically, events can be pushed from an external source and put in a queue of incoming events. However, not all events can be delivered all the time. Some events require dedicated channels for listening to event sources. Alternatively, some external sources require that events will be pulled periodically to avoid the abuse of the network. A common example of the latter involves a more recent technology of RSS feeds, which requires monitoring of pull-based streams.

The pull adaptor, marked 3.1.1 in Figure 5 is in charge of sampling events, or monitoring external queues of pushed events. It serves as a boundary system and transforms technological difficulties between the sampled information and the system technology. To illustrate, consider the case in which the event information is organized as a string formed message. In order to recognize the information, an adapter needs to read the information, detect its content, and parse it into a structured machine-readable format. These specialized adapters wrap one technology and

transform it into another. One can use any existing well-known protocol such as SOAP messages,[4] TCP[5] query, or SNMP[6] for that matter, in order to investigate and query a device, a server, or an explicit event generator.

Event Collection

As can be observed in Figure 4, the Event sources hosting (Figure 5) passes the monitored events to the Events Collection (Figure 6). The Events Collection component is in charge of capturing events by triggering and scheduling an adapter in the event sources hosting. In addition, it has two more main functionalities.

The Load Shedding Filters component is in charge of reducing the number of propagated events into the system based on the system internal logic. For example, if an external event source is known to undergo maintenance, then events from this source are considered garbage and will not be passed on. The shedding policy is configured for a specific CEP unit instance locally, as well as remotely by the Event Situation Management, and its internal CEP grid admin Level 1 component (9.4).

The sampled events are decorated, using the Basic Enrichment (4.3) component, with basic information such as Time Stamp, Event Type, and Event Source, and forwarded to the Event Purification component. It is important to emphasize

Figure 7. Event Purification internals

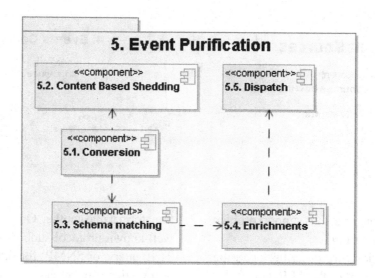

that the Time Stamp is based on a world network clock, to ensure that all events across the grid, will be synchronized and correlated in case of subsequent processing outside the CEP unit itself.

Event Purification

The Event Purification component is in charge of enriching the event with semantic information, as opposed to the minimal, mainly syntactical enrichment that is done in the Event Collection component (see Figure 6). Figure 7 illustrates the internals of this component.

The Conversion component (5.1) retrieves the filtered events from the Event Collection component (4) and classifies them for improved performance. An example might be to organize all the events originated from a specific resource such as a database server. Another classification may be based on origin (using regions of IP addresses) while yet another classification separates slow from fast processed events in order to improve processing batch performance capabilities. After parsing the content of the event, one might apply another level of shedding, using the Content Based Shedding (5.2) Level 1 component for

improved efficiency. The difference between Content Based Shedding and Load Shedding in 4.2, is that the latter does not sample events entirely, eliminating any entry of events to the system by stopping a port from listening to events, or by not running a certain adapter (3.1.1). Such a case might be when a server is restarting due to an upgrade activity. During this time, the server is not in production mode and does not participate in the overall IT system. However, if sampling of events from this server will not be removed by the External Shedding component, it will insert redundant events into the system. In the case of Content Based Shedding, one might consider an a synchronic communication protocol, in which the system would like to receive an event "once and only once," similar to a digital flip-flop gate, protecting the system from reacting to a recurring event, and acting upon it within a reasonable time. In such a case, the system will need to get an event, recognize it as being already processes and prohibit its propagation.

Following this shedding of events, incoming events are normalized, formatted and validated in terms of content, later to be transformed into a new schema using the Schema Matching (5.3) component's mapping rules. Component 5.1 is connected, in parallel, to both 5.2 and 5.3 since Content Based Shedding is infrequently used, and its activation should not be in a sequential manner.

At this point, the system adds missing information using the Enrichment (5.4) component. This is done with the use of external information sources, such as the Configuration Management Database (CMDB, 13.1) or based on pre-defined structured algorithm, configured from the CEP Grid admin remotely (9.4).

Immediately after the enrichment is completed, the basic purified events are sent to either the persistency component (6.1) or external subscribers according to the dynamically configured dispatching information, stored at the Dispatch component (5.5).

Example 3. *To illustrate the functionality of this component we now present an example, based on the Service Availability Management case study.*

Consider an incoming event, captured with the Event Collection (4) Level 0 component, using the Capture (4.1) and Basic Enrichments (4.3) Level 1 components. The data within this event is structured with a string containing "Host: Windows Windows caiSqlA2 Poll Agent:caiSqlA2:sqla2ServerConn OK Critical DB1 Instance MySqlInst". The first task involves parsing this event and converting its unstructured string format into a structured schema that defines fields such as Client, Host, Resource Class, Resource instance and so on. Some of the additional information is provided by the Basic Enrichments (4.3) Level 1 component, identifying the source of the data, the client under investigation, and the unique ID of that resource. Such information might be with values such as "Client=uscigp15" and "Host=usci04", and the time stamp of receiving this event. However, some of the information is not within the content of the event, such as the name of the administrator that is responsible for this SQL DB. This information can be acquired by connecting to a CMDB, and receiving the names of the associated users for this specific SQL instance.

The system is seeking specific information, and therefore not all of the information in the event data will be transformed. In our example, the relevant transformed information fields would be:

- *Client=uscigp15*
- *Host=usci04*
- *ResourceClass=DBMS*
- *ResourceInstance=SqlServer:MySqlInst: DB1*
- *Vendor=Microsoft*
- *User=fred*

An example of detecting a needed call for recovery action is based on the incoming DB event. We can receive events of "IsAlive" using a "ping" TCP command for both client and host, as well as verify that the network is up using network system reports. If all these conditions (a complex situation) are true, the action should be a notification to the DB management services, to improve their load balancing.

This complex event can be calculated by the CEP Execution Engine (7.1) detailed below, requiring a collection of these events, for near real-time inferencing. Another option will be to define a new mapped event, using the Schema Matching component (5.3), creating a compound derived event, aggregated with added data of the statuses of the IT network, client and server from the CMDB health status.

Event Storing

The Persistency (6.1) component, the only Level 1 component of the Event Storing component (Figure 8), enables the warehousing of the received events for immediate processing by the Inferencing component (see Figure 9). Moreover, it temporarily stores the qualitative events, triggered by a specific predetermined time-based rule, and thus enables time-based window processing.

Figure 8. Event storing internals

Figure 9. Event Inferencing internals

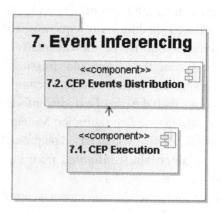

Upon expiration of that window, the events are stored for further analytical processing of future trends and statistics, as well as subsequent archiving and purging of old data. The notion of purified storage is of importance for most of the analytical engines, since in order to correlate events or harvest logs one must have a unified, semantically understood, single format of data. Otherwise, the system will need to apply intense calculation effort for transient translations, to be repeated periodically. It is worth noting that such cascading of persistency types supports real-time (on the fly), near real-time (time-window based), and historical analytics (business intelligence).

Event Inferencing

This module is in charge of the generation of new events, given an existing event. The CEP Execution Engine (7.1) uses purified events, the output of the Persistency Storage (6.1) and runs active situation policies that were provided by the Event Situation Management component (see 9 below). Based on the outcomes of the rules, three major actions can be derived:

1. Additional probing, either direct or deferred, by triggering a pull adapter (3.1.1) instance of the Event Sourcing Level 1 component, to retrieve more events and information. This mechanism of recursive probing reduces the calculation efforts in the system. Consider the case of detecting a degradation in a transaction overall performance from the time a

user submits a query from a web UI, until a response was generated. The root-cause for this might be the load on a Database server, or on a network hub. Alternatively, it may be memory leakage in the Web server. Instead of constantly measuring these three attributed, one can define a rule that triggers CPU performance testing, or memory utilization, only in the case that the overall transactions performance is more than 3 seconds.

2. Sending detected complex events to any regular CEP clients using the CEP Events Distribution component (7.2). For example, a load balancer that instantiates a new server and adds it to the computer farm needs to consider several system conditions. These conditions include the number of running transactions (more than 100) **and** the accumulated free memory (less than 30%) **and** the overall average CPU time within the last 1 hour (more than 70%). The load balancer is required to constantly monitor these aggregated conditions, and act upon the changes in their statuses.

3. Storing the detected complex events. This is done when the CEP Events Distribution component (7.2) sends a complex event to the Pull Adapter (3.1.1) to re-enter the system. Reduced calculation effort is achieved

here using recursive storage. Considering once more the previous example, one might define the first two conditions (transactions and free memory) as an internal rule called "Transaction and Free Memory (TAM)." Consequently, the rule is transformed into TAM AND Average CPU. Moreover, consider that the Database server itself is interested in the TAM rule for self-adaptation needs, in which it will add storage-on-demand to the same server by using remote iSCSI[7] storage memory that enables the utilization of storage over the Internet. In such a case, computing the TAM rule for each complex rule separately is redundant. The system should detect this similarity of the rules, thus define a common granular TAM rule, which will be used by both CEP rule consumers, namely the load balancer, as well as the Database storage Self-Adaptation service.

The Event Inferencing component (7) operates on the active CEP rules only, *i.e.*, those rules to which consumers register. This property increases performance and reduces computation efforts of the overall system.

EVENT SITUATION MANAGEMENT

The Event Situation Management component (9) (see Figure 10) provides CEP service virtualization capabilities, by seamlessly managing the CEP rules and the CEP Grid structure as well as its performance optimization. This capability of centralized design and life cycle management supports compliance regulations of events, and a single dashboard and reporting mechanism to the CEP grid service.

The situation policy/rule construction, design, and execution are managed by the Situation Policy Life Cycle Management component (9.1) that gathers compliance or any other requirements for CEP rules. It uses the Situation Design component

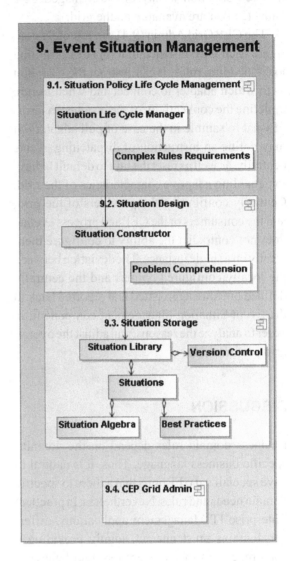

Figure 10. Event Situation Management internals

(9.2) to construct a new situation, based on the problem comprehension. By using the stored situations in the Situation Storage component (9.3), it can either reuse situations as defined by best practices, or construct new ones by using situation algebra. The designer component optimizes the collective rules into aggregated policies, and detects which complex rules should be stored within the system itself to reduce calculations.

After defining a new situation, and detecting the need to deploy it, the Situation Policy Life Cycle Management component (9.1) deploys the required situation in any of the managed CEP Units (2) that are available on the grid.

The CEP Grid Admin (9.4) can remotely configure each CEP instance in terms of submitting load-shedding rules. It can instruct the CEP unit as to which adapter to connect and when to run, or define the connectivity of the instance in terms of what to sample in the case of pull adapters. It may define an instruction of dispatching events centrally, as well as reset the unit to default behavior, enabling adaptive characteristics of the grid. Centrally configuring the providers of the grid, and the consumers of the CEP and process events, does not contradict the ability to configure them locally on a single instance. The deviation between the locally configured options and the centrally defined ones are aggregated and reported back to the Event Situation Management component, in order to analyze the reasons, and adjust the overall grid performance.

DISCUSSION

Business domain rules depend on the domain-specific business language. Thus, it is natural to have specialized CEP units that adhere to specific domain needs and rules. Nevertheless, in practical Enterprise IT Management applications, different domains must interact, jointly providing a comprehension of the overall system behavior, and cross relating services performance and availability measurements across business domains. The CEP grid architecture enables the separation of concerns using specialized CEP units utilizing domain-specific CEP algebra rules. The units are interacting using the grid publish/subscribe and dispatching mechanism, while transforming the event schemas in correlation with the propagation of the inferred events within the grid.

Different clients have different needs. A full-blown solution that might be acceptable for one customer might not be applicable to another. Thus, based on separable Level 1 components,

customers can deploy a partial solution of a CEP unit, with minimal configuration, and later expand the investment and add functionality by means of additional components. Specifically, consider the evolutionary behavior of schema matching, or ones that require central situation management and a repository of best practices, which recommends business CEP rules.

Computation efforts are important for scalable systems, and therefore the easiest method for improving performance is to avoid investing calculation efforts in redundant or repeatable situations. In the proposed reference architecture, scalability is achieved by several means. The first is the reuse of pre-calculated complex rules, *e.g.*, by using aggregated trees. Each node contributes partial calculation, either producing an inferred event that is dispatched to consumers, or serving as a common mathematical effort for another inferred event. Thus, each fundamental situation is calculated only once.

The second mechanism involves the reduction of overall processed data, by either preventing events from entering the grid, or stopping them from propagating into the system (based on content). The specifications of which events should be blocked may be done locally at each CEP unit, or for the grid as a whole.

The third mechanism is a cascade of persistency engines: one that caches and calculates the real-time events as they propagate into the system; the second is the time-window based for longer term events; and the third is an archive for offline analytics. This separation of duty enables the system to divide the computation efforts even for a single CEP unit, and construct a load balancing solution.

Overall system perceived performance, business rules correctness, and system non-functional characteristics are important for enterprise applications. The system's quality of service must be maintained, adhering to regulation and compliance needs. Centrally collecting and managing the lifecycle of CEP rules is critical for validating their current applicability as well as timing the execution. This balance enables regulated management of business rules that are aggregated and reported to executives and other stakeholders while conforming to compliance needs. Moreover, CEP central configuration services provide the ability to mesh-up CEP units during run-time. In addition, it enables high availability of the system in terms of active CEP services, fail-over mechanisms, and accumulation of reusable rules. The interchangeable execution of CEP units provides virtual CEP services that are globally deployed.

CONCLUSION

This chapter focused on the reference architecture of Complex Event Processing (CEP) systems as a generic comprehensive recommendation for event management. The implementation architecture might differ from one solution to another, however, and we consider the fundamental components presented in this chapter as the ones that capture the essence of a CEP system. Given the enterprise environment, one should be cautious when selecting the specific components, since this solution is structured with an EITM (Enterprise IT Management) approach in mind. The complexity and scalability of other systems might differ, thus, only a subset of the recommended architectural parts will be applicable. Since the collected components reflect separable customer needs, one must first examine actual customer needs, and adjust the architecture accordingly.

Separate business languages, performance considerations, scalability, availability, as well as multi-tenancy considerations affect the practical configuration of a CEP grid. These concerns, as well as others, were assimilated into the proposed CEP reference architecture when selecting the modules and deployable units.

We have experience with implementing components of the generic architecture at CA. The outcome was an abstraction of an existing product, evaluating the needs for future possible deliverables.

Given this architecture, we plan on exploring specific elements in it, including the extension of existing works on rule ordering (*e.g.*, (Widom & Ceri 1996)) to a grid environment, the use of existing schema matching techniques (*e.g.*, (Gal 2007)) for semantically enriching events, grid optimization of load shedding, and the development of algebra and data structures for the recursive activation of CEP units.

REFERENCES

Adi, A., & Etzion, O. (2004). Amit - the situation manager. *The VLDB Journal*, *13*(2), 177–203. doi:10.1007/s00778-003-0108-y

Balazinska, M., Khoussainova, N., & Suciu, D. (2008). PEEX: Extracting probabilistic events from rd data. In *Proceedings of ICDE*.

Chakravarthy, S. (1994). Snoop: an expressive event specification language for active databases. *Data & Knowledge Engineering*, *14*(1), 1–26. doi:10.1016/0169-023X(94)90006-X

Dayal, U. (1988). The HiPAC project: Combining active databases and timing constraints. *SIGMOD Record*, *17*(1), 51–70. doi:10.1145/44203.44208

Gal, A. (1999). Semantic Interoperability in Information Services: Experiencing with CoopWARE. *SIGMOD Record*, *28*(1), 68–75. doi:10.1145/309844.310061

Gal, A. (2007). Why is Schema Matching Tough and What Can We Do About It? *SIGMOD Record*, *35*(4), 2–5. doi:10.1145/1228268.1228269

Gehani, N. H., Jagadish, N. H., & Shmueli, O. (1992). Composite event specification in active databases: Model and implementation. In *Proceedings of VLDB*, 23-27.

Hadar, E., & Silberman, G. (2008, October 19-23). Agile Architecture Methodology: Long Term Strategy Interleaved with Short Term Tactics. In *Proceedings of the International Conference on Object Oriented Programming, Systems, Languages and Applications*, (OOPSLA 2008), Nashville, TN.

Haddow, G. D., & Bullock, J. A. (2004). *Introduction to Emergency Management*. Amsterdam: Butterworth-Heinemann.

Hall, D. L., & Llina, J. (2001). *Handbook of Multi-sensor Data Fusion*. Boca Raton, FL: CRC Press.

Luckham, D. (2002). *The Power of Events: An Introduction to Complex Event Processing in Distributed Enterprise Systems*. Reading, MA: Addison-Wesley.

Michelson, B. M. (2006). *Event-Driven Architecture Overview*. Boston, MA: Patricia Seybold Group.

Shmueli, G., & Fienberg, S. (2006). Current and potential statistical methods for monitoring multiple data streams for biosurveillance. In *Statistical Methods in Counterterrorism* (pp. 109–140). Berlin, Germany: Springer Verlag. doi:10.1007/0-387-35209-0_8

Widom, J., & Ceri, S. (Eds.). (1996). *Triggers and Rules for Advanced Database Processing*. San Francisco: Morgan-Kaufmann.

Zimmer, D., & Unland, R. (1999). On the semantics of complex events in active database management systems. In *Proceedings of ICDE*, (pp. 392-399).

ENDNOTES

[1]   http://www.progress.com/psm/apama/event-driven-architecture/index.ssp

[2]   http://soa.omg.org/Uploaded%20Docs/EDA/bda2-2-06cc.pdf

[3]   http://debs08.dis.uniroma1.it/tutorials.php#etzion

[4]   http://en.wikipedia.org/wiki/SOAP

[5]   http://en.wikipedia.org/wiki/Transmission_Control_Protocol

[6]   http://en.wikipedia.org/wiki/SNMP

[7]   http://en.wikipedia.org/wiki/ISCSI

Chapter 2
Event Models in Distributed Event Based Systems

Rolando Blanco
University of Waterloo, Canada

Paulo Alencar
University of Waterloo, Canada

ABSTRACT

The event model of a system determines how events are defined and generated, and how events are notified to interested components. In this chapter we look at key differences between the event model in distributed event based systems (DEBSs) and event models found in other implicit invocation systems. We identify features common to all DEBS event models, and variations within different DEBSs implementations. The main goal of the chapter is to elicit important features in event models that need to be supported in the engineering of DEBS applications.

INTRODUCTION

Applications are regularly developed by composing functionality encapsulated in units of computation. Depending on the type of system these units of computation, *components* for short, include modules, classes, and programs. The functionality provided by components can be composed by procedural abstraction and implicit invocation (Garlan & Shaw, 1994; Dingel et al., 1998; Notkin et al., 1993). When composing functionality by procedural abstraction, also referred to as *explicit invocation*, names that identify a component are statically bound to the component implementing the functionality. This is the case of a function in one module invoking another function in another module, or a program in one computer using a Remote Procedure Call (RPC) to invoke functionality implemented by a different program on another computer. In contrast, when composing functionality by implicit invocation, a component *announces* an event. This event announcement triggers the invocation of functionality implemented by another component. The component announcing the event may or may not be required to know the name nor location of the component triggered by the event.

DOI: 10.4018/978-1-60566-697-6.ch002

Distributed Event Based Systems (DEBSs) are implicit invocation systems (IISs) comprised of distributed components that interact with each other using events only. Events are generated by components called *publishers*. Components interested in the events that have been generated, are called *subscribers*. A subscriber is notified when an event of interest to the component is generated by a publisher.

The term *event model* is used in the area of IISs research to characterize two different things. For some researchers, the event model of a system determines the support that the system provides for structuring event data (Rozsnyai et al., 2007). Here, we take a more general approach, similar to that of Garlan and Scott (1993), and Meier and Cahill (2005). We consider the event model to determine the application-level view that a developer must have in order to develop an event-based application or component. Hence, the event model determines how events are defined, how they are announced to other components, how components manifest their interest in events, and how events are delivered to interested components.

The event model found in DEBSs is frequently referred to as *publish/subscribe* (Eugster et al., 2003). The term publish/subscribe is quite generic and applies, not only to DEBSs, but to any IIS where components invoke a *subscribe* operation to manifest their interest on events, and a *publish* operation to announce events. Since the event model of a system goes beyond the actual mechanism used to announce and subscribe to events, we refrain from using the term publish/subscribe, and refer to the event model found in DEBSs, generically, as the *DEBS event model*.

In this chapter we identify and characterize the features of the DEBS event model, compare the DEBS event model to event models found in other IISs, and illustrate the variations that exist within different DEBSs. The identified features elicit the key aspects that need to be supported by software engineering methodologies, structuring and modularization constructs for DEBSs.

Hence, the DEBS event model here presented can be used to validate the support provided by existent software engineering methodologies for the developments of applications in DEBSs, or it can be used as the starting requirement for new proposals.

Understanding the DEBS event model is essential in the development of these systems and their applications. To sustain the need for our categorization, consider the design and specification methods typically proposed for IISs and applied to DEBSs (e.g., Wieringa, 2002). These specification methods assume event models that are incompatible or do not support all the features in the DEBS event model. For example, Harel statecharts (Harel, 1987) assume instantaneous event processing: the reaction to an event occurs in zero time, upon notification of the event. This is not the case when dealing with DEBSs, where events take time to reach subscribed components. Another issue is the assumption that only one event may happen at a time. UML statecharts (OMG, 2007) do not have this restriction, providing instead a queue of events. Unfortunately, both Harel and UML statecharts assume broadcasting of events, where events are globally visible to all components in the system. In contrast, in DEBSs, events are only notified to subscribed components. The incompatibilities between the DEBS event model and the statechart event model make the use of statecharts to model behaviour in DEBSs and DEBS applications difficult. This problem is not unique to DEBSs and arises when dealing with any IISs that have an event model incompatible with the statechart event model. The effect of this impedance between event models has been the proposal of multiple, different, and sometimes incompatible, statechart variations when attempting their use to model complex IISs (e.g., Beeck, 1994; Leveson et al., 1994; Dias & Vieira, 2000; Barbier & Belloir, 2003; Ryu et al., 2006).

Other specification methods based on finite state machines for representing system behaviour also suffer from event model incompatibilities when applied to IISs (e.g., Bultan et al., 2003). More recently, interface automata (Alfaro & Henzinger, 2001) have been used to describe the behaviour of reactive systems (Veanes et al., 2005; Völgyesi et al., 2005). As with statecharts, the event model assumed in interface automata is different than the event model found in DEBSs. For example, in interface automata, the arrival of a message while on a state not prepared to handle the message, would indicate an incompatibility between the environment and the automaton. In DEBSs, on the other hand, the message would be queued until the component is at a state ready to handle the message.

This chapter does not provide a comprehensive list of existent DEBSs. Instead, we are interested in categorizing variations of the DEBS event model. Specific features of the DEBS event model are illustrated with few sample DEBSs. Therefore, the reader should not assume that the mentioned DEBSs are the only systems that exhibit a particular feature.

We start the chapter by enumerating the key features in the event model common to all DEBSs. We then review related proposals for the categorization of these features. Since there is no agreed-upon categorization of event models for IISs, to be able to compare the DEBS event model with the event models of other IISs, we introduce a general event model categorization, applicable to any IIS. After the related work, each section in this chapter covers one of the main categories: event definition, event attributes, attribute binding, event binding, event announcement, event subscription, event delivery, event persistence, and event notification. In each section, we start by describing the category, first in terms of general IISs, and then specifically for DEBSs and their variations. We conclude the chapter by summarizing the features of the DEBS event model and discussing future trends.

FEATURES OF THE DEBS EVENT MODEL

The following event-related features are common to most DEBSs:

- The kinds of events, components publish and subscribe to, are not predetermined by the system. New kinds of events can be introduced to the system, and existent kinds of events can be removed from it, at run time.
- Event bindings determine which components are to be notified when an event is announced. In DEBSs, bindings between events and components that react to the events can be established or terminated dynamically. Moreover, event bindings are maintained by the DEBS, without the need for publisher components to be aware of which components are interested in their events.
- The announcer of an event does not specify the components that will be notified of the event.
- A component publishes an event by explicitly invoking an announce or publish operation.
- When an event is announced, the component publishing the event continues its execution without being blocked. The publisher component does not wait for subscriber components to be notified or for their reaction to the event.
- Components must register their interest on the events they want to be notified about.
- The system attempts to deliver an event to all the components interested in the event.

There are variations between DEBSs with regards to how events are declared, the form of the subscription operation, whether or not event filtering is supported, the event delivery semantics, and the support for event persistence. Variations

Figure 1. Event models, related work

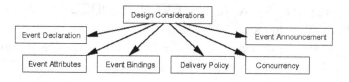

(a) Notkin et al., Design considerations when supporting event–based interactions

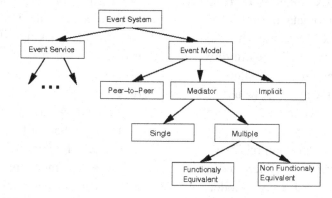

(b) Meier and Cahill's taxonomy of DEBSs, event model dimension adapted from Meier and Cahill (2005)

also exist on how the event model is implemented and its architecture.

We now look at previous works, related to the categorization of these features, not only for DEBSs, but for IISs in general.

RELATED WORK

Most of the research efforts in DEBSs have been oriented towards the development of commercial and prototype systems (e.g., *middleware*) that can efficiently provide the functionality required for the execution of DEBS applications (Oki et al., 1993; Cugola et al., 1998; Carzaniga et al., 2001; Mühl, 2002; Bhola et al.,2002; Pietzuch, 2004). There has also been some interest from the formal methods community, specifically centered in the modeling of event systems (Dingel et al., 1998; Mühl, 2002; Garlan et al., 2003; Fenkam et al., 2004; Cai et al., 2007). Unfortunately, few research works have focused on the engineering of DEBS applications.

The most important works related to the classification of event models in IISs are Notkin et al. (1993) and Meier and Cahill (2005). Notkin et al. (1993) enumerate key design considerations that arise when extending traditional languages with implicit invocation. Meier and Cahill (2005) propose a taxonomy of distributed event-based programming systems.

Although proposed for the addition of implicit invocation to traditional languages, the design considerations enumerated by Notkin et al. (1993), and shown in Figure 1 (a), relate to the event model that will be provided by the extended language. These design considerations are:

- **Event declaration:** what vocabulary is used to define events, when is the event definition done, and whether or not the event vocabulary is extensible.
- **Event parameters (attributes):** what information is associated to events.
- **Event bindings:** how and when are events bound to the components that process them.

- **Event announcement:** whether events are announced explicitly or implicitly. If events are explicitly announced, what procedures exist to announce the events.
- **Delivery policy:** whether events are delivered to one or all components interested in the event.
- **Concurrency:** number of threads of control that exist in the system.

These considerations cover some of the essential properties in an event model, but important issues not related to programming languages, and relevant to distributed systems, are not included. For example, delivery semantics that arise when dealing with unreliable communication, addressability of events, and announcement call models.

Meier and Cahill (2005) take a different approach, and focus instead on the communication and structural properties of the event model in distributed IISs. Meier and Cahill (2005) categorize DEBS based on their event model, and the event system (middleware) implementing the event model. Event models are classified by Meier and Cahill (2005) as *peer-to-peer*, *mediator*, and *implicit* (see Figure 1 (b)). In a peer-to-peer event model, components announcing and consuming events communicate directly with each other. This form of interaction is common in systems implementing the Observer pattern (Gamma et al., 1995). In most Observer pattern implementations, components that are interested in events need to know the components that announce the events, and register their interest directly with the publisher components. Publisher components must keep track of the components that are interested in the events and, when an event is announced, send the event to the interested components.

In our opinion, the peer-to-peer event model classification proposed by Meier and Cahill (2005) does not apply to DEBSs. The main reason is that, as we will see later, in DEBSs the components interested in events (similarly, announcing events) are not required to know the name nor location of the components announcing (similarly, interested in) events. The end result is that applications in DEBSs tend to have reduced coupling when compared to Observer pattern implementations: the providers and users of functionality in DEBSs can be decided at run time, without requiring a priori knowledge of their names nor locations. This low coupling favours the development of components that are autonomous and heterogeneous.

In terms of Eugster et al. (2003), DEBSs provide *time*, *space* and *synchronization* decoupling features that are not always provided by peer-to-peer event models. By time decoupling, Eugster et al. (2003) refer to the fact that interacting components in a DEBS do not need to be actively participating in the interaction at the same time; space decoupling refers to the fact that components do not need to know each other for them to interact; and synchronization decoupling refers to the fact that when a publisher announces an event, the publisher does not wait for the event to reach interested components.

Continuing with the classification of Meier and Cahill (2005), in the mediator event model, the communication is made via one or more mediator (broker) components. The mediator event model is similar to the Event Channel pattern (Buschmann et al., 1996). In the Event Channel pattern, components communicate via events without the need to register with each other, communicating instead, indirectly, via the mediators. Typically different mediators are in charge of different kinds of events. Hence, components interested in a particular kind of event must be able to know and communicate with the mediator (event channel) in charge of disseminating the events of interest.

Finally, according to Meier and Cahill (2005), in an implicit event model, components consuming events subscribe to a particular event type rather than to another component or mediator. In this case the system must keep track of the components' interests and of the delivery of events. The implicit event model can be seen as a mediator event model where the system is the broker for all kinds of events.

As described, most of the work in Meier and Cahill (2005) centers on the architectural aspect of the IIS, while the work in Notkin et al. (1993) centers on some of the design options available to produce the event model for an IIS. Neither work provides a complete categorization of the diverse event models found in IISs. Notkin et al. (1993) lacks categories that are relevant in distributed IISs, while many of the categories in Meier and Cahill (2005) are made from an structural point of view and allow features undesirable in DEBSs. With the purpose of categorizing the different event models in IIS, we combine and complement Notkin et al. (1993) and Meier and Cahill (2005) into an extended categorization summarized in Figure 2, and discussed in the following sections. The extended categorization is key to understanding the differences between DEBSs and other IISs. It also provides a collection of event related features that need to be supported in the engineering of DEBS applications.

As originally done by Notkin et al. (1993), we start the characterization of event models by looking at the support that the system provides for defining new kinds of events.

EVENT DECLARATION

The *event vocabulary* of a system is the collection of the kinds of events that components in the system can announce. Our first criterion for categorizing event models is whether or not this collection is static or dynamic, and if dynamic, whether new kinds of events need to be declared before events can be announced.

A system has *fixed event vocabulary* when the kinds of events that a component can announce are predetermined by the system. In contrast, in a system with *variable event vocabulary*, different kinds of events can be added to the system as required. Aspect Oriented Programming (AOP)

(Kiczales et al., 1997), is an example of an IIS with fixed event vocabulary. In AOP, components called *advice* are invoked when a program reaches certain execution points (e.g. a method call, a method call return, a constructor call, etc). The kinds of execution points where advice can be specified are predetermined by the AOP system.

When new kinds of events can be added to a system, but the set of events must be known before the system operates, the system has *static event declaration*. An example is the extension to Ada and other languages proposed in Garlan and Scott (1993). Specifically, interfaces for Ada packages are extended with the declaration of the events each package generates. The names of the package methods to be invoked when specific events are generated are also part of the implicit invocation declaration. The event declaration in this Ada extension is then precompiled and Ada-only code is generated. An executable program is built by compiling and linking the generated Ada code with the rest of the application code.

When new kinds of events can be added at run time, the system has *dynamic event declaration*. In some IISs, there is no need to declare the kinds of events that will be announced. Such systems are said to have *no event declaration*, and events are typically announced by specifying an arbitrary string or a list of strings. When new kinds of events need to be declared in a system, the declaration may occur at a *specialized centralized* location, or alternatively, they may occur at multiple locations. In this latter case, the system implements *distributed* declaration of events.

Event Declaration in DEBSs

DEBSs commonly have variable event vocabulary: the kinds of events that can be announced by components are not predetermined by the system. Nevertheless, there are variations with regards to whether or not event types need to be declared, whether the declaration is static or dynamic, and the location of the declaration.

Figure 2. Event model categorization for IISs

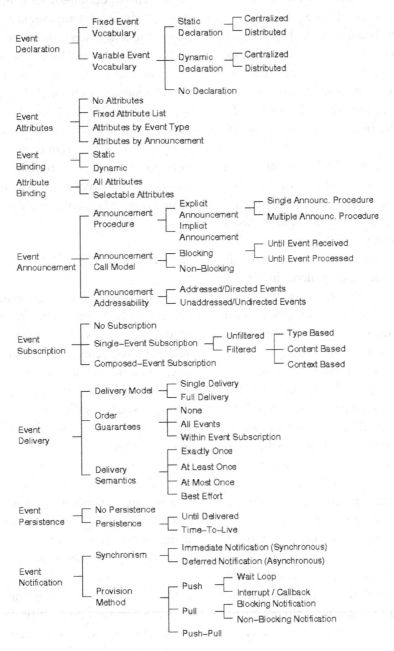

Before an event can be published by a component, most DEBS require that the type of the event be registered in the system. Since the event declaration happens at run time, these DEBSs have dynamic event declaration. For example, in Hermes (Pietzuch, 2004), the components that react to events are programs. Programs subscribe and unsubscribe to events via event brokers. Programs generating events register event types with the event brokers. Once an event type is registered, a program can announce events of the registered event type. Event types can be registered and unregistered dynamically.

A DEBS where new kinds of events are not declared is the Java Event-Based Distribution Architecture (JEDI) (Cugola et al., 1998, 2001). Each event in JEDI corresponds to a list of strings, where the first string is the event name. The other strings in the list are the event attributes. An event subscription in JEDI specifies the name of the event and, possibly, filtering arguments on the event attributes.

A system that supports both dynamic event declarations and no event declarations is SIENA (Carzaniga et al., 2001). SIENA can operate under what the SIENA creators call *subscription-based semantics* and *announcement-based semantics*. Irrespectively of the operation mode, an event is a set of typed attributes. Components in SIENA are objects, and they announce events by invoking a publish call. Under subscription-based semantics, components interested in being notified, subscribe to events by invoking a subscribe call. A filtering expression specifying values for the event attributes is passed as parameter of the subscribe call. A component is notified of the occurrence of an event if the filtering expression specified in the subscription call matches the event notified via a publish call. When operating under subscription-based semantics the SIENA system requires no event declaration. In contrast, when operating under announcement-based semantics, components generating events need to register the events they will be generating by invoking an advertise call. An unadvertise call is used by components to inform that a given event will no longer be generated. Under announcement-based semantics the SIENA system has dynamic event declaration.

With regards to the location of the declaration, an example of a DEBS where the events are declared in a central location is Yeast (Krishnamurthy & Rosenblum, 1995). In Yeast events are generated when object attributes change. Yeast provides a predefined set of objects and attributes (e.g. the object file has attributes file name, creation time, modification time, etc). Components can declare new events by defining objects and their attributes via commands that are executed on the machine where the components run. These declarations are processed and stored by the Yeast server.

Examples of DEBSs with distributed event declaration are Rebeca (Mühl, 2002), Hermes (Pietzuch, 2004), and Gryphon (Bhola et al., 2002). In these DEBSs several event brokers, possibly running at different locations in the system, process the event registration, subscription and unsubscription requests from the components generating and reacting to the events.

EVENT ATTRIBUTES

IISs may allow the association of data attributes to events. These data attributes are also known as the *parameters* of the event. If attributes are not supported, relevant information related to the event must be deduced from the event name or be retrieved, via global variables, shared memory or database space, explicit invocation, or any other means by the component reacting to the event announcement.

Few IISs do not support event attributes. The signal notification system in UNIX (CSRG, 1986) is an example of an IIS where there are no attributes associated with the events. Processes using signals, must develop their own protocols for retrieving any relevant information associated to the occurrence of a signal. Moreover, the protocol used to retrieve the relevant information is separate from the signal notification system itself.

An IIS may associate the same *fixed attribute list* to each event. This is the case of the Java Message Service (SUN-JMS, 2002). All messages in JMS have a message header, a set of message properties, and a message body. Any application information, relevant to the message, must be coded in the message body by the component generating the message. Components receiving the message must know how to decode the information in the message body. JMS supports the codification of data in the message body as a string (for example representing a XML document), a list of attribute/value pairs, a serialized object, a stream of bytes, or a stream of Java primitive data types.

Similarly to JMS, the Corba Event Service (CORBA-ES, 2004) supports a *Generic Event Communication* mode where all events have a single attribute of type *any* corresponding to the event data. Components in the Corba Event Service must agree on how required event information is coded into the event data attribute.

In most recent IISs, events are instances of a certain *event type* or class (the term *event schema* is also used in the area to specify an event type). When event types are used, the set of attributes associated to the event is determined by the type of the event. EAOP, the model and tool described in (Douence & Südholt, 2002) for event-based aspect-oriented programming, is an example of an IIS where the number and type of attributes associated to an event are determined by the type of the event. EAOP has a fixed event vocabulary with four types of events: method call, method return, constructor call, and constructor return. Depending on the type of the event, there are a number of attributes associated to each event. For example, some of the attributes associated to events of type method call are the method name, the values of the arguments passed to the method, and the depth of the execution stack.

Alternatively, any number of attributes and their types may be determined at the time the event is announced. In this case, two different event announcements of events of the same type may have different parameters.

Event Attributes in DEBSs

Event attributes are typically supported by DEBSs. There are variations in DEBSs with regards to whether or not all events have the same fixed attribute list. If different attributes are associated to different kinds of events, there are also variations on whether or not the attributes associated to an event are determined by the type of event.

As previously mentioned, JMS (SUN-JMS, 2002) associates the same fixed attribute list to each event. One of the attributes in the list corresponds to a data attribute that is used by applications to associate any relevant application data to the event. Hence, although JMS based DEBSs could be restricted by the fixed attribute list, they implement a higher level abstraction of events where the data attribute is used to provide support to typed events.

Another example of a DEBS that associates the same fixed attribute list to every event is JINI (SUN-JINI, 2003). JINI allows Java objects to be notified of events occurring on other, possibly remote, objects. Every event has four associated parameters: (1) an attribute identifying the type of event; (2) a reference to the object on which the event occurred; (3) a sequence number identifying the instance of the event type; (4) a hand-back object. The hand-back object is a Java object that was originally specified by the component receiving the event when the component first registered its interest on the event.

Most DEBSs require events to be typed, the event type determining the attributes in an event. For example, Hermes (Pietzuch, 2004) uses XML Schema specifications (W3C, 2004) to represent event type definitions. The event type definitions are then used in Hermes to type check event subscriptions and publications. CAE, the Cambridge Event Architecture (Bacon et al., 2000), is also a DEBS where the event attributes depend on the event type. An event occurrence is represented in CAE as the instance of a given event class. Event types are defined using an Interface Definition Language (IDL). Components interested in events of a specific class, specify a value or wildcard for each attribute of the given event class.

In DEBSs with event attributes dependent on the type of event, every single instance of a given event type has the same attributes. In contrast, in JEDI (Cugola et al., 1998) all events are represented by a list of strings. The first string in the list indicates the event name. Hence, in JEDI there is no guarantee that two events with the same name represent the same event type. Similarly, there is no guarantee that all announcements of events of the same type are done with same-sized lists of strings.

GREEN (Sivaharan et al., 2005) is a DEBSs implemented as a framework (Fayad et al., 1999) that supports different event attribute representations by way of plugins. A plugin is a specific implementation of an aspect of the event model. Plugins exist for representing events attributes as sequences of strings, name value pairs, and objects.

EVENT BINDING

The event binding in an IIS determines which components are to be informed when an event is announced. In *static event binding*, the components that react to an event are predetermined at compile time. An example of static event binding is the proposal in Garlan and Scott (1993) to extend ADA with implicit invocation. In Garlan and Scott (1993) an event specification language is used to indicate, for each package, the events the package wishes to be notified about, and the methods that are to be invoked when the event occurs. Hence, the components generating the events (e.g., packages in the case of Ada), and the components reacting to each event, are known at compile time.

In *dynamic event binding*, bindings between events and components that react to the events can be established or terminated dynamically. Most IISs support dynamic event binding.

Event Binding in DEBSs

In DEBSs, bindings between events and components that react to the events can be established or terminated dynamically. Moreover, event bindings are maintained by DEBSs without the need

for publisher components to be aware of which components are interested in their events. Because of this low coupling between event publishers and subscribers, DEBSs have been proposed for applications with large, autonomous, and possibly heterogeneous components (Weiser, 1993; Geihs, 2001).

ATTRIBUTE BINDING

The attribute binding in an IIS determines the attributes that are passed to an interested component when an event is announced. The majority of IISs pass *all attributes* in the event to the components reacting to the event. A different option is to allow a component to *select the event attributes* of interest. All but one of the reviewed IISs that support event attributes pass all the event attributes to the components reacting to the event. The exception is the Ada extension in Garlan and Scott (1993) where the names of the event attributes to pass to the methods reacting to an event are included as part of the event declaration. Not all event attributes need to be bound to method attributes.

As an illustration, consider an event NewCustomer) with attributes CustomerNum and CustomerName. This event NewCustomer is generated when a new customer is added to an Accounts package. Another BankGateway package, may only be interested in the attribute CustomerNum. Hence in Garlan and Scott (1993) it is possible to bind a package to only the attributes of interest, as illustrated by the method InformBank in the following listing:

```
for Package_Accounts
  declare Event_NewCustomer
    CustomerNum: Integer,
    CustomerName: String (1.. 30)
  when Event_BankPayment =>
Method_ProcessPayment TransId
  end for Package_Accounts
  for Package_BankGateway
```

```
declare Event_BankPayment
   TransId: Integer;
when Event_NewCustomer =>
Method_InformBank CustomerNum
   end for Package_BankGateway
```

Attribute Binding in DEBSs

We are not aware of any DEBSs where components can decide what event attributes to bind. This may change in the future with the addition of role based access controls (RBAC) to DEBSs (Belokosztolszki et al., 2003; Bacon et al., 2008). In general, access roles characterize a set of components (Ferraiolo & Kuhn,1992). Roles are currently used in DEBSs to restrict the types of events that a component can subscribe to (e.g., Belokosztolszki et al., 2003). The granularity of the restriction can, in theory, be increased to impose access controls at the event attribute level. Hence the event attributes to bind in a component being notified of the occurrence of an event would depend on the roles that have been granted (or not granted, if negative expressions on roles are allowed) to the component. When roles are used to restrict attribute access, the binding is then imposed by the system, and it is not a choice made by the subscribed component. This is nevertheless an open area of research.

EVENT ANNOUNCEMENT

Most IISs have *explicit event announcement* procedures that need to be used to generate an event. In some systems there is a unique announce or publish procedure or method, while in other systems there are several announcements methods.

In IISs with *implicit event announcement*, the event is generated as a side effect of executing an instruction or procedure. AOP is an example of an implicit invocation system with implicit event announcement (Kiczales et al., 1997). In AOP, events are generated when the execution of the program reaches certain points (e.g., method calls, control structures, assignments). Another example of implicit announcement is active database systems (Cilia et al., 2003). In active database systems, components, known as database triggers, are invoked as a side effect of the insertion, deletion, or updating of data in the database.

When an event is announced, the execution of the component announcing the event may be blocked *until the event is received* by all components to be notified of the event, or *until the event is processed* by all components receiving the event. Alternatively, the announcement of an event may not block the execution of the component announcing the event.

In general, centralized IISs tend to follow a blocking call model for the announcement of the event, with many of the systems blocking until the components processing the event finish their processing of the event. Most distributed IISs, on the other hand, provide a non-blocking call model.

The model and tool for event-based aspect-oriented programming (EAOP) presented in Douence and Südholt (2002) is an example of an implicit invocation system where the code generating events is suspended until all components, *advice* in this case, process the event. An execution monitor in EAOP tracks the events generated during the execution of a base program on which advice has been defined. When an event is generated, the execution of the base program is suspended and the execution monitor sequentially invokes every advice associated to the given event. Once each advice has executed, the monitor gives control back to the base program.

Most active databases support a blocking call model as well. Programs inserting, deleting, or updating data are blocked until the code reacting to these database operations (*database triggers*) complete their execution. An added feature is that, in active database, the changes made to the data in the database, by the components generating the events and the functional components reacting to the events, are typically part of a single database transaction. Logically, all or none of the modifications to the database are carried out - independently of whether the modification was performed by the component generating the event or the component reacting to the event.

Independently of the announcement method used by the IISs, components may or may not be required to *address* the announced events. Addressing of events, also referred to as directing of events, happens when the announcer of the event specifies the component that will be notified of the event. Dingel et al. (1998) use this type of announcement in their model for the verification of IISs. In Dingel et al. (1998), components, called methods, send events via an announcement call where the first argument of the call is the intended recipient of the event, and the second parameter of the call is the event data itself.

The UNIX signal notification system also requires functional components to direct the events (CSRG, 1986). In UNIX, the functional components generating and processing the signals are called processes. A process is uniquely identified by a numeric process identifier, and a group of processes is uniquely identified by a numeric group identifier. A signal can be directed to a specific process, or to all the processes in the same group as the process sending the signal.

The requirement to address events augments the degree of coupling between event announcers and event consumers. Hence, several IISs have unaddressed event announcement, where the system is in charge of directing the event to the components registered for the event.

Event Announcement in DEBSs

Components in DEBSs explicitly announce events, the operations used to announce the events are non-blocking, and the announced events are unaddressed.

Most DEBSs have only one operation to announce events. An exception is Hermes (Pietzuch, 2004), which provides three announcement meth-ods: publishType, publishTypeAttr and publish. The method publishType announces an event that will trigger the notification of components that have subscribed to events of the given event type (or to any ancestor of the published event type, since Hermes supports inheritance of event types). The second method, publishTypeAttr, triggers the notification of components that have subscribed to events of certain type (or event type ancestor) and that have specified value conditions on the attributes. The third announcement method, publish, has the same effect as invoking both the publishType and publishTypeAttr calls.

EVENT SUBSCRIPTION

Components may or may not be required to register their interest to be notified when events are announced. If there is no requirement to register for events, also referred to as *subscribing* for events, event announcements may be broadcasted to all components in the system. Alternatively, event announcements may be registered in a shared memory space that is accessed by components wishing to inquiry if a certain event has been announced. Linda (Gelernter, 1985) is an example of an IIS where components are not required to announce their interest for events. In Linda, implicit invocation is done via a shared memory region called the *tuple space*. Functional components, processes in the case of Linda, generate tuples that are stored in the tuple space. Other processes monitor the tuple space and can read and, optionally, remove the tuples that have been added to the tuple space.

Most IISs require that components register for the events they wish to be notified about. In systems with *single-event subscription*, there is a subscription procedure that must be invoked for each event of interest. In systems with *composed-event subscription*, a component can express its interest to be notified when a composition of events occurs.

Some IISs with single-event subscription allow the specification of a filtering expression as part of the subscription operation. An event is then delivered to a component only if the component is interested in the event, and the filtering expression associated with the interest of the component for the event holds. If the system supports event types, the expression can be based on the *type* of the event. Recall that when event types are supported, every generated event in the system is an instance of a certain event type. As previously mentioned, the term event schema is also used in the area to refer to event types.

Some systems are described as having *topic* (a.k.a. subject) based subscription (Eugster et al., 2003). Events in these systems are grouped in feeds, and interested components subscribe to the feeds. We will assume that this type of filtering (topic or feed based), is an instance of type based filtering, where a different event type can be defined for each feed of interest.

When a system allows the specification of filtering conditions on the attributes of the event, the system is said to have *content* based filtering. An event is delivered to an interested component if the conditions imposed on the attribute values of the event are met. Recent systems also allow the filtering of events based on contextual information related to the physical location or proximity of the components announcing and subscribed to the event (Meier & Cahill, 2002; Sivaharan et al., 2005).

When IISs support composed-event subscription, a mechanism is provided for components to express event composition conditions. This mechanism is typically implemented as a language that allows the specification of temporal conditions on the event occurrences (Carlson & Lisper, 2004; Konana et al., 2004; Liebig et al., 1999; Mansouri-Samani & Sloman, 1997). For example, the language proposed in Konana et al. (2004) for the specification of composite events allows the identification of a sequence of events that satisfy or violate timing and event attribute-value constraints. Based on real time logic (RTL), the specification of conditions of the type "the third occurrence of the event of type e_i after time t must have a value v for attribute $e_i.attr$" are possible in the proposed language. Konana et al. (2004) also assumes the existence of data repositories in the form of relational databases. Hence, conditions on the data stored in the data repositories are also part of the event composition language.

The actual event filtering may occur at a central location (Konana et al., 2004), at each component (Oki et al., 1993), or at specialized event servers (Carzaniga et al., 2001).

Event Subscription in DEBSs

DEBSs have mandatory event subscription: components must register their interest on the events they want to be notified about. Most DEBSs support single-event subscription and allow the specification of conditions on the values of the event attributes. Hence, most DEBSs support content-based event filtering. For example, in NaradaBrokering (Pallickara & Fox, 2003) filtering conditions are specified as SQL queries on properties contained in JMS messages, as well as XPath queries. In JEDI (Cugola et al., 2001), a filtering expression is specified as an ordered set of strings. Each string in the set represents a simple form of regular expression that is matched against the attribute in the same position in the event of interest. Recall that, in JEDI, an event is represented as a list of strings, where the first string in the list is the event name.

SIENA is an example of an implicit invocation system that supports composed event subscription. A filtering condition f, on the event type and event attribute values, can be specified in SIENA as part of the event subscription call. A pattern f_1, $f_2,...,f_n$ can also be specified, where each filtering condition f_i may apply to a different event type. Such subscription indicates that the functional component running the subscription operation shall be notified if events $e_1, e_2,..., e_n$ are generated, such that:

- e_i occurs after e_{i-1} for all $2 \leq i \leq n$
- The filtering condition f_i is true when evaluated for the event e_i, with $1 \leq i \leq n$

GREEN (Sivaharan et al., 2005) supports multiple filtering models via plugins. Recall that GREEN is implemented as a framework, where different aspects of the event model can have different implementations. When a GREEN system is deployed, application developers instantiate the system with the plugin implementations that meet their requirements. There is support in GREEN for type based, content based and proximity based filtering. Assuming event types represented in XML, an example of a filtering condition is:

```
//RoadTraffic/[%type = $Traf-
ficLight$, colour = $Red$]?#DISTANCE#
< $100$
```

This example, based on an example from Sivaharan et al. (2005), indicates that the component should be notified if an event of type *RoadTraffic*, representing a traffic light changing to colour *Red* within a distance of *100* units.

GREEN also supports composite-event subscription via a plugin that interfaces with CLIPS (C Language Integrated Production System) (NASA Software Technology Branch, 1995). CLIPS allows the specification of Event-Condition-Action rules as filtering conditions in the event subscriptions.

EVENT DELIVERY

Once an event is announced, the system must select the functional components that will receive the event. In *single delivery* of events, an event is delivered to only one of the components interested in the event. Single delivery of events is useful in a pool of servers where only one server is required to attend a request represented by an event. In *full delivery*, an event is delivered to all the components interested in the event. IISs implementing addressed events typically support single delivery of events. For example, the UNIX signal notification system (CSRG, 1986), operates in single delivery mode when a signal is sent to a process. Full delivery operation occurs in UNIX when a signal is addressed to a group of processes.

Linda (Gelernter, 1985), supports both single and full delivery. Events, represented as tuples in Linda, are stored in a tuple space that is accessible to all components. A component reacting to an event has the option of removing the event from the tuple space. To guarantee that only one component accesses the event, semaphores and other process synchronization techniques, can be modeled in Linda.

The delivery semantics of a system determine if announced events are delivered *exactly once*, *at least once*, *at most once*, or in *best effort*, there are no delivery guarantees. Exactly-once and at-most-once delivery are usually more difficult to implement than the other options, since the implementation may require the use of transactional protocols. An important part of the delivery semantics, is whether or not there are order guarantees in the delivery of events. Some systems may provide order guarantees within events of a single event type. In these systems it is possible to identify, for two events of the same type, which one was generated before the other, or whether both were generated at the same time. Other systems may provide system-wide ordering guarantees. In this later case, it would be possible to identify, for any two events, even if not of the same type, which one was generated before the other.

Event Delivery in DEBSs

DEBSs commonly implement full delivery of events: events are delivered to all subscribed components. There are variations with regards to the delivery semantics provided by DEBSs. For example, IBM's Gryphon project (Bhola et al., 2002), implements a protocol that guarantees exactly once delivery if the components being notified of the events maintain their connectivity to the system. The protocol models a knowledge graph where nodes, named routing brokers, represent components in charge of routing events. Arcs in the graph represent filtering conditions on the events. The filtering conditions are used to split the routing of events between routing brokers. The graph is dynamically adjusted in case of node or network failures. Further refinement of the protocol is presented in (Zhao et al., 2004). In this later work, the protocol is extended to guarantee not only exactly once delivery, but ordered delivery of events matching a single subscription. Components, named subscription brokers, receive subscription requests from other functional components in the system wishing to be notified of events. Ordered delivery is accomplished by associating, with each event generated, a vector containing information for each subscription request related to the event. Virtual timers at subscription brokers are used to identify, from a stream of events matching a subscription, the first event in the stream after which every single event is guaranteed to be delivered, in order, to the component that subscribed to the events. To accomplish this functionality, the protocol propagates subscription information from subscription brokers to the functional components generating the events. Event information is propagated from the components generating the events to the subscription brokers. In this refinement of the original protocol, routing brokers are in charge of routing, both, the subscriptions and events.

Another system that provides delivery guarantees is NaradaBrokering (Pallickara & Fox, 2003). Specifically, at-least-once, at-most-once, exactly-once, and ordered delivery are provided via a Web Services Reliable Messaging (WSRM) implementation (OASIS, 2004). For reliable delivery, the system needs to have access to reliable storage or DBMS (Database Management System).

In JEDI (Cugola et al., 2001), events are guaranteed to be received according to the causal relationships that hold among them. In other words, if an event e_2 is generated as a reaction to another, previous event e_1, a component interested in both events is notified of e_1, before it is notified of event e_2.

EVENT PERSISTENCE

When an event is delivered, the intended recipients of the event may not be available to receive the event. In this case, the IIS may choose to save the event and attempt the delivery at a later time, or it may choose to abort the delivery of the event to the component that is unavailable. When *persistence* of events is supported by the system, the event may be maintained in the system until all intended recipients receive the event. Alternatively, the event may be maintained in the system until a *time-to-live* expires. The time-to-live may be the same for all events, it may be determined by the type of event, or it may be specified when the event is announced.

Most IISs deliver events only to functional components available at the time the event is generated. Linda (Gelernter, 1985) and other systems based on data repositories are exceptions. In Linda, events represented as tuples are stored in a common area until they are explicitly removed, either by the component generating the event, or by any functional component reacting to the event. Similarly, in Oracle's Advanced Queuing (Oracle, 2005), events in the form of messages are stored in Oracle's relational database. An expiration interval can be individually associated to each generated event. When no expiration interval is associated with the event, a component must explicitly remove the event from the system.

The delivery semantics of the system highly influence the event persistence supported by the system. With the exception of best effort delivery, the system must support some kind of event persistence to be able to provide delivery guarantees.

Event Persistence in DEBSs

Most DEBSs deliver announced events to the subscribed components that are running at the time of the delivery. A notable exception is Rebeca (Mühl, 2002). In Rebeca, local brokers serve as access points of the system and are in charge of managing clients. Local brokers are connected to router brokers. The router brokers handle event delivery. A special kind of components, named *history clients*, save the events they are subscribed to. When a component submits a subscription request to the system, it can indicate that it wants to be notified of past events that have been saved by history clients. History clients are then notified, and they re-publish the saved events for the interested component, and only for that component. The amount of events stored by history clients is not predetermined by the system and is in general up to the history client to decide.

As previously mentioned, NaradaBrokering (Pallickara & Fox, 2003) also implements event persistence when the system has access to reliable storage or DBMS. A replaying feature in NaradaBrokering allows the replaying of events at any time. This is in contrast with Rebeca where replaying can happen only during the execution of a subscription operation.

EVENT NOTIFICATION

When an event is announced, the components receiving the event may be *immediately notified* or, in *deferred notification*, notified at a later time. When the system provides immediate notification of events, the events are usually pushed to the

components interested in the events. The receiving components may implement a wait loop, or their execution may be interrupted by calling a previously registered callback routine.

When the event notification is deferred, the component may *pull* the system for information about any events of interest that may have occurred since any previous pull call. The pull operation may be *blocking* or *non-blocking*. An example of a system where events are pulled by components reacting to the events is the Corba Event Service Notification CESN (CORBA-ES, 2004). In CESN, a blocking pull call is used by a component to retrieve an event generated by an event producer. If no events have been generated, the execution of the component is suspended until an event is available. An alternate try_pull call in CESN may be used when the functional component reacting to the event does not wish to be blocked in the absence of events. The Java Message Service JMS (SUN-JMS, 2002) also supports blocking event polling. Blocking is supported via a receive method call. A mode of event pushing is also supported in JMS via an extra component called a message listener. Upon arrival of an event, the message listener invokes a previously registered call-back method.

A hybrid push-pull option occurs when subscribed components are signaled that an event of interest has been generated, and the components poll the system to retrieve the event data.

Event Notification in DEBSs

Most DEBSs initiate the notification actions as soon as an event is generated. The notification itself is typically pushed to the components, and a routine registered at the time of the subscription is invoked when the event is received by the component.

JEDI (Cugola et al., 2001) supports, both push and pull notification mechanisms. Events are typically pushed to subscribed components, but

Figure 3. Event model categorization for DEBSs

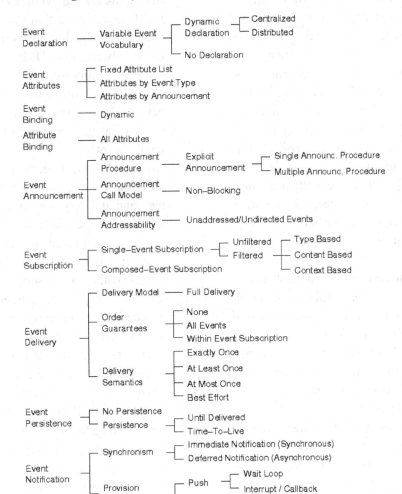

if a subscribed component is not online when an event is generated, the event can later be retrieved by the subscribed component via a getEvent call. A hasEvent call is also used by subscribed components in pull mode to check if there are any outstanding events of interest to be processed by the component. Outstanding events are events that have been generated in the system, but have not been successfully pushed to the component.

DEBS EVENT MODEL: RECAPITULATION

We have categorized the event model in IISs and, for each category, we have discussed the variations in the DEBS event model. In Figure 3, we summarize the DEBS event model properties:

- *Event Declaration*. The kinds of events that components announce in a DEBS are not predetermined by the system. Some DEBSs require that publisher components register the kinds of events they will announce. The actual registration can happen at a central location, or it can happen at multiple locations within the system.

- *Event Attributes*. Data attributes can be associated to events in DEBSs. Some DEBSs allow unstructured lists of strings to be associated with the events. Other systems support attribute name/value pairs, while yet other systems support complex data structures. The actual event attributes may be determined by the type of the event, or in some systems, the attributes may vary by announcement.

- *Dynamic Event Binding*. In all DEBSs, the event bindings between publishers and subscribers can be established and terminated dynamically. The bindings are maintained by the system without the awareness of the publisher or subscriber components.

- *Attribute Binding*. All event attributes are passed to subscriber components upon notification of an event.

- *Event Announcement*. Publisher components must explicitly announce the events by invoking a publish call. Some DEBSs provide more than one publish call. The publisher component is not blocked until subscribed components receive the announced event, or until the interested components react to the event. When an event is announced, the publisher component does not address the event to any particular component.

- *Event Subscription*. Subscriber components are required to inform the system of the kinds of events of their interest via the invocation of a subscribe call. As part of the subscribe call, some systems allow the specification of filtering conditions. An event is then delivered to a subscribed component only if the filtering condi-

tion holds for the event. In some systems, the filtering condition can specify multiple events. Such systems have composed event-subscription. DEBSs may support filtering conditions on the type of event, its attributes, or contextual information. The actual filtering may happen at a central location, at each component, or at specialized servers.

- *Event Delivery*. Events are delivered to all subscribed components. Order guarantees and delivery semantics vary among DEBSs. Some systems guarantee that events are delivered in order, while other systems have order guarantees only for events within a subscription. Most systems do not provide delivery guarantees. The delivery semantics in a system specify whether an event is guaranteed to be delivered exactly once, at least once, at most once, or if there are no guarantees.

- *Event Persistence*. Events may be kept in the system until all interested subscribers are notified, for a time-to-live period, or more frequently, the event is removed after its delivery - even if interested subscribers cannot be notified. The event persistence supported by a DEBS is typically determined by its delivery guarantees.

- *Event Notification*. When an event is announced, the system may attempt to deliver the event immediately, or it may delay its delivery. A subscriber component may be notified of an event via a previously registered call-back routine, or it may need to pull the system for any outstanding events. As part of the pull operation, if there are no outstanding events, the system may block the subscriber component until an event of interest occurs, or it may inform the subscriber of the absence of events.

FUTURE TRENDS

Trends related to the DEBS event model include DEBS frameworks and context aware DEBSs. DEBS frameworks are toolboxes that allow the development of DEBSs. DEBS frameworks abstract functionality implemented by DEBSs into interfaces. When implementations (called plugins) for these interfaces are provided, a running DEBSs is obtained. Since different plugins may exist for the same interface, DEBSs with different characteristics can be obtained by instantiating the framework with different plugins. Examples are REDS (Cugola & Picco, 2006) and GREEN (Sivaharan et al., 2005). In both of these systems different event filtering capabilities are supported by different plugins. In the case of GREEN, different event representations are supported as well (e.g., events as objects, and events as sequences of strings).

The ability to support context-aware applications is a requirement that, we believe, will greatly influence the DEBS event model. Dey (2001) defines context as: any information that can be used to characterize the situation of an entity. An entity is a person, place, or object that is considered relevant to the interaction between a user and an application, including the user and applications themselves. (p. 5)

Context-aware applications are able to alter their behaviour based on patterns of use, location, and timing conditions. In our categorization of event models, we included context in the filtering of events. But context-awareness has the potential of affecting many more aspects of the event model. For example, consider a component that subscribes to events when it is within a certain geographical region, and that unsubscribes when it leaves the region. In fashion with adaptive applications, the subscribe/unsubscribe operations should be implicitly invoked on behalf of the component without the need for the component to be constantly checking its location coordinates. Context-awareness also influences publishers of events. Events may be generated due to particular conditions in the context of the component. This context may in turn be associated to the published event and be transmitted along with the event attributes to interested subscribers. Contextual information for the event itself (e.g., the time the event was generated), may be of interest as well. Research into the development of context-aware applications is very active, including the development of context-models for representing contextual information (Bolchini et al., 2007). Since context-aware applications have been proposed for the same type of applications as DEBSs, we believe that the research areas will eventually intersect.

CONCLUSION

We have provided a categorization of the event model in IISs, summarized in Figure 2. Within this categorization we have identified features in the DEBS event model commonly found in DEBSs: the ability to declare new types of events in the system, at run time; event bindings are maintained by the DEBSs and are established and terminated dynamically; components publishing events do not direct them to subscribed components; components publishing events do not wait for subscriber components to be notified; components must subscribe to the events of interest; published events are delivered to all subscribed components. We have also identified key aspects of the event model in which variations exist within different DEBSs. These variations include the requirement to declare new type of events; diverse event attribute representations; different subscription operations, including filtering capabilities; diverse delivery semantics and ordering guarantees; and support for event persistence.

We have also discussed the impedance between the DEBS event model and the event model of methodologies widely used in the design of IISs. We argue that successful design and specification methods for DEBSs need to be compatible with the DEBS event model. Therefore, the event features here reported can be used to validate candidate design and specification methods, or be included as event feature requirements for new methods.

REFERENCES

W3C. (2004). *W3C xml schema*. Retrieved from http://www.w3.org/XML/Schema

Bacon, J., Eyers, D. M., Peter, R., & an Pietzuch, J. S. (2008, July). Access control in publish/subscribe systems. In *Proceedings of the 2nd International Conference on Distributed Event-based Systems (DEBS'08)* (pp. 23-34). Rome: ACM Press. (Chair Roberto Baldoni)

Bacon, J., Moody, K., Bates, J., Hayton, R., Ma, C., & McNeil, A. (2000). Generic support for distributed applications. *Computer, 33*(3), 68–76.

Barbier, F., & Belloir, N. (2003, April). Component behavior prediction and monitoring through built-in test. In *Proceedings of ECBS '03: The 10th IEEE International Conference and Workshop on the Engineering of Computer-based Systems* (pp. 17-22). Huntsville, AL: IEEE Computer Society.

Belokosztolszki, A., Eyers, D. M., Pietzuch, P. R., Bacon, J., & Moody, K. (2003, June). Role-based access control for publish/subscribe middleware architectures. In Proceedings of the *2nd International Workshop on Distributed Event-based Systems (DEBS'03)*. San Diego, CA: ACM Press. (Program Chair Hans-Arno Jacobsen)

Bhola, S., Strom, R. E., Bagchi, S., Zhao, Y., & Auerbach, J. S. (2002). Exactly-once delivery in a content-based publish-subscribe system. In *Proceedings of DSN '02: The 2002 International Conference on Dependable Systems and Networks* (pp. 7-16). Washington, DC: IEEE Computer Society.

Bolchini, C., Curino, C., Quintarelli, E., Schreiber, F. A., & Tanca, L. (2007). A data-oriented survey of context models. *SIGMOD Record, 36*(4), 19–26.

Bultan, T., Fu, X., Hull, R., & Su, J. (2003). Conversation specification: a new approach to design and analysis of e-service composition. In *Proceedings of WWW '03: The 12th International Conference on World Wide Web* (pp. 403-410). New York: ACM Press.

Buschmann, F., Meunier, R., Rohnert, H., Sommerlad, P., & Stal, M. (1996). *A system of patterns - pattern oriented software architecture*. New York: Wiley.

Cai, L. R., Bradbury, J. S., & Dingel, J. (2007). Verifying distributed, event-based middleware applications using domain-specific software model checking. In Bonsangue, M. M., & Johnsen, E. B. (Eds.), *FMOODS* (*Vol. 4468*, pp. 44–58). Berlin, Germany: Springer-Verlag.

Carlson, J., & Lisper, B. (2004). An event detection algebra for reactive systems. In *Proceedings of EMSOFT '04: The 4th ACM International Conference on Embedded Software* (pp. 147-154). New York: ACM Press.

Carzaniga, A., Rosenblum, D. S., & Wolf, A. L. (2001). Design and evaluation of a wide-area event notification service. *ACM Transactions on Computer Systems, 19*(3), 332–383.

Cilia, M., Haupt, M., Mezini, M., & Buchmann, A. (2003). The convergence of AOP and active databases: towards reactive middleware. In *Proceedings of GPCE '03: The Second International Conference on Generative Programming and Component Engineering* (pp. 169-188). New York: Springer.

CORBA-ES. (2004). *Corba event service, version 1.2*. Retrieved from http://www.omg.org/technology/documents/formal/event service.htm

CSRG. (1986). Unix programmer's reference manual (4.3 BSD ed.) [Computer software manual]. Berkley, CA: Computer Systems Research Group, University of California.

Cugola, G., Nitto, E. D., & Fuggetta, A. (1998). Exploiting an event-based infrastructure to develop complex distributed systems. In *Proceedings of ICSE '98: The 20th International Conference on Software Engineering* (pp. 261-270). Washington, DC: IEEE Computer Society.

Cugola, G., Nitto, E. D., & Fuggetta, A. (2001, September). The JEDI event-based infrastructure and its application to the development of the OPSS WFMS. *IEEE Transactions on Software Engineering* (Vol. 27) (No. 9). Piscataway, NJ: IEEE Press.

Cugola, G., & Picco, G. P. (2006). Reds: a reconfigurable dispatching system. In *Proceedings of SEM '06: The 6th International Workshop on Software Engineering and Middleware* (pp. 9-16). New York: ACM Press.

de Alfaro, L., & Henzinger, T. A. (2001). Interface automata. *ACM SIGSOFT Software Engeneering Notes*, 26(5), 109–120.

Dey, A. K. (2001). Understanding and using context. *Personal and Ubiquitous Computing*, 5(1), 4–7.

Dias, M. S., & Vieira, M. E. R. (2000). Software architecture analysis based on statechart semantics. In *Proceedings of IWSSD '00: The 10th International Workshop on Software Specification and Design* (p. 133). Washington, DC: IEEE Computer Society.

Dingel, J., Garlan, D., Jha, S., & Notkin, D. (1998). Reasoning about implicit invocation. In *Proceedings of SIGSOFT '98/FSE6: The 6th ACM SIGSOFT International Symposium on Foundations of Software Engineering* (pp. 209-221). New York: ACM Press.

Douence, R., & Südholt, M. (2002). *A model and a tool for event-based aspect-oriented programming (EAOP)* (Tech. Rep. No. 02/11/INFO). Ecole des Mines de Nantes.

Eugster, P. T., Felber, P. A., Guerraoui, R., & Kermarrec, A.-M. (2003). The many faces of publish/subscribe. *ACM Computing Surveys*, 35(2), 114–131.

Fayad, M. E., Schmidt, D. C., & Johnson, R. E. (1999). *Implementing application frameworks: Object-oriented frameworks at work*. New York: John Wiley & Sons, Inc.

Fenkam, P., Jazayeri, M., & Reif, G. (2004, May). On methodologies for constructing correct event-based applications. In A. Carzaniga & P. Fenkam (Eds.), *3rd International Workshop on Distributed Event-based Systems (DEBS'04)* (pp. 38-43). Edinburgh, UK: IEEE Computer Society.

Ferraiolo, D., & Kuhn, R. (1992, October). Role-based access controls. In *15th NIST-NCSC National Computer Security Conference* (pp. 554-563). Baltimore, MD.

Gamma, E., Helm, R., Johnson, R., & Vlissides, J. (1995). *Design patterns: elements of reusable object oriented software*. Reading, MA: Addison-Wesley Professional.

Garlan, D., Khersonsky, S., & Kim, J. S. (2003). Model checking publish-subscribe systems. In *Proceedings of the 10th International SPIN Workshop on Model Checking of Software (SPIN 03)*. Portland, OR.

Garlan, D., & Scott, C. (1993). Adding implicit invocation to traditional programming languages. *In Proceedings of ICSE '93: The 15th International Conference on Software Engineering* (pp. 447-455). Los Alamitos, CA: IEEE Computer Society Press.

Garlan, D., & Shaw, M. (1994, January). *An introduction to software architecture* (Tech. Rep. No. CMUCS-94-166).Pittsburgh, PA: School of Computer Science, Carnegie Mellon University.

Geihs, K. (2001). Middleware challenges ahead. *Computer, 34*(6), 24–31.

Gelernter, D. (1985). Generative communication in Linda. *ACM Transactions on Programming Languages and Systems, 7*(1), 80–112.

Harel, D. (1987, June). Statecharts: a visual formalism for complex systems. *Science of Computer Programming, 8*(3), 231–274.

Kiczales, G., Lamping, J., Mendhekar, A., Maeda, C., Lopes, C. V., Loingtier, J.-M., et al. (1997, June). Aspect-oriented programming. In *Proceedings of the 11th European Conference on Object-oriented Programming (ECOOP)*. Berlin, Germany: Springer-Verlag.

Konana, P., Liu, G., Lee, C.-G., & Woo, H. (2004). Specifying timing constraints and composite events: An application in the design of electronic brokerages. *IEEE Transactions on Software Engineering, 30*(12), 841–858.

Krishnamurthy, B., & Rosenblum, D. S. (1995). Yeast: A general purpose event-action system. *IEEE Transactions on Software Engineering, 21*(10), 845–857.

Leveson, N. G., Heimdahl, M. P. E., Hildreth, H., & Reese, J. D. (1994). Requirements specification for process-control systems. *IEEE Transactions on Software Engineering, 20*(9), 684–707.

Liebig, C., Cila, M., & Buchmann, A. (1999). Event Composition in Time-dependent Distributed Systems. In *Proceedings of the 4th IFCIS International Conference on Cooperative Information Systems (CoopIS 99)* (pp. 70-78). Washington, DC: IEEE Computer Society.

Mansouri-Samani, M., & Sloman, M. (1997, June). Gem: A generalised event monitoring language for distributed systems. *IEE/IOP/BCS Distributed Systems Engineering Journal, 4*(2), 96-108.

Meier, R., & Cahill, V. (2002). Steam: Event-based middleware for wireless ad hoc networks. In *Proceedings of the 22nd International Conference on Distributed Computing Systems Workshops* (pp. 639-644). Los Alamitos, CA: IEEE Computer Society.

Meier, R., & Cahill, V. (2005). Taxonomy of distributed event-based programming systems. *The Computer Journal, 48*(5), 602–626.

Mühl, G. (2002). *Large-scale content-based publish/subscribe systems*. Unpublished doctoral dissertation, Technische Universität Darmstadt, Darmstadt, Germany.

NASA Software Technology Branch. L. B. J. S. C. (1995, February). *C language integrated production system*. Retrieved from http://www. cs.cmu.edu/afs/cs/project/ai-repository/ai/areas/ expert /systems/clips/0.html

Notkin, D., Garlan, D., Griswold, W. G., & Sullivan, K. J. (1993). Adding implicit invocation to languages: Three approaches. In *Proceedings of the First JSSST International Symposium on Object Technologies for Advanced Software* (pp. 489-510). London: Springer-Verlag.

OASIS. (2004). *Oasis web services reliable messaging (WSRM) TC*. Retrieved from http://www. oasis-open.org/committees/wsrm/

Oki, B., Pfluegl, M., Siegel, A., & Skeen, D. (1993). The information bus: an architecture for extensible distributed systems. In *Proceedings of SOSP '93: The Fourteenth ACM Symposium on Operating Systems Principles* (pp. 58-68). New York: ACM Press.

OMG. O. M. G. (2007, February). *UML 2.1.1 superstructure specification.* Retrieved from http://www.omg.org/technology/documents/formal/uml.htm

Oracle. (2005, June). *Oracle® streams advanced queuing user's guide and reference 10g Release 2 (10.2), Part number b14257-01.* Retrieved from http://download-east.oracle.com/docs/cd/B1930601/server.102/b14257/toc.htm

Pallickara, S., & Fox, G. (2003). Naradabrokering: A distributed middleware framework and architecture for enabling durable peer-to-peer grids. In *Proceedings of ACM/IFIP/USENIX International Middleware Conference Middleware-2003* (Vol. 2672, pp. 41-61). Berlin, Germany: Springer-Verlag.

Pietzuch, P. R. (2004). *Hermes: A scalable event-based middleware.* (Unpublished doctoral dissertation, University of Cambridge). New York: Queens' College.

Rozsnyai, S., Schiefer, J., & Schatten, A. (2007, June). Concepts and models for typing events for event based systems. In H.-A. Jacobsen, G. Mühl, & M. A. Jaeger (Eds.). In *Proceedings of the Inaugural Conference on Distributed Event-based Systems.* New York: ACM Press. Retrieved from http://debs.msrg.utoronto.ca

Ryu, M., Kim, J., & Maeng, J. C. (2006, June). Reentrant statecharts for concurrent real-time systems. In H. R. Arabnia (Ed.). In *Proceedings of the International Conference on Parallel and Distributed Processing Techniques and Applications & Conference on Real-time Computing Systems and Applications* (pp. 1007- 1013). Las Vegas, NV: CSREA Press.

Sivaharan, T., Blair, G., & Coulson, G. (2005). Green: A configurable and re-configurable publish-subscribe middleware for pervasive computing. In [DOA]. *Proceedings of the International Symposium on Distributed Objects and Applications, 3760,* 732–749.

SUN-JINI. (2003). Jini's distributed events specification, version 1.0. Retrieved from http://java.sun.com/products/jini/2.1/doc/specs/html/event-spec.html

SUN-JMS. (2002). Java message service (JMS) specification, version 1.1. Retrieved from http://java.sun.com/products/jms/docs.html

Veanes, M., Campbell, C., Schulte, W., & Kohli, P. (2005, January). *On-the-fly testing of reactive systems* (Tech. Rep. No. MSR-TR-2005-05). Microsoft Research.

Völgyesi, P., Maróti, M., Dóra, S., & Osses, E., & Lédeczi Ákos. (2005). Software composition and verification for sensor networks. *Science of Computer Programming, 56*(1-2), 191–210.

von der Beeck, M. (1994). A comparison of statecharts variants. In *Proceedings of PROCOS: The Third International Symposium, organized jointly with the Working Group Provably Correct Systems on Formal Techniques in Real-time and Fault-tolerant Systems* (pp. 128-148). London, UK: Springer-Verlag.

Weiser, M. (1993). Some computer science issues in ubiquitous computing. *Communications of the ACM, 36*(7), 75–84.

Wieringa, R. J. (2002). *Design methods for software systems: Yourdon, Statemate and UML.* San Francisco: Morgan Kaufmann Publishers.

Zhao, Y., Sturman, D., & Bhola, S. (2004). Subscription propagation in highly available publish/subscribe middleware. In *Middleware '04: Proceedings of the 5th ACM/IFIP/USENIX International Conference on Middleware* (pp. 274-293). New York: Springer-Verlag New York, Inc.

KEY TERMS AND DEFINITIONS

Event: Happening of interest in the system or the environment.

Explicit Invocation: Architectural style in which functionality provided by a component is directly invoked, via a name bound to the functionality, by interested components.

Implicit Invocation: Architectural style in which functionality provided by a component is executed indirectly as a reaction to the announcement of an event in the system.

Event Vocabulary: Collection of kinds of events that can be announced in a system.

Event Model: Collection of features in an implicit invocation system that determine how events are defined and announced, and how interested components are notified of the events.

Event Binding: Components to be notified of an event occurrence.

Publish/Subscribe: Type of implicit invocation where components manifest their interested in events by invoking a subscribe operation, and components announce events by invoking a publish operation. When an event is published, the event is asynchronously propagated to the components that have subscribed to the event.

Chapter 3
Time/Space Aware Event Correlation

Eiko Yoneki
University of Cambridge, UK

ABSTRACT

Ubiquitous computing, with a dramatic increase in the event monitoring capabilities of wireless devices and sensors, requires more complex temporal and spatial resolution in event correlation. In this chapter, we describe two aspects of event correlation to support such environments: event modelling including spatio-temporal information, and composite event semantics with interval-based semantics for efficient indexing, filtering, and matching. It includes a comparative study of existing event correlation work. The described durative event model, combined with interval-semantics, provides a new vision for data processing in ubiquitous computing. This makes it possible to extend the functionality of simple publish/subscribe filters to stateful subscriptions, parametrisation, and computation of aggregates, while maintaining high scalability.

INTRODUCTION

We envision event correlation in a publish/subscribe system becoming important. A publish/subscribe system may offer content-based filtering, which allows subscribers to declare their interests in the attribute values of events using a flexible subscription language, while event correlation addresses the relationships among event instances of different event types. Especially with a dramatic increase in the event monitoring capabilities of wireless devices and sensors, selective data dissemination in publish/subscribe systems requires more complex temporal and spatial resolution in event correlation.

Sensor technology enables monitoring of the environment that may trigger the operation of subsequent services. An example of such a subscription is 'Notify' the products in the store that become on sale with 30% discount price within 50 minutes after the total number of customers in the store reaches 120. This subscription may

DOI: 10.4018/978-1-60566-697-6.ch003

Figure 1. Filtering, correlation and aggregation

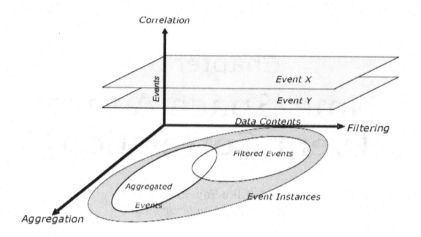

not be described by filtering, where an exact temporal constraint (e.g., time instant, time interval) is expected. A time interval is relative to a time point (e.g., 50 minutes after the number of customers reaches 120). Thus, it is desirable that the semantics of filtering and correlation can be defined together.

Wireless Sensor networks (WSNs) raise new issues to be addressed such as data aggregation and an operation dealing with redundancy, summarisation, and quality control of data in the network. Figure 1 shows the relationships between aggregation, filtering and correlation. For example, some conditions/values being sensed last longer than the sensing interval, which causes repetitive event reports. Similarly, the same events will be reported by several spatial groups, if the size of the detection group is too small. Another characteristic of WSNs is that they aim to link the real and virtual worlds. Events tie the two worlds together, and relevant state changes in the real world have to be detected and signalled to the virtual world by generating appropriate events. Note that sensors can detect state changes in the real world, but low-level sensor information is often not directly useful. Typically, raw sensor input is therefore processed and combined to form high-level events that model real world actions.

We envision supporting more sophisticated and unambiguous event correlation over time and space. Spatio-temporal queries may contain changes in both object and query locations over time. For example:

- Moving queries on stationary objects: *as I am moving along a certain route, show me all gas stations within 3 minutes of my location.*
- Stationary queries on moving objects: *how many speeding cars are within the city boundary.*
- Moving queries on moving objects: *as the President moves make sure that the number of security guards within 50 metres of his/her location is more than 50.*

Event Correlation will be a multi-step operation from event sources to the final subscribers, combining information collected by wireless devices into higher-level information or knowledge. Distributed events are fragmented and dispersed among various devices, services, and agents, and interpretation of events can occur at any time and any location. Events flow based on queries to databases, subscriptions, notifications, and search results over the networks.

In this chapter, we look closely into events and find out how their semantics and representation should be defined. Events contain multidimensional attributes and should be well indexed for searching and complex correlation. Besides the existing attributes of events, it will be crucial to incorporate continuous context information such as time or geographic location within an event model. Defining an unambiguous event model provides a common interface for event correlation. We then introduce a novel and generic interval-based semantics for composite event detection. This precisely defines complex timing constraints among correlated event instances. The described durative event model, combined with interval-semantics, provides a new vision for data processing. This extends the functionality of simple publish/subscribe filters to enable stateful subscriptions, parameterisation, and computation of aggregates.

BACKGROUND

Recent progress has led mobile devices to become ubiquitous and Wireless Sensor Networks (WSNs) are composed of sensor nodes distributed in the environment communicating with the other devices. Automatic, self-organising and self-managing systems will be essential for such ubiquitous environments, where billions of computers are embedded in everyday life. This new dimension of ubiquitous computing requires more complex communication mechanisms and, most importantly, intelligent data processing throughout the networks.

Emergence of Loosely Coupled Communication

In daily life, the synchronous polling mode dominates the search for information on the World Wide Web (WWW). A complementary model, asynchronous publish/subscribe event notification is becoming popular, where a user subscribes to specific events and receives notifications when any of these events are published. Amazon shopping alerts, the auction notification of *eBay*, and stock quotes and news alerts are all examples of this model. In ubiquitous computing scenarios, applications that communicate via WSNs to perform automation tasks will rely on the event notification model. Non-human subscribers increase the scalability and impact the design of publish/subscribe systems, because of the high number of subscribers/publishers, more complex subscriptions, and high rate of event processing.

Publish/Subscribe is a powerful abstraction for building distributed applications. Communication is message-based and can be anonymous, where participants are decoupled giving the advantage of removal of static dependencies in a distributed environment. It is an especially good solution to support highly dynamic, decentralised systems (e.g., wired environments with huge numbers of clients, mobile ad-hoc networks (MANETs), and P2P).

Most distributed event-based middleware contains three main elements: a publisher who publishes events (messages), a subscriber who subscribes his interests to the system, and an event broker network to match and deliver the events to the corresponding subscribers. Event brokers are usually connected in an arbitrary topology.

Subscription Model

Most early event-based middleware systems are based on the concepts of group (channel) or topic communication (i.e., topic-based publish/subscribe). These systems categorise events into predefined groups. Topic-based publish/subscribe is an abstraction of numeric network addressing schemes.

On the other hand, content-based subscription used in SIENA (Carzaniga & Wolf, 2003), Gryphon (IBM Research, 2001), and Elvin (Segall & Arnold, 1998) delivers the messages depending

Figure 2. Content-based model

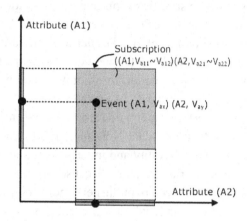

on their content; applications can therefore select different combinations of messages without changing the addressing structure. Content-based subscription extends the capability of event notification with more expressive subscription filters compared to topic-based subscription. The most advanced and expressive form of subscription language is content-based with pattern matching, which is important for event notification. Common topic-based systems arrange topics in hierarchies, but a topic cannot have several super topics. Research is also ongoing to structure complex content-based data models (Muhl, Fiege, & Buchmann, 2002) and reflection-based filters (Eugster, Felber, Guerraoui, & Kermarrec, 2003). Figure 2 depicts a content-based subscription with two attributes, where range values are specified in both attributes. The combination of hierarchical topics and high speed content filtering provides the flexibility necessary to allow applications to evolve beyond their initial design. Content-based publish/subscribe is essential for better (filtered) data flow in mobile computing.

Event Structure

In SIENA, an event is described with simple name and value pairs, while XRoute (Snoeren, Conley, & Gifford, 2001) defines an event type

and attribute in an XML schema. In type-based publish/subscribe (Eugster, Felber, Guerraoui, & Sventek, 2000), the event type model integrates with the type model of an object-oriented programming language.

In a system with large numbers of brokers, events and subscribers, it is necessary to develop efficient data structures and algorithms for subscription propagation and event matching/filtering. Semantically summarising subscriptions would be one approach, which saves network bandwidth and processing cycles for matching.

The subscription can be summarised at the attribute level by a *subsumption* (Borgida & Patel-Schneider, 1994) mechanism among attribute values. The frequency of subsumption therefore makes data structures more compact. A distributed scheme for propagating, merging, and updating the subscription summaries in a network of brokers is useful. Each node summarises the received summary from its neighbours, adds its own subscription summary, and propagates the result to other neighbours. It can create a summary in a hierarchical structure among nodes.

Distributed Stream Data Processing

Stream Data Processing shares many problems with publish/subscribe systems. In many applications such as network management, stock analysis, and Internet-scale news filtering and dissemination, data arrives in a stream. Examples are news-feed data and continuously arriving measurements from sensors. Stream processing systems typically support numbers of continuous queries. Stock analysts continuously monitor incoming stock quotes to discover matching patterns. Classic database systems are optimised for one-shot queries over persistent datasets, and solutions based on database triggers do not scale for stream processing.

In the database community, TelegraphCQ (Chandrasekaran et al., 2003), Aurora (Abadi et al., 2003), Borealis (Abadi et al., 2005), and

STREAM (Arasu et al., 2003) have progressed in providing support for stream data manipulation from a database-centric perspective. Aurora manages continuous data stream for monitoring applications; it focuses on DBMS-like support for applications. Borealis extends Aurora by adding dynamic revision of query results and dynamic query modification. Similarly, the STREAM project (Arasu et al., 2003) views stream processing as the running of continuous queries expressed in a query language (CQL) that includes sliding windows and sampling over the data stream. Queries are converted into an execution plan that includes stream-related operators, data queues, and data stores that manage the sliding windows and samples over data that recently passed through the stream.

Event Correlation

In event-based distributed systems, composite events represent complex patterns of activity from distributed sources. Although composite events have been a useful modelling tool in active database research and telecommunications network monitoring, little progress has been made in using them in large-scale, general-purpose distributed systems.

Much composite event detection work has been done in active database research. COMPOSE (Gupta, Sahin, Agrawal, & Abbadi, 2004) provides expressive composite event operators similar to regular expressions and implements composite events using Finite State Automata (FSA). SAMOS (Gatziu & Dittrichothers, 1994) uses Petri nets, in which event instances are associated with parameter-value pairs. An early language for composite events follows the Event-Condition-Action (ECA) model and resembles database query algebras with an expressive syntax. Snoop (Chakravarthy & Mishra, 1996) is an event specification language for active databases, which informally defines event contexts. The detection mechanism in Snoop is based on trees that express the composite events. Instances of the primitive events are inserted at the leaves.

The transition from centralised to distributed systems led to the need to deal with time. (Chakravarthy & Mishra, 1996) presents an event-based model for specifying timing constraints and to process both asynchronous and synchronous monitoring of real-time constraints. Various event operators have been used for defining composite events. However, composite events specified with these operators are often interpreted differently in terms of their occurrences. (Liu, Mok, & Konana, 1998) proposes an approach that uses the occurrence time of various event instances for time constraint specification. GEM (Mansouri-Samani & Sloman, 1997) allows additional conditions, including timing constraints, to combine with event operators for composite event specification.

Event-based systems provide a way to design large-scale distributed applications and require event detection as a middleware functionality to monitor complex systems. We aim to provide an unified event model, event correlation semantics, and composite event detection including the operations and event contexts so that the system can ensure the resource bounds for event detection in resource constrained environments.

EVENT AND QUERY MODEL

In this section, we emphasise the fundamental design of event representation. Besides existing event attributes, we could incorporate event order and continuous context information, such as time or geographic location, within an event description. We define a novel event model of primitive events and durative events with a time interval. We then briefly introduce a multi-dimensional event representation (i.e., a *hypercube* in an RTree (Guttman, 1984)) for efficient indexing, filtering, and matching in publish/subscribe systems. See (Yoneki, 2006) and (Yoneki & Bacon, 2007) for further details of the hypercube representation.

There are various types of data on the Internet such as unstructured documents, web data, and documents in databases. Sensor captured data are typically name-attribute pairs, and many of these attributes have scalar values. Sensor data may be redundant, and may not be valid after a certain period. A mechanism to discard obsolete data may be necessary, since the data rate could be high and dynamic. The sensor data may be transmitted over the Internet as a real-time data stream, which is a time series of data that arrives in some order. One of the characteristics of stream data is the spatial and temporal information that are annotated to the data.

Event Model

In this section, we aim at modelling events in an unambiguous way to deal with types of events that require integration of multiple continuous attributes (i.e., time, space, etc.). There are two steps in sensor data operations. First, processing sensor data to generate a meaningful event. This involves signal processing and various algorithms such as Bayesian networks, Markov models, rule based, and neural networks. Second, after the generation of primitive events, higher-level information can be computed.

We consider events and services associated with events to be of prime importance for ubiquitous computing and define semantics of events and instances. An event is a message that is generated by an event source and sent to one or more subscribers. Actual event representation (e.g., data structures) may be a structure encoded in binary, a typed object appropriate to a particular object-oriented language, a set of attribute-value pairs, or XML.

Event

The event concept applies to all levels of events from business actions within a workflow to sensing the air temperature. Primitive and composite events are defined as follows:

Definition 3.1 (Primitive Event) *A primitive event is the occurrence of a state transition at a certain point in time. Each occurrence of an event is called an event instance. The primitive event set contains all primitive events within the system.*

Definition 3.2 (Composite Event) *A composite event is defined by composing primitive or composite events with a set of operators. The universal event set E consists of the set of primitive events E_p and the set of composite events E_c.*

The operator $\succ$ identifies the events that contribute to a composite event. $\succ$ is defined: Let $e_1,..., e_n \in E$ be event instances contributing to the composite event $e \in E_c$. $\{e_1,...e_n\} \succ e$ expresses this relation where $\{e_1,...e_n\}$ can be instances of primitive or composite events.

Timestamps

Definition 3.3 (Time) *Time in the real world may be continuous, but we define time as discrete and finite with limited precision. It is assumed that time has a fixed origin and equidistant time domain. Absolute time systems (e.g., 15:00 GMT) can map to the time-axis.*

Definition 3.4 (Timestamp) *Each event has a timestamp associated with the occurrence time. There is uncertainty associated with the values of timestamps in implemented systems. A timestamp is a mandatory attribute of an event defined within a time system, while the event occurrence time is the real-time of the occurrence of the event which itself cannot be known precisely. Thus the timestamp is an approximation of the event occurrence time.*

The accuracy of timestamps depends on the event detection and timestamp model (see (Yoneki, 2006) for the time model). The stages of event capture are:

- Physically capturing an event, i.e., time of occurrence (T_p).
- Recognising it (by the processor), i.e., time of detection (T_d).
- Obtaining a point-based timestamp (T_g) or an interval-based timestamp (T_g^l, T_g^h) with

  l for lower bound and h for upper bound.
- Inserting the event in a communication line.

In most cases, it is expected that $|T_d - T_p|$ to be 0. T_g indicates T_p with the available values within the system. Depending on the time system, the timestamp can be point-based (T_g) or interval-based (T_g^l, T_g^h).

Most point-based timestamps consist of a single value indicating the occurrence time. If there is a virtual global clock, each event can have a global time value for timestamping when the event occurred. Granularity and non-zero precision of a virtual global clock cause errors. For example, two distinct events may have the same timestamp value if issued at different places, or if communication is involved, the timestamp of the message sender could be greater than or equal to the timestamp of the corresponding receiver. Thus, adding the margin by calculating the delay time from the communication with the global clock (e.g. UCT) can be represented in either a point-based timestamp or an interval-based timestamp.

In (Liebig, Cilia, & Buchmann, 1999), the time when an event is detected is given as an interval-based timestamp, which captures clock uncertainty and network delay with two values: the low and high end of the interval. Although an interval format is used, it represents a single point (*point-interval-based timestamp*).

Let e be a primitive instance of an event and $t(e)$ the timestamp for it. There are three possibilities in representing timestamps:

- **Point-based timestamp:** denoted $t_p(e)$.

- **Point-interval-based timestamp:** denoted $t_{pi}(e)_1^h$. It is a point-based timestamp in an interval-based format. An interval represents error margins from event detection delay, processing time, and network delay.
- **Interval-based timestamp:** denoted $t_i(e)_1^h$. Composition of events creates an interval. $t_i(e)^l$ and $t_i(e)^h$ themselves may be represented in either point-based or point-interval-based format. Thus, the timestamp could take a two layer structure of interval values.

Composite events are built up from events occurring at different times, and the associated real-time is usually that of the last of its contributory primitive events. This is natural in a context where the prime focus is on event detection, since typically a composite event will be detected at the time that its last contributory event is detected. However, this does lead to logical difficulties in the case of some composite events. A composite event is defined with duration and given an interval-based timestamp. A point-interval-based timestamp is an accurate representation of a primitive event and is distinct from interval-based timestamps representing the duration of composite events. In theory, for a primitive event, either a point-based or point-interval-based timestamp can be used. However, to focus on the interval semantics for composite events, the point-based timestamp for a primitive event is used throughout this chapter.

The duration of composite events depends on composition semantics and the time system. Figure 3 depicts an interval-based timestamp for composite events in the system. For example, the disjunction operation of event A and B (i.e., $A | B$) detects A as a result of the event composition, and if a point-based time system is used, the timestamp of event A is maintained as a timestamp of the composite event, while the conjunction operation of A and B (i.e., $A + B$) results in the duration of A

Figure 3. Timestamp of composite event

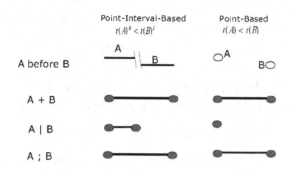

and *B* as a timestamp of the composite event. The sequence operation of event *A* and *B* (i.e., *A; B*) results in the same as the conjunction operation.

Spacestamp

Definition 3.5 (Spacestamp) *A spacestamp is an optional attribute of an event, indicating the location of event occurrence. The location can be absolute location, relative location, and grouping (e.g., position information (x,y,z), postcode, global node id). The Global Positioning System (GPS) (Hoffmann-Wellenhof, Lichtenegger, & Collins, 1994) can provide each node with its location information (latitude, longitude and elevation) with a high degree of accuracy. This information can be used for classifying events within the given space.*

Currently, the semantics of *spacestamp* is dependent on the applications. No global ordering scheme is defined. For example, when *postal codes* are used, the ordering could be simple string order or a specific order derived from the geographic locations. Spacestamp could represent 2-dimensional or 3-dimensional coordinates.

Duration

Definition 3.6 (Durative Event) *Composite events can have duration, where the time of occurrence denotes the duration which binds the*

event instances in the course of the detection of composite events.

A durative event can be seen as an abstraction constructed over two primitive events, which binds its occurrence period instead of start-end instantaneous events. Considering the time of occurrence and the time of detection, these two times in primitive events usually coincide. For composite events, however, the time of occurrence (occurrence period) denotes the span that binds the event instance, while the time of detection is an instant or greater than the last instant of the occurrence period.

(Allen, 1983) argues that all events have duration and considers intervals to be the basic timing concept. A set of 13 relations between intervals is defined, and rules governing the composition of such relations control temporal reasoning.

A durative event can be seen as capturing the uncertainty over the time of occurrence and the time of detection of an event rather than modelling an event that persists over time. In this sense, durative events are akin to the point-interval-based-timestamps described above. The proposed model is based on primitive events that represent instantaneous changes of the system state, with uncertainty over their measurement. We would regard an event that persists over time as akin to a state, with an event at the start and one at the end of the time period. This could also be defined as a composite event. Determination of the duration of composite events requires semantics of composition and time system information. Figure 4 shows 7 temporal relations defined in the system. Complex timing constraints among correlated event instances are precisely defined.

Duplication

It is important to distinguish between multiple instances of a given event type, which may be primitive or composite events.

Figure 4. Interval and point based timestamps over 7 temporal relations

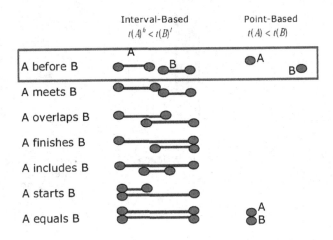

In sensor networks, to avoid loss of events by communication instability, duplicates of events may be produced to increase reliability. Duplicates have to be handled differently depending on the application and contexts within applications. In object tracking, for example, the most recent reading from a sensor is valid, and events prior to that will be obsolete except for the historical record. On the other hand, for a transaction event in which a customer cancels an order, a duplicate event should be ignored as a transaction is being repeated.

Both primitive and composite events are recurrent. An event may occur multiple times, and simply applying event operators on events instead of specific event instances might cause semantic ambiguity. The semantics of event composition have to address handling of duplicates. We propose duplicate handling in two ways: adding a selection operator as an event composition operator and adding subset rules as parameters without loss of meaningful data.

Typed Events

Definition 3.7 (Event Type) *The event type describes the structure of an event.*

Event types can be defined in XML with a certain document type definition, attribute-value pairs with given attributes and value domains or strongly typed objects. For example, event notification from a publisher could be associated with a message m containing a list of tuples <type, attribute name(a), value (v)> in XML format, where type refers to a data type (e.g., float, string). Each subscription s is expressed as a selection of predicates in conjunctive form, i.e., $s = \wedge_{i=1}^{n} P_i$.

Each element P_i of u is expressed as <type; attribute name(a); value range(R) >, where R: (x_i; y_i). P_i is evaluated to be true only for a message that contains $< a_i; v_i >$. A message m matches a subscription s if all the predicates are evaluated to be true based on the content of m.

Events in Hypercube

Event filtering on content-based publish/subscribe can be considered as queries in a high dimensional space, but applying multidimensional index structures to publish/subscribe systems is still unexplored. Ubiquitous computing produces high volumes of sensor data, which may suit multidimensional indexing techniques. We exploit *Hypercube*, a multidimensional indexing

Figure 5. 3-dimensional subscription

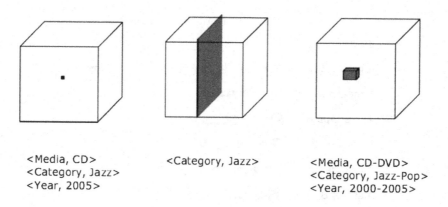

<Media, CD>
<Category, Jazz>
<Year, 2005>

<Category, Jazz>

<Media, CD-DVD>
<Category, Jazz-Pop>
<Year, 2000-2005>

scheme based on an RTree, which provides more effective range queries. Thus, both publication and subscription are modelled as hypercubes in the implementation, where matching is regarded as an intersection query on hypercubes in an n-dimensional space. Point queries on *Hypercube* are transformed into range queries to make use of efficient point access methods for event matching. This corresponds to the realisation of symmetric publish/subscribe, and it automatically provides range queries, nearby queries, and point queries. It is similar to B-trees, and rectangles represent the data spaces, which can be hierarchically nested and overlapping. Search algorithms including intersection, containment, and nearest search use boundary rectangles to decide whether to search further nested rectangles.

Cube Subscription

Events and subscriptions can essentially be described in a symmetric manner with *Hypercube*. Consider an online market of music, where old collections may be on sale. Events represent a cube containing 3 dimensions (i.e., Media, Category, and Year). Subscriptions can be:

- **Point Query:** CDs of Jazz released in 2005
- **Partial Match Query:** Any media of Jazz
- **Range Query:** CDs and DVDs of Jazz and Popular music released between 2000 and 2005

Figure 5 depicts the 3-dimensional *Hypercube* and the above subscriptions are shown accordingly

Expressiveness

Typical examples of queries and subscriptions in geographical information systems are as follows:

- Find objects that crossed through region A at time t_1, came as close as possible to point B at a later time t_2 and then stopped inside circle C some time during interval (t_3, t_4).
- Find objects that first crossed through region A, then passed point B as close as possible and finally stopped inside circle C (the exact time when this occurs is not important).
- Find the object trajectory that crossed to point A as close as possible at time t_1 and then as close as possible from point B at a later time t_2.

Hypercube can express these subscriptions and filtering by use of another dimension with time values. A simple real world example for use of *Hypercube* can be with geographical data coordinates in 2-dimensions. A query such as *Find all book stores within 2 miles of my current location* can be expressed in an RTree with the data splitting space of hierarchically nested, and possibly overlapping, rectangles. See more detail in (Yoneki & Bacon, 2007) and (Yoneki, 2006).

EVENT CORRELATION

In this section, we first discuss the characteristics of composite events and queries, and report a comparative study of existing event correlation work. We then introduce a novel and generic interval-based semantics for composite event detection. This precisely defines complex timing constraints among correlated event instances.

Correlation Definition Language

Traditional publish/subscribe systems are scalable, but the subscription language is often a simple subset of SQL, which is not sufficiently expressive. In contrast, data stream management systems provide an expressive language (e.g., CQL (Arasu, Babu, & Widom, 2002), (Motwani, R., Widom, J., Arasu, A., Babcock, B., Babu, S., Datar, M.,... Varma, R., 2003)) but with only little work on scalability in distributed environments. Another contradicting and complex issue is to balance the expressiveness of event/query with the quality and accuracy of events. The applications that process events could range from medical applications to shopping guidance. An exact event detection mechanism with crucial real-time support may be required for some but not all applications.

Many expressive languages have been introduced in the Active Database community (Widom & Ceri, 1996) for complex event-condition-action (ECA) rules. The composite event definition lan-

guages of Snoop (Chakravarthy & Mishra, 1996) and ODE (Gehani, Jagadish, & Shmueli, 1992) describe composite events in regular expressions using a non-deterministic FSA model. The FSA construction of (Gehani et al., 1992) supports a limited parameterisation for composite events (e.g., equality constraints between attributes). On the other hand, the more expressive event languages (Galton & Augusto, 2002),(Zimmer & Unland, 1999) do not provide clear semantics. Thus, it is not easy to compare the expressiveness of languages.

From a survey on formal methods in event specification and analysis (Babich & Deotto, 2002), most of the popular approaches are based on theoretical models such as finite state machines (FSM), timed automata, process algebra, and Petri-Nets. (Daws, Olivero, Tripakis, & Yovine, 1996) discusses the difficulty of using FSMs to deal with hierarchical modelling and synchronisation. FSMs can be augmented to Timed Automata by incorporating a finite set of real-valued clocks. FSM and Timed Automata contain deterministic models.

In WSNs, near future specifying more complex and *higher-level* events by event-description languages is emerging. For example, SNEDL is based on Petri-Nets (Mansouri-Samani & Sloman, 1997), which provides a system with asynchronous and distributed features, where SNEDL can form a hierarchy of events.

Event Correlation in Middleware

This section shows a comparative study of existing event correlation mechanisms. Event correlation may be deployed as part of applications, as event notification services, or as part of a middleware framework. Definition and detection of composite events vary, especially over distributed environments. Equally, named event composition operators do not necessarily have the same semantics, while similar semantics might be expressed using different operators. Moreover, the exact semantic

Figure 6. Event composition semantics - composition operators

	Operators						
	Conjunction	Disjunction	Sequence	Conc.	Negation	Iteration	Selection
ECCO	$A+B$	$A|B$	$A;B$	$A\|B$	$-A$	A^*	A^N
Opera	$A\|B$	$A|B$	$A; B or AB$	-	$\neg A$	A^*	-
CEA	$A\&B$	$A|B$	$A;B$	-	$-A$	-	-
Schwiderski	A,B	$A|B$	$A;B$	$A\|B$	$NOT A$	$A^* or A^+$	-
A-mediAS	$A\&B$	$A\|B$	$A;B$	-	$-A$	-	$A^{[i]}$
Ready	$A\&\&B$	$A\|B$	$A;B$	-	$not A$	-	-
Eve	$CON(A,B)$	$DEX(A,B)$	$SEQ(A,B)$	$CCR(A,B)$	$NEG(A,B)$	$REP(A,n)$	-
GEM	$A\&B$	$A\|B$	$A;B$	-	$!A$	-	-
Snoop	A,B	$A \vee B$	$A;B$	-	-	A^*	-
Rebeca	$A \wedge B$	$A \vee B$			$\neg A$	A^*	-
SAMOS	A,B	$A|B$	$A;B$	-	$NOT A$	$TIMES(n,A)$	$A^*/last(A)$

description of these operators is rarely explained. Thus, we define the following schema to classify existing operators: conjunction, disjunction, sequence, concurrency, negation, iteration, selection, aggregation, spatial restriction, and temporal restriction. Considering the analysed systems, it becomes clear that it is not sufficient to simply consider the operators to convey the full semantic meaning. Each system offers parameters, which further define/change the operators' semantics. The problem is that the majority of systems reflect parameters within the implementations. Parameters for consumption mode and duplicate handling are rarely explicitly described. Figure 6 and Figure 7 show a comparative study of event composition semantics in eleven existing systems. Note that the list of analysed systems cannot be exhaustive but considers a representative set of selected composition semantics. Recent event correlation services for WSNs or embedded systems provide time restricted operations. Most event notification systems still only support primitive events, and their focus is on efficient filter algorithms.

Figure 6 shows composition operators, while Figure 7 emphasises temporal order related parameters such as consumption modes. Concepts of conjunction, disjunction, sequence, concurrency, negation, and iteration operators are based on an event algebra. Selection defines the event instance selection, by which events qualify for a compos-

ite event, and how duplicate events are handled. Conjunction and disjunction are supported in most systems, some of which implement sequence operators (requiring ordering), while fewer support negation. Selection and concurrency are rarely supported. Concurrency is difficult to determine for distributed systems. Time operators are not always supported, requiring a time handling strategy for distributed systems. Consumption mode and temporal conditions are rarely made explicit. If they are explicit, several options are supported as otherwise they are hard-coded in the system and difficult to determine. The listed systems are notification services, event composition languages, and workflow coordinators, in which common characteristics are fairly complete semantics of event composition.

- **ECCO:** the proposed prototype described in this section (Yoneki, 2006).
- **Opera:** a framework for event composition in a large scale distributed system (Pietzuch, Shand, & Bacon, 2004) aiming at reduction of event traffic by distributed composite event detectors. The language of composite events is based on the finite state automata.
- **CEA:** our group's early work on the Cambridge Event Architecture (CEA) extended the then-predominant, object-

Figure 7. Event composition semantics - time-related parameters

	Time Operator		Timestamp	Composition	T.C.	Consumption			Subset	Detection
	Period	Life				Unrest.	Recent	Chron.		
ECCO	$(A;B)_T, (A+B)_T$	$(A)_T$	P, PI, I	I	×	×	×	×	×	algorithm
Opera	$(A,B)_T$	-	PI	P	-	-	-	-	-	T-FSA
CEA	-	-	P	P	-	×	-	-	×	FSA
Schwiderski	-	-	P	P	-	-	-	-	-	Rule
A-mediAS	$(A,B)_T, (A;B)_T, -A_T$	-	P	P	-	×	×	×	×	T-FSA
Ready	-	-	P	P	-	-	-	-	×	-
Eve	-	-	P	P	-	-	-	×	×	Graph
GEM	$(A+timeperiod)$	-	P	P	×	-	×	-	×	Rule
Snoop	$(A, [tiestring] : param, B)$	-	P	P	-	-	×	×	×	Graph
Rebeca	slide window	-	PI	P	-	-	×	×	-	Rule
SAMOS	-	-	-	-	-	-	-	×	×	PetriNet

Time Operator: Period (between event A and B) Life (valid time for event A)

Timestamp: P (Point-based), PI (Point represented in Interval-based format), and I (Interval-based).

Composition: Event composition semantics. P (Point-based), and I (Interval-based).

Temporal Condition (T.C.): Temporal conditions such as *A before B, A meets B, etc. for event composition.*

Consumption: Event consumption (Unrestricted, Recent, and Chronicle).

Subset: Parameters for selection or subset on duplication handling.

Detection: Implementation methods.

oriented middleware (CORBA and Java) with an *advertise, subscribe, publish, and notify* paradigm (Bacon, Bates, Hayton & Moody, 1995). COBEA (Ma & Bacon, 1998) is an event-based architecture for the management of networks using CEA based composite event operators.

- **Schwiderski:** enhanced distributed event ordering and introduced 2g-precedence-based sequence and concurrency operators for detection of composite events (Schwiderski, 1996).

- **A-mediAS:** an integrating event notification service that is adaptable to different applications specifically on handling composite events and event filtering methods (Hinze & Bittner, 2002).

- **Ready:** an event notification service from AT&T Research similar to SIENA. Ready supports composite events and its grouping functionality can be shared among cli-

ents (Gruber, Krishnamurthy, & Panagos, 1999).

- **EVE:** combines characteristics of active databases and event-based architectures to execute event driven workflows (Gehani et al., 1992).

- **GEM:** GEM (Mansouri-Samani & Sloman, 1997) is an interpreted, generalised, event monitoring, rule-based language.

- **Snoop:** a model-independent event specification language (Chakravarthy & Mishra, 1996) supporting parameter contexts. It supports temporal, explicit and composite events for active databases.

- **Rebeca:** an event-based electronic commerce architecture focusing on event filtering in a distributed environment (Muhl et al., 2002). Temporal delays in event composition have been addressed in (Liebig et al., 1999).

- **SAMOS:** The SAMOS (Swiss Active Mechanism - based Object - oriented Database System) (Gatziu & Dittrichothers, 1994) project addresses the specification of active behaviour and its internal processing, supporting ECA rules. The detection of composite events is implemented based on coloured PetriNets.

Event Correlation Semantics

We define unified composite events by expressions built from primitive and composite events and algebraic operators. Event algebra is an abstract description of event composition independent of the actual composite event definition languages. Parameters including time, selection, consumption policy, subset policy, and precision policy are also supported. Basic operators provide the potential of expressing the required semantics and are capable of restricting expressions by parameters. Defined composite events can be transformed using algebraic laws, into more feasible and implementable forms for resource constrained environments.

Interval-based semantics for event detection is introduced based on (Adaikkalavan & Chakravarthy, 2006) and extended, defining precisely complex timing constraints among correlated event instances (Yoneki, 2006). WSNs produce a high volume of both continuous and discrete data, and both types need to be dealt with efficiently. Timed automata can process time but not space as continuous variables. Also, an interval-semantics supports more sensitive interval relations among events in environments where real-time concerns are more critical such as wireless networks or multi-media systems. The temporal operators introduced in (Allen, 1983) are not uniformly defined in many applications, and precise timing constraints are defined.

Composite Event Operators

An event algebra is used to define the semantics of composite events. A composite event consists of any primitive or composite events. The algebra has a relatively simple declarative semantics, and it also ensures that detection can be implemented with restricted resources.

Event instances are denoted by e, while event types are denoted by E. An event instance e that belongs to an event type E is denoted as $e \in E$. An event type may have subtypes such as $e \in E \subset E_{super}$, where E_{super} indicates a super-class of E. *NULL* denotes an event set NULL = φ with no event instances.

The occurrence time of the composite event depends on the event composition semantics. T refers to the occurrence time defined, based on the time system; T is a time span in reference time units.

Figure 8 defines the timing constraints related to event composition operators in the classification of timing semantics defined in (Allen, 1983).

The event operators are informally defined as follows:

Definition 4.1 (Conjunction A + B) *Events A and B occur in any order. $(A + B)_T$ with a temporal parameter T denotes the maximal length of the interval between the occurrences of A and B. Note that $(A + B)_\infty$ or $(A + B)$ has no temporal restrictions.*

Definition 4.2 (Disjunction A|B) *Event A or B occurs.*

Definition 4.3 (Concatenation A B) *Event A occurs before event B, where timestamp constraints are A meets B, A overlaps B, A finishes B, A includes B, and A starts B.*

Definition 4.4 (Sequence A ; B) *Event A occurs before B, where timestamp constraints are A before B, and A meets B. $(A; B)$ ensures events are disjoint, whereas (AB) allows overlaps between events. $(A; B)_0$ is a special case of A meets B.*

Figure 8. Interval semantics - timestamps for composite events

	Relation	Timestamps of Primitive Events	Point	Interval	Interval/Point	Point/Interval
1	A before B (A + B) (A | B) (A ; B)	P-P: $t_p(A) < t_p(B)$ I-I: $t_i(A)^h < t_i(B)^l$ I-P: $t_i(A)^h < t_p(B)$ P-I: $t_p(A) < t_i(B)^l$	○ A ○ B	○–A–○ ○–B–○	○–A–○ ○ B	○ A ○–B–○
2	A meets B * (A + B) (A | B) (A B) (A ; B)₀	P-P: NA I-I: $t_i(A)^h = t_i(B)^l$ I-P: $t_i(A)^h = t_p(B)$ P-I: $t_p(A) = t_i(B)^l$		○–A–○ ○–B–○	○—A—○ ○ B	○ A ○—B—○
3	A overlaps B (A + B) (A | B) (A B)	P-P: NA I-I: $(t_i(A)^l < t_i(B)^l) \wedge (t_i(A)^h > t_i(B)^l)$ I-P: NA P-I: NA		○—A—○ ○——B——○		
4	A finishes B (A + B) (A | B) (A B)	P-P: NA I-I: $(t_i(A)^l < t_i(B)^l) \wedge (t_i(A)^h = t_i(B)^h)$ I-P: $t_i(A)^h = t_p(B)$ P-I: $t_p(A) = t_i(B)^h)$		○——A——○ ○——B——○	○——A——○ ○ B	○A ○——B——○
5	A includes B (A + B) (A | B) (A B)	P-P: NA I-I: $(t_i(A)^l < t_i(B)^l) \wedge (t_i(A)^h > t_i(B)^h)$ I-P: $(t_i(A)^l < t_p(B)) \wedge (t_i(A)^h > t_p(B))$ P-I: NA		○——A——○ ○–B–○	○——A——○ ○ B	
6	A starts B (A + B) (A | B) (A B)	P-P: NA I-I: $(t_i(A)^l = t_i(B)^l) \wedge (t_i(A)^h < t_i(B)^h)$ I-P: $t_i(A)^l = t_p(B)$ P-I: $t_p(A) = t_i(B)^l$		○—A—○ ○——B——○	○—A—○ ○ B	○ A ○——B——○
7	A equals B (A + B) (A | B) (A || B)	P-P: $t_p(A) = t_p(B)$ I-I: $(t_i(A)^l = t_i(B)^l) \wedge (t_i(A)^h = t_i(B)^h)$ I-P: NA P-I: NA	○ A ○ B	○——A——○ ○——B——○		

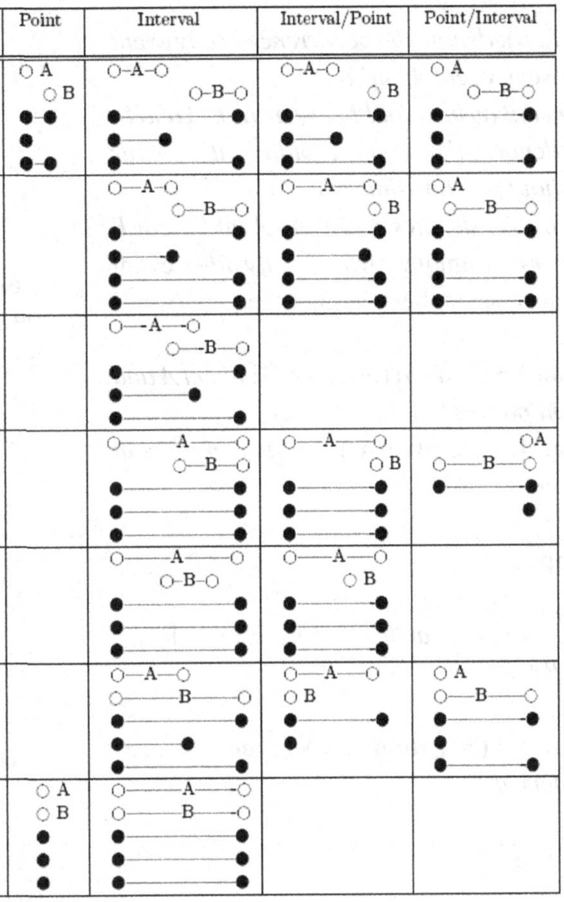

●—● depicts the timestamp for the composite events
* A meets B where $t(A)^h$ and $t(B)^l$ share the same time unit

$t(A)$: timestamp of an event instance A
$t_p(A)$: Point-based timestamp
$t_i(A)_l^h$: Interval-based timestamp from event composition
$t_{pi}(A)_l^h$: Point-interval-based timestamp
(compared as point-based timestamp but using interval comparison)

P-P: Between Point-based and Point-based timestamps
I-I: Between Interval-based and Interval-based timestamps
I-P: Between Interval-based and Point-based timestamps
P-I: Between Point-based and Interval-based timestamps
Real-time Period T:

$$[t(B)^l, t(B)^h] - [t(A)^l, t(A)^h] < T = \begin{cases} YES: & max(t(A)^h, t(B)^h) - min(t(A)^l, t(B)^l) \leq T(1 - \rho) \\ NO: & max(t(A)^l, t(B)^l) - min(t(A)^h, t(B)^h) < T(1 + \rho) \\ MAYBE: & otherwise \end{cases}$$

where ρ is maximum clock skew.

Examples:

(*A*; *NULL*; *B*): *denotes no occurrence of any event between event A and B.*
(*A*;*B*)$_T$: *means an interval T between event A and B.*
(*A*; *B*)$_0$: *denotes that event A and event B occur without any time-interval.*
(*A*; *NULL*; *B*)$_0$: *denotes that event A and event B occur contiguously without any other* event occurrence at the meeting time.

Definition 4.5 (Concurrency ‖ A B) *Event A and B occur in parallel.*
Definition 4.6 (Iteration A*) *Any number of event A occurrences.*

Examples:

A(A|B)*C: would match input such as AAC or AABAC.

Definition 4.7 (Negation-A$_T$) *No event A occurs for an interval T.*

Examples:

(*A* - *B*): *denotes no B starts during A's occurrence.*
(*A* - *B*)$_T$: *denotes no B starts after starting A's occurrence within an interval T.*
(*A*; *B*) - *C*: *denotes that event A is followed by B and there is no C in the duration of (A; B).*

Definition 4.8 (Selection A^N)$_T$ *The selection A^N defines the occurrence defined by the operation N within an interval T, where N ∈ ALL, n, FIRST, LAST,...*

Examples:

A_T^{ALL} : denotes taking *all the event instances during an interval T.*

A_T^n : denotes taking *the nth instance during an interval T.*
A_T^{FIRST} : denotes taking *the oldest instance during an interval T.*
A_T^{LAST} : denotes taking *the most recent instance during an interval T*

Definition 4.9 (Aggregation A^G) *The aggregation A^G defines the operation of aggregation G within an interval T, where G ∈ AVG, MAX, MIN,... The attribute to be used for the aggregation operation is assumed to be implicitly identified.*

Examples:

A_T^{AVG} : denotes taking *the average during an interval T.*
A_T^{MAX} : denotes taking *the maximum during an interval T.*
A_T^{MIN} : denotes taking *the minimum during an interval T.*

Definition 4.10 (Spatial Restriction A$_S$) *Event A occurs within a spatial restriction defined by S such as location identifiers or GPS coordinate values.*

Examples:

A_{CB30FD}: *The area code CB30FD identifies the zone around the Computer Laboratory in Cambridge. Event A is valid only when this spatial condition is satisfied.*

Definition 4.11 (Temporal Restriction A$_T$) *Event A occurs within T.*

Examples:

(*A*; *B*)$_T$: *B occurs within an interval T after A.*
B_T: *B is valid for an interval T.*

For event expression A and B, $A \equiv B$ can be established using the laws of algebra. This can be used for adapting the expression for resource constrained environments. For example, $A; (B|C) \equiv (A; B)|(A; C)$ could be applied when the event rate C is low or $(A - B) - C \equiv A - (B|C)$ to avoid the negation operation as much as possible.

The following examples illustrate the use of the operators to describe composite events.

Example 1: The temperature of rooms with windows facing south is measured every minute and transmitted to a computer placed in the corridor. T denotes a temperature event and T_{30}^{AVG} denotes a composite event of the average temperature during 30 minutes. $\left(T_{room1} + T_{roomT}\right)_{30}^{AVG}$ denotes to take the average of room 1 and 7.

Example 2: Two sensing receivers are placed before and after a stop sign on the street. When car passes, the sensors generate events to the local computer. Suppose the event received before the stop sign is B and after the stop sign is A for a given car. $(B; A)_2$ denotes A occurs 2 seconds after B, and indicates a car did not make a full stop at the stop sign. On the other hand, $((B; A)_{60})^*$ indicates cars may not be flowing in the street, which indicates potential traffic congestion.

Example 3: At a highway entrance, a sensor detects movement of a passing car as event E. The number of cars entering the highway $\left(E_{10}^{SUM}\right)_{HWY1ENT7}$ can be locally calculated at a computer, which can be used to detect traffic congestion on the highway (e.g., a congestion event $C = \left(\left(E_{10}^{SUM}\right)_{HWY1ENT7}\right)^{LESS12}$.

Example 4: At the four roads of a roundabout, sensors capture car movement and produce events: N for movement towards North, E for East, S for South, and W for West. $(N^{n-1}; N^n)_2 + (E^{n-1}; E^n)_2 + (S^{n-1}; S^n)_2 + (W^{n-1}; W^n)_2$ denotes a composite event for any car movement within a 2 second interval. Detection of this composite event indicates traffic

flow, otherwise either no traffic or heavy congestion is assumed in the roundabout area.

Example 5: On a road, sensors are placed every kilometre to detect car movement, and they collect only specific subsets of events $E_{tagxxx001}$. Drivers may be interested in finding the speed of cars ahead of them to find travelling time by use of roadside sensors. This information can be transmitted over the air to drivers who are interested in estimating driving times.

Example 6: Primitive events include strong wind detected W, high humidity detected H, air pressure increase detected P, and above zero degree temperature recorded T. An avalanche alerting system is operated, which detects an event A when strong wind is detected followed by high humidity detection within 3 minutes, unless either *air pressure goes up* or *temperature goes above zero* occurs in between. A set of rules specifying reactions to three primitive events can express the combined behaviour. The above composite event A corresponds to the expression $(S; H)_3 - (P | T)$.

Example 7: *Notify the frozen products that are in the store, when the temperature changes 2% in a 10 second time interval* can be expressed with $_{(s)}((E^{FIRST} + E^{LAST})_{10})$, where s denotes the sub-setting function (i.e., temperature change rate = 2%).

Temporal Conditions

Defining temporal conditions for the semantics of composite events can be tricky, especially when timing constraints are important as in processing transactions. This may cause an incorrect interpretation according to the intuition of the user. Figure 9 depicts a composite event E (*snow storm alert*): during the period when primitive event A (*humidity raises 60%*) occurs followed by primitive event B (*wind blows from north*), if primitive event T (*temperature goes down below zero degrees*) occurs. Two situations are shown. If we follow the interpretation of temporal conditions described in (Gomez & Augusto, 2004), in the first situation, event E is detected and in the second situation,

Figure 9. Composite Event

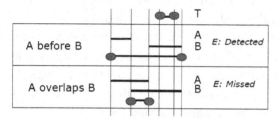

event *E* is missed. In (Gomez & Augusto, 2004), *overlaps* and *during* only comprise the period when two events are simultaneously occurring, while every other operator takes the period over both event occurrences. This inconsistency may cause a problem. On both occasions, the natural interpretation of event *E* is the same. With the proposed definition, both examples will yield consistent results.

Interval Semantics

In most event algebras, an individual primitive and/or composite event occurrence is detected at a point time (i.e., the time of detection of the end of the occurrence). This may cause incorrect detection of operator combinations. The main reason for such problems is that composite events are defined for detection conditions but not in terms of occurrence conditions. Examples are nested sequence operators.

Instead, the detection time of composite events, using the interval of the occurrence, will solve this issue. In (Allen, 1983), 13 basic interval relations are shown. Figure 8 summarises and maps to the interpretation of interval relations for composite event operators.

In (Galton & Augusto, 2002) and (Carlson & Lisper, 2003), the ambiguity of the detection condition is described, which is illustrated in Figure

10. When a point timestamp mechanism is used, an event *B*; (*A*; *C*) is detected, although the actual order of event occurrences is *A*, *B*, and *C*. This is because detection of *B*; (*A*; *C*) is dependent on *A*; *C*. Interval semantics with precise timing conditions solves this problem. In this example, the sequence *B* followed by *A*; *C* is a relation of *overlap*, which is not considered as a *sequence*. Thus, *B*; (*A*; *C*) would not be detected.

Event Context

Adding a policy defining the constraints provides a way to create modified operator semantics. This parameter-dependent algebra can accommodate different policies on event consumption or subsetting. Each operator is given a principal definition of the operation. Then, the various event contexts can be defined, acting as modifiers to the native operator semantics. Selection mechanism of event instances may be defined by this, creating a unique operator for each composite event. This helps resource constrained environments, e.g., keeping the most recent instance for future use instead all instances. Event filtering can be incorporated with event context as parameterisations.

Consumption Policy

For event consumption policy, five contexts can be defined: *unrestricted, recent, chronicle, continuous* and *cumulative*. Snoop (Chakravarthy & Mishra, 1996) uses these contexts but is not capable of applying an individual context to different event operators. The parameterised algebra solves this problem. The following gives an informal definition for detecting sequence operation *A*; *B*. Figure 11 illustrates the sequence operator with these contexts.

- **Unrestricted:** All combinations of instances of A and B.
- **Recent:** The most recent instance of *A* is used for composition.

Figure 10. Point and interval semantics

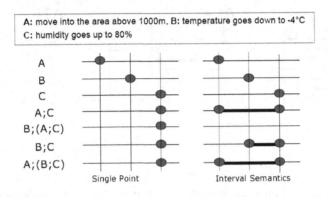

Figure 11. Event consumption policy

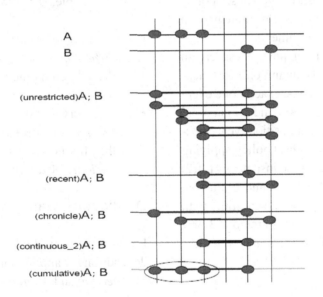

- **Chronicle:** Let A be the initiator and B be the terminator. The initiator and terminator are paired uniquely in occurrence order. Thus, the oldest initiator pairs with the oldest terminator.
- **Continuous:** A moving time window determines the selection of A. In Figure 11, the window size is 2 and only the most recent A is used for composition.

- **Cumulative:** All instances of A are grouped that are used for composition

Subset Policy

The subset policy defines the subset of events to detect. When the subset policy is applied for efficient use of resources, the subset policy should not give crucial impact to unrestricted semantics.

Figure 12. Subset policy: Valid restrictions of A

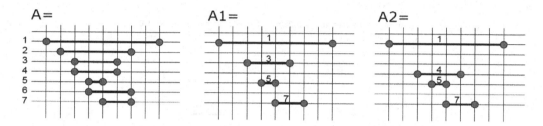

At the same time, efficient detection without wasting resources should be realised. For example, conjunction and sequence operations should be able to detect non-valid instances as early as possible. The main task of the subset policy is to make an effective algebra to implement in resource-constrained environments.

For example, the subset policy restricts an event stream, and a subset contains only instances with the same end-time. Thus, exactly one event instance with the same end-time and the maximal start-time is stored for the composition event detection. Figure 12 shows the subset policy applying to an event stream A. From the instances of $A(2, 6, 7)$ with the same end-time, 2 and 6 must be removed. 5 and 7 do not share the end-time with the others. The choice whether to remove 3 or 4 results in two restricted event streams A, named $A1$ and $A2$.

Precision Policy

The precision Policy defines the required precision of the events to be detected. Dynamic spatial-temporal data from sensor networks is generated at a rapid rate and all the generated data may not arrive at the event correlation node over the network due to its lossy/faulty. Various techniques, including compression and model adaptation, have provided certain levels of guarantee (Lazaridis & Mehrotra, 2003). On the other hand, defining the available precision becomes important if some

imprecision of the collected data can be tolerated by the application. Parameters can be based on position, periodicity, urgency, relative deadline, or timeliness.

For example, *High, Default, Low* can map to:

- the ratio of sensor nodes that are awake: *80%, 20%, 5%*
- the delivered time-series data: *100%, 70%, 50%*
- the interval of data collection: *1 second, 10 seconds, 60 seconds*
- the frequency of data report: *Urgent, Periodic, Available.*

Duplication Handling

Interpretation of duplication and redundancy depends on the application, and it can be handled by *Selection* and *Aggregation* operators together with an event consumption policy. This enables an individual duplicate handling policy for the event stream.

A duplicate set of event instances D for event type A within the time period T can be defined by the composite event A_T^{ALL} or A_T^{LAST}. An aggregation operator can be used, e.g., A_T^{MAX}. A consumption policy can be applied over the composite event, e.g., $_{(recent)}A_T^{LAST}$.

Adaptation to Resource-Constrained Environments

We have used an event algebra for the definition of composite events. An event algebra helps us to use properties such as the laws of distributivity to transform one form of event algebra to another to make detection more efficient or with fewer resources. As a simple and intuitive example, an expression $(A; B)_t$ can be transformed to $(A; B_t)_t$. This will restrict the buffer size for B before $(A; B)$ is evaluated.

Event Stack Size: Consider a composite event with a sequence operator, $A; B$, where the length of the instance of B never exceeds t (i.e., $B \equiv B_t$). The event stack size for A can be optimised by only storing A with maximum t time unit earlier. Furthermore, when the consumption policy defines A with maximum start-time for the composition, only a single instance of A needs to be stored. Without this mechanism, multiple instances of A must be stored.

The event algebra operation itself should not be restricted, and event contexts (e.g., consumption, subset, and precision policies) can be defined for restricting each operator. Wireless devices are resource constrained, and the algebra's properties can be used to determine necessary resources for detection of composite events. An expensive operation such as keeping event instances for a specific event type for infinite time can be prevented by applying rules such as event context so that detection may become possible. If the event expressions are known before execution, a canned detection component can be created for common use. Once the semantics are defined in an event algebra, the implementation can be done by any means such as FSA or even by rule-based approaches (i.e., ECA).

Event Detection

Defining composite events in an event algebra achieves simplicity, but implementing the defined algebra in an efficient manner is difficult. After investigating different approaches (Chakravarthy & Mishra, 1996), (Cayuga Project, 2005), (Hayton, 1996), (Carlson & Lisper, 2003), and (Hinze & Bittner, 2002), we take a simple imperative algorithm for the detection based on combining the approaches of (Pietzuch et al., 2004) and (Carlson & Lisper, 2003).

Detection Algorithm

The event detection process consists of three stages: applying event context policies to the incoming stream (e.g., consumption, subset, and precision policies), matching the operations on event types, and event composition operations. Figure 13 shows the detection algorithm of event composition operations. When this algorithm is executed, the matching operations on the target event types, and restriction on the event streams (i.e., step1 and 2), have already been applied. Depending on the consumption policy, duplicated events may be input to the detection.

Let E be a composite event expression to be detected, and let us index sub-expressions of E from 1 to k in bottom-up order. Thus, the final result of the operation is $E^k(= E)$, and an initial expression is a primitive event $E^1 \in P$, where P is a finite set of primitive events. The main loop selects sub-expressions dynamically and calculates the current composite event instance from the current event and stored information. This operation loops over every time instant.

Let E^i be an output of a composite event in an individual step, E^x a left hand side event, and E^y a right hand side event. Each operation in the expression needs its own indexed state variables (e.g., past events, time instant, and spatial information). Each operator in the composite event expression requires its own variables to keep the state, and thus indexed variables are denoted from 1 to k. Let v_i be a variable of the output composite event instance of E^i, and thus the final v_k contains the result of the algorithm.

Figure 13. Event detection algorithm

```
FOR i = 1 to k
    CASE E^i OF
        [E^i ∈ P]: v_i = E^i ∨ φ
        [E^x + E^y]:
            CASE timestamp OF
                [x_i(start-time) ¡ v_x(start-time)]: x_i = v_x
                [y_i(start-time) ¡ v_y(start-time)]: y_i = v_y
                [v_x(start-time) ≤ v_y(start-time)]: v_i = x_i ∪ y_i
                [OTHERS]: v_i = v_x ∪ y_i
        [E^x|E^y]:
            CASE timestamp OF
                [v_x(start-time) < v_y(start-time)]: v_i = v_x
                [OTHERS]: v_i = v_y
        [E^x E^y]:
            v_i = φ
            IF v_x = φ THEN
                REPEAT event α in z_i
                    IF (α_(end-time) = v_y(start-time)) THEN v_i = α
            IF (v_i = φ) THEN v_i = v_i ∪ v_y
        [E^x : E^y]:
            v_i = φ
            IF v_x = φ THEN
                REPEAT event α in z_i
                    IF (α_(end-time) ≤ v_y(start-time)) THEN v_i = α
            IF (v_i = φ) THEN v_i = v_i ∪ v_y
        [E^x||E^k]:
            v_i = φ
            IF v_x = φ THEN
                REPEAT event α in z_i
                    IF ((α_(start-time) = v_y(start-time)) ∧ (α_(end-time) = v_y(end-time))) THEN v_i = α
            IF (v_i = φ) THEN v_i = v_i ∪ v_y
        [E^x·]: v_i = v_x ∪ y_i ; x_i = v_i
        [E^x - E^y]:
            CASE timestamp OF
                [t_i ¡ v_y(start-time)]: t_i = v_y(start-time)
                [t_i ¡ v_x(start-timte)]: v_i = v_x
                [OTHERS]: v_i = φ
        [E^xN]: v_i = funcN(x_i, v_x) ; x_i = v_i
        [E^xG]: v_i = funcG(x_i, v_x) ; x_i = v_i
        [E_s^x]:
            CASE spacestamp OF
                [(v_x(spacestamp)) ≡ s]: v_i = v_x
                [OTHERS]: v_i = φ
        [E_t^x]:
            CASE timestamp OF
                [(v_x(end-time) - v_x(start-time)) ≤ t]: v_i = v_x
                [OTHERS]: v_i = φ
ENDFOR
```

Let *x* be a variable for storing the left-hand side of a past event, *y* a variable for the right-hand side of a past event, *t* a time instant, *s* a spatial instant, and *z* a set of instances of the event.

The variables *x, y, t, s, z* are used for storing past information for each step of the composition operation. $e_{(start-time)}$ and $e_{(end-time)}$ denote the start- or end-time of the interval timestamp in the event *e.* Creation of new interval timestamps after the detection of the target composite events follows the definition in Figure 8.

In the main loop, when one of the event instances is empty (e.g., E^x happens to be φ in conjunction with operation $E^x + E^y$) or both are empty, the existing event is retained out for the operation or it becomes automatically empty. To highlight the main algorithm, these intuitive operations are not shown in Figure 13. When the composite event expressions are static and can be defined before the system deployment, the operation can be statically determined as the canned operation. See further details in (Yoneki, 2006).

Experiments

Event detection by imperative algorithms is implemented as a prototype with limited functions. Thus, experiments in this section highlight the resource efficiency of controlled event consumption mechanisms such as time restriction, subset policies, and use of durative events. This significantly reduces the event buffer for composite event detection. Object tracking with real Active BAT data is described in the next section.

Prototype Implementation

The prototype has about a 20KB class library in Java with JDK 1.5 SE and J2ME CDC Profile, which supports resource constrained devices. The prototype currently does not provide a high level language interface, and composite events need to be described as event expressions with the operators described in Figure 14. For example, a composite event expression A; B; C must be described as (A; B); C or A; (B; C). The prototype implements a subset of operators. Event expressions are parsed and kept. Each valid expression is associated with a correlation listener that operates the correlation process. The primitive events are processed every 0.1 seconds. The experiment expects deterministic results. Each experiment is

Figure 14. Operators in the prototype

+	Conjunction
|	Disjunction
=	Concatenation
;	Sequence
:	Concurrency
*	Iteration
-	Negation
"	Selection
@	Spatial Restriction
!	Temporal Restriction

repeated a minimum of 3 times and it produced the expected deterministic results. The event operators for composite event expressions are shown in Figure 14. An example of a valid expression is ((A; B)5)|((A; B) - B), and several experimental results are shown as follows.

Example 1

Figure 15 shows the basic operation of two different composite event detections.

Example 2

Figure 16 shows that an ordered set of primitive events (i.e., input stream) should trigger both specified composite events (e.g., simultaneous Disjunction operation). The composite event detection is reported at the time-line described.

Example 3

Figure 17 shows the associativity of sequence.

Figure 15. Composite event detection: Multiple composite event detection

```
Timeline:           1 2 3 4 5 6 7 8
----------------------------------------
Primitive Event:    A C B C D B A D
----------------------------------------
Composite Event Expression: A;C
                            A;D

At 2:   (Detected A;C)
At 4:   (Detected A;C)
At 5:   (Detected A;D)
At 8:   (Detected A;D)
```

Figure 16. Composite event detection: Simultaneous disjunction

```
Timeline:          1 2 3 4 5 6 7 8 9 10 11 12 13 14 15 16 17
------------------------------------------------------------------
Primitive Event:   A B C D A C E D C B  A  D  B  A  C  B  D
------------------------------------------------------------------
Composite Event Expression: A;((B;D)|(C;D))
                            A;((C;D)|(B;D))

At 4:   (Detected A;((B;D)|(C;D)) A;((C;D)|(B;D)))
At 8:   (Detected A;((B;D)|(C;D)) A;((C;D)|(B;D)))
At 12:  (Detected A;((B;D)|(C;D)) A;((C;D)|(B;D)))
At 17:  (Detected A;((B;D)|(C;D)) A;((C;D)|(B;D)))
```

Figure 17. Composite event detection: Associativity of sequence

```
Timeline:          1 2 3 4 5 6 7 8 9 10 11 12 13 14 15 16 17 18
------------------------------------------------------------------
Primitive Event:   A B C D E B B D A E  D  C  B  A  E  D  C  D
------------------------------------------------------------------
Composite Event Expression: (A;B);(C;D)

At 4:   (Detected (A;B);(C;D))
At 7:   (Detected (A;B);(C;D))
At 8:   (Detected (A;B);(C;D))
At 11:  (Detected (A;B);(C;D))
At 13:  (Detected (A;B);(C;D))
At 16:  (Detected (A;B);(C;D))
At 18:  (Detected (A;B);(C;D))
```

Figure 18. Composite event detection: Temporal restriction

```
Timeline:           1 2 3 4 5 6 7 8 9
----------------------------------------
Primitive Event:  A A B B B B B B B
----------------------------------------
Composite Event Expression: (A_5);B

At 3:   (Detected (A_5);B)
At 4:   (Detected (A_5);B)
At 5:   (Detected (A_5);B)
At 6:   (Detected (A_5);B)
At 7:   (Detected (A_5);B)
```

Figure 19. Composite event detection: Negation

```
Timeline:          1 2 3 4 5 6 7 8 9 10 11 12
-------------------------------------------------
Primitive Event:   A A B A A C C B C A  D  B
-------------------------------------------------
Composite Event Expression: (A;B)-C

At 3:   (Detected (A;B)-C)
At 12:  (Detected (A;B)-C)
```

Example 4

Figure 18 shows the use of temporal restriction. The composite event (A_5); B denotes that a primitive event A is valid for 5 time units and, under that condition, event B follows A must be detected. The target composite event is detected up to the time line 7 in Figure 18, however after time unit 8, an event A is no longer valid.

Example 5

Figure 19 shows the use of the negation operator. The composite event (A; B) - C denotes that a primitive event A is followed by B, but event C must not occur between the occurrences of A and B. The result shows the correct detection.

Object Tracking with Active BAT

Sentient computing is ubiquitous computing using sensors to perceive the environment and react accordingly. One research prototype of a sentient computing system was developed at AT&T Research in the 1990s, and this research continues at the University of Cambridge as the Active BAT system (Harter, Hopper, Stegglesand, Ward, & Webster, 2002). It is a low-power, wireless indoor location system, which is accurate up to 3 cm. It uses an ultrasound time-of-flight trilateration technique to provide more accurate physical positioning. Users and objects carry Active BAT tags. In response to a request that the controller sends to a BAT via short-range radio, a BAT emits an ultrasonic pulse to a ceiling-mounted receiver grid. At the same time, the central controller sends the radio frequency request packet to the ceiling sensors to reset the baseline via a wired serial network. Each ceiling sensor computes the distance to the BAT from the time interval from the baseline to ultrasonic pulse arrival from the BAT. This distance value is forwarded to the central controller, where the trilateration computation is performed for object location.

Distributed Gateways

The current Active BAT system employs a centralised architecture, and all the data are gathered in the database. The Active BAT system, as described, is expensive to implement in that it requires large installations, and a centralised structure. We envision a future version of the Active BAT system should be based on real-time mobile ad hoc communications and distributed coordination must be supported, so that mobile device users can promptly subscribe to certain information. In this experiment, using real Active BAT data, a hypothetical environment described below is created that regenerates events of the Active BAT data. It is simulated that each room and corridors hold gateway nodes, which are able to participate in event broker grids. Thus, each local gateway node performs event filtering and correlation by registering the service that associates subscriptions with abstractions such as Andy in room SN04. These subscriptions are decomposed to the units operable by the event broker grid, where event correlation, aggregation and filtering are supported.

Figure 20. Events (Andy, Brian)

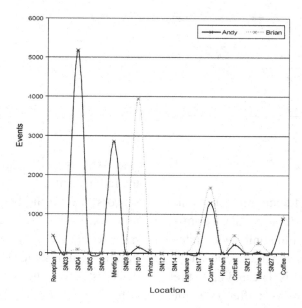

Figure 21. Durative events (Andy, Brian)

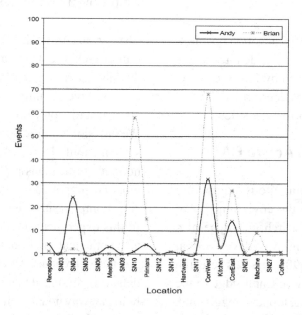

Durative Event Efficiency

Several event correlations are performed. Among them, the use of durative events is shown below. During this experiment, 21 BATs participated.

Figure 20 and Figure 21 depict the events identified on the BAT holders Andy and Brian. Andy's office (room SN04) is most likely the location where the highest number of events was recorded. Figure 20 and Figure 21 show the events over

Figure 22. Events over time

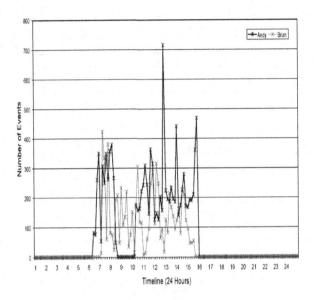

Figure 23. Durative events over time

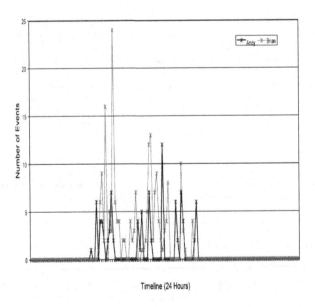

the location, but they do not indicate when they occurred. Figure 22 and Figure 23 depict the events over the time-line (24 hours). Most activities are recorded during day time. Durative event composition over the time-line (24 hours) shows a significant reduction in the number of events

Event Correlation

Figure 24 shows a specific period, when Andy and Brian were positioned in *corridor west*. The composite event (*Brian*; *Andy*)$_{corrwest}$ is detected at time unit 2564. In Figure 25, the detection

Figure 24. Composite event – (Brian; Andy)$_{corrwest}$

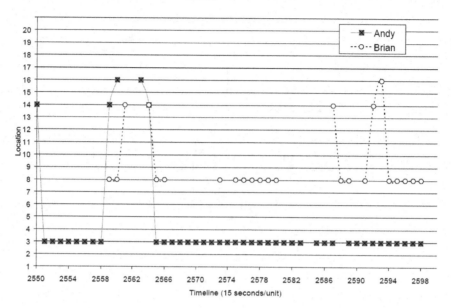

Figure 25. Composite event – (Andy + Brian)$_{machine}$

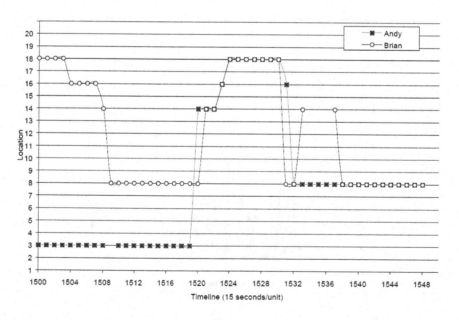

of composite event $(Andy + Brian)_{SN25(machine)}$ is shown at the time unit between 1523 and 1529. A local gateway can detect this correlation, if the composite event is subscribed to through the event broker. This composition could be a part of services provided by the service grid.

FUTURE TRENDS

The work presented in this chapter provides an important first step into data-centric networking research. Major research challenges in this area are information diffusion over wireless mobile network environments, event aggregation, filtering, and correlation over networks with selective routing. We describe the future trends for specific topics in this section

Fuzzy Semantic Query

Current search engines such as Google use keyword-based queries to help users find relevant information. Traditional tools assume that the parameters of a query model represent exactly the features of the modelled objects. However, some query processes are uncertain and hard to express in the form of traditional query languages. In publish/subscribe systems, the subscribers may not know exactly what to subscribe to when issuing subscriptions. For example, if the subscriber is interested in any abnormality in a time series of data, the pattern of the composite event is unknown. This requires a new type of subscription definition (i.e., fuzzy subscription), using a more ambiguous subscription language based on natural language or a series of keywords.

Fuzziness is often used to express the uncertainty or vagueness concerning the semantics of events, phenomena, or statements. The power of range matching may not be sufficient for many applications, as users may not have the knowledge to formulate precise queries. It is desirable to apply fuzzy logic to model publish/subscribe including approximate matching for the uncertainty of wireless sensor data. In (Liu & Jacobsen, 2004), subscriptions are expressed in a formula using a fuzzy set to integrate with publish/subscribe systems. The concept of fuzzy subscription is not explored in depth in publish/subscribe systems yet, and will be interesting future work.

High Level Language for Event Correlation Semantics

We have described the semantics of event correlation with a parameterised event algebra. A high level language is desirable, which may combine subscription and correlation languages. Transformation of correlation semantics to π-$calculus$ can automatically provide the capability to compile into Java code or other conventional programming languages from the definition. Formalisation and transformation will help to create implementable, reliable, and QOS-guaranteed event detection in various resource equipped devices.

Programmable Networks

The majority of programming for WSNs follows a component-based event-driven programming model, where an event triggers the operation to migrate data from one node to another. TAG (Madden, Franklin, Hellerstein, & Hong, 2002) considers WSNs as a database and provides a high-level SQL-like language for queries. The programmability of sensor networks has received attention recently. (Nagpal, 2002) introduces a high-level language inspired by biology, where program agents are considered as cells to form a globally specified shape via local computation and communication.

Currently most sensor network applications are implemented as complex, low level programs that specify the behaviour of individual sensor nodes. Programming for WSNs raises an issue on programming abstractions, where abstractions of sensors and sensor data need to be defined. Programming abstractions fall into two categories: application level and system level. The former defines and manipulates events at the desired level of semantic abstraction (e.g., latest position of target). The latter precisely specifies distributed computation and communication (e.g., apply f(x) to all x, send d to the 10 nearest nodes). The tradeoffs between these two models are expressiveness, efficiency, reusability and automation. The programmable network research area is still at an early stage and widely open to new developments.

CONCLUSION

In this chapter, we placed an emphasis on the design of the event itself. Besides the existing attributes of events, continuous context information such as time or geographic location is incorporated within an event description. The semantics and representation of events and queries will be even more important in future ubiquitous computing, where events flow over heterogeneous networks. We have briefly shown that multidimensional event indexing with an RTree enables efficient indexing, matching, and scalability. Applications are now dependent on XML and RDF for data representation, and there is a growing need for efficient data centric routing protocols that can support query language expressiveness.

Secondly we introduced a generic interval-based semantics for composite event detection. This allows the precise definition of complex timing constraints among correlated event instances, which adds a new dimension to event processing in global computing, understanding event aggregation, filtering and correlation. The sensed data will be aggregated and combined into higher-level information at appropriate points, while flowing over heterogeneous networks. Our work on 'Time/Space Aware Event Correlation' provides the way to define temporal and spatial relationships among correlated events from heterogeneous network environments.

Recent evolution in WSNs has meant that ubiquitous computing must now play an important role in data processing, where sensors are used to gather high volumes of data, to feed as context to a wide range of applications. This data processing requires novel mechanisms that make intelligent use of diverse, simple data to create meaningful information. Complex data management must be integrated within the context of data-centric networking. We hope this work represents a step towards accomplishing a universal event correlation framework and routing network.

REFERENCES

Abadi, D. J., Ahmad, Y., Balazinska, M., Cetintemel, U., Cherniack, M., Hwang, J., & Zdonik, S. (2005). The design of the Borealis stream processing engine. In *Proceedings of CIDR*, (pp. 277–289).

Abadi, D. J., Carney, D., Cetintemel, U., Cherniack, M., Convey, C., Lee, S., & Zdonik, S. (2003). Aurora: A new model and architecture for data stream management. *The VLDB Journal, 12*(2), 120–139. doi:10.1007/s00778-003-0095-z

Adaikkalavan, R., & Chakravarthy, S. (2006). SnoopIB: Interval-based event specification and detection for active databases. *Data & Knowledge Engineering, 59*(1), 139–165. doi:10.1016/j.datak.2005.07.009

Allen, J. (1983). Maintaining knowledge about temporal intervals. *CACM, 26*(1), 832–843.

Arasu, A., Babcock, B., Babu, S., Datar, M., Ito, K., & Nishizawa, I. Widom, J. (2003). Stream: The Stanford stream data manager. In *Proceedings of ACM SIGMOD*, (pp. 665).

Arasu, A., Babu, S., & Widom, J. (2002). *An abstract semantics and concrete language for continuous queries over streams and relations.* (Technical Report 2002-57). Palo Alto, CA: Stanford University.

Babich, F., & Deotto, L. (2002). Formal methods for specification and analysis of communication protocols. *IEEE Communication Survey and Tutorials, 4*(1), 2–20. doi:10.1109/COMST.2002.5341329

Bacon, J., Bates, J., Hayton, R., & Moody, K. (1995). Using events to build distributed applications. In *Proc. IEEE SDNE*, (pp.148–155).

Borgida, A., & Patel-Schneider, P. (1994). A semantics and complete algorithm for subsumption in the classic description logic. *Journal of Artificial Intelligence Research*, 277–308.

Carlson, J., & Lisper, B. (2003). An interval-based algebra for restricted event detection. In *Proceedings of FORMATS*, (pp. 121–133).

Carzaniga, A., & Wolf, A. L. (2003). Forwarding in a content-based network. In *Proceedings of ACM SIGCOMM*, (pp. 163–174).

Cayuga Project. (2005). *Cayuga technical report*, (Technical Report). Ithica, NY: Cornell, University. Retrieved from http://www.cs.cornell.edu/mshong/cayuga-techreport.pdf

Chakravarthy, S., & Mishra, D. (1996). Snoop: An expressive event specification language for active databases. *Data & Knowledge Engineering*, *14*(1), 1–26. doi:10.1016/0169-023X(94)90006-X

Chandrasekaran, S., Cooper, O., Deshpande, A., Franklin, M., Hellerstein, J., Hong, W., & Shah, M. (2003). Continuous dataflow processing for an uncertain world. In *Proceedings of Innovative Data System Research* (pp. 269–280). TelegraphCQ.

Daws, C., Olivero, A., Tripakis, S., & Yovine, S. (1996). *The tool kronos in Hybrid Systems III*. (LNCS 1066), (pp. 208-219). New York: Springer.

Eugster, P., Felber, P., Guerraoui, R., & Kermarrec, A. (2003). The many faces of publish/subscribe. *Computing Surveys*, *35*(2), 114–131. doi:10.1145/857076.857078

Eugster, P., Felber, P., Guerraoui, R., & Sventek, J. (2000). *Type-based publish/subscribe. (Technical report)*. Lausanne, Switzerland: EPFL.

Galton, A., & Augusto, J. C. (2002). Two approaches to event definition. In *Proceedings of DEXA*, (pp. 547–556).

Gatziu, S., & Dittrichothers, K. R. (1994). Detecting composite events in active database systems using Petri Nets. In *Proceedings of RIDE-AIDS*, (pp. 2–9).

Gehani, N. H., Jagadish, H. V., & Shmueli, O. (1992). Composite event specification in active databases: Model and implementation. In *Proceedings of VLDB*, (pp. 327–338).

Gomez, R., & Augusto, J. C. (2004). Durative events in active databases. In *Proceedings of ICEIS*, (pp. 511–516).

Gruber, B., Krishnamurthy, B., & Panagos, E. (1999). The architecture of the READY event notification service. In *Proceedings of ICDCS Workshop on Electronic Commerce and Web-Based Applications*, (pp. 1–8).

Gupta, A., Sahin, O. D., Agrawal, D., & Abbadi, A. E. (2004). Meghdoot: content-based publish/subscribe over P2P networks. In *Proceedings of ACM/IFIP/USENIX Middleware*, (pp. 254–273).

Guttman, A. (1984). R-trees: A dynamic index structure for spatial searching. In *Proceedings of ACM SIGMOD*, (pp. 47–57).

Harter, A., Hopper, A., Stegglesand, P., Ward, A., & Webster, P. (2002). The anatomy of a context-aware application. *Springer Wireless Networks*, *8*(2-3), 187–197. doi:10.1023/A:1013767926256

Hayton, R. (1996). *OASIS: An Open architecture for Secure Inter-working Services*. (PhD thesis), Cambridge, UK: University of Cambridge.

Hinze, A., & Bittner, S. (2002). Efficient distribution-based event filtering. In Proceedings of Distributed Event Based Systems, (pp. 525–532).

Hoffmann-Wellenhof, B. H., Lichtenegger, H., & Collins, J. (1994). *GPS: Theory and Practice*. New York: Springer.

Lazaridis, I., & Mehrotra, S. (2003). Capturing sensor-generated time series with quality guarantees. In *Proceedings of ICDE*, (pp. 429–440).

Liebig, C., Cilia, M., & Buchmann, A. (1999). Event composition in time-dependent distributed systems. In *Proceedings of IFIP CoopIS*, (pp. 70–78).

Liu, G., Mok, A. K., & Konana, P. (1998). A unified approach for specifying timing constraints and composite events in active real-time database systems. In *Proceedings of the Real-Time Technology and Applications Symposium*, (pp. 2–9).

Liu, H., & Jacobsen, H. (2004). Modeling uncertainties in publish/subscribe system. In *Proceedings of ICDE*, (pp. 510–521).

Ma, C., & Bacon, J. (1998). COBEA: A corba-based event architecture. In *Proceedings of COOTS*, (pp. 117–132).

Madden, S., Franklin, M. J., Hellerstein, J., & Hong, W. (2002). TAG: a tiny aggregation tree for ad-hoc sensor networks. In *Proceedings of USENIX Symposium on Operating Systems Design and Implementation*, (pp. 131–146).

Mansouri-Samani, M., & Sloman, M. (1997). Gem: A generalized event monitoring language for distributed systems. *IEE/IOP/BCS Distributed systems Engineering Journal, 4*(2), 96–108.

Motwani, R., Widom, J., Arasu, A., Babcock, B., Babu, S., Datar, M., & Varma, R. (2003). Query processing, approximation, and resource management in a data stream management system. In Proceedings of Innovative Data Systems Research, (pp. 245–256).

Muhl, G., Fiege, L., & Buchmann, A. (2002). Filter similarities in content-based publish/subscribe systems. In *Proceedings of ARCS*, (pp. 224–238).

Nagpal, R. (2002). Programmable self-assembly using biologically inspired multiagent control. In *Proceedings of AAMAS*, (pp. 418–425).

Pietzuch, P., Shand, B., & Bacon, J. (2004). Composite event detection as a generic middleware extension. *IEEE Network, 18*(1), 44–55. doi:10.1109/MNET.2004.1265833

Research, I. B. M. (2001). *Gryphon: Publish/Subscribe over public networks*. Retrieved from http://researchweb.watson.ibm.com/grypohn/Gryphon/gryphon.html

Schwiderski, S. (1996). *Monitoring the Behavior of Distributed Systems*. (PhD thesis), Cambridge, UK: University of Cambridge.

Segall, B., & Arnold, D. (1998). Elvin has left the building: A publish/subscribe notification service with quenching. In *Proceedings of AUUG*. Retrieved from http://www.dtsc.edu.au/

Snoeren, A., Conley, K., & Gifford, D. (2001). Mesh based content routing using XML. In *Proc. ACM SOSP*, (pp. 160–173).

Widom, J., & Ceri, S. (1996). *Active Database Systems: Triggers and Rules For Advanced Database Processing*. San Francisco: Morgan Kaufmann Publishers.

Yoneki, E. (2006). *ECCO: Data Centric Asynchronous Communication*. (PhD thesis), (Technical Report UCAM-CL-TR677), Cambridge, UK: University of Cambridge.

Yoneki, E., & Bacon, J. (2007). ecube: Hypercube event for efficient filtering in content-based routing. In *Proceedings of the International Conference on Grid computing, High-performance and Distributed Applications*, (LNCS 4804, pp. 1244–1263).

Zimmer, D., & Unland, R. (1999). On the semantics of complex events in active database management systems. In *Proceedings of ICDE*, (pp. 392–399).

Chapter 4
Temporal Perspectives in Event Processing

Opher Etzion
IBM Haifa Research Lab, Israel

ABSTRACT

One of the major characteristics of event processing is its strong relationship to the notion of time, yet some of the temporal aspects of event processing still issue challenges to the implementations of event processing tools. This paper provides an overview of the notion of "event processing network" as the underlying model behind event processing; maps the temporal aspects, and discuss each of them. The temporal aspects that are discussed are: temporal dimensions of events, time granularities, temporal context, temporal patterns, event order, and retrospective and proactive operations.

INTRODUCTION

Event Processing is an emerging area in the IT industry, evident by the burst of products, and attention given by analysts, venture capitals and enterprises. One of the notable characteristics of event processing is its close relation to temporal aspects. One can view event processing as getting a decision that is based on looking at the history of transitions in the domain of discourse. This glance on the event history involves multiple aspects of temporal operations; In this section we'll explain briefly the main concepts and architecture

of event processing, and show the touch points between event processing and temporal aspects, the rest of the paper will deal in details with the various issues.

Introduction to Event Processing

Event denotes *something that happens in reality* (Luckham, 2002). While data-item typically denotes the state of some entity, events denote transition between these states. There is a semantic overload in the term "event", since it reflects both the event occurred in the reality and its representation in the computer domain (event message or event object). Event is being processed by the

DOI: 10.4018/978-1-60566-697-6.ch004

Figure 1. Event processing network example

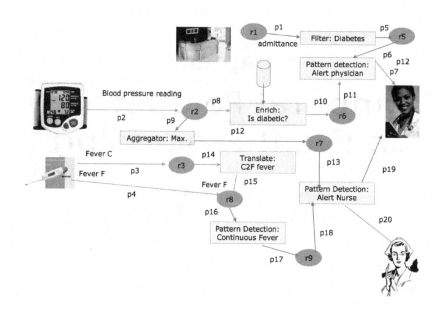

means of "event processing network" (Sharon, 2008).

The life-cycle of event processing application is detailed below (some of it in iterative way):

1. Event schemas are defined.
2. Event producers devise mechanism to emit events: in push, periodic pull, or on-demand pull. Event producer examples are: instrumented program, state observer in business process management system, sensors.
3. The event is published on a channel and routed through an "event pipe" to one or more "event processing agents" which – validate, transform, enrich, filter or detect patterns and create more events that are in turn published on another channel and routed to more event processing agents.
4. At the edge of the event processing network the created events called "situation events" and are routed to "event consumers"

Figure 1 shows an EPN example, for a monitored patient in the hospital.

Event producers are:

- Admittance office that records admittance and release of patients.
- Blood pressure meter
- Fever meter

Agents are performing: filtering, aggregation, enrichment, transformation and pattern detection; routing channels are designated in r1… r9 and event pipes are designated in p1… p20. We shall use this patient monitoring simple example to demonstrate all concepts. The patterns that need to be detected are:

- **Ascending Fever:** Fever of more than 39 degrees Celsius is consistent over three hours and is constantly ascending, this is calculated every three hours – in 00:00, 03:00, 06:00, 09:00, 12:00, 15:00, 18:00, 21:00.
- **Hypotension:** The average systolic blood pressure over three consecutive measurements is less than 85

- Ascending Fever and later Hypotension.
- Systolic Blood Pressure is more than 180 and patient is diabetic within the first day after admittance.

Temporal Aspects of Event Processing

The temporal aspects of event processing are realized in various ways:

- The temporal characteristics of a single event: in temporal database, the time perspectives are transaction time and valid time, what are the equivalent perspectives for events? Are they the same? Section 2 deals with these issues and look also at time-point based semantics vs. interval-based semantics and time granularities.
- The notion of context (Adi, 2003) is one of the fundamental concepts of event processing. One of the major dimensions is temporal. Section 3 discusses the notion of temporal context, from its simple incarnation of "time window" until more sophisticated contexts.
- Section 4 deals with temporal event patterns
- Section 5 deals with extending the event perspectives to "derived events".
- Section 6 deals with an issue that relate to both the notion of context and patterns, and discusses the case in which events arrive "out of order".
- Section 7 deals with the impact of distribution on the temporal aspects
- Section 8 summarizes this paper.

When trying to check whether transition semantics has any distinction from state semantics, we shall try to look at the following issues:

1. Are the temporal dimensions proposed for state semantics also applicable for transition semantics?

2. Does transition occur over a time-point or an interval?
3. Can we conclude anything about event processing from these observations?

TEMPORAL DIMENSIONS OF EVENTS

Temporal Dimensions of States

Temporal database represents the history of states in a certain domain (Jensen, 1999) and consists of snapshot semantics and state history semantics.

Snapshot semantics, as captured by classical database processing, deals with the representing and processing of a **single** state of the universe. Queries that relate to this state enable the processing of this snapshot, for example, the query: "How many daily flights there are from Boston to London" refers to a single (typically the current) state of the universe.

State history semantics, as captured by temporal database processing, deals with the processing of multiple states. Each data-item may have two temporal dimensions:

- *Transaction Time (Tx):* The time-stamp that denotes the commit of the transaction that updated this data-item in the database. This is important to ask queries "as-of": what is the answer to a certain query if asked "as-of" the observation point of June 1, 2008, answer: all data-items whose Tx < June 1, 2008 are counted).
- *Valid Time (Tv):* A time-interval in which the fact stated by the data-item is true in reality.

Example: the fact that there is a "flight BA0238 flies from Boston to London and leaves in 8:15AM" is recorded in the database in May 25, 2008 (Tx), and is valid from June 1, 2008 until September 30, 2008 (Tv).

Temporal Dimensions of Transitions

Looking at the relevance of the temporal dimensions of state, the following observations can be made about transitions; following the observation that an event can be mapped into transition among states (Howe, 1997).

Detection Time: *The timestamp in which the message representing the event has been detected by the processing system.* This definition is related to the execution of the event processing system.

Example: the event "flight BA0238 landed in 10:39" this has been reported in 10:40, which is the detection time of this event.

Occurrence time:*The time-stamp (or interval) in which the event occurred in the real universe.*

In many applications the occurrence time is the correct reference time for the order notion, rather than the detection time. In temporal state processing this time dimension has been mentioned as: event time or decision time (Etzion, 1994)), but was not considered as a major dimension

Valid Time: State spans over time, and may have expiration semantics as the one described by valid time (e.g. the flight schedule is valid between June and September), events typically happen and do not expire, however the notion of "valid time" maybe useful to denote that an event is not relevant anymore for processing (e.g. event is an order, and a "retraction event" of order cancellation has been issued), or that this event is relevant to operations only in the next ten minutes, later than that it expires. However, it is not a major dimension in event processing, and more used for management and auditing purposes.

Conclusion: the main temporal dimensions between temporal databases and event processing are slightly different. Instead of: transaction time and valid time which are the major dimensions of temporal databases, in event processing detection time and occurrence time are the major time dimensions, Detection time is determined by the event processing system, thus, obtaining an accurate value of detection time can be achieved;

contrary to that occurrence time is reported by the producer, and thus there are some issues related to its accuracy and synchronization that are discussed in the sequel.

The "Instantaneous" Question: Does Event Occur Over a Time-Point or Over an Interval?

Transition between states is typically considered as instantaneous (Iwasaki, 1995): that is, occurs in a specific point in time, or has "zero duration". This raises the question, whether an event adopts this property, and is instantaneous, or event occurs over a time interval. Here are three examples that demonstrate the ambiguity of the term "event duration".

The event: Flight BA0238 landed (two hours late). Landing is a process that starts with the descending of the aircraft, and ends when the aircraft parks at the gate. This is clearly an interval. One can argue that "landing" is a state and not a transition that starts at a single time-point and ends at a single time point, and the "landed" is semantically refers to the end of the process when the aircraft arrives at the gate.

The derived event: "alert physician" defined as blood pressure is constantly up within 2 hours period and fever $> 39\circ$ during this period. Here again one interpretation can be that this event occurs during two hours interval and another interpretation is that the event occurs when the derived event is detected (typically some time offset after the two hours interval ended).

The complex event "the financial crisis of 2008", is an event that consists of many atomic events (thus the name complex), and spans over an interval that has not ended when this paper has been written.

In reality, events typically occur over some interval, contrary to the common definition of transitions; thus an event may have a start timestamp and an end timestamp. For computation purposes it is typically easier to deal with events that occur over a single time point, and the most typical operations on events involved questions about their order. The bridge between the reality, and the computational convenience is done in many cases, by selecting the granularity of the time point (see discussion below), and approximating the time interval to such time point.

Time Granularity

Different applications are interested in different time granularities. A salary system may be interested in a granularity of a month, while a missile tracking system may be interested in a granularity of micro-second. In general we can use the term *Chronon that* has been proposed by the temporal database community (Anrung, 2008), and denotes the application-dependent "atomic unit of time". Chronon can denote different time granularities.

Conclusion: While there are some commonalities between the temporal concepts in event processing and the area of temporal databases, however, there are some differences which stem from the fact that temporal databases deal with the processing of states, while event processing with the processing of transitions. The differences are in abstractions and the semantics of transition duration.

Examples in the Patient Monitoring Story

For the event Blood-Pressure-Reading there are two dimensions, assuming that the report is off-line and the nurse has to fill a form on her desktop. It is possible that the nurse records the blood pressure a few minutes after the measurement. The occurrence time is the time in which the measurement has been executed, while the detection time is the time in which the nurse reported the measurement results. In this application a granularity of a minute is sufficient, thus time stamp is in the form YYYYMMDD:HHMM Which denotes year in four digits, month, hour and minute (all in two digits).

TEMPORAL CONTEXTS

The notion of context, as a generic term, has been defined as: *"Context is any information that can be used to characterize the situation of an entity. An entity is a person, place, or object that is considered relevant to the interaction between a user and an application, including the user and applications themselves."* (Dey, 2000). Putting events in contexts have been discussed by several studies like (Adi, 2002) which represents context as a major abstraction of event processing, (Shaw, 2008) which talks about Spatio-temporal contexts, and (Shevade, 2007) who discusses context of events within social networks. All the various context definitions have the same practical role: partitioning the "cloud" of events into various partitions, such that an event processing operation operates on a single partition, i.e. the events in different partitions are not mixed in processing. A partition can be semantic (all events that relate to a single customer, or to platinum customer are group together as input for a single processing), can be based on spatial properties (all events within 1KM from here are those who relevant for processing), and can be based on time. Of course, it can be based on all the above creating a multi-dimensional context. In this section we discuss temporal contexts, which denote temporal partition on events that are being processed together; the notion of temporal context subsumes the notion of "time window" that exists in stream processing (Patroumpas, 2006).

The role of a temporal context is to group together events that need to be processed within a single agent instance. Some examples:

- If the operation is enforcing a regulation that a person cannot withdraw more than three times from ATM machines within **a single day**, then each day (starting in midnight) is a separate context, and the relevant events are associated with it;

- If the regulation is modified such that a person cannot withdraw more than three times from ATM machine within a 24 hours period, then each time that a person withdraws from an ATM machine, the temporal context now starts whenever a customer withdraws money from an ATM machine, and ends 24 hours later. A temporal context can be one of the three types:
  1. Fixed Interval - may be periodic
  2. Event Interval: An interval that starts by event and ends by event.
  3. Sliding Interval: An interval with fixed length that "slides" over the time.

Fixed Interval

Fixed interval is an interval that has a fixed length; it may be repeating, example: *every working day between 8 - 10 AM,* or can be ad-hoc and start explicitly, example: *from now until now + 3 hours.* A fixed interval is used frequently for periodic aggregations, such as, aggregation of events to compare against key performance indicators (KPI). Examples:

- Calculate aggregations about purchases from a web store for every hour separately.
- Monitor patient for a fixed time since some health related event has occurred.
- Calculates some indications about the behavior of the stock market within a single day.

In all these cases – event that have occurred during this fixed period serve as input to a single processing agent.

Event Interval

Event Interval is an interval that starts with the occurrence of event and ends with the occurrence of another event. Example: *From patient admittance to patient release.* Note that event can also be calendar event set by fixed time, or offset from the start event, and that multiple events can start or terminate the interval. There are several variations of this one:

- The event interval starts by the occurrence of a pre-specified event (called: **initiator**). This can be a single event, or a member of collection of events.
- The event interval terminates by the occurrence of a pre-specified event (called: **terminator**). This can be again a single event, or a member of a collection of events.
- The event interval may expire even if a terminator has not arrived, either after some time offsets, or when a certain amount of events of a certain type have been detected.
- If an instance of an initiator event occurs while an event interval is open, the behavior can be set up by policies that determine the behavior, example for such policies are:
  ○ Open another event interval in parallel.
  ○ Ignore the second initiator.
  ○ Close and re-open the event interval
- The policy is set according to the semantics of the application.

Some examples:

- Monitor all movements of a package since it starts the journey until it is delivered (in this case a second initiator does not make sense and should be ignored).
- Monitor a stock to observe if there are at least three big transactions (> $1M value of trade) within a single hour. In this case whenever a big transaction occurs a temporal context instance is opened, and closed after an hour. Note that several intervals can be opened in parallel, since each can start a one hour period in which the condition is satisfied.

Sliding Interval

Sliding interval is an interval with fixed size that continuously slides on the time axis. Sliding intervals may be overlapping or non-overlapping. Overlapping interval example is: *Every 10 minutes start a sliding interval of one hour.* In this case, there will be six partially overlapping intervals in every single time-point. Non-overlapping interval example is: *Every hour start a sliding window of an hour.* At any point in time there is a single sliding interval. Typically sliding interval is being used for aggregative operations, e.g. counting the events of a certain type that occurs in the sliding interval. A sliding interval may count number of events and not time.

As a conclusion, the notion of temporal context is a major abstraction that enables to group input events according to time characteristics. As noted before, a context may be multi-dimensional and include a spatial dimension and semantic dimension. A composite context for a real estate dealer can be: events that relate to real estate transactions within the city of Winchester starting from the time that a house that costs more than $2M has been put on the market, until the same house is being sold.

Examples in the Patients Monitoring Story

- A sliding interval is an interval of three hours that starts and ends every three hours, thus it is non overlapping sliding interval, this interval is the context for the Ascending Fever detection.
- An event interval that starts by event and expires after a certain amount of event instances is the case in which we are looking for three consecutive measurements of blood pressure.
- An event interval that starts by event and expires after certain amount of time is the case in which we are looking for hyper-

tension for diabetic patient in the first day after admittance. Note that patient release serves as a terminator.

TEMPORAL PATTERNS

The notion of pattern is central to complex event processing (Sharon, 2008). The goal is to find patterns in the event history in order to identify high order events (Luckham, 2002). There are various types of event processing patterns. Pattern may or may not be temporal in nature. Examples of patterns that are not related to time are:

- The sales person visited in all cities in the list, where the order is not important;
- Five members of the team have already logged into the system, again the order of these events is not important

In this Section we concentrate on patterns that are temporal based. We revisit the different time perspectives and explains their role in the pattern matching; we then survey patterns related to "event at a time" processing, and survey patterns related to batches of events. This Section relates to an event processing generic model, there were specific work in variations of event processing, such as: (Bry, 2008) that deal with temporal aspects of stream queries, and temporal aspects of event processing based on RETE rules are discussed in (Waltzer, 2008).

Processing Based on the Different Time Perspectives

We have discussed two dimensions: detection time and occurrence time. The semantics of event processing temporal patterns can follow each of them:

- **Detection time semantics:** all the temporal operators are applied on the detection time of events, i.e. the order of events is considered to be the order that is imposed by the detection time timestamps.

- **Occurrence time semantics:** all the temporal operators are applied on the occurrence time of events, i.e. the order of event is determined according to the occurrence time.

Many of the event processing products follow the detection time semantics. It is much easier to implement, since the time-stamp is created by the processing system, thus it is not dependent upon the event producer, its semantics is well defined (see Section 5 for discussion of semantics of derived events to notice that its semantics is not obvious), and it does not require time synchronization between various event producers. While it is easier to implement the detection time semantics, alas, there is one fundamental problem with this solution; it does not guarantee correct results. The reason is that the order of events according to detection time semantics may be different from the order of events according to the occurrence time semantics, and the occurrence time semantics which reflects the orders of events in reality. Section 6 deals with the order issue, and discusses cases where the correct order is required, and the difficulties in maintaining such an order.

Event-at-a-Time Patterns

Event-at-a-time is a processing mode, by which each event is evaluated when it is detected relative to the appropriate context. When an event arrives to a specific agent, it checks the status of a pattern detection operation, and if the pattern is being matched then it acts accordingly. While there are variety of patterns that are not temporal in nature, in the scope of this paper we concentrate on two temporal patterns:

- **Temporal Sequence pattern:** the pattern Temporal-Sequence (E1, ..., En) with predicate P and context C is satisfied (Ei

is an event type, P is a predicate on some or all of these event types) when there are events $e1 \in E1,..., en \in En$ such that, in the temporal order, $e1 < e2 <...< en$ and P is evaluated to true on the relevant members of the events $e1,...,en$. Example: **The temporal pattern** *Speculation-Detection* **occurs when the same customer buys and later sells the same stock in value of more than $1M during the same day,** this is expressed as: Temporal-Sequence (Buy, Sell) where:

> Buy. Value > $1M and Sell. Value > $1M; the context C partitions all events relating to the same stock and same customer to a single agent.

- **Time-out pattern:** The pattern Time-out (E) within context C (AKA as absence pattern) is satisfied if there is no instance of event-type E occurs within the context C. Example: **The pattern** *Shipment-not-arrived* **occurs when a shipment does not arrive within its delivery due period.** This is express as time-out (shipment-arrive), where the context is = (T1, T2) the period during which the shipment should arrive.

Set-at-a-Time Patterns

Set-at-a-time patterns are patterns that are being evaluated after all the set of events have arrived. The set may be a time series (Povinelli, 2003) in which events arrive at fixed rate, or a collection of events of certain type that occurred within some temporal context. The evaluation is typically done at the end of the temporal context. There are several aggregation patterns, in the scope of this paper, we'll concentrate in temporal patterns of the type of trend. A trend temporal pattern is of the form: *<Trend-Name E A>*, where:

- Trend-Name is one of the following values: increasing, strictly-increasing, decreasing, strictly-decreasing, and mixed.
- E is an event name
- A; is a name of an attribute of E..

The trends are defined according to the temporal order (detection time or occurrence time), let the event instances according to this order denoted as e1 < e2 < ... < en.

- **Increasing:** $e1.A \leq e2.A \leq ... \leq en. A$
- **Strictly-Increasing:** $e1.A < e2.A < ... < en. A$
- **Decreasing:** $e1.A \geq e2.A \geq ... \geq en. A$
- **Strictly-Decreasing:** $e1.A > e2.A > ... > en. A$
- **Mixed:** when none of the above is satisfied.

Examples:

- The pattern *Increasing-Blood-Pressure* is defined as blood pressure is strictly increasing for a patient over 24 hours (assuming there is a time series of measurements).
- The pattern *decreasing-stock* is defined as the value of a certain stock at the end of the day is decreasing over a month.

Examples in the Patient Monitoring Story

- The temporal sequence pattern is **Ascending Fever and later Hypotension,** this is an example of two derived events that should occur within a certain order; Section 5 discusses the notion of when a derived event has occurred.
- **Ascending Fever** is a trend oriented pattern, in this case the trend is "ascending" and is calculated at the end of the time context.

TEMPORAL PROPERTIES OF DERIVED EVENTS

Derived event (Etzion, 2005) (AKA virtual event, synthetic event) is an event that is derived from other events using *Event Processing agent* which takes as input one or more events, and produces as an output one or more events. Note that in the case of *raw events,* an event occurs in reality, and its occurrence results in the introduction of an event 8in the computer domain, while in the case of derived event, the process is reversed. The event is produced in the computer domain, and is interpreted as an *event* in the conceptual domain, using its semantic interpretation. Events can be processed in "event at a time" mode (the derived event is derived by analyzing single event) or "set at a time" mode (the derived event is derived by analyzing a sequence of events of the same type, e.g. when the complex event is an aggregation). We demonstrate the issue using several examples in the rest of this section.

Sequence Example

The derived event *Speculation-Detection* occurs when the same customer buys and later sells the same stock in value of more than $1M during the same day.

Analysis of the temporal characteristics, using instance examples (See Table 1).

Speculation-Detection: assume that the events in Table 1 occurred.

In this case, we assume that every 30 minutes the input is a time series, i.e. there is a feed of filtered events that may be relevant to the processor which creates the *Speculation-Detection.* We assume that there are defined event-types for the raw events of Buy and Sell, and to the derived event Speculation-Detection. Further assume that the processor that detects the "speculation-detection" emits this speculation-detection in 11:32. We need to determine --- when the event "Speculation Detection" occurred in the pragmatic domain.

Table 1. Speculation-detection events

Event-Id	Event-type	Customer	Stock	Value	Occurrence Time	Detection Time
E11283	Buy	John Smith	MCD	$1.2M	09:17	09:30
E23034	Sell	John Smith	MCD	$1.1M	11:27	11:30
E26587	Speculation-Detection	John Smith	MCD		?	11:32

Table 2. Shipments have not arrived

Event-Id	Event-type	Shipment-Id	Due-Date	Occurrence Time	Detection Time
E35243	Shipment	S1004	Aug-3-2008	Aug-1-2008 8:30AM	Aug-1-2008 8:30AM
E9765	Shipment-Not-Arrived	S1004	Aug-3-2008	?	Aug-4-2008 12:01AM

The question is what is the occurrence time of the derived event? The answer here is not unique, we can think of three possible interpretations:

- The occurrence time = detection time = 11:32. Rationale: since the event occurs in the "computer domain", its occurrence time is identical to the time it is detected.
- The occurrence time = occurrence time of the last event that completed the pattern = 11:27.
- The occurrence time occurs over the interval of all events participated in the derivation = (09:17, 11:27)

We cannot say that one of the interpretations is more valid than the two others – thus the decision about the occurrence time semantics should be left to the system designer as a policy decision with a default.

Time Out Pattern Example

The derived event *Shipment-not-arrived* occurs when a shipment was due to arrive in a certain day, and at the end of the working day it did not arrive (See Table 2).

In this case, the "end of business day" is considered as 11:59 in August 3rd. The actual processing may occur later (in this case it occurs in 2:00AM). There are, again, three possible interpretations:

- Detection time interpretation which will make the occurrence-time = Aug-4-2008 12:01AM
- Due time by which the shipment did not arrive = Aug-3-2008 11:59PM (assuming this is the end of the day).
- Shipment interval (Aug-1-2008 8:30AM, Aug-3-2008 11:59PM)

We see that in this example we are getting similar possibilities to the previous one.

Aggregation Example

The derived event *maximal-bid* is derived when an auction bid time ends, and the different bid events are analyzed to find the maximal one (note that the pattern itself is not temporal) (See Table 3).

We can have again the similar three interpretations:

Table 3. Maximal-Bid

Event-Id	Event-type	Customer	Bid Value	Occurrence Time	Detection Time
E13456	Start-Bid			10:00	10:00
E14564	Bid	John	$50,000	10:10	10:10
E26587	Bid	Hillary	$53,000	10:30	10:30
E34345	Bid	Janet	$51,000	10:44	10:44
E38229	Bid	Mark	$45,000	10:50	10:50
E40023	End-Bid			11:00	10:00
E40024	Maximal-Bid	Hillary	$53,000	?	11:00

- The Maximal-Bid occurs when it is detected, at the interval end = 11:00.
- The Maximal-Bid occurs when the raw event of the maximum occurs = 10:30
- The Maximal-Bid occurs within the interval (10:00, 11:00).

Note that in some aggregation functions (e.g. count, average), that do not select a single event, the second interpretation is not valid.

Occurrence Time Policies

Looking at these three examples we can conclude that there are three common interpretations for occurrence time of derived events:

- The detection time of the derived event.
- The occurrence time of the event that has completed the pattern (or event that has been selected).
- The temporal interval in which the operations has been evaluated.

The decision which policy should be used is a function of the semantics of the application and joins various other policies that enable to fine tune the semantic; a thorough discussion about methodology for using these policies is beyond the scope of this paper.

THE ORDER OF EVENTS

This section deals with the issue of guaranteeing that events are processed in the correct order.

The Need for Event Ordering

The processing of events needs to be order-aware (in the sense of "order in reality") in various cases:

- An event is part of a time-series event-stream, and there is some trend-oriented pattern that is based on the order of events as occurred in reality.
- The event processing has different results when orders of event change. Some examples:
  a. If all the work-stations in a computer lab are occupied, start a "time sharing enablement system" that times-out users.
- The events are: work-station becomes free, work-station becomes occupied, and work-station becomes offline. The relative timing of these events can determine whether the time sharing enablement system is activated or not.
  b. For on-line auctions, fairness criteria are applied, processing bids according to their relative order.
- There are various bid events that were issued by different users, processing in dif-

ferent order can determine the results of the auction.

c. The event processing itself is performed within a *temporal context*. There are cases in which the event processing is bounded to a temporal context. An example is:

- **Between takeoff and landing, process sensor events, to obtain diagnostics.**

In this case it is important to determine the takeoff and landing times, and determine if a sensor reading was done before or after the takeoff (note that the interpretation of the landing event here may be different than in the context of flight schedule!).

Ordering in the Event Source Level

In a distributed system, there is no guarantee that the order of event *occurrences* is identical with the order of *event detections* for the same event. Recall that *occurrence time* reflects the event ordering in reality, while *detection time* reflects the orders in which the corresponding event becomes available for processing. In some cases, the *detection time* can be a good enough approximation and decisions can be based on it; In other cases, where the "order in reality" is critical, there is a need to use the *occurrence time* as a measurement. This, of course, raises the known issues of clock synchronization in distributed systems, if the measurement is taken from computer clock at the source. See further discussion in Section 7.

Time-Out Reordering

Time-out reordering is based on two assumptions:

a. The order of "event arrival" may not be identical to the order of occurrence time, but the delay time is relatively small and usually bounded by a time-out τ.

b. It is possibly to ignore events that arrive out of order.

Based on these two assumptions, the time-out reordering is performed by putting the incoming events to a queue and delaying their detection by the processing system. Since the maximal delay is τ, let e be an event an Toc(e) be the occurrence time of the event, then in Toc (e) + τ, it is safe to assume that each event whose occurrence time is earlier than then the events already arrived, until the safe time, event e waits in a queue, and if event e' such that, Toc(e') < Toc (e) arrives, then the queue will be reordered, and becomes a priority queue based on the Toc. If an event e'' arrives later than Toc (e'') + τ then it is ignored.

This is similar to buffering techniques in data stream processing (Babcock, 2002). Its main benefit is that it guarantees that the detection time order will be the same as the occurrence time order. This method has two main deficiencies that may or may not be important for various applications:

- Detection time is delayed, thus the end to end latency of the processing system gets higher.
- Events may be ignored due to late arrival.

Retrospective Updates

One way to handle out of order events is to handle them when they arrive, and compensate for each decision that should have taken that event into account. A similar model has been introduced in the past in the context of active databases (Etzion, 1994). This can be achieved in a technique similar to "truth maintenance systems" (Hindi, 1994). This is more complex for implementation, both due its high computational overhead, and that actions taken as a result of processing event may be irreversible; thus the idea still has to be investigated in the context of event processing.

THE IMPACT OF DISTRIBUTION ON THE TEMPORAL ASPECTS

The discussion so far has been orthogonal to distribution issues; this section discusses several issues that are derived from distribution. The discussion will relate to time zones, time synchronization and time validity for mobile consumers.

Time Zones

In distributed systems we may have a cross time-zone event processing system; the cross time-zone can be reflected in various cases:

- The event producers reside in multiple time zones
- The event consumers reside in multiple time zones
- An event processing agent that reside in different time zone relative to one or more event producer
- An event processing agent that reside in different time zone relative to one or more event consumers
- The event processing network is being executed on a collection of processors that span different time zones.

All of these cases may require processing of events with timestamps that occur in different time zones. This can be resolved in two ways:

- Convert all time stamps to a canonical zone time stamp, which can be the most common time zone for this application, or some global time zone like Greenwitch Meantime.
- Keep the timestamp as a pair <timestamp, time zone>

The first alternative requires each actor to convert the time stamp in its own time zone from and to the canonical time zone, while the second one requires each actor to convert the input only, but the conversion is more complicated, especially where daylight saving time is involved. The trade-off here depends on the measure of the distribution, when most of the time stamps are within the same time zone, it is reasonable to use this time zone as the canonical one. There have been some works related to the processing of time zones such as (Broden, 2004) and the time stamp semantics (Dyreson, 1993).

Time Synchronization

The problem of time synchronization is the hardest problem of distributed event processing. While time zone diversity can be bridged, time synchronization. The problem, already mentioned in Section 6.2, relates to the fact that the internal clock of the various machines may not be consistent with each other. There are two ways to cope with this problem: clock synchronization and using a time server.

- **Clock synchronization:** Clock synchronization is a well researched topic in distributed computing, starting with works like (Lamport, 1978) and newer works such as (Liebig, 1999) and (Kshemkalyani, 2007). These methods are aimed to ensure that the clocks of all sources will be synchronized.
- **Time Server:** While the synchronization may be a valid solution in some cases, there is a problem to implement them when the collection of sources is large, not well defined or dynamic. The solution taken in these cases is that all sources will take the timestamp from a common time server and not from the machine clock.

There are various such servers available through the internets, and some organizations provide time servers in their intranet. this solution may have some latency, and may not be applicable when the chronon granularity is small, typically the "time-server" solution is considered good enough, and indeed being used by various of event processing implementations.

Time Validity for Mobile Consumers

The result of an event processing networks may be sending a derived event to some consumer. However, when the consumer is mobile, thus, may be disconnected for a long time (Cao, 2007), In regular Email systems, all mail is waiting for the consumer to connect, and the question which Email message is still relevant is determined manually by the consumer. Event processing systems may provide semantics for validity of reported events; for example, if the consumer subscribes to the derived event that is created wherever there is an arbitrage opportunity (the difference between the quote of the same stock in two different stock exchanges exceeds some threshold). However, if the consumer was disconnected, this information may be irrelevant to this consumer, since this situation does not remain for long. This can be reflected by adding relevance time to each derived event (Workman, 2005) which, in essence, gets us back to the notion of valid time in temporal databases, discussed before. There is also a possibility to define policy about the expired events distribution, possible values can be: discard, store for browsing on request, and aggregate (e.g. showing count only).

CONCLUSION

Event processing has strong relations with temporal aspects, this paper has discussed the notion of occurrence time and detection time, and their effect on processing events, and discussed temporal patterns, contexts and temporal order issues. Further work will continue the thinking to handle retrospective events (events about the past) and predictive events (events about the future), another threads of work is looking at temporal pattern in stored events, and temporal reasoning about the execution of event processing – causality and lineage issues.

Other areas of research are: extending the set of temporal patterns to cover both space and time, thus supporting patterns like "moving north", and investigate the relations between uncertainties in event processing (Wasserkrug, 2008) with the temporal aspect.

REFERENCES

Adi, A., Biger, A., Botzer, D., Etzion, O., & Sommer, Z. (2003). *Context Awareness in Amit* (pp. 160–167). Actuve Middleware Services.

Allen, J. F. (1983). Maintaining Knowledge about Temporal Intervals. *Communications of the ACM*, *26*(11), 832–843. doi:10.1145/182.358434

Anurag, D., & Anu, P. K. (2008). The Chronon Based Model for Temporal Databases. (DASFAA) (pp. 461-469)

Babcock, B., Babu, S., Datar, M., Motwani, R., & Widom, J. (2002). *Models and Issues in Data Stream Systems* (pp. 1–16). PODS.

Brodén, B., Hammar, M., & Nilsson, B. J. (2004). Online and Offline Algorithms for the Time-Dependent TSP with Time Zones. *Algorithmica*, *39*(4), 299–319. doi:10.1007/s00453-004-1088-z

Bry, F., & Eckert, M. (2008). *On static determination of temporal relevance for incremental evaluation of complex event queries* (pp. 289–300). DEBS.

Cao, X., & Shen, C.-C. (2007). Subscription-aware publish/subscribe tree construction in mobile ad hoc networks. In *Proceedings of ICPADS*, (pp. 1-9)

Dey, A. K., & Abowd, G. D. (2000). Towards a Better Understanding of Context and Context-Awareness. In: CHI 2000 Workshop on the What, Who, Where, When, and How of Context-Awareness, Dyreson, C. E., & Snodgrass, R.T. (1993). Timestamp semantics and representation. *Inf. Syst. (IS) 18*(3), 143-166)

el Hindi, K., & Lings, B. (1994). Using Truth Maintenance Systems to Solve the Data Consistency Problem. In Proceedings of CoopIS (pp. 192-201).

Etzion, O. (2005). Towards an Event-Driven Architecture: An Infrastructure for Event Processing Position Paper. In. *Proceedings of RuleML, 2005*, 1–7.

Etzion, O., Gal, A., & Segev, A. (1994). Retroactive and Proactive Database processing. In *Proceedings of RIDE-ADS*, (pp. 126-131)

Howe, A. E., & Somlo, G. (1997). Modelling Discrete Event Sequences as State Transition Diagrams. In *Proceedings of IDA* (pp. 573-584)

Iwasaki, Y., Farquhar, A., Saraswat, V. A., Bobrow, D. G., & Gupta, V. (1995). Modeling Time in Hybrid Systems: How Fast Is "Instantaneous"? In *Proceedings of the IJCAI* (pp. 1773-1781).

Jensen, C. S., & Snodgrass, R. T. (1999). Temporal Data Management. [TKDE]. *IEEE Transactions on Knowledge and Data Engineering, 11*(1), 36–44. doi:10.1109/69.755613

Kshemkalyani, A. D. (2007). Temporal Predicate Detection Using Synchronized Clocks. [TC]. *IEEE Transactions on Computers, 56*(11), 1578–1584. doi:10.1109/TC.2007.70749

Lamport, L. (1978). Time, Clocks, and the Ordering of Events in a Distributed System. [CACM]. *Communications of the ACM, 21*(7), 558–565. doi:10.1145/359545.359563

Liebig, C., Cilia, M., & Buchmann, A. P. (1999). Event Composition in Time-dependent Distributed Systems. In Proceedings of CoopIS (pp. 70-78).

Luckham, D. C. (2002). *The Power of Events: An Introduction to Complex Event Processing in Distributed Enterprise Systems*. Reading, MA: Addison-Wesley.

Patroumpas, K., & Sellis, T. K. (2006). Window Specification over Data Streams. In *Proceedings of the EDBT Workshops* (pp. 445-464).

Povinelli, R. J., & Feng, X. (2003). A New Temporal Pattern Identification Method for Characterization and Prediction of Complex Time Series Events. [TKDE]. *IEEE Transactions on Knowledge and Data Engineering, 15*(2), 339–352. doi:10.1109/TKDE.2003.1185838

Sharon, G., & Etzion, O. (2008). Event Processing Networks – model and implementation. *IBM Systems Journal, 47*(2), 321–334. doi:10.1147/sj.472.0321

Shaw, R., & Larson, R. R. (2008). Event Representation in Temporal and Geographic Context. In *Proceedings of ECDL* (pp. 415-418).

Shevade, B., Sundaram, H., & Xie, L. (2007). Exploiting Personal and Social Network Context for Event Annotation. In *Proceedings of ICME* (pp. 835-838)

Walzer, K., Breddin, T., & Groch, M. (2008). Relative temporal constraints in the Rete algorithm for complex event detection. In *Proceedings of DEBS* (pp. 147-155).

Wasserkrug, S., Gal, A., Etzion, O., & Turchin, Y. (2008). Complex event processing over uncertain data. In *Proceedings of DEBS* (pp. 253-264)

Workman, S., Parr, G., Morrow, P. J., & Charles, D. (2005). Relevance-Based Adaptive Event Communication for Mobile Environments with Variable QoS Capabilities. In *Proceedings of MMNS* (pp. 59-70).

Chapter 5
Serendipity Reloaded:
Fair Loading in Event-Based Messaging

Daniel Cutting
University of Sydney, Australia

Aaron Quigley
University College Dublin, Ireland

ABSTRACT

Client/server approaches to event-based message can scale to millions of users, but at great administrative and financial cost. By contrast, distributed peer-to-peer (P2P) systems offer the promise of smooth scalability from small to large numbers of participants without dedicated infrastructure. Some forms of event-based messaging, such as publish/subscribe, require events to be delivered to groups of consumers based upon their characteristics or interests. Such groups are undefined until the moment of publication and may be very large, posing significant delivery and load distribution problems in P2P environments. This chapter presents ICE, a structured P2P overlay design with scale-free properties that can be used to construct fairly loaded and efficient event-based messaging architectures.

INTRODUCTION

A community is an enduring group of people defined by common characteristics. Historically, a community has described a group of people living together in a particular geographic area, or a social group within a larger society. Formal associations are another form of community but typically members of an association share a common purpose. More recently, Communities of Practice (CoP) (Lave & Wenger, 1991)

have described groups of people participating in shared social or business settings. CoP has been extended to several specific types of communities including Communities of Interest (people who share a common interest or goal) (Fischer, 2001) and Communities of Circumstance (people with illnesses or minority groups, for example).

These notions of community have progressed into the online domain. First used by Rheingold in relation to the WELL web site in the early nineties, the term "virtual community" (Rheingold, 2000) described physically distant members bound intellectually, socially and technically.

DOI: 10.4018/978-1-60566-697-6.ch005

Today the Internet has grown to be a dominant medium for social interaction with more than *1.4 billion* people already online (Miniwatts Marketing Group, 2008). More broadly speaking the term "community" now refers to practically any online shared social activity such as frequent visitors to web sites, chat rooms, and online games (Zhang & Weiss, 2003). Online communities typically require their members to explicitly visit (e.g., websites, portals and member sites), join (e.g., Facebook (Facebook, 2008) and LinkedIn (LinkedIn, 2008)) or opt-in (e.g., mailing lists and RSS feeds). Such approaches often rely on client/ server models and sites such as Google (Google Inc., 2008b) have shown they can scale to handle hundreds of millions of users daily. However, this comes at a great administrative and financial cost, and repeated incidents of outage or the scaling problems which plague services such as Twitter (Twitter, 2008) are common.

In parallel to the growth of such explicit communities has been the rise of countless content-oriented tools such as YouTube, Blogger (Google Inc., 2008a) and Flickr (Yahoo! Inc., 2008) which rely on large server farms to deliver their services. These tools have significantly lowered the technical difficulty of producing and publishing content. Combined with increasingly ubiquitous broadband connectivity (some two-thirds of all users in 2006, based on forecasts from 2004 data) (Organisation for Economic Co-operation and Development (OECD), 2008), these tools have driven the proliferation of specialised weblogs (*blogs*), films, technical documents and podcasts (Pew Internet & American Life Project, 2004, Lindahl & Blount, 2003, Kumar, Novak, Raghavan, & Tomkins, 2003). By 2008, the web site Technorati (Technorati Inc., 2008) was tracking more than 133 million interconnected blog records in the "blogosphere". Four of the top ten entertainment sites accessed are blogs according to a comScore Media Metrix study in July of 2008 (comScore, 2008). "Open source journalism" is taking root (OhmyNews, 2008) and the concurrent evolution

of connected mobile devices such as the Apple iPhone (iPhone, 2008) enables people to publish and receive photographs and other content "on-the-go" as participants in mobile virtual communities (Rheingold, 2002).

However, the Internet is now a mass medium facilitating tremendous social interaction on a scale difficult for an individual to fully grasp or appreciate. While there are hundreds of blogs, thousands of websites and millions of pictures which might be of interest to an individual over the course of their lives, how would one ever find them? As a result, people flock to well known websites or limit themselves to content others have placed into well-defined communities that they opt into over time. The question becomes, how can one serendipitously be exposed to information when online interaction is inherently designed around getting specifically what is requested?

One approach to serendipity relies on the fact that publishers create content with a specific audience in mind. When publishing, publishers could specify the consumer demographic for each article of content. This demographic defines an *implicit group* or a set of consumers that have some inherent features in common. An implicit group is defined by the characteristics of its members, rather than their explicit names. Implicit groups are wider in scope than traditional communities. Although they can describe communities of people with shared interests or circumstances, they are, more specifically, subsets of a population of objects selected by the requirements for membership. They may be enduring or ephemeral, generic or highly specific, social or technical. Implicit groups may specify groups not considered to be communities in the traditional sense, such as a set of computers with specific functionality and available resources. Sending messages to these sorts of groups is called implicit group messaging (IGM) (Cutting, 2007, Cutting, Quigley, & Landfeldt, 2007a, 2007b). Some examples include: the PEACH project (Busetta, Merzi, Rossi, & Zancanaro, 2003) which uses the concept of "implicit organization" in order to communicate between various components of an interactive museum guide; and Chambel et al (Chambel, Moreno, Guimaraes, & Antunes, 1994) which describes the use of implicit groups to aid navigation of and retrieval of data from large distributed "hyperbases".

Another approach to serendipitously delivering content is traditional publish/subscribe messaging (Eugster, Felber, Guerraoui, & Kermarrec, 2003) which allows consumers to create "standing queries" that push matching content to them as it becomes available. Consumers select the set of messages they receive using a subscription language to group messages. Many publish/subscribe systems have been proposed, including SIENA (Carzaniga, Rosenblum, & Wolf, 2001), Elvin (Segall & Arnold, 1997), REBECA (Mühl, 2002), Gryphon (Bhola, Strom, Bagchi, Zhao, & Auerbach, 2002) and INS/Twine (Balazinska, Balakrishnan, & Karger, 2002).

In both IGM and publish/subscribe, the implicit group of consumers varies from message to message, based either on the publisher's description of its content or its target audience. Hence, the final recipients are not known until the point of publication and messages targeted at a large audience may place a significant strain on the messaging infrastructure.

THE LOADING PROBLEM

When groups of consumers wish to repeatedly receive messages from publishers, they may explicitly join specific groups. In these cases, prior construction of messaging infrastructure can be used to achieved efficient distribution. For example, IP multicast (Deering, 1989) allows hosts to explicitly join a group and receive all messages sent to the group address by constructing an efficient spanning tree between all members. This tree can be source-rooted in the case of a single producer, or rooted at a "rendezvous point" when there are multiple producers. An advantage of

multicasting at the IP level in the network stack is that redundant packets can be avoided, optimising the usage of physical network links. Similar application-level approaches have also been explored (Chu, Rao, & Zhang, 2000, Chawathe, 2000, Francis, 2000), designed around hubs that route messages between participants. Although unable to offer network utilisation as optimal as IP multicast, they have the advantage of not requiring additional network support. A number of similar multicast schemes constructed over peer-to-peer (P2P) networks have also been proposed, such as SCRIBE (Castro, Druschel, Kermarrec, & Rowstron, 2002), CAN multicast (S. Ratnasamy, Handley, Karp, & Shenker, 2001) and Bayeux (Zhuang, Zhao, Joseph, Katz, & Kubiatowicz, 2001).

The technical designs usually employed for these kinds of techniques are generally inappropriate for publish/subscribe and IGM because they only need to deliver messages to participants known in advance. In particular, these approaches are not directly applicable to serendipitous messaging as they would require the creation and maintenance of separate multicast groups for each possible implicit group, of which there are a potentially unlimited number.

When considered in the context of serendipitous content publication over the Internet, IGM and publish/subscribe systems must support thousands to millions of participants with very frequent publications to both small and large groups. They must be scalable in terms of the number of participants, the sizes of implicit groups and the number of publications. Potentially ephemeral implicit groups may be messaged frequently, rarely, or not at all, with membership varying as new participants join, leave or change.

The decoupled flexibility of the implicit groups concept poses inherent design problems. They are unbounded in size, undefined until the moment of publication, and may cross-cut the population of consumers along any number of dimensions. Implementations may therefore be subjected to extremely variable loading characteristics. For example, the server in a centralised implementation will receive many messages (high incoming load) and must forward many duplicates for each (extreme outgoing load). Underlying network links will also be subjected to this strain. In distributed implementations, non-uniform group selections by publishers and skewed distributions of consumer features may cause points of heavy loading in the system, either during individual publications, or manifest only after many.

Social interaction over the Internet is inherently decentralised, egalitarian and many-to-many. A solution to the loading problem may seek to embrace these same principles for reasons both technical and social. Distributed peer-to-peer systems offer the promise of smooth scalability from small to large numbers of participants without dedicated infrastructure (Stoica, Morris, Karger, Kaashoek, & Balakrishnan, 2001, S. P. Ratnasamy, 2002, Pyun & Reeves, 2004). P2P systems may also avoid centralised bottlenecks and points of failure, and eliminate inherent bias or censorship of content, whether intentional (Ding, Chi, Deng, & Dong, 1999) or not (Mowshowitz & Kawaguchi, 2002). Of importance to some applications is that distributed systems may also be less vulnerable to legal or technical attacks (Lee, 2003). However, it is important in such systems to ensure the load placed on peers is "fair", precisely because no single peer claims responsibility for the whole system. Here, *fairness* is defined as the degree to which all participants contribute equally to the operation of the system. A system is *efficient* when the maximum contribution of any peer is low relative to the total work that must be done.

The purpose of this chapter is to outline the loading and scaling difficulties faced by event-based P2P messaging systems, and suggest a structured overlay design called ICE that enables architectures to overcome these problems. A sample IGM architecture built on ICE is used to demonstrate its features throughout the chapter. A more complete treatment of this sample architecture can be found in related work (Cutting, 2007, Cutting et al., 2007a, 2007b).

PEER-TO-PEER NETWORKING

P2P is a style of networking characterised by its focus on many participants contributing to the operation or maintenance of the system as a whole. Although most well-known for file sharing applications, P2P is applicable to many domains such as resilient data backup (Cox, Murray, & Noble, 2002, Landers, Zhang, & Tan, 2004), distributed file systems (Muthitacharoen, Morris, Gil, & Chen, 2002), workflow management (Fakas & Karakostas, 2004), distributed search (Waterhouse:01, 2001), web caching (Eaton, 2002), email (Mislove et al.., 2003), video streaming (Kulkarni, 2006), and publish/subscribe messaging (Bhola et al.., 2002, Mühl, 2002, Pietzuch & Bacon, 2002, Gupta, Sahin, Agrawal, & Abbadi, 2004, Xue, Feng, & Zhang, 2004, P.-A. Chirita, Idreos, Koubarakis, & Nejdl, 2004). In a loose sense, P2P has existed since the beginning of the Internet, and by many definitions includes heterogeneous directory-based approaches such as the original Napster. In this discussion, however, P2P refers to systems where all peers have homogeneous roles. Of this type of P2P there are two main approaches: *unstructured* and *structured*.

Unstructured P2P

In unstructured P2P, peers connect to one another randomly or according to certain heuristics, in order to produce small-world networks (Mitre & Navarro-Moldes, 2004, P. A. Chirita, Damian, Nejdl, & Siberski, 2005) or groupings clustered by interest (Khambatti, 2003). The approach is exemplified by early file sharing networks such as Gnutella.

Applications built on unstructured networks must be designed to work with the limited assumptions that can be made. Since each peer only connects to its direct neighbours and there are no central directories available for global system communication, it can be difficult to implement functionality that requires guarantees. Searching for objects that are replicated throughout a network (such as in file sharing applications) is an appropriate application for unstructured P2P as in these cases it is sufficient to find just one or a few objects matching the search criteria. Queries are typically flooded from the source to a certain depth, although some advanced techniques for limiting or directing these floods have been explored (Sarshar, Boykin, & Roychowdhury, 2004, Rhea & Kubiatowicz, 2002, Yang & Garcia-Molina, 2002).

Applications that are required to find unique instances of objects, or every instance of an object are less suitable for such networks. For instance, a publish/subscribe messaging system is designed to deliver a message to all consumers interested in it. It is not sufficient to find only consumers near the publisher. In order to remain relatively efficient at large scales (i.e., without needing to flood messages to all peers or store excessive state at each peer), a network could be seeded with hints that improve the likelihood of finding selected peers. Such techniques include gossiping (Hedetniemi, Hedetniemi, & Liestman, 1988), Rumor Routing (Braginsky & Estrin, 2002), Directed Diffusion (Intanagonwiwat, Govindan, & Estrin, 2000), interest-based communities (Khambatti, 2003), percolation search (Sarshar et al.., 2004) and routing indices (Crespo & Garcia-Molina, 2002). Even so, it is difficult to guarantee that significant fractions of groups will receive messages, and finding all members of very small, scattered groups is infeasible.

Structured P2P

A structured P2P network permits stronger guarantees for delivering messages to all relevant consumers. Structured P2P networks take a very different approach to connect neighbours. Peers are organised according to an abstract, regular topology supporting various routing guarantees which allow certain design assumptions absent from unstructured P2P. A peer is usually assigned a unique address or range of addresses. Routing rules are defined such that a message can be deterministically forwarded from peer to peer towards the peer that is nearest to a destination address.

A panoply of structured P2P designs exist, based on various routing schemes such as Plaxton routing (Pastry (Rowstron & Druschel, 2001), Tapestry (Zhao et al.., 2004)), a one-dimensional ring (Chord (Stoica et al.., 2001)) or a d-dimensional Cartesian surface (CAN (S. P. Ratnasamy, 2002)). Many of these have $O(\log n)$ storage and routing costs (e.g., Chord, Pastry and Tapestry).

Collectively, these are often referred to as Distributed Hash Tables (DHTs) or Distributed Object Location and Routing (DOLR) designs. A DHT conceptually operates like a classical hash table, storing and retrieving objects over the network via fixed length keys. Peers are typically responsible for a portion of the address space and store and serve the objects that are hashed to it. DOLRs are very similar but support routing of arbitrary messages to objects or nodes in the overlay, rather than just storage and lookups.

CAN is of particular relevance to this chapter as the structured overlay introduced below, ICE, bears some similarities. CAN addressing and routing occurs on a d-dimensional plane, which is divided among peers. Peers' regions are specified using Cartesian coordinates, and data is logically stored at points on the surface by routing it to the peer that owns the region containing it. The point-to-point routing available in CAN incurs higher routing costs than many structured networks, $O(dn^{1/d})$, but benefits from just $O(d)$ storage complexity.

DHTs and DOLRs are often used as a substrate for more advanced applications, such as storage or messaging systems. For instance, to create a traditional multicast group, an identifier may be hashed to the addressing system of the underlying DHT and a creation message routed to the nearest node. This node may then act as the root of the multicast group. When a node wishes to join the group, it too hashes the group identifier and routes a join message to the group root, thereby including itself in further group messages. Some systems may use the group root to store a list of all members, while others may take an approach similar to more traditional IP multicast and create a tree from the root to all members along the route of nodes in the DHT from the source to the root (e.g., SCRIBE (Castro et al.., 2002)).

LOAD BALANCING

When messages are delivered from publishers to consumers, some degree of load is placed on the message delivery infrastructure. Some of this load is in the form of state that must be retained by the infrastructure. Other load includes instantaneous or ongoing stress on various components, such as network links forwarding packets, peers receiving incoming packets, and servers transmitting multiple copies of messages. This section is specifically concerned with loading in structured P2P designs for messaging infrastructure.

Storage/Registration Distribution

Storage load is caused by many objects hashing to the same location in a structured overlay, and hence the same peer. Ordinarily, fair storage distribution is trivially achieved with a consistent hashing function which uniformly distributes keys over an identifier range. If the identifier range is also uniformly partitioned across peers, then the load is fairly distributed.

However, some systems require support for range queries. These are usually implemented by replacing consistent hashing with locality-preserving functions which can lead to imbalance when many data items have similar values or ranges. This is often handled by partitioning the identifier range non-uniformly to relieve load (Ganesan, Yang, & Garcia-Molina, 2004, Aspnes,

Kirsch, & Krishnamurthy, 2004), or by mapping the skewed data items to a uniform range (Choi & Park, 2005). For instance, a CAN-like surface may be uniformly partitioned by default, but a load-balanced version could instead split a region such that half of the data objects currently stored there remain in each region. Such an approach would result in a balanced structure, but since the surface would not be uniformly distributed, additional techniques for distributing routing load may be required. Ongoing adaptation could also be difficult since overlays are partitioned according to the loading profile of particular points in time.

Mercury (Bharambe, Agrawal, & Seshan, 2004) is a protocol for supporting multi-attribute range queries and supports load balancing of attributes that are popularly registered. Mercury peers participate in *attribute hubs* which are logical rings of nodes pertaining to a single attribute. Each peer must link to every hub, so the number of hubs (and thus attributes) cannot be particularly great. Peers register data items in the hub for each attribute they contain. They are stored around the hub using a locality-preserving function such that range queries can be resolved. Conjunctive range queries are routed to a single hub (for one attribute from the query) which then resolves the whole query. Popular attributes are handled by moving lightly loaded hub nodes to regions which are heavily loaded such that data items can be partitioned.

Mirinae (Choi & Park, 2005) is a content-based publish/subscribe system that uses a hypercube overlay routing network to group similar subscriptions. Subscriptions are mapped to a corner of the hypercube based upon an application schema, and events are mapped to partial identifiers according to the same schema and sent along the edges to corners that cover it. This scheme works by clustering subscriptions in the topology of the hypercube that are semantically close to one another.

Such storage distribution techniques may be required for publish/subscribe messaging systems built on structured overlays because many peers will likely register similar subscriptions, leading to imbalance. Such systems have an additional caveat. Unlike search engines, for example, which resolve queries and return results to a single peer, these systems must forward publications to many consumers. If a P2P messaging system has a storage loading problem, it will also have an outgoing load problem, since the peer storing the data will also need to notify many consumers. Storage distribution is thus of particular importance in such systems.

Access/Query Distribution

The other major type of loading is access or query loading. In this instance, a peer must frequently respond to queries for objects (i.e., it has a high incoming load). Query loading may also be a problem in publish/subscribe systems where publications do not follow uniform distributions. In such cases, some points in the system will need to handle a large volume of publications.

One technique for handling this problem, described in relation to Tapestry (Zhao et al.., 2004), Chord (Stoica et al.., 2001) and CAN (S. P. Ratnasamy, 2002), uses multiple hash functions to map the same object to many locations in the overlay, any of which can be used to retrieve it.

A related technique is symmetric replication (Ghodsi, Alima, & Haridi, 2005b, Leslie, Davies, & Huffman, 2006). Instead of storing a registration at a single point in the overlay, it is stored at several, symmetrically placed points (by rotating the hashed result, for example). Queries may then be routed to any of the points, which can be calculated deterministically by the querying peer, dividing access load over a number of peers over time. HotRoD (Pitoura, Ntarmos, & Triantafillou, 2006) is a ring-based DHT supporting range queries that uses symmetric replication in the form of multiple virtual rings corresponding to increased levels of replication. When a range of peers becomes "hot" they are replicated and rotated to another region of a virtual ring. Queries can be routed to either the original or any of the replicas to be resolved.

Wang et al (Wang, Alqaralleh, Zhou, Brites, & Zomaya, 2006) addresses the access loading problem for DHTs by replicating data to the neighbours of the peers that are ordinarily required to forward the query. A similar approach is used by Beehive (Ramasubramanian & Sirer, 2004), LAR (Gopalakrishnan, Silaghi, Bhattacharjee, & Keleher, 2004), P-RLS (Cai, Chervenak, & Frank, 2004), Tapestry (Zhao et al.., 2004), and the Chord DHT (Stoica et al.., 2001) which permits replicas of objects to be stored at a number of successor nodes in a peer's routing table. These neighbour-based replication approaches, which may also serve to improve data resilience, typically exploit the routing algorithms of DHTs which route queries through peers on the way to the original. If a replica is encountered first, the original does not need to be contacted.

Other varieties of data replication can also be used to reduce access load. For some applications, ideas from unstructured P2P and sensor networks may be applied, such as owner, path and random replication (Lv, Cao, Cohen, Li, & Shenker, 2002, Braginsky & Estrin, 2002). In such schemes, objects are cached at various points, such as on the peers that request objects, or peers along the path in between. Queries can then be routed as normal through the structured overlay until they encounter a replica by chance or reach the original. Alternatively, structured routing could be replaced by or combined with a random walk or expanding ring search, for instance. These ad hoc replication strategies are suitable for immutable objects such as popular music files, but are less suited to database records that are regularly modified, as it is difficult to ensure that all copies have been updated. Event-based messaging systems which rely on such records are more suited to symmetric or neighbour-based replication techniques so that replicas can be more easily updated.

Comprehensive Load Distribution

Some systems aim to solve both storage and access loading problems. Meghdoot (Gupta et al.., 2004) is a content-based publish/subscribe system based on a structured overlay network similar to CAN. It provides support for arbitrary-range subscriptions, but requires the pre-definition of a schema defining the names, types and limits of possible fields, which may limit its applicability in some domains. A $2k$-dimensional surface akin to CAN is formed and maintained by peers (where k is the number of fields in the schema). Consequently, for a schema of any appreciable complexity, the cost of surface maintenance is relatively high. Subscriptions are installed on the surface by mapping them to a $2k$-dimensional hyperplane, bounded by the ranges specified in the subscription. Events are mapped to a single point on the surface, then broadcast to all neighbours within the region near the point that contains all possible matching subscriptions. To balance storage load, newly arriving peers split the regions of overloaded peers and take half of their subscriptions. When a peer is processing too many events, a new peer replicates the content and takes joint ownership of the region. When messages are routed through that region, the precise peer contacted is chosen at random. Meghdoot claims good and fair performance for scaling up to 10 000 peers, even with heavily skewed subscription and event sets.

Gao's DHT-based Content Discovery System (CDS) (Gao, 2004) is essentially a distributed search engine that returns a matching set for a query. Although CDS addresses *content discovery* (as distinct from content delivery), it does provide a novel load distribution technique called the Load Balancing Matrix (LBM) which manages skewed storage and query loads by increasing the number of peers responsible for storing and responding to queries.

The LBM is an adaptive 2-dimensional matrix that replicates and partitions data at disparate parts of the surface using a hash function. It stores registrations of content, where columns partition registrations and rows replicate the partitions. Queries are resolved by routing to one member of each column (in any row) and collating results at the querying peer. Each registration contains enough information to resolve conjunctive queries without needing to contact other matrices.

However, each partitioned peer must be individually contacted to resolve a query, resulting in potentially many distant routes over the DHT. Furthermore, LBM uses a hash function that does not colocate partitioned peers, so amortised routing cannot be used to reduce loading. The LBM also requires that replica nodes store the same registrations, which does not permit the possibility of peers with different capabilities handling more or fewer than other peers.

EVENT-BASED MESSAGING OVER P2P

Both publish/subscribe messaging and implicit group messaging require infrastructure to support the delivery of messages from publishers to their consumers. As previously noted, the cost and scaling problems of centralised servers require us to explore options such as server federation or looser arrangements such as a homogeneous network of peers that cooperatively serve consumers. Our approach is to develop a structured P2P substrate called ICE which provides a "surface" upon which various algorithms for event-based messaging may be founded. By exploiting various design features of ICE, advanced distribution and replication techniques may be used to ensure that the participating peers perform only limited work and store limited state information, independent of the number of peers, popularity of subscriptions, or frequency of messages. This limited work per peer ensures that ICE maintains a fair distribution of load. This is essential in P2P systems, or certain nodes can become de facto servers whether they wish to or not.

The capabilities of the ICE overlay can be demonstrated by developing a sample P2P-based IGM messaging system. In this section, we model the system using a simple DHT without any load balancing features. The following section discusses ICE in detail and presents results that confirm its basic properties. Finally, we show how it can be used as the basis of a fair and efficient IGM messaging system, extending the model in this section.

All peers in the system partake in a DHT. To register a set of interest "tags", each peer inserts a reference to itself in the DHT at the hashed address of each tag, termed *rendezvous points* (RPs). To send, or "cast", a message, a publisher retrieves across the DHT all peer references stored for each of the tags in an expression describing the target audience for the message. It then combines and intersects them according to the expression before delivering the message to the resultant set of consumer peers. Such an approach has much in common with distributed search indices for document collections which employ "vertical" or "keyword partitioning" (Reynolds & Vahdat, 2003, Zhong, Moore, Shen, & Murphy, 2005).

A problem with this naïve *iterative* approach is that considerable intermediate data must be returned to the publisher before it can determine the selected consumers. A *recursive* alternative is to route the cast to the RPs in a chained sequence rather than returning results immediately to the publisher, as applied in the distributed search indices used in Panaché (Lu, Sinha, & Sudan, 2002) and RDFPeers (Cai & Frank, 2004), for example. At each point, the intersection of results can be calculated and forwarded to the next RP, reducing the amount of data that needs to eventually return to the publisher.

However, the network traffic cost for lookups is non-negligible, whether iterative or recursive. The number of lookups can be reduced by each peer storing a complete list of its tags at the RP for each of its tags. A lookup then returns all tags for a peer and allows the set of selected consumers to be calculated without having to lookup every tag individually. A somewhat similar strategy is employed in keyword fusion (Liu, Ryu, & Lee, 2004) and Keyword-Set Search System (Gnawali, 2002).

The RP for each tag must be informed when a participant's registration is altered, which may be a reasonably expensive operation in terms of network traffic. However, IGM registrations represent interests or inherent attributes of a participant which presumably will not change as quickly as casts are published. Since casts may occur frequently and to potentially very large groups, their efficiency is a priority. Thus, this design is optimised for casts, and adopts the technique of storing all of a peer's tags at the RP for each (Figure 1(a)).

Even with this improvement, results are still returned to the publisher before consumers are notified of a message. Since all the information necessary to find selected peers is stored at RPs, a simple further optimisation is for the peer responsible for the RPs to forward casts directly to consumers instead, reducing the total amount of data transferred around the network (Figure 1(b)). Because other peers in the network must now participate in the actual delivery of casts, such an approach requires a DOLR (where peers execute specific algorithms upon receipt of messages) as opposed to a pure lookup-based DHT.

Although conceptually straightforward, this approach does not work well when implicit groups are very large or frequently selected, as expected in systems that support implicit groups. This sample model illustrates the need for mechanisms that distribute the various types of load precipitated by event-based messaging in P2P systems.

The following section presents ICE, a DOLR design with properties that enable construction of load balanced event-based messaging systems. Its capabilities are subsequently demonstrated by exploiting them in an adapted version of the model in this section.

Figure 1. Schematic of basic event-based messaging on a DOLR

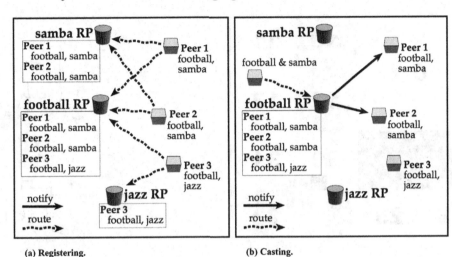

(a) Registering.   (b) Casting.

ICE: A TESSERAL P2P SUBSTRATE

This section introduces a novel P2P substrate called ICE based around tesseral addressing and an efficient amortised multicast routing algorithm. ICE is a useful, lightweight fundament for building load-balanced P2P messaging systems and has just two functions: to organise the peers into a structured overlay over the physical network; and to provide an efficient routing algorithm for delivering messages from source peers to multiple destination peers.

Peers are organised on the d-dimensional surface of a d-torus. The entire surface is claimed by peers; there are no "holes". Each peer "owns" an exclusive region of the surface and communicates directly only with peers that own bordering regions (maintained in a *neighbourhood* table). Messages are routed across the surface by passing them from neighbour to neighbour in the direction of the destination. Figure 2(a) shows the 2-dimensional surface of a 2-torus. The projected surface is considered to adjoin continuously on all edges.

Peers become part of the surface by routing a join request to a chosen region on the surface from an arbitrary bootstrap peer. The peer that owns the chosen region handles the join request by halving

its region, returning a join acknowledgement to the new peer, and informing its neighbours.

The only state information stored by an ICE peer is its neighbour table. The number of neighbours a peer has is dependent on the dimensionality of the surface. Because the surface is randomly partitioned between peers, each peer has approximately one bordering neighbour for each face of its region, or two neighbours for each axis of surface dimensionality. Thus, the storage complexity of ICE is related only to the dimensionality of the surface, i.e., $O(d)$. The number of peers does not affect the state stored by any individual peer.

The ICE surface is superficially similar to CAN (S. P. Ratnasamy, 2002). However, unlike CAN which uses a Cartesian coordinate space, ICE surfaces use hierarchical tesseral addressing (described below). The distinction is motivated by the intended usage: CAN is designed to support point-to-point routing whereas ICE is inherently a multipoint routing substrate, especially designed to enable various load distribution features.

Tesseral Addressing

Hierarchical tesseral addressing is a compact and elegant addressing scheme that describes regions

Figure 2. The ICE surface and multipoint routing

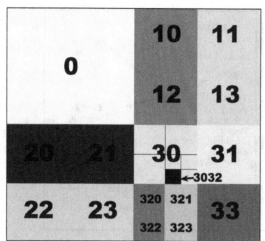

(a) A projection of the surface of a 2-torus.

(b) A message routed from top-right to shaded destination tract.

of space instead of Cartesian points. It is particularly useful for arbitrarily decomposing a space to different granularities and is used to decompose the ICE surface.

The general idea is to efficiently decompose a plane by subdividing it into a regular tessellated pattern, so that each element can completely contain another such division. Square and triangular elements have this property. Figure 2(a) illustrates a hierarchical tesseral address space based on squares; such a structure is often referred to as a quadtree.

Hierarchical tesseral addressing has been applied to many domains including geographic data storage and querying, computer graphics and robotics (Samet, 1984). It has not been used to decompose the address space of a structured P2P network, although some systems have stored such structures over traditional structured networks (Tanin, Harwood, & Samet, 2005).

ICE uses square elements. Regions are addressed by strings of d-bit digits (of base 2^d). These regions are termed *extents*, and each digit represents a progressive index into the hierarchically addressed surface. Extents may be large and coarse-grained or extremely small and fine-

grained. The *depth* of an extent is the number of digits in its address. Deeper extents are necessarily smaller than shallow extents. The solitary extent of zero length is the *universal extent* (denoted U) and represents the entire surface.

Figure 2(a) illustrates the addressing scheme, ordered left-to-right and top-to-bottom. For example, the address 0 specifies the top-left quadrant of a 2-dimensional surface, and address 3032 specifies a small extent towards the bottom-right corner. This mapping approach can be applied to surfaces of arbitrary dimensionality without loss of generality.

A set of extents is called a *tract*. Tracts can represent arbitrary regions of the surface by composing several extents, both contiguous and disparate. For example, the tract {0,2} is the entire left half of a 2-dimensional surface. Every instance of a tract has a definite representation (i.e., a specific set of extents), although there are infinitely many such sets which can define the same tract. Tracts can be subtracted from one another but unlike ordinary set difference, the result is a new tract that describes the remaining region of the surface. For example, on a 2-dimensional surface, {0,2}−{20}={0,21,22,23}.

Figure 3. ICE point-to-point routing

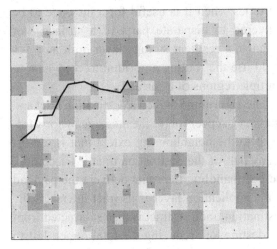

(a) Routing across a 2D surface.

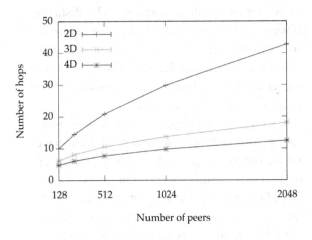

(b) Point-to-point routing is $O(dn^{1/d})$ overlay hops.

Hierarchical tesseral addressing has some important features that make it suitable for describing an overlay surface, and in particular for supporting load distribution. Its hierarchical nature allows a deep address to naturally *scale* to increasingly large covering extents simply by omitting digits from its tail. Similarly, its self-similar properties means an address can easily be *translated* across the surface by replacing its head with digits from another extent.

It is important to note that although this addressing scheme is hierarchical, there is no need for a global data structure to be maintained across peers. In particular there is no "root" peer. The scheme simply affords compact descriptions of large and small regions of the surface, and is a convenient way for peers to store surface information and communicate with one another. For instance, the region of the surface that each peer owns is represented as a tract. Likewise, peers' neighbourhood tables map extents adjacent to their tract to the peers that own them. The versatile hierarchical and self-similar properties of tesseral addressing are integral to load distribution, and to the multicast routing algorithm described below.

Amortised Routing

Besides tesseral addressing, the other major component of ICE is an efficient amortised multicast routing algorithm. Payloads are routed from a source peer to a *destination tract* by passing messages from neighbour to neighbour. All peers that own tracts intersecting the destination tract receive a copy of the message.

Point-to-Point Routing

ICE can emulate point-to-point routing by delivering to a very small destination tract that is covered by the tract of a single peer. Messages are routed geometrically across the surface between neighbouring peers. At each hop, peers forward the message to a neighbour that owns an extent nearer to the destination. Figure 3(a) shows an example of a message routed across a 2-dimensional surface.

For the purposes of finding a neighbour closer to the destination, a distance metric between extents is required. This is achieved by mapping extents from their tesseral addresses to bounding hypercubes in a Cartesian space. To do this, the surface is assumed to be wholly contained within a unit hypercube, within which extents are converted to sub-hypercubes. The distance between two extents is defined as the length of the minimum interval connecting their bounding hypercubes. For example, the distance between extents 3032 and 23 on a 2-dimensional surface is 0.125 (see Figure 2(a)).

The message complexity of point-to-point routing depends on the dimensionality of the surface and the number of peers. An equally partitioned surface has $n^{1/d}$ distinct peers arranged along each edge of the surface. The longest route a message can take is from a corner of the surface to its centre (due to the surface wrapping on all sides). The surface is "quantised" by the peers so the message takes a Manhattan ("city block") path. Point-to-point routing over the ICE surface is thus $O(dn^{1/d})$ overlay hops. Figure 3(b) confirms this in simulation by routing messages between two random extents. As expected, higher dimensionality requires fewer hops to route between two parts of the surface.

In practice, actual routing performance can be improved by considering the physical link latency between neighbours when selecting the next hop, in addition to the distance from the destination. The "best" hop is chosen to be the one with the greatest distance progress to link latency ratio, which tends to favour low latency links. The technique is one of the routing optimisations used in CAN (S. P. Ratnasamy, 2002), but other routing possibilities exist. For example, peers could attempt to minimise total load on neighbours over time, avoid known malicious peers, or monitor the reliability of neighbours in order to bias route selection.

Multipoint Routing

The ICE routing algorithm is not restricted to point-to-point routing; it delivers copies of a message to all peers with tracts intersecting a destination tract. The destination tract may be a small contiguous region of the surface but this is not a requirement. A tract can specify *any* region

of the surface whether contiguous or not. There are times when it is useful for a message to be delivered to disparate parts of a surface. For instance, a resilient storage layer built atop ICE could use the routing algorithm to copy data to multiple regions of a surface.

The algorithm is given a destination tract and a payload. A peer routes the message by forwarding it to the neighbouring extent that is closest to any part of the destination tract. At each hop, the peer handling the route delivers the payload to the application layer if its tract intersects the destination tract. Its tract is then subtracted from the destination, reducing the total region left to visit. The remaining tract is then recursively visited in the same way.

When a tract is not contiguous, it is quicker to route separate copies of a message to each region instead of visiting each in turn. However, routing separate copies of a message from the source is inefficient if part of the route is the same for each copy. Because ICE is a structured overlay and the surface is a geometric construct, the routing algorithm clusters parts of the destination tract based on their direction away from the source in order to amortise the routing cost. If several lie in the same direction away from a source, it is sufficient for a single copy to be routed that way.

This technique is recursively applied. Not only is the destination tract clustered from the initial source, it is clustered again at each hop. This results in messages taking a tree-like route from the source to all peers with tracts that intersect the destination tract. Initially the route has few branches but as the message approaches individual peers it branches as necessary to balance the total number of messages against direct individual routes. Figure 2(b) shows an example of amortised routing delivering a message to a highlighted destination tract (comprising two distinct regions).

The branching points are determined by a divisive hierarchical clustering algorithm that is applied to the remaining destination tract at each hop, resulting in a set of clustered sub-tracts. Each of these clusters is then individually routed a copy of the message. Thus, the message branches when there is more than one cluster but continues in a point-to-point fashion otherwise.

Divisive clustering is a technique that transforms a single set of objects into a number of clustered sets. It works by first finding the maximum distance between two elements in a set and creating a cluster for each. Each remaining element is then assigned to the cluster to which it is closest. This continues until each element is in its own cluster, or the maximum distance between every element is below a threshold.

For example, this set of five numbers is clustered according to their difference, resulting in first level clusters of two and three elements, and so on until each is individually clustered.

```
   13        568
  13        568
 1    3    56    8
 1    3    5    6    8
```

Divisive clustering is used in the ICE P2P substrate to cluster the parts of the network that remain to be visited by the routing algorithm. Each cluster is then visited independently. By setting a suitable threshold value, this approach permits the routing algorithm to reduce the number of messages that must be routed over the surface.

Recall that a tract is simply expressed as a set of extents; the clustering algorithm groups these extents according to the angular difference between their centres as measured from the point of view of the current peer's tract. Those with an angular difference larger than a threshold are placed in separate clusters. This threshold is called the *branch factor*, expressed in radians and taking a value in the range 0–π. Due to parallax, distant extents are more likely to be clustered together than nearby extents.

A branch factor of 0 means messages are never amortised; every extent in a destination tract is considered its own cluster and individually approached using point-to-point routing. A value of π means the entire destination tract is always treated as one cluster such that the message never branches but visits each extent in turn. The advantage of a high branch factor is that a minimal total number of messages are needed, which in turn reduces the incoming and outgoing load on peers and network links. The drawback is that messages take longer to reach their destinations on average, because more circuitous routes are taken.

The effect of surface dimensionality and branch factor on multipoint routing are investigated via simulation. 1000 messages are routed from random sources to random destination tracts intersecting ≈ 500 peers of 1000 on an ICE surface. The total number of messages needed for each delivery and the average number of hops to each destination peer per message are measured.

Figure 4(a) shows that the branch factor has a very significant effect on the total number of messages. By increasing amortisation of packets (with a higher branch factor), the number drops to approximately 1000 irrespective of surface dimensionality, just twice the minimum number of messages needed to directly contact each destination peer from the source. Increasing the surface dimensionality generally increases the total number of messages because there are intuitively more directions in higher dimensions and destinations are less likely to be clustered. However, by increasing the branch factor beyond approximately $\pi/3$ the total number of messages can be reduced even in these higher dimensions because the algorithm is more permissive in what constitutes a "similar" direction.

Higher dimensionality also dramatically reduces the average number of hops to destinations (Figure 4(b)) because the number of hops between two parts of the surface is reduced (as shown previously in Figure 3(b)). A lower branch factor also delivers messages in fewer hops since they do not follow longer generalised routes to many

Figure 4. Multipoint ICE routing

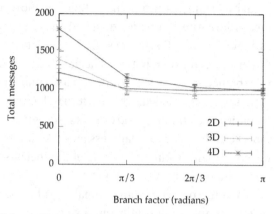

(a) Increased branch factor reduces the total number of messages needed.

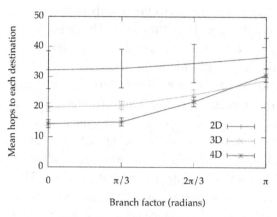

(b) Increased branch factor increases the number of hops to individual peers.

destinations. A branch factor of less than $\pi/3$ tends to produce the most direct routes to destinations.

Based on these results, a surface dimensionality of 3 and branch factor of $\pi/3$ offer the best compromise between low total messages and average hops to each destination.

Overlay Mapping

It is theoretically appealing to arrange peers randomly in the address space of a structured overlay, as uniform randomness simplifies design and permits assumptions of system behaviour and costs at the overlay level. However, such randomness does not consider physical network locality. Because neighbouring peers may be physically separated by intercontinental distances, such overlays may exhibit appalling performance. Therefore, it is beneficial for the overlay surface to map well to the physical underlying network, although this must be tempered by the desirable properties of random placement. Many systems offer functionality to aid creation of overlays that map well to physical networks. These are generally based upon measuring the latency of connections to other peers, or to a set of well-known "landmarks" around the network and include the ping-based Vivaldi (Dabek, Cox, Kaashoek, & Morris, 2004),

Madhyastha et al's traceroute-based approach (Madhyastha, Anderson, Krishnamurthy, Spring, & Venkataramani, 2006) and CAN's landmark binning scheme (S. Ratnasamy, Francis, Handley, Karp, & Shenker, 2001).

This section briefly discusses applying a simple landmark binning scheme to ICE. The guiding principle is to attempt to place physically close peers near one another on the overlay, yet ensure the surface is uniformly partitioned among peers.

l well-known landmarks are randomly placed around the physical network and each is associated with an equally sized extent of the ICE surface. Each peer joining the system measures the round trip time (RTT) to each landmark. It then attempts to join at a random location within the extent associated with the closest landmark. If a landmark is the closest choice for more than n/l peers, the next closest landmark is chosen by a newly joining peer (such information can be determined by consulting with landmarks). This method ensures that each part of the overlay has a similar number of peers, and thus that the ICE surface is uniformly partitioned.

The complementary problem of mapping landmarks to extents on the surface is achieved by finding their minimal spanning tree according to the RTTs between them. The order of landmarks resulting from a depth-first traversal of this tree may then be mapped to a Morton linearisation (Morton, 1966) of the ICE surface. This strategy tends to associate physically nearby landmarks with extents that are nearby on the surface, and since peers within each extent are physically nearby, the overlay as a whole is better able to map to the physical network.

Self-Stabilisation

The unreliable nature of peers in a P2P system guarantees that some peers will unexpectedly fail or leave the system over time. When designing a structured overlay such as the ICE surface, it is important to consider how it will cope with structural damage when it is encountered and how it will repair itself despite peer flux. ICE does not inherently support data storage—it is implemented by higher layers if needed—so the only impact of a failed peer is routing over the surface. To ensure messages are not interrupted by *holes* in the ICE surface (unclaimed tracts), the routing algorithms must be capable of routing around failed peers. Additionally, peers should detect and repair holes. This section sketches how the ICE surface can be made "self-stabilising".

Self-stabilisation (Schneider, 1993) is the process of converging a system to a stable state from an arbitrary starting state. This can be expressed as a set of invariants and rules that move the system state towards satisfying these invariants. In ICE, a stable surface is defined by three invariants. Firstly, the entire surface must be claimed by peers. Secondly, the tracts claimed by peers must not intersect. Thirdly, every extent adjacent to a peer's tract must be described by exactly one mapping in its neighbourhood table.

ICE is suited to the *correction-on-change* and *correction-on-use* approaches to self-stabilisation (Ghodsi, Alima, & Haridi, 2005a). In this technique, problems are corrected as they are detected during normal routing operations, rather than through the use of periodic beacons. This greatly limits the amount of maintenance traffic required and is especially applicable to ICE because the failure of a peer affects only $O(d)$ neighbours.

Correction-on-use dictates that each message routed across the surface must include maintenance information about intermediate source and destination peers. This allows peers to verify that their neighbourhood state agrees with the sender's. Deterministic agreements can ensure that any conflicts arising (such as two peers claiming intersecting extents) are resolved. Furthermore, the very act of receiving a route from a peer indicates the sender's existence and allows the recipient to ensure neighbourhood information is correct.

When routing messages, peers will occasionally detect failed peers. In these cases the message must be alternately routed. A message may be greedily routed around the periphery of a hole on an ICE surface by probing the "next best" neighbour, or more general techniques from sensor networking or geocasting may be applied, such as the FACE protocols (Bose, Morin, Stojmenović, & Urrutia, 2001). In addition to routing around holes, the correction-on-change protocol repairs the hole by deterministically assigning it to a new peer. Detections of the hole by different peers routing messages result in a deterministic marshalling peer being informed. This peer can then claim the hole itself, or delegate it to a neighbour. Finally, update information can be delivered to neighbours so they can update their neighbourhood tables. If a marshalling peer has also failed, the technique may be recursively applied.

Note that a stabilised ICE surface does not negate the need for applications using ICE (such as our IGM sample) to employ their own data integrity algorithms. ICE's only guarantee is to route messages to all peers within a destination tract. Any application-specific stabilisation must build on this guarantee. The application of ICE to the IGM sample in the next section includes a discussion of how registration information is maintained.

APPLYING ICE

We now return to the sample IGM messaging system discussed earlier, and show how it can be adapted to use the scale-free features of ICE to improve its fairness and efficiency. Similarly to the previous generic DOLR design, consumers register at the rendezvous points (RPs) for each tag on the ICE surface and publishers route casts to these RPs which notify all selected consumers. DHTs typically use consistent hashing to map objects to identifiers in the DHT address space (Maymounkov & Mazieres, 2002, Stoica et al.., 2001, Zhao et al.., 2004, S. Ratnasamy, Francis, et al.., 2001, Rowstron & Druschel, 2001). The number of bits in the address space is often fixed at 128 (suitable for the MD5 hash function) or 160 (suitable for SHA-1). Due to ICE's tesseral addressing, the rendezvous points for tags are in fact deep extents and are interchangeably referred to as *rendezvous extents*. The address of a rendezvous extent is found by hashing a tag, then taking d bits at a time for each digit.

Table 1 enumerates four extreme points on the continuum of possible casts and shows their effect on the incoming and outgoing loads of RPs. The work of any one peer should be minimised despite this variety of loading possibilities. For large groups, notification places a high outgoing load on RPs and the links surrounding them, so a better solution is to spread the notification load to other peers. RP peers for tags frequently appearing in casts may have a high incoming load. This incoming cost should be shared with other peers. Two new parameters are needed to capture these principles. The *storage limit* (SL) is the maximum number of registrations a rendezvous peer is willing to store for a particular registry. The *frequency limit* (FL) is the maximum number of casts a rendezvous peer is prepared to service per second. The intention is to reduce storage and outgoing load on RPs that hold registries for common tags, and reduce incoming load on RPs

Table 1. Various types of casts that cause load imbalance

	Frequent cast	Seldom cast
Common tag	↑ incoming, outgoing	↑ outgoing
Rare tag	↑ incoming	↓ incoming, outgoing

that hold registries for tags that appear frequently in target expressions.

The storage and frequency limits guide the operation of two complementary load distribution techniques called *registry distribution* and *registry replication*. The former adaptively distributes the registrations held in a registry over many peers surrounding the original RP, such that each is required to store and forward only a fraction of the total outgoing load for a cast. The latter makes complete copies of registries at other parts of the surface that can be independently used to resolve casts, thereby reducing the total incoming load on any single RP. These techniques can be combined for the registries of tags that are both commonly registered by peers and frequently used in target expressions.

Registry Distribution

The basic registration algorithm stores registries at deep rendezvous extents on the ICE surface. When many peers register the same tags, these registries become large and the RPs responsible for them must store many registrations and notify many consumers.

Registry distribution is designed to reduce the storage and outgoing load of these RPs. It works by incrementally scaling the address of where a registry is stored from the initially deep rendezvous extent to the entire ICE surface, using the hierarchical property of the tesseral addressing scheme. The initial rendezvous extent found by hashing a tag can be thought of as the deepest possible *container* for all registrations. At this depth it is covered by a single peer which is

Figure 5. Registry distribution—a registry's rendezvous extent scales as it is overloaded with registrations. Extent 210 scales to 21, then 2, and finally U

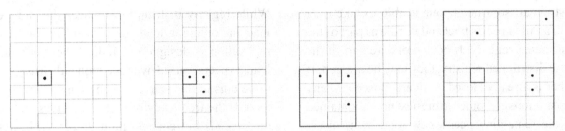

solely responsible for the registry. When an RP reaches its storage limit, the rendezvous extent is scaled by omitting a digit from the end of its address. The new container is 2^d times the size and intersects with the tracts of more peers, each of which becomes an RP for that registry and is responsible for storing a fraction of the registrations. The number of times a rendezvous extent is scaled is called the *notch* of the registry.

Take Figure 5 as an example. Suppose that on a 2-dimensional surface the jazz tag hashes to the rendezvous extent 210 (the small highlighted extent in the leftmost subfigure). This is referred to as notch 0. All peers that register jazz route their registrations to the RP that owns that extent. If the tag is common, the storage limit of the RP will soon be reached so it will increase the rendezvous extent to 21 by omitting the trailing 0. This extent, at notch 1, is $2^d = 4$ times the size and encompasses four peers (assuming the ICE surface is equally partitioned), each of which becomes an RP for the jazz registry and stores a fraction of the registrations. This process can continue as needed to comfortably accommodate all registrations stored in the registry (a peer's portion of the registry is marked as *closed* when it has reached its storage limit). Eventually the rendezvous extent can reach the universal extent which contains the entire surface, such that every peer in the system may store a fraction of the registrations. The benefit of this approach is that the available storage naturally scales with the number of peers registering in the system.

When a cast arrives at the original rendezvous peer, it is efficiently routed to peers at the next notch using the ICE routing algorithm. This continues until all registrations are found, whereupon consumers may be notified as they normally would. By distributing the registrations over larger extents, individual peers need store fewer registrations, and perform fewer notifications.

Peer Flux and Self-Stabilisation

A new peer claiming part of a tract owned by a peer participating in a distributed registry is handled by copying all closed registry portions to the new peer, and splitting any open registry entries equally. Each peer is then capable of handling casts as normal.

Registrations can be removed by routing unregister requests to an RP, which then routes it through the distributed registry in the same ways as casts. To improve efficiency, RPs may periodically forward these casts in bundles, or piggy back them on casts, rather than route each as it arrives. Alternatively, registrations could be *leased* meaning they automatically expire after a period of time. Unregistration would not be required, but consumers would need to reregister periodically.

The departure or failure of a peer that is part of a distributed registry may be handled using neighbour-based replication. Peers are seeded with copies of neighbours' data when registrations first occur, so a neighbour is able to take over both a failed peer's tract and its role as part of the distributed registry. If both peers are part of the same distributed registry, they may be merged. If either is already closed, both are closed and any registrations are routed to the next notch contained by the remaining peer's new total tract. In essence, local repairs to the surface and delegation of roles result in the remaining peer acting as if it had always been the only peer at that location.

General peer flux and the removal of registrations by consumers may eventually lead to distributed registries becoming fragmented and only partially full. It is inefficient for casts to be routed to all RPs when a lesser number would suffice. For this reason, peers can initiate a redistribution process if the number of registrations they store falls below a threshold. A peer does this by routing a redistribution command to those RPs above it, which propagates to the registry's leaves in the same way as a cast. Upon receipt of this command, each RP routes its registrations to the original RP where they are treated as new registrations and redistributed as normal. During this period, old RPs maintain copies of the registrations so that casts can continue to be resolved before they have been redistributed. After a short period of time, the registrations are presumed to have been stored and the old RPs delete their registries.

This algorithm potentially permits recipients to receive duplicate casts or miss casts completely which can be remedied by using a distributed commit protocol between the old RPs and the original. Only once the original has agreed to take responsibility for consumer notifications does the old RP delete its registry. Note that although consumers will not miss any casts, this still permits some to be notified twice if a cast occurs during this commit. Such duplicates can be detected by code at the recipient and eliminated to ensure applications do not receive them.

Registry Replication

While registry distribution reduces the storage load and outgoing load per cast on RPs, registry replication is designed to reduce the incoming load on peers caused by casts frequently using the same tags in their target expressions. Each peer records the frequency of casts it receives. When this exceeds the frequency limit, the peer replicates its registrations to other parts of the surface using the ICE routing algorithm. These replicas then resolve a fraction of new casts in the same way as the original RP, reducing its incoming load.

A form of symmetric replication is used: the extents to which registries are copied are found by replacing the head of the rendezvous extent address with a number of "wildcards". For example, Figure 6 shows a 2-dimensional surface with the RP at 210 that holds the registry for the jazz tag handling many casts (at left). Level 1 replication of this rendezvous extent, *10, copies the registry to corresponding extents in the other three quadrants, i.e., 010, 110 and 310. Level 2 replication results in a total of 16 copies at corresponding extents within the next deepest set of extents, **0.

When a peer determines that its frequency limit is exceeded by incoming casts, it begins the replication process. If the registry to be replicated is not distributed, the peer simply routes a copy of it to the set of extents that will act as the roots of the replicas. Casts may be resolved by the replicas once the routed registry is received and stored by the new RPs. If the registry to be replicated is also distributed, a replication command traverses the whole registry in the same way as casts. Each distributed RP that receives the command routes its fraction of the registry to the set of new replica roots. As the new replicas receive the casts, registry distribution is employed as normal to create newly distributed replicas of the registry.

Every level of replication reduces the probability of a particular replica being chosen by a factor of 2^d, since replication replaces the first digits of an address with wildcards and is thus dependent

Figure 6. Registry replication—0, 1 and 2 levels of replication. Registries are replicated to extents corresponding to the original RP

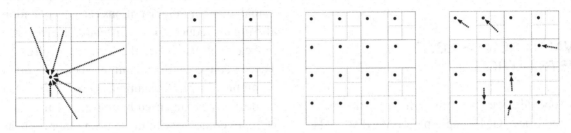

Peer Flux and Self-Stabilisation

on the surface dimensionality. To actually spread load to the replicas, it is necessary to adjust the cast algorithm slightly. Ordinarily, casts are routed from the publisher to the distant rendezvous extents of tags in the target expression. Replicas may be found during this route by perturbing slightly from the most direct path so as to pass through extents where a replica may exist. This is practical since it is possible to calculate the locations of potential replicas from the tag alone. If a replica is found, the route is terminated, resulting in a shortened path and reduced total incoming load on the original rendezvous peer. If no replicas exist for a tag, the message will still reach the original rendezvous extent, having deviated only slightly from the optimum route.

Specifically, the cast is first routed to the nearest extent that may hold the most heavily replicated version of the registrations. If no replica is found at this extent upon arrival, it is known that the registrations have not been replicated to that level, so a lesser level of replication is selected and the process is repeated. For those rendezvous extents which have been replicated, this modification to the cast algorithm means that casts will always be delivered to the closest possible replica (illustrated in Figure 6). Therefore, replication combined with the modified cast algorithm divides the amount of incoming traffic for a tag evenly over all replicas, assuming publishers are randomly spread through the network.

Registry replication can be applied to both distributed and non-distributed registries. Each replica behaves exactly like the original registry, distributing as needed to efficiently store incoming registrations. The algorithms for maintaining distributed registries described earlier apply equally to all replicas.

A new peer may register at any replica. Registrations should be eventually, but not necessarily immediately, active. Thus it is sufficient for replicas to be weakly consistent. I.e., registrations can be collected for a period before they are routed to the other replica points. Similarly, requests to unregister can be collected and copied to other replicas periodically. Note, however, that since a peer that has received a cast should receive all subsequent casts selecting the same registration, a registration should be fully replicated at all replica points before any cast targeting it is honoured.

When a replica's incoming cast frequency drops below a threshold, it tries to reduce the level of replication in the system to avoid the maintenance traffic required to synchronise registries. It does this by routing a suggestion to all other replicas that the replication level be reduced. If no replica vetoes the suggestion within a certain period, each replica reduces its level. In most cases this means the registries stored by the replica can be deleted, although replicas situated at the rendezvous extents for the reduced replica level continue to resolve casts.

EVALUATION OF LOAD DISTRIBUTION

To conclude, we show through simulation how the modified IGM sample performs when 1024 messages are cast to the same large implicit group comprising approximately one quarter of all peers in a network of 4096. A more comprehensive superset of these results can be found in related work (Cutting et al.., 2007b).

As a quarter of peers register at the same RP peer, it will be subjected to high storage costs. Furthermore, without distribution or replication, the same peer is required to handle and notify all group members for every cast. In effect, it is behaving like a server and incurs a high total incoming number of messages (referred to as TIN_M in the following discussion), and a high number of outgoing messages per cast, and in total ($POUT_M$ and $TOUT_M$, respectively). Indeed, it is required to receive 1024 casts and forward more than a million copies to recipients. Fairness is measured using an inequality metric called the Gini coefficient (G). At zero, this indicates that all members of a population have similar values. As the degree of inequality across the members rises, it increases to a maximum value of one. The unfair distribution of total outgoing load across all peers is referred to as $TOUT_G$.

The storage limit (SL) of all peers is varied from 4096 (no registry distribution) to 16 (heavy distribution) and the frequency limit (FL) from 2.56 casts/s (no registry replication) to 0.005 casts/s (level 3 replication). Note that when SL and FL are high, the model behaves akin to the original IGM sample architecture without load balancing. The degree of distribution increases from right to left in these plots as the storage limit decreases.

Figure 7(a) shows that $TOUT_M$ is worst when there is no replication or distribution. A very large

reduction in $TOUT_M$ is observed by reducing the storage limit, because individual peers store only a small number of registrations each. Replication also greatly reduces $TOUT_M$ because each registry replica is only used to resolve a fraction of all casts. $TOUT_M$ converges to a minimum when replication is combined with distribution, because the scaled rendezvous extents used by registry distribution begin to coincide with those from other replicas and load the same peers.

Figure 7(b) shows that TIN_M is lowest when there is no distribution. This is a consequence of the routing algorithm; by not needing to route a cast to all RPs in a distributed registry, fewer total messages are needed.

Figure 7(c) shows that registry distribution is capable of greatly reducing per cast outgoing load (in addition to total outgoing load), at the cost of slightly increased per cast incoming load. The slight increase in PIN_M is due to a small number of duplicate messages routed through the same RPs when searching for registrations in distributed registries. As these metrics measure per cast load, registry replication does not affect them.

It is very difficult to achieve global fairness with an extremely skewed cast loading such as this, but the combination of registry distribution and replication is effective (Figure 7(d)). The original architecture with no distribution or replication has $TOUT_G$ of 0.99 (i.e., one peer is doing all the work). With heavy distribution and replication, $TOUT_G$ is reduced to the far more equitable value of 0.21.

CONCLUSION

This chapter has presented the design of ICE, a distributed, structured P2P scale-free overlay that can efficiently and fairly support event-based messaging paradigms like IGM and publish/subscribe. These styles of messaging require messages to be dispatched to groups of consumers that cannot be determined before the moment

Figure 7. Extreme cast loading

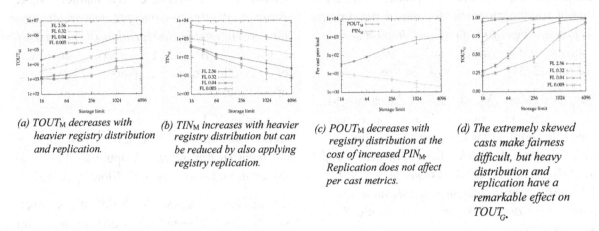

(a) $TOUT_M$ decreases with heavier registry distribution and replication.

(b) TIN_M increases with heavier registry distribution but can be reduced by also applying registry replication.

(c) $POUT_M$ decreases with registry distribution at the cost of increased PIN_M. Replication does not affect per cast metrics.

(d) The extremely skewed casts make fairness difficult, but heavy distribution and replication have a remarkable effect on $TOUT_G$.

of publication and may change from message to message. Using a distributed IGM architecture as an example, the ICE substrate is shown to permit very fairly distributed incoming, outgoing and storage loads over all peers without excessively loading any single peer, by providing the novel properties of tesseral addressing and amortised multipoint routing.

ACKNOWLEDGMENT

The authors wish to thank Dr. Björn Landfeldt and the anonymous reviewers who greatly improved this chapter.

REFERENCES

Aspnes, J., Kirsch, J., & Krishnamurthy, A. (2004, July). Load balancing and locality in range-queriable data structures. In *PODC '04: Proceedings of the twenty-third annual ACM symposium on principles of distributed computing* (pp. 115–124). New York: ACM Press.

Balazinska, M., Balakrishnan, H., & Karger, D. (2002, August). INS/Twine: A scalable peer-to-peer architecture for intentional resource discovery. In *Pervasive '02: Proceedings of the first international conference on pervasive computing* (pp. 195–210). London: Springer-Verlag.

Bharambe, A. R., Agrawal, M., & Seshan, S. (2004, October). Mercury: supporting scalable multi-attribute range queries. *SIGCOMM Computer Communication Review, 34*(4), 353–366. doi:10.1145/1030194.1015507

Bhola, S., Strom, R. E., Bagchi, S., Zhao, Y., & Auerbach, J. S. (2002, June). Exactly-once delivery in a content-based publish-subscribe system. In *Dsn '02: Proceedings of the 2002 international conference on dependable systems and networks* (pp. 7–16). Washington, DC: IEEE Computer Society.

Bose, P., Morin, P., Stojmenović, I., & Urrutia, J. (2001, November). Routing with guaranteed delivery in ad hoc wireless networks. *ACM Wireless Networks, 7*(6), 609–616. doi:10.1023/A:1012319418150

Braginsky, D., & Estrin, D. (2002, October). Rumor routing algorithm for sensor networks. In *Proceedings of the first workshop on sensor networks and applications (WSNA '02), Atlanta, GA* (pp. 22–31). New York: ACM Press.

Busetta, P., Merzi, M., Rossi, S., & Zancanaro, M. (2003, October). Group communication for real-time role coordination and ambient intelligence. In A. Kruger & R. Malaka (Eds.), *Proceedings of workshop on AI in mobile systems (AIMS 2003), 5th international conference on ubiquitous computing (UbiComp 2003)* (pp. 9–16). Seattle, WA: Springer-Verlag.

Cai, M., Chervenak, A., & Frank, M. (2004, November). A peer-to-peer replica location service based on a distributed hash table. In *Proceedings of the ACM/IEEE SC2004 conference on high performance networking and computing, Pittsburgh, PA* (pp. 56–56). Washington, DC: IEEE Computer Society.

Cai, M., & Frank, M. (2004, May). RDFPeers: a scalable distributed RDF repository based on a structured peer-to-peer network. In *Proceedings of the 13th international conference on world wide web (WWW 2004), New York, NY, USA* (pp. 650–657). New York: ACM Press.

Carzaniga, A., Rosenblum, D. S., & Wolf, A. L. (2001, August). Design and evaluation of a wide-area event notification service. *ACM Transactions on Computer Systems, 19*(3), 332–383. doi:10.1145/380749.380767

Castro, M., Druschel, P., Kermarrec, A., & Rowstron, A. (2002, October). SCRIBE: A large-scale and decentralized application-level multicast infrastructure. [JSAC]. *IEEE Journal on Selected Areas in Communications, 20*(8), 1489–1499. doi:10.1109/JSAC.2002.803069

Chambel, T., Moreno, C., Guimaraes, N., & Antunes, P. (1994). *Concepts and architecture for loosely coupled integration of hyperbases.* In (Vol. 3, 4). Broadcast Secretariat, Department of Computing Science, University of Newcastle-upon-Tyne, UK.

Chawathe, Y. D. (2000). *Scattercast: An architecture for Internet broadcast distribution as an infrastructure service.* Unpublished doctoral dissertation, University of California, Berkeley.

Chirita, P. A., Damian, A., Nejdl, W., & Siberski, W. (2005, November). Search strategies for scientific collaboration networks. In *P2PIR '05: Proceedings of the 2005 ACM workshop on information retrieval in peer-to-peer networks* (pp. 33–40). New York: ACM Press.

Chirita, P.-A., Idreos, S., Koubarakis, M., & Nejdl, W. (2004, May). Publish/subscribe for RDF-based P2P networks. In *Proceedings of the European semantic web symposium.* Heraklion, Greece.

Choi, Y., & Park, D. (2005, June). Mirinae: A peer-to-peer overlay network for large-scale content-based publish/subscribe systems. In *NOSSDAV '05: Proceedings of the international workshop on network and operating systems support for digital audio and video* (pp. 105–110). New York: ACM Press.

Chu, Y. hua, Rao, S. G., & Zhang, H. (2000, June). A case for end system multicast (keynote address). In *Sigmetrics '00: Proceedings of the 2000 ACM SIGMETRICS international conference on measurement and modeling of computer systems* (pp. 1–12). New York: ACM Press.

comScore. (2008, December). *comScore Media Metrix.* Retrieved December, 2008 from http://comscore.com/metrix

Cox, L. P., Murray, C. D., & Noble, B. D. (2002, Winter). Pastiche: making backup cheap and easy. *SIGOPS Operating Systems Review, 36*(SI), 285–298.

Crespo, A., & Garcia-Molina, H. (2002, July). Routing indices for peer-to-peer systems. In *Proceedings of the 22nd international conference on distributed computing systems (ICDCS'02)* (pp. 23–32). Washington, DC: IEEE Computer Society.

Cutting, D. (2007). *Balancing implicit group messaging over peer-to-peer networks*. Unpublished doctoral dissertation, School of Information Technologies, University of Sydney.

Cutting, D., Quigley, A., & Landfeldt, B. (2007a). *Special Interest Messaging: A Comparison of IGM Approaches*. The Computer Journal.

Cutting, D., Quigley, A., & Landfeldt, B. (2007b). SPICE: Scalable P2P implicit group messaging. *Computer Communications*.

Dabek, F., Cox, R., Kaashoek, F., & Morris, R. (2004, August). Vivaldi: A decentralized network coordinate system. In *Proceedings of the ACM SIGCOMM '04 conference*. Portland, Oregon, USA (pp. 15–26). New York: ACM Press.

Deering, S. (1989, August). *Host extensions for IP multicasting*. Stanford University: Network Working Group, Internet Engineering Task Force. Retrieved March, 2007 from http://www.ietf.org/rfc/rfc3170.txt

Ding, C., Chi, C.-H., Deng, J., & Dong, C.-L. (1999, October). Centralized content-based web filtering and blocking: How far can it go? In IEEE international conference on systems, man and cybernetics (pp. 115–119). SMC.

Eaton, P. R. (2002, November). *Caching the web with OceanStore* (Tech. Rep. No. UCB/CSD-02-1212). UC Berkeley, Computer Science Division.

Eugster, P. T., Felber, P., Guerraoui, R., & Kermarrec, A. (2003, June). The many faces of publish/subscribe. *ACM Computing Surveys, 35*(2), 114–131. doi:10.1145/857076.857078

Facebook. (2008, December). *Facebook*. Retrieved December, 2008 from http://facebook.com

Fakas, G. J., & Karakostas, B. (2004). A peer to peer architecture for dynamic workflow management using web services. *Information and Software Technology Journal, 46*(6), 423–431. doi:10.1016/j.infsof.2003.09.015

Fischer, G. (2001, August). Communities of interest: Learning through the interaction of multiple knowledge systems. In A. M. S. Bjornestad R. Moe & A. Opdahl (Eds.), *24th annual information systems research seminar in Scandinavia (IRIS'24)*, Ulvik, Hardanger Fjord, Norway (pp. 1–14).

Francis, P. (2000, April). *Yoid: Extending the Internet multicast architecture*. Retrieved March, 2007 from http://www.icir.org/yoid/docs/index.html

Ganesan, P., Yang, B., & Garcia-Molina, H. (2004, June). One torus to rule them all: multidimensional queries in P2P systems. In *WEBDB '04: Proceedings of the 7th international workshop on the web and databases, Paris, France* (pp. 19–24). New York: ACM Press.

Gao, J. (2004). *A distributed and scalable peer-to-peer content discovery system supporting complex queries*. Unpublished doctoral dissertation, Computer Science Department, Carnegie Mellon University. (CMU-CS-04-170)

Ghodsi, A., Alima, L. O., & Haridi, S. (2005a, January). Low-bandwidth topology maintenance for robustness in structured overlay networks. In *Proceedings of the 38th annual Hawaii international conference on system sciences (HICSS'05), Big Island, HI, USA*. Washington, DC: IEEE Computer Society.

Ghodsi, A., Alima, L. O., & Haridi, S. (2005b). *Symmetric replication for structured peer-to-peer systems. In 3rd intl. workshop on databases, information systems and peer-to-peer computing* (pp. 74–85).

Gnawali, O. (2002). *A keyword set search system for peer-to-peer networks*. Unpublished master's thesis. Cambridge, MA: Massachusetts Institute of Technology (MIT).

Google Inc. (2008a, October). *Blogger*. Retrieved October, 2008 from http://blogger.com

Google Inc. (2008b, October). *Google*. Retrieved October, 2008 from http://google.com

Gopalakrishnan, V., Silaghi, B., Bhattacharjee, B., & Keleher, P. (2004, March). Adaptive replication in peer-to-peer systems. In *Proceedings of the 24th international conference on distributed computing systems (ICDCS'04), Hachioji, Tokyo, Japan* (pp. 360–369). Washington, DC: IEEE Computer Society.

Gupta, A., Sahin, O. D., Agrawal, D., & Abbadi, A. E. (2004, October). Meghdoot: Content-based publish/subscribe over P2P networks. In H.-A. Jacobsen (Ed.), Proceedings of Middleware 2004, ACM/IFIP/USENIX international middleware conference, Toronto, Canada (p. 254-273). New York: Springer.

Hedetniemi, S. M., Hedetniemi, S. T., & Liestman, A. L. (1988). A survey of gossiping and broadcasting in communication networks. *Networks, 18*(4), 319–349. doi:10.1002/net.3230180406

Intanagonwiwat, C., Govindan, R., & Estrin, D. (2000, August). Directed diffusion: a scalable and robust communication paradigm for sensor networks. In *Proceedings of the 6th annual international conference on mobile computing and networking, Boston* (pp. 56–67). New York: ACM Press.

iPhone. (2008, December). *Apple iPhone*. Retrieved December, 2008 from http://apple.com/iphone

Khambatti, M. (2003). *Peer-to-peer communities: architecture, information and trust management*. Unpublished doctoral dissertation, Arizona State University.

Kulkarni, S. (2006, September). Video streaming on the Internet using split and merge multicast. In *P2P '06: Proceedings of the sixth IEEE international conference on peer-to-peer computing* (pp. 221–222). Washington, DC: IEEE Computer Society.

Kumar, R., Novak, J., Raghavan, P., & Tomkins, A. (2003, May). On the bursty evolution of blogspace. In *WWW '03: Proceedings of the 12th international conference on world wide web* (pp. 568–576). New York: ACM Press.

Landers, M., Zhang, H., & Tan, K.-L. (2004, August). Peerstore: Better performance by relaxing in peer-to-peer backup. In 4th international conference on peer-to-peer computing (P2P 2004), Zurich, Switzerland (p. 72-79). IEEE Computer Society.

Lave, J., & Wenger, E. (1991). *Situated learning: Legitimate peripheral participation (learning in doing: Social, cognitive & computational perspectives)*. Cambridge, UK: Cambridge University Press.

Lee, J. (2003, February). An end-user perspective on file-sharing systems. *Communications of the ACM, 46*(2), 49–53. doi:10.1145/606272.606300

Leslie, M., Davies, J., & Huffman, T. (2006, April). Replication strategies for reliable decentralised storage. *The First International Conference on Availability, Reliability and Security (ARES 2006)*.

Lindahl, C., & Blount, E. (2003, November). Weblogs: simplifying web publishing. *Computer, 36*(11), 114–116. doi:10.1109/MC.2003.1244542

LinkedIn. (2008, December). *LinkedIn*. Retrieved December, 2008 from http://linkedin.com

Liu, L., Ryu, K. D., & Lee, K.-W. (2004, April). *Keyword fusion to support efficient keyword-based search in peer-to-peer file sharing.* Presented at the Fourth IEEE International Symposium on Cluster Computing and the Grid (CCGrid'04) Chicago, IL.

Lu, T., Sinha, S., & Sudan, A. (2002). Panaché: A scalable distributed index for keyword search (Tech. Rep.). Cambridge, MA: Massachusetts Institute of Technology (MIT).

Lv, Q., Cao, P., Cohen, E., Li, K., & Shenker, S. (2002). Search and replication in unstructured peer-to-peer networks. In *ICS '02: Proceedings of the 16th international conference on supercomputing* (pp. 84–95). New York: ACM.

Madhyastha, H. V., Anderson, T., Krishnamurthy, A., Spring, N., & Venkataramani, A. (2006, October). A structural approach to latency prediction. In *IMC '06: Proceedings of the 6th ACM SIGCOMM on Internet measurement* (pp. 99–104). New York: ACM Press.

Maymounkov, P., & Mazieres, D. (2002, March). Kademlia: A peer-to-peer information system based on the XOR metric. In Druschel, P., Kaashoek, M. F., & Rowstron, A. I. T. (Eds.), *First international workshop on peer-to-peer systems at IPTPS 2002, Cambridge, MA, USA* (pp. 53–65). New York: Springer.

Miniwatts Marketing Group. (2008, October). *Internet usage statistics. Internet World Stats.* RetrievedOctober, 2008 from http://www.internetworldstats.com/stats.htm

Mislove, A., Post, A., Reis, C., Willmann, P., Druschel, P., & Wallach, D. S. (2003, May). POST: A secure, resilient, cooperative messaging system. In M. B. Jones (Ed.), *Proceedings of HOTOS'03: 9th workshop on hot topics in operating systems, Lihue (Kauai), Hawaii, USA* (pp. 61–66). USENIX.

Mitre, J., & Navarro-Moldes, L. (2004, June). P2P architecture for scientific collaboration. In *WETICE '04: Proceedings of the 13th ieee international workshops on enabling technologies: Infrastructure for collaborative enterprises (WETICE '04)* (pp. 95–100). Washington, DC: IEEE Computer Society.

Morton, G. M. (1966). *A computer oriented geodetic data base; and a new technique in file sequencing (Tech. Rep.).* Ottawa, Canada: IBM Canada Ltd.

Mowshowitz, A., & Kawaguchi, A. (2002, September). Bias on the web. *Communications of the ACM, 45*(9), 56–60. doi:10.1145/567498.567527

Mühl, G. (2002). *Large-scale content-based publish/subscribe systems.* Unpublished doctoral dissertation, Darmstadt, Germany: Darmstadt University of Technology.

Muthitacharoen, A., Morris, R., Gil, T. M., & Chen, B. (2002, Winter). Ivy: a read/write peer-to-peer file system. *SIGOPS Operating Systems Review, 36*(SI), 31–44.

OhmyNews. (2008, October). *OhmyNews International.* Retrieved October, 2008 from http://english.ohmynews.com

Organisation for Economic Co-operation and Development (OECD). (2008, October). *OECD Broadband Portal.* OECD RetrievedOctober, 2008 from http://www.oecd.org/sti/ict/broadband

Pew Internet & American Life Project. (2004, February). *Content creation online: 44% of U.S. Internet users have contributed their thoughts and their files to the online world.* Pew Internet. Retrieved from March, 2007 from http://www.pewinternet.org/PPF/r/113/report_display.asp

Pietzuch, P. R., & Bacon, J. (2002, July). Hermes: A distributed event-based middleware architecture. In *Proceedings of the 22nd international conference on distributed computing systems workshops (ICDCSW '02), Vienna, Austria* (pp. 611–618). Washington, DC: IEEE Computer Society.

Pitoura, T., Ntarmos, N., & Triantafillou, P. (2006, March). Replication, load balancing and efficient range query processing in DHTs. In Y. E. Ioannidis et al. (Eds.), *Proceedings of 10th international conference on extending database technology (EDBT06)* (pp. 131–148). Berlin, Germany: Springer.

Pyun, Y. J., & Reeves, D. S. (2004, August). Constructing a balanced, (log(N)/loglog(N))-diameter super-peer topology for scalable P2P systems. In 4th international conference on peer-to-peer computing (P2P 2004), Zurich, Switzerland (p. 210-218). IEEE Computer Society.

Ramasubramanian, V., & Sirer, E. G. (2004, March). Beehive: O(1) lookup performance for power-law query distributions in peer-to-peer overlays. In *Nsdi '04: Proceedings of the 1st conference on symposium on networked systems design and implementation* (pp. 99–112). Berkeley, CA: USENIX Association.

Ratnasamy, S., Francis, P., Handley, M., Karp, R., & Shenker, S. (2001, August). A scalable content-addressable network. In *Sigcomm '01: Proceedings of the 2001 conference on applications, technologies, architectures, and protocols for computer communications* (pp. 161–172). New York: ACM Press.

Ratnasamy, S., Handley, M., Karp, R. M., & Shenker, S. (2001, November). Application-level multicast using content-addressable networks. In *NGC '01: Proceedings of the third international COST264 workshop on networked group communication* (pp. 14–29). London: Springer-Verlag.

Ratnasamy, S. P. (2002). *A scalable content-addressable network*. Unpublished doctoral dissertation, Berkely, CA: University of California at Berkeley.

Reynolds, P., & Vahdat, A. (2003, June). Efficient peer-to-peer keyword searching. In *Proceedings of acm/ifip/usenix international middleware conference (middleware 2003)* (pp. 21–40). Rio de Janeiro, Brazil: Springer.

Rhea, S. C., & Kubiatowicz, J. (2002, June). Probabilistic location and routing. In *Proceedings of INFOCOM 2002: Twenty-first annual joint conference of the IEEE computer and communications societies, New York, USA* (pp. 1248–1257).

Rheingold, H. (2000). *The virtual community: Homesteading on the electronic frontier* (Revised ed.). Cambridge, MA: MIT Press.

Rheingold, H. (2002). *Smart mobs: The next social revolution*. New York: Perseus Books Group.

Rowstron, A., & Druschel, P. (2001, November). Scalable, decentralized object location, and routing for large-scale peer-to-peer systems. In IFIP/ACM international conference on distributed systems platforms (middleware) (pp. 329–350). Heidelberg, Germany: Pastry.

Samet, H. (1984, June). The quadtree and related hierarchical data structures. *ACM Computing Surveys, 16*(2), 187–260. doi:10.1145/356924.356930

Sarshar, N., Boykin, P. O., & Roychowdhury, V. P. (2004, August). Percolation search in power law networks: Making unstructured peer-to-peer networks scalable. In 4th international conference on peer-to-peer computing (p. 2P). Zurich, Switzerland: IEEE Computer Society.

Schneider, M. (1993, March). Self-stabilization. [CSUR]. *ACM Computing Surveys, 25*(1), 45–67. doi:10.1145/151254.151256

Segall, B., & Arnold, D. (1997, September). Elvin has left the building: A publish/subscribe notification service with quenching. In *Proceedings of Australian UNIX and open systems user group conference (AUUG 97), Brisbane, Australia.*

Stoica, I., Morris, R., Karger, D., Kaashoek, M. F., & Balakrishnan, H. (2001, August). Chord: A scalable peer-to-peer lookup service for Internet applications. In *Proceedings of the 2001 conference on applications, technologies, architectures, and protocols for computer communications* (pp. 149–160). New York: ACM Press.

Tanin, E., Harwood, A., & Samet, H. (2005, April). A distributed quadtree index for peer-to-peer settings. In *Proceedings of the 21st international conference on data engineering (ICDE '05), Tokyo, Japan* (pp. 254–255). Washington, DC: IEEE Computer Society.

Technorati Inc. (2008, October). *State of the blogosphere / 2008.* Technorati. Retrieved August, 2008 from http://www.technorati.com/blogging/state-of-the-blogosphere/

Twitter. (2008, December). *Twitter.* Retrieved December, 2008 from http://twitter.com

Wang, C., Alqaralleh, B. A., Zhou, B. B., Brites, F., & Zomaya, A. Y. (2006, September). Self-organizing content distribution in a data indexed DHT network. In *P2P '06: Proceedings of the sixth IEEE international conference on peer-to-peer computing* (pp. 241–248). Washington, DC: IEEE Computer Society.

Waterhouse, S. (2001, May). *JXTA search: Distributed search for distributed networks* (Tech. Rep.). Palo Alto, CA: Sun Microsystems, Inc. Retrieved January, 2008 from http://gnunet.org/papers/JXTAsearch.pdf

Xue, T., Feng, B., & Zhang, Z. (2004, October). P2PENS: Content-based publish-subscribe over peer-to-peer network. In H. Jin, Y. Pan, N. Xiao, & J. Sun (Eds.), *Proceedings of third international conference on grid and cooperative computing (GCC 2004), Wuhan, China* (pp. 583–590). Berlin, Germany: Springer-Verlag.

Yahoo. Inc. (2008, October). *Flickr.* Retrieved October, 2008 from http://flickr.com

Yang, B., & Garcia-Molina, H. (2002, July). Efficient search in peer-to-peer networks. In *Proceedings of the 22nd international conference on distributed computing systems.* Vienna, Austria.

Zhang, Y., & Weiss, M. (2003, Fall). Virtual communities and team formation. *Crossroads, 10*(1), 5. doi:10.1145/973381.973386

Zhao, B. Y., Huang, L., Stribling, J., Rhea, S. C., Joseph, A. D., & Kubiatowicz, J. (2004, January). Tapestry: A resilient global-scale overlay for service deployment. *IEEE Journal on Selected Areas in Communications, 22*(1), 41–53. doi:10.1109/JSAC.2003.818784

Zhong, M., Moore, J., Shen, K., & Murphy, A. L. (2005, June). An evaluation and comparison of current peer-to-peer full-text keyword search techniques. In *Proceedings of the 8th International Workshop on the Web & Databases* (WebDB 2005), ACM SIGMOD/PODS 2005 Conference. Baltimore, MD.

Zhuang, S. Q., Zhao, B. Y., Joseph, A. D., Katz, R. H., & Kubiatowicz, J. D. (2001, June). Bayeux: An architecture for scalable and fault-tolerant wide-area data dissemination. In *NOSSDAV '01: Proceedings of the 11th international workshop on network and operating systems support for digital audio and video* (pp. 11–20). New York: ACM Press.

KEY TERMS AND DEFINITIONS

Serendipitous Messaging: A form of messaging where recipients receive relevant messages when and as they are published.

Implicit Group: A group defined by the characteristics of its members, rather than their explicit names.

Publish/Subscribe Messaging: An event-based messaging paradigm where recipients select the messages they receive as an implicit group.

Implicit Group Messaging (IGM): An event-based messaging paradigm where publishers select recipients as an implicit group.

Ice: A lightweight overlay for building load-balanced P2P messaging systems.

Hierarchical Tesseral Addressing: A compact addressing scheme used in ICE that describes regions and subregions of space.

Amortised Routing: A geometric multipoint routing algorithm used in ICE that clusters destinations in similar directions in the overlay.

Chapter 6
Securing Event–Based Systems

Jean Bacon
University of Cambridge, UK

David Eyers
University of Cambridge, UK

Jatinder Singh
University of Cambridge, UK

ABSTRACT

The scalability properties of event-based communication paradigms make them suitable for building large-scale distributed systems. For effective management at the application level, such systems often comprise multiple administrative domains, although their underlying communication infrastructure can be shared. Examples of such systems include those required by government and public bodies for domains such as healthcare, police, transport and environmental monitoring. We investigate how to build security into these systems. We outline point-to-point and publish/subscribe event-based communication, and examine security implications in each. Publish/subscribe decouples communicating entities. This allows for efficient event dissemination, however it makes controlling data visibility more difficult. Some data is sensitive and must be protected for personal and legal reasons. Large pub/sub systems distribute events using intermediate broker nodes. Some brokers may not be fully trusted. We discuss how selective encryption can effect security without impacting on content-based routing, and the implications of federated multi-domain systems. We discuss the specification of policy using role-based access control, and demonstrate how to enforce the security of the communications API and the broker network.

INTRODUCTION

This chapter examines techniques for securing various types of event-based systems. The first section discusses typical application requirements. The following section examines specific event

DOI: 10.4018/978-1-60566-697-6.ch006

dissemination approaches. Applying application-level security to event-based systems is introduced at first, along with an overview of Role-Based Access Control. Application-level security is a perimeter defence for an event-based system. The next section demonstrates mechanisms for implementing security within a large-scale distributed broker network, including discussion of distributed cryptographic key management. The following section explores a number of broker-specific policy enforcement mechanisms: where context-sensitive event processing allows more than binary (permit/deny) access control. The chapter conclusion highlights some of the open issues in this area of work.

APPLICATION REQUIREMENTS

Many large-scale distributed applications are best modelled as a federation of domains. A domain is defined as an independently administered unit in which a domain manager has, or may delegate, responsibility for naming and policy specification. Although most communication is likely to be within a domain, there is also a clear need for inter-domain communication. Some examples of application domains and communication requirements are as follows:

Police infrastructure: A number of regional (e.g. UK counties') police domains need support for intra- and inter-domain messages. Incident reports may be sent within and between domains for real-time response and may also be stored as part of an audit or record-keeping process. Databases for court records and the licensing of drivers of vehicles are accessible from all domains.

Healthcare systems: A national health service comprises many independently administered hospitals, clinics, primary-care practices, etc. The care of a patient may move from primary care to treatment in hospital. Specialists may be associated with more than one domain, such as hospitals and clinics. Caring for post-operative or elderly patients in their homes involves carers from many domains. This includes sharing information with various care providers (Singh, Vargas, & Bacon, 2008), making aspects of patient information persistent in centralised health record services (Moody, 2000), auditing data flows, to monitor compliance with procedures (Singh, Bacon, & Moody, 2007), and investigating anomalies.

Communication within and between domains must therefore be supported but because of the sensitivity of the data must be strictly controlled.

Environmental monitoring: Traffic, noise, pollution, and weather conditions are monitored in a city to provide real-time information for citizens (Bacon, Beresford, et al., 2008). All data is recorded for historical analysis to aid prediction and for use by Local Government for planning purposes.

In such applications the need to communicate is driven by the actions of people, emergency situations, and the sensing and reporting of environmental conditions. Communication is naturally event-driven, requiring the transmission of data that captures the nature of some occurrence according to application-specific event-naming, specification and management, see Section 2.

Traditional access control mechanisms tend to focus on client authentication and authorisation specifics. In highly dynamic, distributed applications, context becomes increasingly important; that is, what are the *circumstances* in which access is appropriate. Often data is highly sensitive, and must be protected, yet must also be delivered in a timely manner to those parties that need-to-know. Data must be protected, not only from inappropriate clients, but also from other components of infrastructure (e.g. brokers). Policy must therefore be specified and enforced on how data is transmitted within and between domains, see Section 4.

EVENT DISSEMINATION INFRASTRUCTURE

As motivated above, events represent incidents. An event is a data-rich occurrence, encapsulating a particular semantic. Typically, an event instance consists of a set of (attribute, value) pairs, conforming to a named event type definition. A typical example of an event would be when a sensor device reports a reading from its sensor. Another often used example is each constituent update within a "stock ticker" – i.e. reporting the change of a stock price. Event-based communication is inherently asynchronous, and is most commonly either point-to-point or publish/subscribe.

Point-to-Point

Point-to-point, asynchronous communication implies that endpoint addresses are connected directly. A static structure may exist where senders and receivers communicate via named channels. This scheme originated from database servers and clients, as in IBM MQSeries (now Websphere) (IBM, 2002). For more dynamic point-to-point communication, middleware services must exist, equivalent to those familiar from synchronous communication systems. These originated as remote procedure call (RPC) and evolved towards object oriented schemes (CORBA (OMG, 2002), Java RMI, etc.) and Web Services (Gottschalk, Graham, Kreger, & Snell, 2002). Support services such as naming and directories (Yellow Pages) allow potential services to advertise their availability, and details sufficient for invoking them. Potential clients of these services or objects can look up these details of the interfaces and locations. The clients can then bind to and invoke the services or objects.

Similarly, for asynchronous messaging, potential senders can advertise the events they are prepared to publish and their locations, while potential receivers can connect to them and request to receive publications. For large-scale systems, with many publishers and subscribers, this myriad of point-to-point direct connections is not a scalable solution. Also, application-level clients are interested in events, embodied as data, and should not need to know the address of every possible publisher in order to connect to them all. These issues have been tackled at both the client and communication substrate levels through the publish/subscribe paradigm.

It should be noted however that the mutual knowledge of senders and receivers, with direct connections between them, has advantages from a security point of view. Senders know exactly where the data is to be sent and thus security policy is relatively easy to enforce.

Publish/Subscribe

Publish/subscribe (Eugster, Felber, Guerraoui, & Kermarrec, 2003) is emerging as an appropriate communication paradigm for large-scale systems. It allows loose coupling between mutually anonymous components and supports many-to-many communication. In the publish/subscribe paradigm, a principal takes the role of a publisher and/or a subscriber. Principals connect to the publish/subscribe middleware in order to communicate. Publishers advertise the events they are prepared to publish. Subscribers register their interest in receiving events through a subscription that the middleware handles. Publishers produce events without any dependence on subscribers. This occurs through an event broker, which routes—typically in cooperation with other brokers—events from publishers to subscribers. An event is delivered to a subscriber if it matches a subscription. This process is termed notification.

Publish/subscribe systems are classified as type/topic or content/attribute-based (Eugster et al., 2003). Topic-based publish/subscribe involves the association of an event channel with a particular named topic/type. Producers publish events to the appropriate channel, while subscribers express their interest in receiving messages of a certain type. Content-based publish/subscribe considers message content: a subscriber defines their interest in receiving particular events based on the type and attribute values of the event instance.

Large-scale, publish/subscribe messaging systems often comprise a network of dedicated brokers that provide a communication service, and lightweight clients that use the service to advertise, subscribe to and publish messages (Carzaniga, Rosenblum, & Wolf, 2001; Banavar et al., 1999). A broker network can have a static topology e.g. Siena (Carzaniga et al., 2001) and Gryphon (P. R. Pietzuch & Bhola, 2003) or a dynamic topology, e.g. Scribe (Castro, Druschel, Kermarrec, & Rowstron, 2002) and Hermes (P. R. Pietzuch & Bacon, 2002). A static topology enables the system administrator to build trusted domains and in that way improve the efficiency of routing by avoiding unnecessary encryptions (see Section 4), which is more difficult with a dynamic topology. On the other hand, a dynamic topology allows the broker network to dynamically re-balance itself when brokers join or leave the network either in a controlled fashion or as a result of a network or node failure.

Broker networks are subject to failures of nodes and links, and brokers may join and leave dynamically. Thus, a communication service must be robust under these conditions, fault-tolerant and dynamically reconfigurable. For this reason the message brokers may exploit an overlay network (P. R. Pietzuch & Bacon, 2003), since peer-to-peer naming and protocols provide the necessary robustness.

The alternative to dedicated brokers is a peer-to-peer arrangement where the middleware components coexist in the same machines as pub/sub clients. This lack of separation of clients and middleware has security implications. It is more difficult to guarantee that clients cannot see more than they should of the data being sent, received and routed.

It is already standard practice to protect confidential data on the wire by means of encryption, when underlying network segments are considered to be vulnerable to listeners. In addition, we observe that the brokers in a large, widely distributed network are not likely to be trusted universally with all data. In Section 4 we show how both static and dynamic topologies can be secured by encryption, taking into account the trustworthiness of the brokers.

Note that in publish/subscribe we have deliberately moved from the mutual knowledge of publishers and subscribers. For highly sensitive

data it may be essential that the sender knows the destination domain of the data. We revisit these ideas in Section 5. At the communication level, we have replaced the myriad point-to-point direct connections with shared channels and content-based routing. Control of where data travels has been forfeited for efficient routing. Even with a dedicated broker network, for a large-scale system we must be concerned about the trustworthiness of every broker.

Hermes Publish/Subscribe

Although our approach is generally applicable, we have used specific prototypes for our research. We have used Hermes publish/subscribe for our experiments on securing event-based communication.

Hermes (P. R. Pietzuch & Bacon, 2003; P. R. Pietzuch & Bacon, 2002) is an architecture for distributed, content-based publish/subscribe with an integrated programming model. It consists of two kinds of component: event brokers and event clients, the latter being publishers and subscribers. Event clients publish, or subscribe to, events in the system. An event client has to maintain a connection to a local event broker, which then becomes publisher-hosting, subscriber-hosting, or both. We assume that clients fully trust this broker. A local broker is usually either part of the same domain as the client, or it is owned by a service provider trusted by the client. An event broker without connected clients is called an intermediate broker.

Event brokers form the application-level overlay network that performs event propagation by means of a content-based routing algorithm. Most publish/subscribe systems, including Hermes, optimise content-based routing of events with a subscription coverage relation, which states which subscriptions are subsumed by others (Carzaniga et al., 2001). This allows brokers to reduce the number of events sent through the system by enabling them to filter non-matching events as close as possible to the publisher; these filters become increasingly specific as events approach subscribers.

Hermes supports strong event typing: every published event (publication) in Hermes is an instance of an event type. An event type has an owner, a name and a list of typed attributes so that, at runtime, publications and subscriptions can be type-checked by the system. Subscribers express their interest in the form of subscriptions that specify the desired event type and a conjunction of (content-based) filter expressions over the attributes of this event type. Hermes event types are organised into inheritance hierarchies, but our work does not depend on this. Inheritance can be used within domains when it is available, see Section 4.

Section 3 describes how event types are named, specified and registered in a multi-domain system. Registration of event types causes encryption status and keys to be set up and delegated to authorised domains for authorised clients.

Before a publisher can publish an event instance, it must submit an advertisement to its local event broker, indicating the event type that it wishes to publish. A Hermes publication consists of an event type identifier and a set of attribute value pairs. The type identifier is the SHA-1 hash of the name of the event type. It is used to route the publication through the event broker network. It conveniently hides the type of the publication, i.e. brokers are prevented from seeing which events are flowing through them unless they are aware of the specific event type name and identifier.

In Hermes each event type has an associated *rendezvous node* in the broker network for routing purposes. The rendezvous node is selected by hashing the type name to a broker identifier—an operation that is supported by the peer-to-peer routing substrate (Rowstron & Druschel, 2001). In order to build an event dissemination tree for each event type, advertisements and subscriptions are each routed along paths that meet at the rendezvous node. These paths set up routing and filtering state in each broker node that they

pass through that is not already a member of the dissemination tree.

For reliability reasons, rendezvous nodes are replicated for each event type. In Hermes, a rendezvous node keeps an authoritative copy of the event type definition, which is cached at other brokers throughout the system for type-checking advertisements, subscriptions, and publications. From a security viewpoint, note that selection of the rendezvous node is arbitrary. In our recent work, authoritative, domain-specific type information is digitally signed and stored within the originating domain and rendezvous nodes hold a copy.

APPLICATION-LEVEL DATA DEFINITION AND ACCESS CONTROL

In a large scale multi-domain system the creation, definition and management of event types must be supported and controlled by policy. Also, policy for access to event instance data by clients must be specified and enforced. We use the roles that principals may inhabit in order to carry out various actions in the system as the basis for policy specification.

Role-Based Access Control

Role-Based Access Control (RBAC) (R. Sandhu, Coyne, Feinstein, & Youman, 1996; R. S. Sandhu, Ferraiolo, & Kuhn, 2000) is an established technique for simplifying scalable security administration by introducing roles as an indirection between principals (i.e. users and their agents) and privileges. Privileges, such as the right to use a service or to access an object managed by a service, are assigned to roles. Separately, principals are associated with roles. This separates the administration of people, and their association with roles, from the control of privileges for the use of services (including service-managed data). The motivation is that users join, leave and change roles in an organization frequently, and the policy of services is independent of such changes. Service developers need only be concerned with specifying access policy in terms of roles, and not with individual users.

RBAC is suitable particularly for securing event-based systems because the process of agreeing the notions of role between decoupled participants within an event-based system closely parallels the process by which those same decoupled participants must agree on how to interpret events. Both are well suited to operation in widely distributed systems. Here we focus on securing access to the communication service using RBAC. Authentication into roles must be securely enforced to control the use of all protected services and access to the data they manage (Bacon, Eyers, Moody, & Pesonen, 2005). Domain managers, or their delegates, specify communication policy in terms of message types and roles; that is, which roles may create, advertise, send and receive which types of message. Inter-domain communication is achieved through negotiated agreements, expressed as access control policy, on which roles of one domain may receive (which attributes of) which types of message of another.

The notion of role is ideally suited to a multi-cast communication style. For example, a police notification service may define a role officer-on-duty and message topics such as burglary and traffic-accident with associated attributes. Officers on duty can subscribe to receive notifications of incidents for which they should take responsibility.

RBAC causes principals to be anonymous (i.e. the privileges available to role holders do not depend on their identity), whereas parametrised RBAC gives the option of anonymity or identification, for example officer-on-duty(*station-ID*, *police-ID*). The use of parametrised roles can also help to avoid an explosion in the number of roles required when RBAC is used in large systems. For the communication service, RBAC policy indicates the visibility (to roles, intra- and inter-domain) of specified attributes of message types.

The fact that advertisement is required before messages can be published, and that both are RBAC-controlled, prevents the spam that pervades email communication between humans. Without such control denial-of-service through publication or subscription flooding could degrade large-scale inter-software communication in the same way that it consumes resources in email management. With our approach, a spammer could only be an authorised, authenticated member of a role and therefore could be held accountable.

Management of Event Names, Types and Policies

When constructing policy-secured, multi-domain pub/sub systems, a mechanism is needed to agree on the naming of event types. We assume that domains are allocated unique names within the system as a whole and those roles are named and managed within a domain. Each domain provides a management interface through which role activation policies and service authorization policies can be specified and maintained.

A group of domains may have a parent domain from which an initial set of role names and policies is obtained. For example, county police domains may agree to use a nationally defined set of police roles; health service domains may start from an initial national role-set. The domain management interface allows local additions and updates, for example when national government policy needs to be customised for implementation regionally. As mentioned above, parametrised roles allow domain-specific parameters, for example sergeant(*domain-ID*, *police-ID*). This allows relationships to be captured as well as avoiding excessive numbers of roles in large-scale systems.

In an evolution from a single-domain Hermes pub/sub system, we introduce a format of event type definition that binds the type name and definition together in a secure manner. Public key cryptography is used to guarantee the authenticity and integrity of this type information. Thus we

Table 1.

Name tuple:	1. Type-issuer's public key
	2. User-friendly name
	3. Version number
Body:	4. Attributes
Digital signature:	5. Delegation certificates
	6. Digital signature

protect the system against forged or tampered event type definitions. We reduce the chance of accidental name collisions, and provide a unique handle through which policy can refer to the names of types and attributes.

We require that all participating brokers in the pub/sub system have a key-pair. We can thus require that event-type issuers incorporate this public key into the type name. This facilitates an event naming scope for each particular type issuer. Since event type names include a public key, it is intuitive that the event type definitions should be signed by the corresponding private key. This binds the type definition to the type name, and facilitates verification of event type integrity and issuer authenticity.

The six items that make up a secure event type definition are shown in Table 1. Items 1-3 identify the name of the type. The Attributes item 4 indicates the core event type definition, and items 5 and 6 contain a digital signature of the event type.

Adding a public key to the type name eliminates event name conflicts. While this is desirable, a user-defined name is also maintained. The user-defined name is able to encode useful aspects, such as hierarchical naming. This enables administrative grouping of type definitions across multiple event type owners. Orthogonal to both of those concerns is type evolution, hence the provision of a version number. Releasing a new version of an event type definition will not conflict with previous instances still in use within the pub/sub system. Indeed, to avoid race conditions for version numbers when multiple type managers

are releasing updated event definitions, a UUID scheme is used for version numbers. UUIDs are 128-bit values that are coupled with a practically collision free generation algorithm.

Item 4 in the event type definition describes the event type structure. Each attribute definition itself consists of a user-defined name, a unique identifier (UUID), and an attribute type identifier. The set of types supported depends on the subscription filter language used. User-defined names for attributes are intended to be used by clients of the pub/sub system, whereas the UUID is used by intermediate brokers during distributed event routing. The user-defined names only need to be unique within the context of one particular version of an event type definition. When a publisher-hosting broker receives an event to route, it looks up the user-defined names used by its client using the publisher's event type definition. This allows the UUID fields to be correctly populated. The reverse of this process occurs at subscriber-hosting brokers. The UUIDs allow multiple versions of an event type to exist within the pub/sub system at a point in time.

The digital signature of an event type provides a guarantee of the authenticity and integrity of the type definition. The signature is calculated over all the type definition items except the signature itself: items 1–4. It thus binds the type definition and the name tuple.

The delegation certificates (item 5) facilitate Internet-scale management of event types. Since key-pairs are involved in signing event types, without delegation certificates the type owner would need to re-sign all updates to the type. Delegation certificates facilitate a digitally-signed path of trust from the original event type owner to type managers: parties that are allowed to update event types on their behalf. The delivery of delegation certificates to type managers can be performed out-of-band. The delegation certificates also provide the means to specify fine-grained access rights. Our prototype implementations have typically supported rights such as those used to change the number of attributes (e.g. addAttribute and removeAttribute), as well as rights that change the nature of the existing attributes (e.g. editAttributeName and editAttributeType).

We have investigated fine-grained security, where message attributes are encrypted selectively, with key management transparent to the client level. Encryption overhead *per se* does not need to be justified, and our evaluation indicates that this approach can often incur less overhead than using whole-message encryption. A summary of the approach is given in Section 4, and further details can be found in (Pesonen & Bacon, 2005; Pesonen, Eyers, & Bacon, 2006; Pesonen & Eyers, 2007) and an overview in (Bacon, Eyers, Singh, & Pietzuch, 2008).

OASIS Role-Based Access Control

In this section we provide a brief introduction to the Open Architecture for Secure Interworking Services (OASIS) (Bacon, Moody, & Yao, 2001; Bacon, Moody, & Yao, 2002). OASIS is the particular RBAC implementation that we have used in our research: it provides a comprehensive rule-based means to check that users can only acquire the privileges that authorise them to use services by activating appropriate roles. Although we will refer to OASIS policy in subsequent sections of this chapter, we aim to ensure that the fundamental design principles we illustrate will be generally applicable.

A role activation policy comprises a set of rules, where a role activation rule for a role r takes the form:

$$r_1,..,r_n,a_1,..,a_m,e_1,..,e_l \quad r$$

where r_i are prerequisite roles, a_i are appointment certificates (most often persistent credentials) and e_i are environmental constraints. The latter allow restrictions to be imposed on when and where roles can be activated (and privileges exercised),

for example at restricted times or from restricted computers. Any predicate that must remain true for the principal to remain active in the role is tagged as a role membership condition. Such predicates are monitored, and their violation triggers revocation of the role and related privileges from the principal.

An authorisation rule for some privilege p takes the form

$$r,e_1,..,e_l \quad p$$

An authorisation policy comprises a set of such rules. OASIS has no negative rules, and satisfying any one rule indicates success.

OASIS roles and rules are parametrised. As mentioned previously, this allows fine-grained policy requirements to be expressed and enforced, such as exclusion of individuals and relationships between principals, for example treating-doctor(doctor-ID, patient-ID).

Access Control Policy for Publish/Subscribe Clients

The most general access control requirement in pub/sub will relate to how clients connect to the pub/sub service (e.g. a local broker of a distributed broker network), and make requests using its API. This implements security at the pub/sub network edge.

Many pub/sub systems (P. Pietzuch, Eyers, Kounev, & Shand, 2007) include the following service methods in one form or another:

- define(*message-type*)
  - ◦ Registers a message type of a particular description.
- advertise(*message-type*)
  - ◦ Announces the potential publication of messages of the specified type.
- publish(*message-type, attribute-values*)

- ○ Publishes an instance of the message-type with the given attribute values. This message is routed to the appropriate subscriptions.
- subscribe(*message-type, filter-expression-on-attributes*)
  - ○ Creates a subscription channel for a principal to receive messages of the particular type. The filter expression specifies preference, limiting the messages delivered based upon content.

Some policy languages will only be able to provide a coarse specification of the client privileges required to use the API. In OASIS RBAC, the authorisation policy for any service specifies how it can be used in terms of roles and environmental constraints, and parametrised roles can be used to effect fine-grained control.

OASIS policy indicates, for each method, the role credentials, each with associated environmental constraints, that authorise invocation. OASIS role parameters can be used to limit privileges to particular message types. The define method is used to register a message type with the service and specify its security requirements at the granularity of attributes. On advertise, publish and subscribe, these requirements are enforced. We can therefore support secure publish/subscribe within a domain in which roles are named, activated and administered.

Domain Security Architecture

A domain-structured OASIS system is engineered with a per-domain, secure OASIS server, as described in (Bacon et al., 2001), and a per-domain policy store containing all the role activation and service-specific authorisation policies. This avoids the need for small services to perform authentication and secure role activation. The domain's OASIS server carries out all per-domain role activation and monitors the role membership rule conditions while the roles are active.

This optimisation concentrates role dependency maintenance within a single server and provides a single, per-domain, secure service for managing inter-domain authorisation policy specification and enforcement.

SECURING COMMUNICATION IN LARGE-SCALE, DEDICATED BROKER NETWORKS

In a distributed pub/sub system there will usually be many more clients (publishers and subscribers) than brokers. One approach to making the implementation of data security more easy, is to avoid clients of the pub/sub system managing encryption. Instead, the clients can be required to work through their local event broker. In general this helps ensure that access control policies will be enforced regardless of the specific clients in the system.

Typically, access control mechanisms tend to be client-centric. In this section, we consider broker networks of sufficient scale that the brokers cannot all be trusted unconditionally. By 'trust' here, we mean that the brokers can be relied upon to route messages, but that there may be a precautionary requirements in place (e.g. legal conditions) that these 'untrusted' brokers are not given open access to the event data they are routing (In many ways this is analogous to the role of today's IP network routers.).

We will assume that link-level security has been catered for already: it is not the focus of our work here. In our implementations, all of the connections between brokers use Transport Layer Security (TLS) (Dierks & Allen, 1999) in order to prevent unauthorised access to data within lower layers of the network stack.

The level of data visibility within brokers will impact directly on their ability to decrypt events for the sake of performing content-based routing. In order to support this efficient routing, but without releasing sensitive information, we introduce attribute encryption to allow some event data to remain opaque to some brokers. Decoupling the encryption from the whole event requires secure associations between type names and encryption keys. The secure event types presented in section 3 are used to make these secure associations, for more detail see (Pesonen & Bacon, 2005; Pesonen et al., 2006; Pesonen & Eyers, 2007).

The likely effectiveness of security measures depends on the threat model employed. For example, if properties of the network behaviour of brokers can be monitored by an external agent, it may be possible to use statistical traffic analysis to determine properties about particular event streams. We have not focused on this type of security threat, although security measures can be escalated: for example, dummy events can be inserted to obfuscate event streams.

The rest of this section examines how to build up a secure, distributed infrastructure. The first step is to control broker membership; authorised event-brokers can be allowed to join the network.

Building the Domain Structure

For scalability it is likely that an access control manager will be included in each domain. As described in section 2, each domain contains a number of pub/sub clients and an access control manager. The access control manager is responsible for granting privileges to brokers and clients within the domain, according to the access control policy of the domain.

In order to oversee the multi-domain system, one of the domains within the wide-area distributed system should be designated as the *coordinating domain*. Other domains can then be invited into the shared pub/sub system by the coordinating domain.

Joining a shared pub/sub infrastructure provides a domain two main benefits. First, wide-area distributed applications can communicate across domains. Second, pooling brokers will increase both the resilience and the coverage of the broker

network. Domains that do not have complete trust in each other must at least trust that access control policy will be enforced, as a condition of their cooperation.

Four main aspects of system behaviour need to be controlled:

- Brokers and clients joining and leaving the pub/sub network.
- Definition of event types and topics.
- Modifications to event types and topics.
- Clients accessing the API of the shared pub/sub network.

All four of the above aspects involve a resource owner. In the first two cases, the coordinating domain will be the resource owner. In the latter two cases, the event type or topic owner will be the resource owner.

Delegating Authority

Some mechanism of privilege delegation will be required in order for the system to scale appropriately. Our model employs delegation certificates to allow resource owners to entrust privileges to access control managers of target domains. An access control manager can thus be authorised to perform further intra-domain delegations as it sees fit. The intra-domain policy representation can be as expressive as the access control manager is willing to support.

Because of the separation between inter- and intra-domain concerns, resource owners must trust that access control managers within a domain will behave appropriately. The resource owner will not have fine grained control over the certificates issued by a domain's access control manager to its members. Of course there is still the option of an access control manager having its authority revoked completely. These safeguards will ensure that the inter- and intra-domain management separation will not cause problems.

In our systems, when a client requests an action, it will present a capability that it has been issued by the domain's access control manager. It will also provide a chain of delegation certificates that link the access control manager back to the resource owner. The key-pair of the access control manager is the link between the capability and delegation certificate chains. Verifying a client request involves traversal of the combined certificate chain.

A number of access control actions can be requested—we illustrate these actions with reference to the design of our prototype. Multiple authorities can be represented in a single certificate if appropriate. A coordinating domain can issue a certificate granting connect and install rights to an access control manager with action: connect | install. Authorities can also contain wild-cards. For example, action:* will enable all possible actions. Naturally a resource owner can only grant authority to their own resources.

Broker Network Access

Authority for nodes joining the broker network is rooted at the coordinating domain: it is the owner of the shared pub/sub system. Each request will be provided a Boolean response, that is coupled with the name of the pub/sub network. Verification should be two way: the brokers joining the network should analyse the coordinating domain's credentials and *vice versa*.

Introducing New Types/Topics

The authority to install new types or topics into a shared pub/sub system is also under the control of the coordinating domain. Similar to the case for broker network access, a binary response coupled with the name of the network will be returned to the requester.

Extending Types/Topics

Event type owners are the source of authorisations for extending types and topics. Type extensions inherit properties from their parent types. Authorities are Boolean responses as to whether a particular event type can be extended by some principal. Wild-cards in the type name can be used to aggregate authorisation from event type owners to principals extending those types. The extended type's definition must include the delegation certificate granting its inheritance from its parent type.

Accessing the Pub/Sub API

The highest number of requests will come from clients accessing the pub/sub API. The use of RBAC to secure client access is introduced in section 3. Any publication request will be related to some event type or topic. Thus, the resource owner of that type or topic will issue delegation certificates to the necessary access control managers.

A pub/sub API will usually have rights that need to be controlled in a fine-grained manner. Clients will have a wide range of privileges over a large number of different event types and attributes. Further, independent access control will be needed both for publication and subscription. Our prototypes have allowed a publisher-hosting broker to need to force an attribute to a particular value, and for subscriber-hosting brokers to filter subscriptions to particular attributes.

Credential Propagation

When a client wishes to use the pub/sub API, it will present its credentials to its local broker. The local broker will assess whether the requested action is authorised or not. The client's credentials will need to be passed through the pub/sub network alongside event data: in a wide-area pub/

sub system, intermediate brokers may need to independently verify a clients' access rights. It is likely that some mechanism for avoiding replay attacks will be needed—for example the client could sign their credentials and a time-stamp to indicate the time they made the request.

Brokers can select an appropriate balance between verification and speed. Caution will dictate that a larger proportion of the events that pass through a broker will need to be verified. It is assumed that the brokers can be trusted to pass client credentials through the network.

Our implementations use soft-state within the broker network to store some aspects of security privileges. This has been shown to be more scalable than explicit state management: soft-state simply removes data if it is not refreshed in a sufficiently timely manner.

The capability-based access control mechanism described above can be extended to use its certificates to control access to data encryption keys. Publishers that are authorised to publish events of a particular event type will be able to access (perhaps by proxy) the encryption keys used to protect the data fields of events of that type.

Encrypting Event Content

The use of large-scale broker networks may make it necessary that encryption be used to enforce protection of event data—limiting their plaintext visibility to certain subsets of brokers. We thus need to manage the encryption keys that are employed. Since local brokers need to be trusted by clients anyway, their local brokers can manage the encryption of event content on the clients' behalf. Such an approach reduces the number of nodes that need to access encryption keys. In order to maintain security, the specific keys used need to be rotated in response to changes in the topology of the broker network. Keys should also be rotated periodically to avoid too much data being encrypted by one key. Advantages of this approach include:

- Confidential encryption keys being trusted to a smaller number of parties (the brokers). Thus, the probability of accidental key disclosure is reduced
- Key refresh operations incur lower overheads since they involve fewer nodes
- Any local broker only needs to decrypt data once for all of its local subscribers

If data encryption is employed in a given application, a fundamental design decision is whether to apply encryption to overall event data, or to encrypt the data in event attributes separately. The encryption of whole events is easier to implement, involves fewer cryptographic keys, and requires a smaller number of cryptographic operations. However attribute encryption allows finer-grained enforcement of event data protection, and may be very useful in cases where intermediate brokers can perform selective event processing based on particular unencrypted attributes.

Whole-Event Encryption

As mentioned, the simplest approach to event data protection is to encrypt the whole event. Nonetheless, the event type identifier needs to be left intact and unencrypted for routing purposes. While in transit, events will consist of a tuple that contains the event type identifier, a publication time-stamp, ciphertext and an overall message authentication tag. Note that a separate encryption key will be associated with each event type.

Brokers that are authorised to access event data will be able to acquire the current key, perform decryption, and thus also perform content-based routing. Non-authorised event brokers will have to route the event based on its event type alone—this may reduce the quality of routing decisions significantly.

Only one encryption is required per publication, and it is performed by the publisher's local broker. Each intermediate broker that performs filtering must do a decryption, and any subscriber-hosting broker must also perform a decryption.

Attribute Encryption

Associating an independent encryption key with each attribute facilitates finer-grained data protection and control. In our prototypes the encryption key in use will be selected through the attribute's UUID. Event type identifiers need to be left intact to allow all brokers to perform at least some degree of event routing that is more sophisticated than flooding. Brokers authorised to access attribute data will be able to acquire the appropriate keys for each attribute and carry out fine-grained content-based routing.

Events in transit contain an event type identifier, a publication time-stamp and a set of attribute tuples. The attribute tuples each contain an attribute identifier, some ciphertext and an authentication tag. Our prototype implementation uses an attribute identifier formed by the SHA-1 hash of the attribute name (as used within the event type definition). Using such a hash prevents unauthorised parties from learning which attributes have been included in a given event.

Attribute encryption usually causes higher computational overhead than whole-event encryption. This is because there are a number of repeated encryption initialisations: an expensive phase of most cryptographic algorithms. Initialisation usually dominates the total time taken to encrypt attributes that only contain small amounts of data, and has to be repeated for each attribute to be encrypted. These performance issues have been demonstrated in (Pesonen & Bacon, 2005; Pesonen et al., 2006).

Attribute encryption allows type owners to enforce different levels of client access to the same event type. It also increases the proportion of intermediate brokers that are likely to be able to perform content-based routing. This will reduce the number of messages sent between brokers in cases where content-based filtering is important.

An attempt can be made to emulate attribute encryption by introducing new event types for the different subscriber authorisation levels. Publishers will then need to publish multiple events: in effect these are views on a single source event. Beyond the risk of consistency failures, this approach will quickly become unwieldy as either the number attributes, or the number of different levels of subscriber authorisation grows.

Encrypting Subscriptions

In environments where brokers are not considered completely trustworthy, it is likely that event subscriptions will need to be encrypted also. This way only authorised brokers can issue subscriptions to the pub/sub network. It is important that unauthorised brokers do not gain information when establishing subscription paths.

Under whole-event encryption schemes, subscription filters should be encrypted. The event type identifier within the subscription should be left intact so that type-based routing can still be performed by unauthorised brokers. All other parts of the subscription will need to be assumed to be unfiltered.

When using attribute-based encryption, all of the attribute filters are encrypted individually. The attribute identifiers will still be left in plaintext, however.

Avoiding Unnecessary Encryption

Brokers that have compatible privilege levels need not encrypt the data travelling between them (apart from at a transport level). Such brokers will frequently arise as a set within a given domain, for example. Brokers must examine each others' credentials at connection time, and retain this information in their routing tables. Note that in our systems a publisher-hosting broker will always encrypt content: overall it is cheaper computationally to do so once for the entire event dissemination tree.

When brokers are of a compatible privilege level, a plaintext cache can be attached to the events being disseminated. Such a cache will be formed on demand, and discarded whenever the next routing hop brings events to a broker without such compatibility. The reduction in the computational cost of cryptographic operations will usually outweigh the increased communication cost caused by addition of the cache.

Key Management

In our model, encrypted data, be it event or attribute, always has a UUID. This identifier determines the cryptographic key in use. Controlling access to the encryption key effects enforcement of access control over the event data.

Our implementations collect brokers into key groups. These key groups are issued and re-issued the actual encryption keys in use. Each key group has a key group manager that is responsible for checking that brokers are authorised to join that key group. The event type owners must trust that the key group managers will enforce access control appropriately. The key group manager is either a trusted third party, or a member of the event type owner's own domain. The capability structures used for managing pub/sub requests (see Section 4) can be used for authorising membership of key groups also; the mechanisms used to enforce access control are different however.

Any client request will require both the client and that client's local broker to have sufficient authorisation. The local broker is responsible for checking the client's credentials, however the local broker needs its own credentials to be acceptable in order to acquire the necessary encryption keys.

Secure Group Communication

Decentralised, multi-domain, encrypted pub/sub communication is a type of secure group communication (in the sense of networking research). Secure group communication mechanisms require a key management system that must scale well in the number of clients. Desirable properties are that the communication mechanism operates efficiently over widely-distributed parties, and that high rates of node churn (i.e. participants joining and leaving) do not affect safety or liveness of the mechanism.

Several scalable key management protocols are surveyed in (Rafaeli & Hutchison, 2003). The one-way function tree (OFT) (McGrew & Sherman, 1998) mechanism was used in our prototype implementation. OFTs are binary trees that place participants at the leaves. The algorithm's per-node storage requirement, and computational and communication costs all scale in $log_2 n$ for n participants. This property was experimentally verified in (Pesonen & Bacon, 2005; Pesonen et al., 2006; Pesonen & Eyers, 2007). The OFT mechanism is reasonably straightforward to implement, but still performs well. The OFT communications can re-use the structured overlay network used for inter-broker event communication.

Key Refreshing

Group key management schemes regenerate encryption keys for two primary reasons. First, regenerating keys periodically avoids large volumes of data being encrypted with one key. Thus should a key be compromised, the data exposure will be limited. Second, keys are generated when the group membership changes, so as to provide forward and backward secrecy. In other words, members who have left the group cannot access new data, and those members who have recently joined, cannot access old data. A broker in a key group will hold a capability that supports their membership.

Even state-of-the-art key management protocols cannot provide a cheap re-keying operation. The extra network traffic and distributed coordination is best avoided where possible. A simple approach to ensure that brokers' capabilities are fresh is to limit their validity periods. Rather than having

Figure 1. A key refresh schedule adjusted based on the validity periods of brokers' capabilities

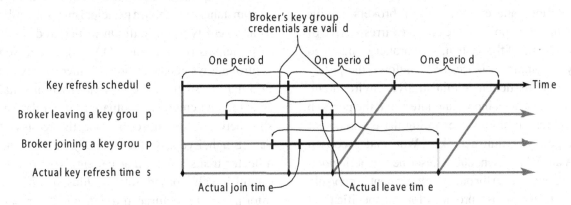

BROKER-SPECIFIC POLICY ENFORCEMENT

to always re-key when brokers join and leave the group, the key group manager can check the earliest validity of each joining broker's capability. If the capability was valid prior to the most recent re-key, another re-key is not required. Similarly, when a broker leaves, re-keying can be deferred until that broker's capability expires. These situations are illustrated in Figure 1. The key refresh schedule line shows the times, if there were no group membership changes, at which re-keying would occur. The second series from the top shows a broker whose capability expires after they have left the group: the key refresh skew is only to the point at which their capability expires. The third series shows a broker joining a key group after a key refresh, however their credentials were valid before the preceding key refresh, and therefore no additional key refresh occurs. The validity of brokers' capabilities should, in general, be far less dynamic than the capabilities held by the clients of the pub/subsystem.

Data may still be in transit within the pub/sub network after a key refresh has occurred. Thus, the old keys will need to be retained for that margin of time. New encryption keys are tagged with a time-stamp so that the correct encryption key can be employed on the basis of an event's publication time-stamp. The period of time for which to retain old keys will depend on properties of the pub/sub network.

Security models must reflect the structure and requirements of the application environment. In situations of federated administrative policy, accountability becomes a consideration (Weitzner et al., 2008). This makes domains *responsible* for the information they hold and share. Domains maintain their own policies, procedures, practices and requirements concerning data management. Further, they have control over the brokers in their local environment. This section describes broker-specific access control mechanisms that allow individual domains to meet their data management responsibilities.

As discussed, the decoupling of entities in event-driven communication suits cross-domain interaction. For example, the events in a distributed healthcare environment may be of interest to several domains. Such events include actions (e.g. administration of a treatment, swallowing of a pill), observations (e.g. sensor readings) and state transitions (e.g. emergency situations). These must be delivered to a number of domains, where each provides a particular service. Example domains from healthcare include surgeries, specialists, outsourced care agencies, billing services, and centralised statistical services.

In pub/sub systems it is appropriate that policy is defined and enforced within brokers, as this ensures that policies are enforced irrespective of the clients of the system. As a broker is managed by a domain, it is trusted to enforce the policy of that particular administrative environment. There has been increasing interest in the pub/sub research community regarding the enforcement of policy within a broker. Wun and Jacobsen (Wun & Jacobsen, 2007) describe a generic *post-matching* model for policy enforcement in pub/sub. Our research also provides for the specification and enforcement of policy within brokers, but is focused specifically on the data-control aspect. This notion of *interaction control* is described in the following section.

Interaction Control

Pub/sub interaction control (Bacon, Eyers, et al., 2008) provides the means for data to be monitored and transformed as part of the event dissemination process. In our prototypes, it functions by loading policy rules into a pub/sub broker to give context-aware control over event distribution. This control allows, in addition to restricting connections and event instances, transformations to be made to event instances. That is, an event instance can be perturbed, fuzzified, summarised, enriched or translated into another format, either on publication or on delivery to a particular subscriber. As such, interaction control provides more than just binary (permit/deny) access control by allowing content to be customised to circumstance.

In our work, interaction control provides the following types of policy rules. These are described in detail in (Singh, Vargas, Bacon, & Moody, 2008).

Client Restrictions

Two types of restrictions can be imposed on a client.

Subscription authorisation rules define the circumstances in which a principal may subscribe to an event type. In addition to the credentials of principals (e.g. through OASIS), interaction control allows consideration of other contextual information such as the current environmental state (e.g. an emergency situation) or a relationship between the subscriber and the requested data (Singh, Eyers, & Bacon, 2008), i.e. ensuring a doctor treats a particular patient. Note that to establish value-based relationships, additional information is required through mandatory attributes. See (Singh, Eyers, & Bacon, 2008) for details.

Imposed conditions restrict the flow of event instances to a particular subscriber. They function similarly to a subscriber-specified filter, except they are defined by policy rules. In addition to event content, these filters may also reference other aspects of context. Imposed conditions are imposed silently, so that a subscriber is unaware that their event-stream is filtered. This avoids revealing any sensitive information encoded in the restriction itself.

Transformations

Transformations serve to customise an event instance to context. Transformations rules are conditional, defining the circumstances in which a transform occurs. The transformation is effected through a function, which may alter the values of an event instance, perhaps enriching or degrading event content, or convert an event into another type to encapsulate a different semantic. Rather than a permit/deny access control scheme, transformation allows the data released to be tailored to circumstance.

Interaction control rules are context-aware, defined with reference to circumstance. Our implementation builds policy enforcement mechanisms into PostgreSQLPS (Vargas, Bacon, & Moody, 2008), an integrated publish/subscribe and database system. In this environment, each broker is also a database instance. This provides a broker with a rich representation of context, while coupling delivery and storage operations under a common administrative interface. The effect is that interaction control allows *fine-grained* control over event dissemination, where policy rules can reference detailed information regarding principals, current environmental state, event content and stored data (through queries). Context can be managed by different entities, where credentials might be maintained in centralised (OASIS) services, and environmental state in a local domain's datastore. In addition to flexibility, this separation suits real-world distributed environments where various services and aspects of state might be managed by different domains.

Pub/Sub Data Control in Distributed Broker Networks

Data control mechanisms apply intuitively where principals' interconnections can form only using broker(s) within the domain dealing in particular information. The question arises of how to manage data in distributed broker networks, where brokers deal in information from other domains. This is non-problematic if brokers are trusted. However, as discussed, in large-scale application environments with federated control, it is unrealistic to assume that a broker network is completely trusted.

Several pub/sub control models allow publishers to define delivery constraints, that are predefined (Opyrchal, Prakash, & Agrawal, 2007), propagated with the publication (Tomasic, Garrod, & Popendorf, 2006), or for (general) policy to be attached to the messages of pub/sub operations (Wun & Jacobsen, 2007). Such approaches assume that brokers will properly enforce policy.

We have described the use of encryption to protect information in untrusted broker networks. This allows sensitive attributes, or indeed whole event instances, to be encrypted when passing through brokers of untrusted domains. However, encryption methods are not always suitable. They

are useful where only particular information (type/attributes) requires protection. In healthcare *all* personal health information is sensitive—(generally) data cannot be disclosed without consent. There are penalties, not only for misuse, but for inadequate protection. Such a scenario requires whole-event encryption to prevent brokers outside the publisher hosting domain from accessing event content. Encryption is particularly appropriate for information that is only temporarily sensitive. For example, battlefield data only requires protection for the duration of an attack. Health information, however, remains sensitive over long periods of time. Thus, encrypted events may leak information if a key is compromised at *any* point in the future.

Is there a use for content-based publish/subscribe in environments of highly sensitive information? The issue is that 'content-based' implies that messages are readable to some degree (by brokers) during the dissemination process. Clearly this is inappropriate where much information is sensitive. Further, if interaction control mechanisms are employed, domains have fine-grained control over the information released. In such scenarios, routing concerns credentials (and context) over content. Consider two identical subscriptions from two principals at a single broker. Each might receive a different event instance, due to restriction/transformation rules that reference the credentials of the principals and the context of the situation. For quality and audit purposes, domains managing information must be able to record where data flows, to identify situations of data leakage. Such information is harder to track in distributed broker environments.

A point-to-point model of communication facilitates data management as communication occurs directly between sources and sinks. However, as mentioned, this burdens principals (clients) with addressing specifics, and also means that each producer must specify its dissemination policy. Apart from issues of scalability, relying on clients to appropriately enforce security policy is

Figure 2. An illustration of a hierarchical broker network

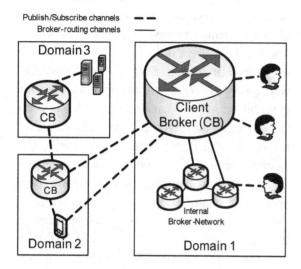

appropriate broker—see Figure 2. Domain 1 in the figure has four brokers, but only one of them is permitted to form links to brokers in domain 2 and domain 3.

This approach maintains indirection, meaning clients are not burdened with the addressing specifics of each data source, nor with the details of dissemination policy. Instead, the middleware layer enforces domain policy, thereby ensuring adherence. As domains deal directly with information consumers, they have precise control over the disclosure of event content. This facilitates information governance, allowing domains to meet their data management responsibilities.

CONCLUSIONS AND OPEN ISSUES

A realistic system architecture for large-scale systems comprises multiple administration domains sharing a dedicated event-broker network. We have found this to be appropriate for many applications. We also assume a secure server per domain that manages credentials and activates roles according to policy. With access control functionality located in the client-hosting brokers, we are able to enforce RBAC on the pub/sub clients. In pub/sub systems it is appropriate that policy is defined and enforced within brokers, as this ensures that policies are enforced irrespective of the clients of the system. In general, separating event-management functionality into a dedicated event service makes access control easier to enforce than in a peer-to-peer approach where the client and event service are co-located. The latter seems inappropriate for applications transmitting sensitive data.

We have assumed content-based routing, for efficiency of communication, rather than broadcast or gossip-based routing. When some brokers are not trusted to see certain sensitive data this style of routing can still be used, with the modifications we describe. But maintaining the required confidentiality of data depends on those brokers

intuitively dangerous, increasing potential points of failure.

A pragmatic approach to managing inter-domain communication in environments of highly sensitive data is to leverage the advantages of the pub/sub and point-to-point communication paradigms. One technique to achieve this would be to prevent the formation of certain links between brokers. In the extreme, a (logical) hierarchical broker network could be formed that has a single, external-facing broker that routes relevant requests to domain-internal brokers.

Partitioning off subsets of brokers in this way will lose the resilience and scalability advantages of an application-wide, shared broker network. Nonetheless, internal broker networks can still assist in load balancing and facilitating domain-internal scalability.

Importantly, trust becomes less of a problem, since the brokers of a domain are managed under a common administrative policy—no information is routed through external brokers (this does not preclude external brokers from acting as publishers or subscribers, though in this capacity they function as principals). Clients producing information within (or relevant to) a domain publish to its given access to keys continuing to be trustworthy and reliable. For sensitive data that persists long term this may not be a sufficient guarantee; some data may remain sensitive over the timescale of a human lifetime or longer. For some data, routing other than point-to-point may be out of the question.

The domain that creates and owns the data may have legal obligations relating to its transmission. Also, it may be that entire domains have reduced requirements on the data, known to the data source. In this case, data transformation can be used to augment the security mechanisms that are used when domains are fully trusted and have an established need for full access to data.

REFERENCES

Bacon, J., Beresford, A., Evans, D., Ingram, D., Trigoni, N., Guitton, A., et al. (2008, January). Time: An Open Platform for Capturing, Processing and Delivering Transport-Related Data. In *Proceedings of the fifth IEEE consumer communications and networking conference (CCNC)* (pp. 687–691). Las Vegas, NV: IEEE Press. (Session on Sensor Networks in Intelligent Transportation Systems)

Bacon, J., Eyers, D. M., Moody, K., & Pesonen, L. I. W. (2005, November). Securing publish/subscribe for multi-domain systems. In G. Alonso (Ed.), Middleware (Vol. 3790, pp. 1–20). Grenoble, France: Springer Verlag.doi:10.1007/11587552_1

Bacon, J., Eyers, D. M., Singh, J., & Pietzuch, P. R. (2008). Access control in publish/subscribe systems. In Baldoni, R. (Ed.), *DEBS* (*Vol. 332*, pp. 23–34). New York: ACM.

Bacon, J., Moody, K., & Yao, W. (2001, November). Access control and trust in the use of widely distributed services. In Middleware '01, IFIP/ACM international conference on distributed systems platforms (Vol. 2218, pp. 295–310). Heidelberg, Germany: Springer Verlag.

Bacon, J., Moody, K., & Yao, W. (2002, November). A model of OASIS role-based access control and its support for active security. [TISSEC]. *ACM Transactions on Information and System Security*, *5*(4), 492–540.doi:10.1145/581271.581276

Banavar, G., Kaplan, M., Shaw, K., Strom, R. E., Sturman, D. C., & Tao, W. (1999). Information flow based event distribution middleware. In *Electronic commerce and web-based applications/middleware workshop at the international conference on distributed computing systems 1999*. Austin, TX: IEEE.

Carzaniga, A., Rosenblum, D. S., & Wolf, A. L. (2001, August). Design and evaluation of a wide-area event notification service. *ACM Transactions on Computer Systems*, *19*(3), 332–383. doi:10.1145/380749.380767

Castro, M., Druschel, P., Kermarrec, A., & Rowstron, A. (2002, October). Scribe: A large-scale and decentralized application-level multicast infrastructure. [JSAC]. *IEEE Journal on Selected Areas in Communications*, *20*(8), 1489–1499. doi:10.1109/JSAC.2002.803069

Corporation, I. B. M. (2002, May). *WebSphere MQ Event Broker*. Retrieved from http://www.ibm.com/software/integration/mqfamily/eventbroker/

Dierks, T., & Allen, C. (1999, January). The TLS protocol version 1.0. *RFC 2246*.

Eugster, P. T., Felber, P. A., Guerraoui, R., & Kermarrec, A.-M. (2003). The many faces of publish/subscribe. *ACM Computing Surveys*, *35*(2), 114–131.doi:10.1145/857076.857078

Gottschalk, K., Graham, S., Kreger, H., & Snell, J. (2002). Introduction to Web services architecture. *IBM Systems Journal*, *41*(2), 170–177. doi:10.1147/sj.412.0170

Jacobsen, H.-A., Mühl, G., & Jaeger, M. A. (Eds.). (2007, June). *Proceedings of the inaugural conference on distributed event-based systems (DEBS'07)*. New York: ACM Press. Retrieved from http://debs.msrg.utoronto.ca

McGrew, D., & Sherman, A. (1998, May). *Key establishment in large dynamic groups using one-way function trees* (Tech. Rep.). Glenwood, MD: TIS Labs at Network Associates, Inc.

Moody, K. (2000, August). Coordinating policy for federated applications. In *14th IFIP WG3 working conference on databases and application security* (pp. 127–134). Schoorl, The Netherlands: Kluwer.

Object Management Group (2002, December) *The Common Object Request Broker Architecture: Core Specification, Revision 3.0.* Needham, MA: OMG.

Opyrchal, L., Prakash, A., & Agrawal, A. (2007). Supporting privacy policies in a publish-subscribe substrate for pervasive environments. *JNW, 2*(1), 17–26.doi:10.4304/jnw.2.1.17-26

Pesonen, L. I. W., & Bacon, J. (2005, September). Secure Event Types in Content-Based, Multi-domain Publish/Subscribe Systems. In *SEM '05: Proceedings of the 5th international workshop on Software Engineering and Middleware* (pp. 98-105). New York: ACM Press.

Pesonen, L. I. W., & Eyers, D. M. (2007, June). Encryption-Enforced Access Control in Dynamic Multi-Domain Publish/Subscribe Networks. In H.-A. Jacobsen, G. Mühl, & M. A. Jaeger (Eds.), *Proceedings of the inaugural conference on distributed event-based systems* (pp. 104–115). New York: ACM Press. Available from http://debs.msrg.utoronto.ca

Pesonen, L. I. W., Eyers, D. M., & Bacon, J. (2006, January). A capabilities-based access control architecture for multi-domain publish/subscribe systems. In *Proceedings of the symposium on applications and the internet (SAINT 2006)* (pp. 222–228). Phoenix, AZ: IEEE.

Pietzuch, P., Eyers, D., Kounev, S., & Shand, B. (2007, June). Towards a Common API for Publish/Subscribe. In H.-A. Jacobsen, G. Mühl, & M. A. Jaeger (Eds.), *Proceedings of the inaugural conference on distributed event-based systems* (pp. 152–157). New York: ACM Press. Retrieved from http://debs.msrg.utoronto.ca

Pietzuch, P. R., & Bacon, J. M. (2002, July). Hermes: A distributed event-based middleware architecture. In *1st international workshop on distributed event-based systems (DEBS'02)* (pp. 611–618). Vienna, Austria: IEEE Press.

Pietzuch, P. R., & Bacon, J. M. (2003, June). Peer-to-peer overlay broker networks in an event-based middleware. In *Proceedings of the 2nd international workshop on distributed event-based systems (DEBS'03)*. New York: ACM SIGMOD.

Pietzuch, P. R., & Bhola, S. (2003, June). Congestion Control in a Reliable Scalable Message-Oriented Middleware. In M. Endler & D. Schmidt (Eds.), In *Proceedings of the 4th int. conf. on middleware (Middleware '03)* (pp. 202–221). Rio de Janeiro, Brazil: Springer.

Rafaeli, S., & Hutchison, D. (2003). A survey of key management for group communication. *ACM Computing Surveys, 35*(3), 309–329. doi:10.1145/937503.937506

Rowstron, A., & Druschel, P. (2001, November). Pastry: Scalable, decentralized object location and routing for large-scale peer-to-peer systems. In Middleware '01, IFIP/ACM international conference on distributed systems platforms (pp. 329–350). Berlin / Heidelberg, Germany: Springer Verlag.

Sandhu, R., Coyne, E., Feinstein, H. L., & Youman, C. E. (1996). Role-based access control models. *IEEE Computer, 29*(2), 38–47.

Sandhu, R. S., Ferraiolo, D. F., & Kuhn, R. (2000). The NIST model for role-based access control: towards a unified standard. In *Rbac '00: Proceedings of the fifth ACM workshop on role-based access control* (pp. 47–63). New York: ACM Press.

Singh, J., Bacon, J., & Moody, K. (2007, April). Dynamic trust domains for secure, private, technology-assisted living. In *Proceedings of the the second international conference on availability, reliability and security (ARES'07)* (pp. 27-34). Vienna, Austria: IEEE Computer Society.

Singh, J., Eyers, D. M., & Bacon, J. (2008). Decemberin press). Credential management in event-driven healthcare systems. In *Middleware*. Leuven, Belgium: Springer Verlag.

Singh, J., Vargas, L., & Bacon, J. (2008, January). A Model for Controlling Data Flow in Distributed Healthcare Environments. In Pervasive Health 2008: Second international conference on pervasive computing technologies for healthcare. Tampere, Finland: IEEE Press.

Singh, J., Vargas, L., Bacon, J., & Moody, K. (2008, June). Policy-based information sharing in publish/subscribe middleware. In *IEEE workshop on policies for distributed systems and networks (Policy 2008). IBM Palisades*. New York: IEEE Press.

Tomasic, A., Garrod, C., & Popendorf, K. (2006). *Symmetric publish/subscribe via constraint publication* (Tech. Rep. No. CMU-CS-06-129R). Pittsburgh, PA: Carnegie Mellon University.

Vargas, L., Bacon, J., & Moody, K. (2008). Event-Driven Database Information Sharing. In *British national conference on databases (BNCOD)* (*Vol. 5071*, pp. 113–125). Cardiff, UK: Springer.

Weitzner, D. J., Abelson, H., Berners-Lee, T., Feigenbaum, J., Hendler, J., & Sussman, G. J. (2008). Information accountability. *Communications of the ACM, 51*(6), 82–87. doi:10.1145/1349026.1349043

Wun, A., & Jacobsen, H.-A. (2007). A policy management framework for content-based publish/subscribe. In *Middleware '07* (pp. 368–388). Newport Beach, CA: Springer.

Chapter 7
Automating Integration Testing of Large-Scale Publish/Subscribe Systems

Éric Piel
Delft University of Technology, The Netherlands

Alberto González
Delft University of Technology, The Netherlands

Hans-Gerhard Gross
Delft University of Technology, The Netherlands

ABSTRACT

Publish/subscribe systems are event-based systems separated into several components which publish and subscribe events that correspond to data types. Testing each component individually is not sufficient for testing the whole system; it also requires testing the integration of those components together. In this chapter, first we identify the specificities and difficulties of integration testing of publish/subscribe systems. Afterwards, two different and complementary techniques to test the integration are presented. One is based on the random generation of a high number of event sequences and on generic oracles, in order to find a malfunctioning state of the system. The second one uses a limited number of predefined data-flows which must respect a precise behaviour, implementable with the same mechanism as unit-testing. As event-based systems are well fitted for runtime modification, the particularities of runtime testing are also introduced, and the usage in the context of integration testing is detailed. A case study presents an example of integration testing on a small system inspired by the systems used in the maritime safety and security domain.

INTRODUCTION

Our research focuses on the fast integration of systems-of-systems (SoS). In collaboration with industrial partners, we ensure that the results can be applied for Maritime Safety and Security (MSS) systems, which are typical event-based systems and are often built on top of a publish/subscribe architecture. An important aspect of this research covers the validation through checking the cor-

DOI: 10.4018/978-1-60566-697-6.ch007

rect integration of the system in order to ensure their reliability.

MSS systems are large-scale distributed SoS in which the sub-components are elaborate and complex systems in their own right. The primary tasks of MSS SoS are sensing issues at sea, analysing these issues, thus, forming a situational awareness, and initiating appropriate action, in case of serious issues (EU Commission, 2007; Thales Group, 2007; Lockheed Martin, 2008; Embedded Systems Institute, 2007). Provision of situational awareness requires analysis and synthesis of huge volumes of data coming from the various types of sensor components such as vessel-tracking (AIS) systems, satellite monitoring, radar systems, or sonar systems. A distinguishing characteristic of MSS systems-of-systems is their data-centric, distributed and event-based nature (Muhl, Fiege, & Pietzuch, 2006), which means that the publish/subscribe paradigm is well fitted, and it is readily being used as underlying system infrastructure.

In their landmark paper on "The Many Faces of Publish/Subscribe", Eugster et al. (2003) present an array of advantages of the publish/subscribe paradigm as a cure for developing highly dynamic large-scale systems. However, the advantages of fully decoupling the communicating entities in publish/subscribe platforms in terms of time, space and synchronization, can also be seen as a curse when it comes to runtime evolution and ensuring a system's integrity after dynamic updates. In this chapter, we will discuss the issues of integration and acceptance testing arising from the loose coupling of communicating entities advocated by event-based systems. Because we are dealing with highly dynamic systems that have to provide constant operational readiness, we will concentrate on the challenges system engineers are facing when it comes to testing and accepting dynamic system reconfigurations.

First in section *Properties of Publish/Subscribe Platforms*, we will present the testing requirements in publish/subscribe systems and highlight the specific challenges of this context. In Section *Integration Testing in Event-based Systems*, we will describe the usage of testing methods in order to implement integration testing in such dynamic and decoupled environment. The Section *Runtime Testing for Component-based Platforms* provides details on the specificities of handling testing at runtime. An example of usage of the methods previously defined will then be presented in Section *Runtime Integration Testing: A Controlled Experiment*, using a simplified version of an MSS system. An overview of the future research to come in the domain of testing and runtime evolution will be given in Section *Future Trends*. Finally, Section *Summary and Conclusion* will summarize and conclude this chapter.

PROPERTIES OF PUBLISH/ SUBSCRIBE PLATFORMS

The quality of a software system is a compromise between the cost of an error happening and the cost of improving the quality (cost being used with a large meaning such as time, money or physical damage). In order to improve the quality of a software system, the most commonly used technique is *software testing*. Due to the nature of their applications, systems based on event-driven platforms, and publish/subscribe platforms in particular, tend to have high needs for software quality and testing. Systems designed on event-driven platforms can have all kinds of applications, but some typical classes of applications are:

- Embedded software, which controls physical devices such as cars, music players, factory robots, etc. Sensors generate input data and events that are treated by software components which, in turn, send output data to controller devices.
- Web applications, which react to the requests from web users, process them and send the results.

- Graphical User Interfaces (GUI), where events such as clicks or key presses trigger specific code which modifies the state of the application and eventually leads to producing outputs.

The first two application classes often have strong *reliability requirements*: incorrect behaviour or lack of reaction could have very fatal consequences, either in terms of cost or safety.

Moreover, the event-driven approach permits to separate the code in several small entities dedicated to a specific task. A formal way to distinguish those entities is the notion of a *component*. This notion allows to handle the complexity of the software system as it is possible to concentrate only on a small part of the functionality at a time. However, each component will still interact with the other components, either directly by sending or receiving data, or indirectly by affecting the state of a third component of the system. Testing each of the component separately is not sufficient (Gao, Tsao & Wu, 2003), integration problems only occur when the entities interact with each other, due to missing/wrong interactions, or because one component does not behave as another one expects it to. It is, therefore, necessary to perform *integration testing*.

The decomposition into components provides flexibility in the system architecture. Publish/subscribe architectures permit to keep components very loosely coupled as, in contrast to other architectures (client/server, peer-to-peer) the components are not directly bound to each other, but only bound to data or event types. This eases the *runtime reconfiguration* of a system, as components can be added or removed independently. Runtime reconfiguration is useful in the context of high availability systems, especially for the large-scale and complex ones which cannot be duplicated. Often duplication is not possible because of the cost of duplicating the hardware (such as having a entire boat available just for testing purpose), but also due to license cost for

both a production and a test system, or in case of cooperation between different actors not willing to share full control of their components between each other. With respect to reliability, reconfiguration has to be treated carefully, as the original behaviour might be affected. Therefore, integration testing must also be done during reconfiguration. It is important to pay special attention to this case, where testing happens at runtime.

In the following, we highlight the issues encountered during integration testing when dealing with a publish/subscribe architecture, which are not so problematic in other runtime architectures such as client/server.

Explicit vs. Implicit Dependencies

Most service-centric architectural models are based on binding required to provided interfaces. Bindings make dependencies and communication between components explicit (Oreizy, Medvidovic, & Taylor, 1998), i.e., components are constrained to interact with a known set of other components, defined by the system architect. When a part of the system is modified, obtaining the list of components affected by the reconfiguration is a matter of identifying the modified components and bindings, and following the dependency graph derived from the bindings. Then, the test cases that exercised the modified parts have to be re-run (Orso, Do, Rothermel, Harrold, & Rosenblum, 2007).

In contrast, Publish/subscribe systems are characterized by loose coupling of their components. Each component listens to data messages of specific types and generates other data messages. There are no explicit dependencies between the components themselves. Dependencies are rather implicit in the data types they publish, or subscribe to. This simplifies the integration process considerably, but on the other hand, it makes the task of finding dependencies between component instances more difficult and inhibits the testing after a reconfiguration of the system.

Ways to handle the dependencies is the various cases are presented in Section *Integration Testing in Event-based Systems*

Rendez-Vous vs. Persistent Messages

In architectures where components are linked directly, such as Remote Procedure Call (RPC) based systems, the communication between components happens only when they are simultaneously present (rendez-vous). Either the interaction between the components is immediate, or it does not happen. On the other hand, in publish/subscribe platforms, communication between components can be time-decoupled. When a component emits data, no assumption is done about the presence of components that may receive them. The data might be used immediately, they might never be used, or they might be used later when a component interested by the data appears, even if the emitting component has, since then, been removed. The data might even be first used immediately and also at later time by another component. This *persistence* is provided by the platform.

Persistence of data or events leads to useful features such as better reliability, or complete independence between the components, but it also brings difficulties in testing: when a new component is added to the system, not only interactions with the current components have to be tested but also the effect of the components previously removed. As we will see in Section *Integration Testing in Event-based Systems*, this can be considered as special case of component dependencies.

Synchronous vs. Asynchronous Interaction

In some architectures, when a component requests the service of another component, it is blocked until the second component finishes. This ensures that the operation has been completed before the first component resumes. In other architectures, and often in event-driven platforms, component interactions are asynchronous, meaning that a component is still active while the service is processed. When the called component finishes the processing, it can alert the first component through some event.

Asynchrony is advantageous in terms of performance and independence, but it also introduces complexity when testing (Michlmayr, Fenkam, & Dustdar, 2006b). Because the reply event can occur at various times during the execution of the first component, additional care must be taken to ensure that all interaction combinations have been adequately assessed. In other words, it introduces an observability issue while testing: comparing the execution trace with the expected result would lead to an inconclusive test result.

Several proposals have already been introduced to handle asynchrony in requirement specification and in test cases. For instance TTCN-3 (Schieferdecker, 2007), a test case description language, permits to specify separately the messages sent, the values returned, and also the minimum and maximum time that the result may take to be answered.

Call-Reply vs. Data-Flow

In a service-centric architecture, an activity originates from a component which "pulls" a result from another component by requesting the processing of some data. The system follows what could be called a *call-reply* model, the basis of a client-server architecture. A component needs a service (the client) and calls a component that provides this service (the server) with the specific data to be treated. Once the processing is finished, the server returns the result. This organisation is presented in Figure 1. The curved line represents the information flow. Component *A* calls component *B*, which replies. After that, component *A* calls component *C*, etc. This scheme can be recursive: a server can be client of other components. Writing integration

Figure 1. Example of system organised following the call-reply model

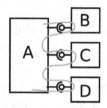

Figure 2. Example of system organised following the data-flow model

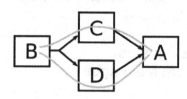

test cases for such architecture can be done at the development time of each component, without requiring knowledge of the system in which it will take part. There is a direct causality between what a component sends and what it receives. From the client side, it is possible to express both the input and the expected output of the server, because the component (and the developer) has an expectation towards the service that can be translated into test cases.

In contrast, in a data-centric architecture, the data is "pushed" from one component to another. The system is better understood as a series of *data-flows*. A component receives data from the previous component in the flow, processes the data and sends the result to the next component in the flow. An example of this type of organisation is shown in Figure 2. In this component model, integration test cases cannot be associated with one specific component alone, because components do not have any requirements on their predecessors or successors; they only know about the data.

Therefore, integration tests are not associated with pairs of components but with either complete or partial *data-flows*. The integration test suite has to verify that the data is correctly processed. In Section *Integration Testing in Event-based Systems* a method is presented to implement such testing which can also work in case of reconfiguration.

Runtime Testing

In case of runtime reconfiguration, testing of the new configuration has to be performed while the

system is in operation (with the previous configuration). Ideally, this can be done by duplicating the entire system. Unfortunately the harsh reality often has limitation on the physical resources. The larger the system, the more likely are there limitations. For instance, it is not feasible to duplicate an entire ship whenever one of the software components of a radar must be updated. When the system cannot be duplicated, it is possible to rely on *runtime testing* (Vincent, King, Lay, & Kinghorn, 2002).

There are two main difficulties with using a component at the same time in the context of normal operation (production) and in the context of testing. First, testing involves interactions with the System Under Test (SUT), i.e. sending stimuli to verify that the SUT responds as expected. Runtime Testing bears the danger that testing interaction with the component will affect its other users. Test operations and data must be ensured to stay in the testing realm and not affect the other clients of a component or sub-system. This characteristic is known as *test isolation* (Suliman et al., 2006). Second, when components are tested which have effect outside of the system (i.e. they produce the output of the system) some operations might not be safe to test at runtime. For instance, it might not be possible to write in a database containing bank account information, or controlling the actuators in a robot while they are already being controlled following a different control algorithm.

Similarly, it can also happen that an input resource is unique. For example, in some frameworks, if a web service listens on one TCP/IP port, it is not possible to run another web service on the same port simultaneously. There are several possible solutions, such as setting up a simulator of the outside world, sharing the resource between the component under test and the component in production, or refraining from executing the test cases. Nevertheless, whichever approach is taken, first it is necessary to be able to define the *test sensitivity* of the component. The platform must provide a way for the component to "tell the difference" between test and non-test data or event, which permits the components to be what we call *test-aware*.

Handling test isolation and test sensitivity is described in Section *Runtime Testing for Component-based Platforms*.

INTEGRATION TESTING IN EVENT-BASED SYSTEMS

Event-based platforms allow decoupling of components that are handling specific operations on data. Unit-testing of the components is not sufficient in order to validate the correctness of the entire system made of such components. In addition, it is necessary to check that the components interact correctly with each other. This validation is the goal of integration testing.

Background

Assumptions on the System

Before detailing further some approaches to test the integration of an event-based system, it is important to define some basic assumptions and expectations on the system under consideration.

First, we consider that the system is component-based (Szyperski, 1998), meaning it is made of separate software units which can only interact with each other via predefined interfaces. In the context of publish/subscribe platforms, an inter-face is defined by the data types which will be received or sent (a simple *event* being represented as a data type which contains no data). Components may be hierarchically defined: a component can be composed out of several other sub-components. Components need not be *pure*[1]: they can have state and interact with the context. They can also be black, or "grey" boxes, i.e. their specification is known but their implementation is not.

Second, each component is considered to have already passed the unit-tests, and, therefore, implements its own specification correctly. The test cases should be available. This can be done either through providing inputs and an oracle, or by providing a specification of the component and a component capable of generating test cases out of it. For example Michlmayr et al. have proposed a framework (2006a; 2006b) to specifically unit-test publish/subscribe applications using Linear Temporal Logic (LTL): depending on the input data received it can define the type and order of the output data.

In most systems, there exist components which, from the point of view of the publish/subscribe framework, do not have inputs or do not have outputs. They correspond to *sensors* (obtain data from a hardware component, read data from a file, etc.) and *actuators* (display data on a screen, write data to a file, etc.). We will not include those special components in the rest of the discussion on integration testing as the information about which events they publish or subscribe to is not sufficient to test them. In practice, test cases should be adapted specifically to fit a component, and they should be able to handle the input and output correctly. Care should be taken to prohibit the sensor components from generating inputs which could interfere with the inputs provided by the test cases. They should either be stopped or their data should separated from the rest of the data using the same mechanisms which will be presented for runtime testing in Section *Runtime Testing for Component-based Platforms*.

Verification and Validation

Verification and validation are the two major processes to ensure the quality of the system. Those processes are complementary. Verification consists of ensuring that every properties and behaviour defined at a given level of abstraction still holds true at a lower level of abstraction. For instance, it might consist of manually comparing one requirement defined in English in the project agreement with the UML model of the system. *Formal* verification checks that a certain property is true using formal methods. For example, it can consist in automatically analysing a source code to verify that the transitions between the states can only follow the transitions defined in a model of the system in a Finite State Machine. The validation process is concerned by evaluating the implementation with respect to the expected behaviour of the system. It can be for instance done using test cases which contains the description of the inputs sent to the implementation and the expected resulting behaviour of the system.

For ensuring the quality of the integration of the system, not only the interaction of the components together has to be taken into account but also the framework on which the components will be executed. While both verification and validation are useful for this task, verification tends to be difficult because it requires also the model of the framework in addition to the models of the components themselves, and because the increase of complexity of the entire system often resolves in exponentially increasing complexity of the proof. This is one of the reasons we are focusing only on validation in this chapter. Nevertheless, it is worth mentioning the work by Garlan et al. (2003) on modelling a generic publish/subscribe architecture for verifying via LTL properties systems based on this type of architecture. Extending this work, Baresi et al. (2007) have proposed special language support for modelling and verifying the system efficiently thanks to the reduction of the number of states to explore. Zhang et al. (2006)

have defined a modelling language dedicated at event-based systems which, using source code transformations, can both generate code used for the verification (using LTL properties associated with the original model) and execution traces of the implementation which can be used for the validation.

Built-In Testing and Beyond

Built-In Testing (BIT) is a useful paradigm in order to test a dynamic component-based system (Vincent et al., 2002; Suliman et al., 2006). BIT refers to any technique used for equipping components with the ability to check their execution environment, and their ability to be checked by their execution environment (Gross & Mayer, 2004), before or during runtime. It aims at a better maintainability of the testing aspect surrounding each component.

BIT has two facets. The first is concerned with the testability of the component. Components can be equipped with special interfaces in order to facilitate the testing. For instance, this can be the ability to control the component's internal state in order to set up quickly the context of a test case. It can also be the ability to let the component become aware of the fact that testing is happening, in case it is test-sensitive. This aspect will be treated more in depth in the next section, concerning runtime testing.

The second facet of BIT is concerned with the association of test cases with the component, whose original goal was to allow the component to carry out all or part of the integration testing by itself. The requirements of the component on its execution environment (i.e.: the platform on which it is running, the physical components with which it is linked, the components on which it relies for providing specific services) can be validated using test cases contained in the component. That way, the components can perform much of the required system validation effort automatically and by "themselves" (Gross, 2005; Brenner et al.,

Figure 3. Built-in Testing in Action

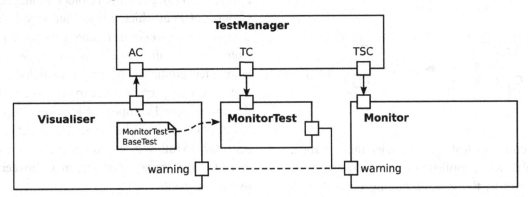

2007). Figure 3 presents schematically the validation using the BIT infrastructure. The *Visualiser* component needs to test the Monitor component on which it depends. Via the AC interface, one of the BIT interfaces, the *Visualiser* contacts the *TestManager* component, providing the *MonitorTest* test. *TestManager* takes care of the connection between the test and the *Monitor* component, let the *Monitor* component know that a test is taking place via the TSC interface (another of the BIT interfaces), and finally report the result of the test to the *Visualiser*.

In the case of a data-flow organisation, only some tests might be possibly done in this way (those directly validating the platform compatibility and the hardware). As seen in the previous section, it is not possible to provide tests to ensure that the requirements on the other components are fulfilled, simply because components do not have any expectations on the other components. Nevertheless, we are going to present approaches that rely on the basic idea of embedding test cases with components and see how this may facilitate testing and maintenance of the system.

This distribution of the responsibility of validating the component's environment to the components themselves is very interesting for large-scale dynamic systems. It can help to maintain the independence of each of the participating components which are likely developed by different teams as tests can be decentralised and

associated with each component. Another interesting property of distributing the test definition is that it helps to keep the tests synchronised with the respective version of the components. For instance, when a component is updated to support additional functionality, the tests to validate the functionality must be updated simultaneously. Associating the tests with components also means that the testing infrastructure can automatically benefit from the dynamicity of the component infrastructure. Updating the testing information alone can be useful because test cases themselves can be erroneous, or for extending the test coverage while keeping the system running.

Finding Dependencies between Components

As previously mentioned in Section *Properties of Publish/Subscribe Platforms*, and as we will see in more details later on, one important piece of information needed to perform integration testing efficiently is represented by the dependencies between the components. Usually in publish/subscribe systems, and opposed to other component-based architectures, the connections between components are not expressed explicitly, although one way to handle this lack of information would be to be very pessimistic and consider that every component interacts with every other component, the combinational explosion of the

Figure 4. Example of dependency graph computation

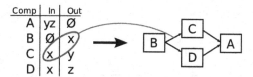

interactions to test would render this approach unusable when applied to large-scale system. In other words, the goal of finding dependencies between the components is to allow a better *test selection*. The dependency information can also be useful when the system is modified, so that the testing happens only on the parts which could be affected by the modification, and not on the whole system.

In order to define a component dependency graph of the whole system, for each component one has to determine the components that might be triggered (executed) after the component has been triggered. In publish/subscribe platforms, components' interfaces are identified by the type of data which will be passed. An output interface corresponds to the generation of a given type of data. An input interface corresponds to a type of data which will trigger the component's execution. Therefore, constructing the dependency graph translates into listing every component of the system, for each of them list the types of data they listen to, and consider them dependent on every component generating events of this type of data. Figure 4 shows an example of the dependency graph computation. The table on the left-hand side represents the information on the components obtained from the architectural description. As component *B* generates data of type *x*, both components *C* and *D* are dependant on this component.

In some platforms, the input might be more precisely defined, for instance by the use of rules which filter the input data also according to their content. This is the case in the Data Distribution

Service (DDS) standard (Object Management Group, 2007) in which a *topic* can be refined by using *topic_expressions* (using a syntax derived from SQL). In this case, it might happen that a component produces a type of data which interests a second component but, due to the content of the data, this second component is never triggered. Unless this case can be detected statically, it is necessary to follow a pessimistic approach which considers that there is a dependency between the two components.

When reconfiguring the system at runtime by introducing a new component, it can happen that some components which are no longer part of the system have left data affecting this component. It is important to detect this situation during the acceptance process of the new configuration. Nevertheless, these components should not be included in the dependency graph because once the persistent data is received the actual behaviour (no more data) would be different from the behaviour of the component being active. The way to handle this persistent data will not be different from the rest of the persistent data during runtime testing, which is described in Section *Runtime Testing for Component-based Platforms*.

One of our assumptions on the system is that it is possible to statically know the full interface of the components. Studies have also been done on dependency computation of even-driven systems where the interfaces are implicit. This happens when the components are not explicitly defined, for instance when event filtering is done within the component via manually-written code, or when the events generated are not declared beforehand. Holzmann and Smith have proposed a method to compute the dependencies statically (1999), given that the source code is annotated (using the *@-format*). Memon et al. have introduced a tool (2003) to obtain this information by running the system and actively triggering every possible event (the tool is focused on GUI application based on Java).

We will now discuss two different methods for testing the integration of an event-based system.

Testing Random Event Sequences

Unit-testing checks each of the components in isolation. However, the integration of the components brings intrinsic problems which have to be detected. Once the system is entirely constituted, an event (or a data) can affect several components simultaneously, in which case they might be executed in any arbitrary order, which might cause unexpected behaviour. Components may not support data with specific values, as generated from another component. Moreover, components having a state (i.e.: non pure components) can be affected by the order in which the data is received. In order to detect such cases, a method that can be used is generating a random sequence of data and verifying that the system reacts correctly to it.

Defining random sequences of events is straightforward. Much more difficult is it to know how to generate the data associated with the event and to know whether the system reacted correctly to them. In case the input data is just events (without containing more information), it is possible to generate them automatically, but if the input contains complex data, this cannot be created randomly in an easy way. In order to generate the input data, a solution consists of relying on the unit-tests of each component. If the unit-tests are specified via a high abstraction model, it might be possible to use the specification to directly generate the input data. When this approach is not possible to obtain the input data it is necessary to execute the test cases entirely. Once the input data has been generated, the components triggered by these inputs generate outputs, which will, in turn, trigger other components. Eventually, from the generated inputs, all the components are triggered.

For every instance of generated input, the sequence of events increases, and every component which has received input must be validated. An oracle compatible with the component and adapted to the sequence of events must be available. There are several complementary ways to obtain oracles:

- The unit-test of a component might have an oracle valid for any input data. Typically, the oracle asserts some invariant properties specific to the component. This type of validation might also be found as *runtime monitors*, which are used to estimate the system health. If such monitoring components are present they can be also used are oracle.

- Some oracles are generic by focusing on common properties that hold true for any component. For instance an oracle might verify that the function terminates within a bounded time, that there is no memory leak or, as used in (Yuan & Memon, 2008), that the component does not crash (a non-handled Java exception).

- In the context of runtime testing it is possible to compare the outputs of the production configuration with those of the configuration under test. Different outputs indicate a potential failure. Whether this failure is in the production configuration or in the one under test has to be decided by the system integrator.

Figure 5 presents an example of one of the test case generated by the method of random event sequences. Three input data are inserted, in the order m, s, m. After each input, the data flow along the components as in the final system. The third insertion of input leads to a crash in component E. This crash is detected by a generic oracle, which, therefore, detects a failure in the test case.

A difficulty with the generation of random sequences of input data is the high number of integration tests created: it is infinite[2]. Depending on the context, it might be possible to simply decide that enough tests has been performed after a

Figure 5. Example of random sequence testing. 3 inputs are generated, which leads to a crash in component E. The crash is detected as a failure to pass the test case by the oracle

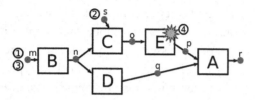

certain time span as passed, but it is still desirable to first execute the tests which are more likely to expose errors. A selection on the sequences can be done by limiting the sequence size to a maximum number of events. Moreover, as the integration testing focuses on the behaviour when multiple components are involved, it is useful to restrain the number of times a type of data is generated (for instance one or two times), so that more interaction happens and the number of possible combinations is reduced. Going further, Yuan and Memon have proposed a test selection technique (2008) which takes into account the interdependencies between events. They propose to concatenate only events which involve different components but for which the components have either a direct interaction or an indirect interaction (defined by *predicates*, which informally correspond to interactions with a common third component). This method has the advantage that if sufficiently many unit-tests are available, it can execute a very large number of test cases. However, as the oracles have to be broad enough to support any order of inputs to the components, this method does not allow to detect if the behaviour of the integration is incorrect, i.e. that for each specific input sequence the correct outputs were generated. This is why it is interesting to combine this method with a second method dedicated to test some typical interactions between the components. We will discuss such a method in the following sub-section.

Testing Specific Data-Flows

In some cases the integration of components might be incorrect without leading to strong misbehaviour of each component. For instance, two components which are supposed to communicate together might actually be publishing and subscribing to different types of data, leading to the subscriber not being triggered at all. It can also happen that due to a misunderstanding during the specification phase, the data types are not processed in the same way (e.g.: X and Y co-ordinates inverted). In order to detect these errors, specific test cases validating the behaviour of the integrated system must be executed. When an error on a running system is found and fixed, similarly, one would need to be able to specify a specific sequence of events with a specific oracle for regression testing.

If the system is entirely based on the call-reply model, it is possible to define a test case for the client component (from which the call is originated), so that the complete interaction sequence may be tested: every client component has explicit requirements on its servers, so it can have test cases validating those requirements on its own. As can be seen in Section *Properties of Publish/ Subscribe Platforms*, in the case of data-flow model, defining such test cases requires knowledge of the integration by the system tester. Moreover, it is important that the test cases written can be reused as much as possible, even if the system is updated or reconfigured.

Our approach consists of defining *virtual components* which correspond to a specific data-flow. The virtual component is only delimited by a set of inputs (data types) and a set of outputs (other data types). In contrast to the usual components, there might be inputs or outputs going into or out of the virtual component without being part of its interfaces. For instance, in Figure 6 the type *n* goes from the inner component *B* to the outer component *D*. This essentially means that some

Figure 6. Example of a virtual component. The virtual component defined solely by its inputs (m) and outputs(p) and automatically encompasses the behaviour of the components B, C, and E

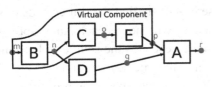

Figure 7. Testing a virtual component consists of executing unit tests associated with the component via the testing interface provided by BIT

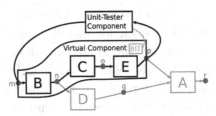

inputs or outputs will never be used to validate the behaviour of this specific flow. As seen in Figure 6, the virtual component inherently encompasses all the components which are used in the flow of data between the input and the output events. Compare to normal composite component, this permits a high flexibility in case of modification around the virtual component, i.e. if components are added or removed within the data-flow the set of inner components will evolve automatically. Referring to the example displayed in Figure 6, if the components *C* and *E* were replaced by three components *X*, *Y*, and *Z* having the same functionality, the virtual component would still be valid and pass the tests. In case the component *C* is mistakenly removed, the virtual component would still be valid, but some test cases would likely not pass, as *E* would never receive any data.

Another advantage of using virtual components over normal composite components is that it is possible to define several overlapping flows. In Figure 6, it would, for example, be possible to define a flow between *m* and *q*, although this would signify that component B is contained simultaneously in several components, which is a forbidden property in component-based systems.

The virtual component does not have any content, it is used only has a placeholder to represent a specific functionality that the testers would like to test, and to associate with the flow a list of test cases. This association can be done by using a BIT infrastructure, as the virtual component is just another component, after all. Each test case

is written as a unit-test for the virtual component, which can be easily handled since it is similar to the unit tests for the normal components, as shown in Figure 7. This is a very interesting effect of our approach: integration testing becomes in practice very similar to unit-testing. This means a large part of the infrastructure used to run unit-tests (associating test cases with components, executing the test cases, etc.) can be reused. Even more important, this signifies that the engineer in charge of the testing can apply his knowledge of the usual unit-testing to integration testing.

Effectively, the unit-tests are executed on the sequence of components encompassed by the virtual component. The execution takes place in the integrated system, resulting in the real components *B*, *C*, and *E* being tested. The computation to determine which component is part of the virtual component is not necessary for running the test cases, this is done implicitly by the flow of data going from the inputs until the expected outputs.

Typically, this approach allows defining a handful of complex interactions in the system which must be tested extensively. The test cases can sequences of events of arbitrary length, and the oracle can precisely verify the output data conform to the specifications. It is *complementary* to the previous method which generates random sequences of events.

After having reviewed techniques to handle the integration testing in publish/subscribe architectures, we will concentrate on the specific difficulties coming from runtime testing.

RUNTIME TESTING FOR COMPONENT-BASED PLATFORMS

Publish/subscribe systems are based on components, which facilitate the modification of just one part of the system. In addition, those components are loosely coupled, so it is technically simple to reconfigure the system at runtime. The reconfiguration itself is straightforward: the components are simply added to, or removed from, the pool of components running in the system. As the component connections are not explicit, but implicit in the event types that are published and subscribed to, there is no need to let the other components know about the modification. This is one of the reasons why publish/subscribe runtime architectures are readily used in systems with high availability requirements. Such systems cannot be stopped only because of a correction being introduced, a new feature being added, or an existing function being removed because the hardware supporting it is about to be removed. Therefore runtime reconfiguration plays a big role in the context of high availability.

However, the fact that a new configuration will work seamlessly, is not guaranteed a priori. Every re-configuration activity should follow the same verification and validation processes, in order to provide the same confidence in the new configuration as it was the case for the old system. Moreover, due to resource limitation, it is not always possible to duplicate the system for testing the new configuration while the previous one is used in production. This is a very likely situation with large-scale systems. To test the new configuration it is then possible to use runtime testing: execute test cases on some components while the whole production system keeps running normally.

Background

Some studies on the topic of *regression testing* can be related to this problem. The main question of this topic concerns the validation of a system after some modifications. There are alternatives to runtime testing. As one of our assumptions is that components can be black or grey boxes, because it is likely that the source code of some components will not be available, approaches that require access to the source code cannot be applied in our context.

Although model-based approaches can be a solution (Orso et al., 2007; Muccini, Dias, & Richardson, 2005), the complexity of the models of the components to be integrated, can often amount to intractable resulting combined models. A solution that can be used when a model is not available, or the available model is too complex, is to derive state models dynamically from usage traces (Mariani, Papagiannakis, & Pezze, 2007). As an advantage, these models will be smaller, as they will be restricted to the way the component has been used earlier in the system, thus, leaving out all the features of the component not relevant in the current context. There exist also some methods (Stuckenholz & Zwintzscher, 2004), that address this problem from a formal point of view, providing a way to measure system updates and finding conflict-free configurations.

Runtime Testability

As seen in Section *Properties of Publish/Subscribe Platforms*, in order to be able to implement runtime testing in a reliable fashion, the middleware needs to provide test-isolation and test-awareness. Let us have a deeper look at the difficulties coming along with runtime testing.

There are two orthogonal dimensions that lead to problems during runtime testing:

- **Component state.** If the component has states, or in other words its outputs depend not only on the current input but also on the previous ones, the testing data might interfere with the production data. For instance in a simple component providing as output a number corresponding to the sum of all the inputs it has received, a test case running concurrently to the normal usage will likely lead to wrong outputs both for the test case and for the normal usage.

- **Component side effects.** If a component's behaviour does not only involve its inputs and outputs, but also has side effects on the ``outside world'', it might be unsafe to perform testing at runtime. For instance, it is not possible to test at runtime a component driving the arm of a robot while the robot has to continue moving normally. As another example, the side effect could be the communication with a database in an external system. It would be a rather bad idea to execute test cases modifying the database of real bank accounts!

These two dimensions corresponds to the notion of *purity* in functional programming. A component which has either states or side effects is considered *test-sensitive*. Unfortunately, often components in real systems are test-sensitive. We could even go further and say that a system which has no component with side effect is useless, as it does not communicate with the outside world.

To overcome the limitations brought up by the test-sensitiveness of the components, the notion of test-awareness is introduced. Test-awareness allows being sure that a component can distinguish the events corresponding to the testing process from the events of the production process. In case of state, components can store a second state in order to handle the testing events separately. In case of side effect, the components' developers can choose either not to perform the function requested at all, or to use a simulation of the real world.

Still, not all components might be able to provide appropriate test-awareness functionality (e.g. because of lack of development time or difficulty of the implementation), or some functions might be considered too dangerous to try even if the test-awareness is implemented (because, after

all, some things might go wrong during the test). In order to allow the testing framework to handle those cases, a test-control interface is needed for each component, that gives the component the possibility to restrain the testing to only some test cases at runtime, and allows more when the testing is done off-line in a controlled environment.

From this information, the *runtime testability ratio* metric can be derived. It is the ratio between the percentage of the system being testable at development time versus the percentage of the system being testable at runtime. The actual value depends on the way the testing coverage is measured, such as code coverage, model coverage, or coverage of usage scenarios. Nevertheless the runtime testability measurement gives an indication on how much is lost on the acceptance quality when testing has to be done at runtime.

Avoiding Interferences Between Production and Testing Domains

In order to maximise the runtime testability ratio, one has to maximise the number of test cases that can be performed during runtime. This can be done through minimizing the interferences between the running system and the tests being executed concurrently during runtime.

One way of minimizing interference is to add a special flag to the testing data in order to be able to discriminate it from production data. We call this approach *tagging*. During testing, components subscribe to events with this specific flag, and *every* output event generated when handling this data must also be tagged with the test flag. The principal advantage of this method is that one component can receive production as well as testing data and process outputs accordingly. In case of testing, a different component state can be used, or some side effect actions omitted. Components which are test insensitive can be used unmodified with this scheme if subscription to the test data and setting the test flag on the outputs is managed automatically by the framework or via

Figure 8. Example of data isolation for runtime testing

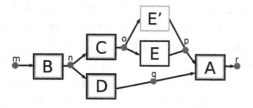

a simple wrapper. For the components with more complex functionality, it is not possible to verify that the output data depend exclusively on input data from the production world or exclusively on input data from the test domain. It is up to the component to separate the handling correctly. Nevertheless, the framework can do some sanity checking by verifying that the output data is from the same domain as the input data that triggered the function call.

Figure 8 shows the usage of tags for data isolation on a system for which the reconfiguration consist of updating the component *E* by *E'*. Components with full black lines process the production data. Components with dotted outlines process the testing data. Components which are not modified process data for both production and testing. Components which are going to be removed, only process the data from the old configuration, the production data. Components which will be added, only process data from the future configuration, the testing data.

In case data persistence exists in the publish/subscribe architecture, the testing could also be done by using the data left by previous components. This is persistent data that has the same type as the inputs expected for the testing, and which has been produced by components during runtime until the beginning of the validation process. As this data will potentially be read by the new components when they will be started in the production mode, starting them during testing with the same data available approximates their real behaviour better. So, before starting the test,

the persistent data which is used as input of one of the components under test should be copied along with the test tag.

Another approach consists of a technique to allow the component to be aware of the testing (Suliman et al., 2006; González, Piel, & Gross, 2008). Obviously, this awareness has to be used with parsimony because it adds more burden to the component developer to handle the various cases, and also with care because the component's behaviour, when under test, must differ as little as possible from the normal behaviour. The basic information sent to the component concerns the beginning of a test and its end. This enables the component to duplicate its state or to set up a simulator for its outputs. The provided information should also detail whether the component is going to be operating exclusively for the test, or whether it will also have to respond to production events. In particular, when updating a component, during the test the previous version of the component must only handle the production data, while the newest version must handle only the testing data.

It is important to mention that interference between production operation and testing can also occur at the non-functional properties level, i.e. there might be shortage of available processing power, memory usage, communication bandwidth, etc. Due to lack of space it is not possible to treat in detail this broad subject. Nevertheless, this is also an aspect that must be taken into account during runtime testing. The techniques such as Quality of Service or system health monitoring can be used in order to ensure that the test case execution never prevents the production operation to be executed normally and to schedule the testing at moments when the system's load is low.

Reducing the Amount of Test Cases

The testing process can be costly in terms of time and resources. When a modification of the system configuration is requested, it is best to reduce the delay to a minimum before it takes place. A typical modification affects only a very constrained part of the whole system, in terms of components as well as in terms of behaviour. For example, updating a component comprises only removing the component and replacing it by another component which has a similar interface and a close behaviour. Moreover, the previous configuration has passed the testing process. Therefore, it seems rather logical that not the whole testing process has to be repeated. Only the tests which might have a different outcome should be re-executed (and the newly introduced tests).

The general idea is to re-test only the parts which are affected by the changes. In integration testing, this means every interaction that involves a component which is affected by the changes.

When using the *random event sequence* technique, only sequences which involve an event affected by the modification have to be re-tested. An event (or data type) is affected when one of the subscribers has changed and, therefore, the reaction of the system to it might change. The fact that a producer has been modified must not necessarily be taken into account, as either the event is explicitly generated during the test using the test cases (so the modified component is not involved) or the event is generated through the modified component, in which case the sequence would be tested anyway, as the component has subscribed to the previous event.

With the validations based on *virtual components*, similarly, every virtual component containing a component affected by the modification will have to be revalidated, because the output data might be affected by the removed or added component. If a virtual component either contains a component that will be removed, or will contain a new component, it will have to be retested. It is important to mention that if the virtual component contains only components with dependencies on a modified component, it is not necessary to retest the data-flow as the interaction with the modified component would not be triggered anyway. Of course, new virtual components should be vali-

dated, and similarly, if a virtual component has new test cases, they should be executed before accepting the new configuration.

In case the architecture is hierarchical, in other words, there are components which are composed by a set of sub-component, the re-testing has to be done recursively. Each composite component containing a modified component has to be considered as modified as well. The testing technique is applied starting from the modified component, going through each parent component, until the topmost level is reached. Theoretically, there is no order in which the levels of hierarchy have to be re-tested, because if a test case can detect a wrong behaviour it will be eventually run, whichever is the execution order. Nevertheless, a problem at a low level of component hierarchy can lead to test cases failing at this level but also at all the higher levels. Therefore, starting from the lowest level (the level at which the modification took place) allows to pinpoint more directly to where the error comes from.

RUNTIME INTEGRATION TESTING: A CONTROLLED EXPERIMENT

In order to illustrate the methods outlined so far, as well as to summarize the whole process of validating the integration, we present the usage of those methods on a simplified sub-system of the MSS domain. This example is an experiment of reduced scale in which we apply and assess the key concepts of the presented methods. This scenario involves validating the integration of the assembled system, and keeping the quality at the same level after an update of the system has taken place.

AIS: Automatic Identification System

The Automatic Identification System (AIS) is a worldwide adopted standard used for vessel identification (International Telecommunication Union,

2001). Ships must broadcast over radio their static voyage data (cargo type, destination, etc.) as well as their dynamic positioning data with a variable report rate that depends on the ship's speed and rate of turn. These messages are received by other ships and by the coast authorities, who can use the data for traffic control, collision avoidance, and assistance. In order to provide a full coverage of the Dutch coastal waters, many AIS base stations are distributed along the coast of The Netherlands as well as in buoys at sea. The messages received by these stations are then relayed to the central coastguard centre, and displayed on the screen of the control room.

The amount of information received cannot be grasped by humans. At any given time, more than 800 AIS identified ships can be sailing in Dutch waters, sending signals every so many seconds. Therefore, automated processing is essential. One particular automatic task is the monitoring of the messages to identify ships with a malfunctioning AIS transponder, or ships whose captain has forgotten to correctly adjust the transmitted information. The detected errors are shown as warnings to the operator next to the ship's icon on the display to indicate that the information is uncertain. Moreover, because AIS messages are broadcasted and can be received by many different base stations in range at the same time, multiple occurrences of the same message will be received in the central data centre with some seconds of delay in between, requiring some initial filtering and cleaning before the data is displayed.

The Initial System

The global organisation of the system is shown in Figure 9. The *World* component generates the same data as boats would in reality. It would obviously not be present in a real system, but would be replaced by the ships' AIS emitters. It actually replays a record of the AIS messages received by the Dutch coastguards during a week of normal operation. The *LS* (Local Station) components simulate the individual AIS receivers by transmitting only messages happening in the coverage zone specified as a parameter. The *Merger* component receives the data from the several *LS* components and removes the duplicated messages (which happen when a boat is in a zone covered by several receivers). The *Tracker* component converts the messages into a database containing the information about each ship. This information can be accessed via a specific protocol based on receiving commands and answering with tracks. The *Monitor* component passively observes all the data and detects inconsistencies in the information sent by the ships. In case a problem is detected, a warning is generated. The *Plotter* component displays on a map the ships, their associated information, and the warnings. The publish/subscribe framework (OpenSplice[3]) is based on the DDS standard. The components are implemented in Java. They are also described using a specific Architecture Description Language (ADL), in order to define explicitly the inputs and outputs, and to associate them with test cases.

First Integration Acceptance

Before the system is started, and after every component has passed the unit-testing phase, the validation of the integration of the components must take place. During this validation the *World* component is not executed. Virtual components can be used to test some specific interactions. For instance a virtual component can be set between uniq_ais and warning to verify that *Monitor* and *Tracker* are compatible at the protocol level. A sequence of uniq_ais is sent at a high frequency (which is not authorised in the AIS protocol). The oracle verifies that the correct warning is generated.

It is also possible to test the core of the flow: *Merger* followed by *Tracker*. To do so, a virtual component is set up with ais and command as

Figure 9. Global view of the AIS processing system

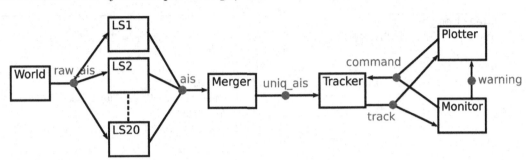

inputs, and track as output. Several test cases can be associated. The first test case has as input a series of identical messages and a command requesting the information for the ship specified in the messages. The oracle checks that only one data is returned. A second test case has as input several different messages, corresponding to several ships. The command input is more complex than in the first test case: it requests information about several ships, including several times the same ship and a non-present ship. The oracle validates whether the information in the track is adequate with respect to the requests.

As an example, Figure 10 is an abridged version of the ADL file that can be used with our framework for specifying the second virtual component. The three interface definitions specify the inputs and outputs of the component. The test definition corresponds to one of the BIT interfaces, and associates test cases with the component. The provider JUnitProviderFlow indicates a class used to run this kind of test (it is generic to all unit-tests for virtual components). The definition argument indicates a JUnit[4] class in charge of the test case. As we can see, the notion of virtual component conveniently allows us to reuse the usual infrastructure for unit-testing.

An extract of the unit-test TestDuplicate referenced in the ADL file is presented in Figure 11. The Java class extends the helper class JUnitFlowTest which automatically sets up the component in the system with the topics specified in the ADL file.

The setUp method is executed at the beginning of the test case to complete the initialisation. The method noDuplicateMessageInRow corresponds to one test case. It emits 10 times a identical AIS message, and reads the tracks received during one second. The reception of the tracks is handled by implementing the TrackListener interface (the methods corresponding to this interface are not shown here for brevity).

Update of the System

Once the system is running, comes a moment when it is reconfigured. On the system, we update the *Monitor* component in order to detect more type of inconsistencies in the AIS data sent. To do so, within the "acceptance perspective", the *Monitor* component is removed, and a *Monitor2* component is added, with the same inputs and outputs. Before the new version of the system can be transferred to the "production perspective", it must pass the validation phase. All the virtual components which are affected by the modification are tested. Here, only the first virtual component, between uniq_ais and warning, is affected.

The *Tracker* component is instructed to use both the testing data and normal data via its testing interface. The *Monitor2* component is instantiated, instructed to use testing data only, and started. The other components, such as the *Monitor* or the *Plotter* do not receive any specific instructions, so they continue running normally,

Figure 10. Definition of a virtual component with two test cases in an ADL file

```
<virtual-composite name="flowCore">
  <interface name="AISIn" role="subscribe" signature="ais" />
  <interface name="CmdIn" role="subscribe" signature="command" />
  <interface name="TrackOut" role="publish" signature="track" />
  <test provider="JUnitProviderFlow" name="Dup"
                             definition="TestDuplicate" />
  <test provider="JUnitProviderFlow" name="MS"
                             definition="TestMultiShip" />
</virtual-composite>
```

Figure 11. Definition of a test case for validating the flow

```
public class TestDuplicate extends JUnitFlowTest
                       implements TrackListener {
  private GenericDataWriter<AM_ais_msgDataWriter> aisW;
  ...
  @Before
  public void setUp() {
    aisW = writers.get("ais");
    readers.get("track").bind(this);
    ...
  }

  @Test
  public void noDuplicateMessageInRow() {
    AISMessage[] result;
    AISMessage in = new AISMessage("11OGQ0?0EpeVNE2:Hjakf0");

    for (int i=0; i < 10; i++) {
 int ret = aisW.w.write(in, DDS.HANDLE_NIL.value);
 Assert.assertTrue("Error while writing message.",
                       ret == RETCODE_OK.value);
    }

    result = getReceivedTracks(1.0f);
    Assert.assertTrue("Duplicated message not detected.",
           warnings.length == 1);
  }
  ...
}
```

using only the production data. The unit-test component is then executed. The input data is generated with the "test" tag (corresponding to a *domain* in OpenSplice), and similarly the component subscribes to the test data only for receiving the output of the data-flow. If one test case fails, an error is displayed to the system integrator, who will have to provide a new version of the component which passes the test. Once the test case is passed, the transfer to production can be executed: the new components are restarted, and the deleted components are stopped.

FUTURE TRENDS

In this chapter we have presented approaches to validate the integration of publish/subscribe systems in an initial configuration and at runtime. One of the assumptions we have taken is that the test cases were either already available or would be provided by the system integrator. A trend in the testing area seems to emerge concerning *automatic test generation*. The goal is to reduce the burden of writing test cases manually by using high level information regarding the expected behaviour of the system (expressed as a model) and transforming it into test cases. Ideally, the behaviour can be captured in models which are useful also for the rest of the development process, such as the documentation in UML, or the implementation of a component with a Finite State Machine. This has been considered for instance in the works of Abdurazik and Offutt (2000), Chandler et al. (2005), Hartman et al. (2005). The development of model transformation techniques should help to go towards this direction.

As this chapter was focused on testing, the subject of *verification of the integration* has not been treated here. Complimentary to the validation, the verification is another effective way to improve the quality of the system. One of the major problems encountered in this domain is when dealing with complex systems the number of states explodes. This is especially noticeable when verifying the component interaction in publish/subscribe systems because of non-determinism due to the concurrent execution of components. The latest works in this domain, such as presented by Baresi et al. (2007) have focused on reducing this combinational explosion, so that it can become a more usable approach. Moreover, there has also recently been work to allow verification with runtime evolution (Deveaux & Collet, 2006).

Concerning the validation aspect, it might be worthy to explore the *synergy between runtime evolution and validation*. Based on approaches used for regression testing, we have seen that it is possible to reduce the amount of test cases to execute, because it is known that the running configuration has been validated. It might be possible to reduce the number even further by observing the running system and discarding test cases validating non-used data types, or non-used components. It could also be possible to leverage the fact that the two configurations are running simultaneously to compare the behaviour of the configuration under test against the validated configuration. The configuration under test would receive real inputs, but the outputs would be only compared instead of being really transmitted.

Finally, it is important to mention that all these discussed approaches will be possible to be applied only once a *user interface* provides to the system integrator the functionalities. Following the status of each test case (i.e. failed or passed) for every component and interaction between the components has to be straightforward. We have started work in this direction for a client-server framework (González, Piel & Gross, 2008). Defining virtual components by selecting inputs and outputs and associating test cases with it should also be an easy operation. It should also facilitate the runtime evolution of the system by allowing the user to view the current configuration and in parallel deriving it into another valid configuration.

SUMMARY AND CONCLUSION

A large number of event-based systems have high availability and high reliability requirements. Ensuring the high software quality of such system is a must, in its initial configuration and also whenever the configuration evolves. Validating each component individually is not sufficient for testing the whole system; it also requires validating the integration of those components together. In event-based systems the dependencies between the components are not explicit, and, when the processing model follows the typical data-flow model, components have no expectations on the behaviour of the other components. As components for event-based architectures are loosely coupled, this type of architecture is well adapted for runtime reconfiguration. Moreover, when the system evolves at runtime and if due to its high complexity it can not be duplicated, the testing must also take place at runtime. In this situation, the testing must not interfere with the normal operation.

In this chapter, we have presented two different and complementary methods to test the integration. One is based on the random generation of a high number of event sequences and on already available oracles, in order to find a state of system malfunctioning. The second one uses a limited number of predefined data-flows which must respect a precise behaviour. It introduces the notion of virtual components, which have the advantages over usual components that they can overlap each other, and the exact set of inner component automatically follows the system's evolution. The two presented methods rely on several known and proven techniques such as unit-testing or Built-In Testing.

The usage of Built-In Testing allows keeping with each component a particular set of test cases. Thanks to this, at any time during the life cycle of the system, the test cases of any component are available, even at runtime. In this chapter, the specific aspects of runtime testing have been presented. First, components require to be assessed by the developer in order to identify state-dependant behaviour and resources outside the component which should not be accessed during testing or cannot be shared. Test-sensitive components can then make use of a special test interface to be aware of the testing process. At reconfiguration time, before a modification is accepted, the parts of the system which might behave differently must be re-tested for integration using this set-up and the integration testing techniques previously introduced. This ensures the high quality of the system at any given time of its lifespan.

ACKNOWLEDGMENT

This work has been carried out as part of the Poseidon project under the responsibility of the Embedded Systems Institute. This project is partially supported by the Dutch Ministry of Economic Affairs under the BSIK03021 program.

REFERENCES

Abdurazik, A., & Offutt, J. (2000, October). Using UML collaboration diagrams for static checking and test generation. In A. Evans, S. Kent, & B. Selic (Eds.), *Proceedings of UML 2000 - the unifed modeling language. advancing the standard. third international conference*. York, UK, (Vol. 1939, pp. 383-395). London: Springer.

Baresi, L., Ghezzi, C., & Mottola, L. (2007). On accurate automatic verification of publish-subscribe architectures. In *ICSE '07: Proceedings of the 29th international conference on software engineering* (pp. 199–208). Washington, DC: IEEE Computer Society.

Brenner, D., Atkinson, C., Malaka, R., Merdes, M., Paech, B., & Suliman, D. (2007). Reducing verification effort in component-based software engineering through built-in testing. *Information Systems Frontiers*, *9*(2-3), 151–162. doi:10.1007/s10796-007-9029-4

Chandler, R., Lam, C. P., & Li, H. (2005). Ad2us: An automated approach to generating usage scenarios from uml activity diagrams. In *Apsec '05: Proceedings of the 12th Asia-Pacific software engineering conference* (apsec'05) (pp. 9-16). Washington, DC: IEEE Computer Society.

EU Commission. (2007, October). *An integrated maritime policy for the European Union*. Brussels, Belgium: European Commission, Maritime Affairs.

Deveaux, D., & Collet, P. (2006). *Specification of a contract based built-in test framework for fractal.*

Embedded Systems Institute. (2007). *The Poseidon project.* Retreived from http://www.esi.nl/poseidon

Eugster, P. T., Felber, P. A., Guerraoui, R., & Kermarrec, A.-M. (2003). The many faces of publish/subscribe. *ACM Computing Surveys, 35*(2), 114–131. doi:10.1145/857076.857078

Gao, J. Z., Tsao, H. J., & Wu, Y. (2003). *Testing and quality assurance for component-based software.* Norwood, MA: Artech House.

Garlan, D., Khersonsky, S., & Kim, J. S. (2003, May). Model checking publish-subscribe systems. In *Proceedings of the International spin workshop on model checking of software* (pp. 166–180). Portland, OR.

González, A., Piel, É., & Gross, H.-G. (2008, September). Architecture support for runtime integration and verification of component-based systems of systems. In *Proceedings of the 1st international workshop on automated engineering of autonomous and run-time evolving systems* (ARAMIS 2008). L'Aquila, Italy: IEEE Computer Society.

Gross, H.-G. (2005). *Component-based software testing with UML.* Heidelberg, Germany: Springer.

Gross, H.-G., & Mayer, N. (2004). Built-in contract testing in component integration testing. *Electronic Notes in Theoretical Computer Science, 82*(6), 22–32. doi:10.1016/S1571-0661(04)81022-3

Hartmann, J., Vieira, M., Foster, H., & Ruder, A. (2005). A UML-based approach to system testing. *Innovations in Systems and Software Engineering, 1*(1), 12–24. doi:10.1007/s11334-005-0006-0

Holzmann, G. J., & Smith, M. H. (1999). A practical method for verifying event-driven software. In *ICSE '99: Proceedings of the 21st international conference on software engineering* (pp. 597-607). Los Alamitos, CA: IEEE Computer Society Press.

International Telecommunication Union. (2001). *Recommendation ITU-R M.1371-1.* Geneva, Switzerland: ITU.

Lockheed Martin. (2008). *Maritime safety, security & surveillance integrated systems for monitoring ports, waterways and coastlines.* Retrieved from http://www.lockheedmartin.com

Mariani, L., Papagiannakis, S., & Pezze, M. (2007). Compatibility and regression testing of COTS-Component-Based software. In *ICSE '07: Proceedings of the 29th international conference on software engineering* (pp. 85-95). Washington, DC: IEEE Computer Society.

Memon, A., Banerjee, I., & Nagarajan, A. (2003, November). GUI ripping: reverse engineering of graphical user interfaces for testing. In *Proceedings of the 10th working conference on reverse engineering* (pp. 260-269). Piscataway, NJ.

Michlmayr, A., Fenkam, P., & Dustdar, S. (2006a, September). Architecting a testing framework for publish/subscribe applications. In *Proceedings of the 30th annual international computer software and applications conference* (compsac'06), (Vol. 1, pp. 467-474).

Michlmayr, A., Fenkam, P., & Dustdar, S. (2006b, July). Specification based unit testing of publish/subscribe applications. In *Proceedings of the 26th IEEE international conference on distributed computing systems workshops* (ICDCSW '06), (p. 34).

Muccini, H., Dias, M., & Richardson, D. J. (2005). Reasoning about software architecture-based regression testing through a case study. In *Proceedings of the 29th annual international computer software and applications conference* (Vol. 2, pp. 189-195). Washington, DC: IEEE Computer Society.

Muhl, G., Fiege, L., & Pietzuch, P. (2006). *Distributed event-based systems*. Berlin, Germany: Springer.

Object Management Group. (2007). *Data distribution service for real-time systems, v1.2*. Needham, MA: OMG.

Oreizy, P., Medvidovic, N., & Taylor, R. N. (1998). Architecture-based runtime software evolution. In *ICSE '98: Proceedings of the 20th international conference on software engineering* (pp. 177-186). Washington, DC: IEEE Computer Society.

Orso, A., Do, H., Rothermel, G., Harrold, M. J., & Rosenblum, D. S. (2007). Using component metadata to regression test component-based software. *Software Testing. Verification and Reliability, 17*(2), 61–94. doi:10.1002/stvr.344

Schieferdecker, I. (2007, February). *The Testing and Test Control Notation version 3*; Core language (Tech. Rep). Sophia-Antipolis Cedex, France: European Telecommunications Standards Institute.

Stuckenholz, A., & Zwintzscher, O. (2004). Compatible component upgrades through smart component swapping. In Reussner, R. H., Stafford, J. A., & Szyperski, C. A. (Eds.), *Architecting systems with trustworthy components* (*Vol. 3938*, pp. 216–226). Berlin, Germany: Springer. doi:10.1007/11786160_12

Suliman, D., Paech, B., Borner, L., Atkinson, C., Brenner, D., Merdes, M., et al. (2006, September). The MORABIT approach to runtime component testing. In *Proceedings of the 30th annual international computer software and applications conference* (Vol. 2, pp. 171-176).

Szyperski, C. (1998). *Component software: Beyond object-oriented programming*. New York: ACM Press.

Thales Group. (2007). *Maritime safety and security*. Retrieved from http://shield.thalesgroup.com/offering/port_maritime.php

Vincent, J., King, G., Lay, P., & Kinghorn, J. (2002). Principles of Built-In-Test for Run-Time-Testability in component-based software systems. *Software Quality Journal, 10*(2), 115–133. doi:10.1023/A:1020571806877

Yuan, X., & Memon, A. M. (2008, August). Alternating GUI test generation and execution. In Testing: Academic and industry conference - practice and research techniques (taic part'08) (pp. 23{32). Windsor, UK: IEEE Computer Society.

Zhang, H., Bradbury, J. S., Cordy, J. R., & Dingel, J. (2006). Using source transformation to test and model check implicit-invocation systems. Science of Computer Programming, *62*(3), 209 - 227. (Special issue on Source code analysis and manipulation (SCAM 2005))

KEY TERMS AND DEFINITIONS

Component: An abstract and unique compositional unit of high cohesion and low coupling with contractually specified interfaces for external communication and an execution context.

Call-Reply: The model of event sequence when the communication between two components is based on a request and a response.

Data-Flow: The model of event sequence when the communication between two components is based on the data transmission only in one direction.

Validation: Development phase during which one ensures that an implementation does behave as defined in the requirements.

Verification: Development phase during which one ensures that a model does conform to all the restrictions defined in the models of higher level of abstraction.

Integration Testing: Testing of the behaviour of a set of components assembled in their final configuration.

Test Sensitivity: The property of a component that indicates to which extend the component can be tested without unwanted side-effects.

Built-In Testing: Testing technique used for equipping components with the ability to check their execution environment, and their ability to be checked by their execution environment.

ENDNOTES

[1]   Pure is used here with the same meaning as in functional programming: no side effects are created as a result of an operation.

[2]   Because each input event could be repeated any number of times.

[3]   http://www.prismtechnologies.com/opensplice-dds

[4]   http://www.junit.org

164

Chapter 8
The PADRES Publish/ Subscribe System

Hans-Arno Jacobsen
University of Toronto, Canada

Alex Cheung
University of Toronto, Canada

Guoli Li
University of Toronto, Canada

Balasubramaneyam Maniymaran
University of Toronto, Canada

Vinod Muthusamy
University of Toronto, Canada

Reza Sherafat Kazemzadeh
University of Toronto, Canada

ABSTRACT

This chapter introduces PADRES, the publish/subscribe model with the capability to correlate events, uniformly access data produced in the past and future, balance the traffic load among brokers, and handle network failures. The new model can filter, aggregate, correlate and project any combination of historic and future data. A flexible architecture is proposed consisting of distributed and replicated data repositories that can be provisioned in ways to tradeoff availability, storage overhead, query overhead, query delay, load distribution, parallelism, redundancy and locality. This chapter gives a detailed overview of the PADRES content-based publish/subscribe system. Several applications are presented in detail that can benefit from the content-based nature of the publish/subscribe paradigm and take advantage of its scalability and robustness features. A list of example applications are discussed that can benefit from the content-based nature of publish/subscribe paradigm and take advantage of its scalability and robustness features.

DOI: 10.4018/978-1-60566-697-6.ch008

INTRODUCTION

The publish/subscribe paradigm provides a simple and effective method for disseminating data while maintaining a clean decoupling of data sources and sinks (Cugola, 2001; Fabret, 2001; Castro, 2002;Fiege, 2002; Carzaniga, 2003; Eugster, 2003; Li, 2005; Ostrowski, 2006; Rose, 2007). This decoupling can enable the design of large, distributed, and loosely coupled systems that interoperate through simple publish and subscribe invocations. While there are many applications such as information dissemination (Liu, 2004; Nayate, 2004; Liu, 2005) based on group communication (Birman, 1999) and topic-based publish/subscribe protocols (Castro, 2002; Ostrowski, 2006), a large variety of emerging applications benefit from the expressiveness, filtering, distributed event correlation, and complex event processing capabilities of *content-based publish/subscribe systems*. These applications include RSS feed filtering (Rose, 2007), stock-market monitoring engines (Tock, 2005), system and network management and monitoring (Mukherjee, 1994; Fawcett, 1999), algorithmic trading with complex event processing (Keonig, 2007), business process management and execution (Schuler, 2001; Andrews, 2003;), business activity monitoring (Fawcett, 1999), workflow management (Cugola, 2001), and service discovery (Hu, 2008).

Typically, a distributed content-based publish/subscribe systems is built as an application-level overlay of content-based publish/subscribe brokers, with publishing data sources and subscribing data sinks connecting to the broker overlay as clients. In a content-based publish/subscribe system, message routing decisions are not based on destination IP-addresses but on the content of messages and the locations of data sinks that have expressed an interest in that content.

To make the publish/subscribe paradigm a viable solution for the above applications, additional features must be added. This includes support for *composite subscriptions* to model and detect com-posite events, and to enable event correlation and in-network event filtering to reduce the amount of data transferred across the network.

Furthermore, the publish/subscribe substrate that carries and delivers messages must be robust against non-uniform workloads, node failures, and network congestions. In PADRES[1], robustness is achieved by supporting alternate message routing paths, load balancing techniques to distribute load, and fault resilience techniques to react to broker failures.

It is also essential for a publish/subscribe system to provide tools to perform monitoring, deployment, and management tasks. Monitoring is required throughout the system to oversee the actual message routing, the operation of content-based brokers, and the interaction of applications via the publish/subscribe substrate. Deployment support is required to bring up large broker federations, orchestrate composite applications, support composition of services and business processes, and to conduct controlled experiments. Management support is required to inspect and control live brokers.

This chapter presents the PADRES content-based publish/subscribe system developed by the *Middleware Systems Research Group* at the University of Toronto. The PADRES system incorporates many unique features that address the above concerns and thereby enable a broad class of applications. The remainder of this chapter beings with a description of the PADRES language model, network architecture and routing protocol in Section 1. This is followed by an outline of the PADRES load balancing capabilities whereby the system can automatically relocate subscribers in order to avoid processing or routing hotspots among the network of brokers. The section then addresses failure resilience describing how the PADRES routing protocols are able to guarantee message delivery despite a configurable number of concurrent crash-stop node failures. Some of the PADRES distributed management features are presented, including topology monitoring and deployment tools. Next, the section discusses a wide variety of applications and illustrates how the features of the PADRES system enable or support the development of these applications. Finally, a survey of related publish/subscribe projects and the contributions of the PADRES project are presented, followed by some concluding remarks.

MESSAGE ROUTING

All interactions in the PADRES distributed content-based publish/subscribe system are performed by routing four messages: advertisements, subscriptions, publications, and notifications. This section outlines the format of each of these messages, then describes how these messages are routed in the PADRES network.

Language Model

The PADRES language model is based on the traditional [attribute, operator, value] predicates used in several other content-based publish/sub-scribes systems (Opyrchal, 2000; Carzaniga, 2001; Cugola, 2001; Fabret, 2001; Mühl, 2002; Bittner, 2007). In PADRES, each message consists of a message header and a message body. The header includes a unique message identifier, the message type (publication, advertisement, subscription, or notification), the last and next hops of the message, and a timestamp that records when the message was generated. The content and formats of each message type are detailed below.

Publications

Data producers, or *publishers*, encapsulate their data in *publication* messages which consist of a comma separated set of [attribute, value] pairs. Each publication message includes a mandatory tuple describing the class of the message. The class attribute provides a guaranteed selective predicate for matching, similar to the topic in topic-based

publish/subscribe systems.[2] A publication that conveys information about a stock listing may look as follows:

```
P: [class, 'STOCK'],
   [symbol, 'YHOO'],
   [open, 25.2], [high, 43.0], [low,
   24.5], [close, 33.0], [volume,
   170300],
   [date, '12-Apr-96']
```

A publication is allowed to traverse the system only if there are data sinks, or *subscribers*, who are interested in the data. Subscribers indicate their interest using subscription messages which are detailed below. If there are no interested subscribers, the publication is dropped. A publication may also contain an optional *payload*, which is a blob of binary data. The payload is delivered to subscribers, but cannot be referenced in a subscription constraint.

Advertisements

Before a publisher can issue publications, it must supply a template that specifies constraints on the publications it will produce. These templates are expressed via *advertisement* messages. In a sense, an advertisement is analogous to a database schema or a programming language type, and can specify the type and ranges for each attribute as shown in the following example:

```
A: [class, eq³, 'STOCK'],
   [symbol, isPresent, @STRING],
   [open, >, 0.0],
   [high, >, 0.0],
   [low, >, 0.0],
   [close, >, 0.0],
   [volume, >, 0],
   [date, isPresent, @DATE]
```

The above advertisement indicates that the publisher will publish only STOCK data with any symbol. The *isPresent* operator allows an attribute to have any value in the domain of the specified type.

An advertisement is said to *induce* publications: the attribute set of an induced publication is a subset of attributes defined in the associated advertisement, and the values of each attribute in an induced publication must satisfy the predicate constraint defined in the advertisement. Note that a publisher may only issue publications that are induced by an advertisement it has sent. Two possible publications P1 and P2 induced by the above advertisement are listed below, while P3 is not induced by the advertisement due to the extra attribute company.

```
P1:  [class, 'STOCK'],
     [symbol, 'YHOO'],
     [open, 25.25],
     [high, 43.00],[low, 24.50]
P2:  [class, 'STOCK'],
     [symbol, 'IBM'],
     [open, 45.25]
P3:  [class, 'STOCK'],
     [symbol, 'IBM'],
     [company, 'IBM']
```

Subscriptions

Subscribers express their interests in receiving publication messages by issuing *subscriptions* which specify predicate constraints on matching publications. PADRES not only allows subscribers to subscribe to individual publications, but also allows correlations or joins across multiple publications. Subscriptions are classified into *atomic* and *composite subscriptions*.

An atomic subscription is a conjunction of predicates. For example, below is a subscription for Yahoo stock quotes.

```
S:  [class, eq, 'STOCK'],
    [symbol, eq, 'YHOO'],
```

```
[open, isPresent, @FLOAT]
```

The commas between predicates indicate the conjunction relation. Similar to publications, each subscription message has a mandatory predicate specifying the class of the message, with the remaining predicates specyfing constraints on other attributes.

A publication is said to *match* a subscription, if all predicates in the subscription are satisfied by some [attribute, value] pair in the publication. For instance, the above subscription is matched by publications of all YHOO stock quotes with an open value. A subscription is said to *cover* another subscription, if and only if any publication that matches the latter also matches the former. That is, the set of publications matching the covering subscription is a superset of those matching the covered subscription.

Composite subscriptions consist of atomic subscriptions linked by logical or temporal operators, and can be used to express interest in composite events. A composite subscription is matched only after all component atomic subscriptions are satisfied. For example, the following subscription detects when Yahoo's stock opens at less than 22, and Microsoft's at greater than 31. Parenthesis are used to specify the priority of operators.

```
CS:  ([class, eq, 'STOCK'],
      [symbol, eq, 'YHOO'],
      [open, <, 22.0]) &&
     ([class, eq, 'STOCK'],
      [symbol, eq, 'MSFT'],
      [open, >, 31.0])
```

Moreover, unlike the traditional publish/subscribe model, PADRES can deliver not only those publications produced after a subscription has been issued, but also those published before a subscription was issued. That is, PADRES realizes a publish/subscribe model to query both the future and the past (Li, 2007; Li, 2008). In this model, data from the past can be correlated with data from the future. Composite subscriptions that allow correlations across publications continue to work with future data, and also with any combination of historic and future data. In that sense, subscriptions can be classified into *future subscriptions*, *historical subscriptions* and *hybrids* of the two.

For example, the following subscription is satisfied if during the period Aug. 12 to Aug. 24, 2008, MSFT's opening price was lower than the current YHOO opening price. The variable $X correlates the opening price in the two stock quotes. This is an example of a hybrid subscription.

```
CS:  ([class, eq, 'STOCK'],
      [symbol, eq, 'YHOO'],
      [open, eq, $X] &&
      [class, eq, 'STOCK'],
      [symbol, eq, 'MSFT'],
      [open, >, $X],
      [_start_time, eq,
       '12-Aug-08'],
      [_end_time, eq,
       '24-Aug-08'])
```

PADRES also provides an SQL-like language called PSQL (PADRES SQL) (Li, 2008), which has the same expressiveness as described above and allows users to uniformly access data produced in the past and future. The PSQL language supports the ability to specify the notification semantic, and it can filter, aggregate, correlate, and project any combination of historic and future data as described below.

In PSQL, subscribers issue SQL-like SELECT statements to query both historic and future publications. Within a SELECT statement, the SELECT clause specifies the set of attributes or aggregation functions to include in the notifications of matching publications, the WHERE clause indicates the predicate constraints to apply to matching publications, and the optional FROM and HAVING clauses help express joins and aggregations.

```
SELECT  [ attr | function ],...
        [FROM src,...]
WHERE   attr op val,...
        [HAVING function,...]
```

The above composite subscription is translated as follows in PSQL.

```
SELECT  src1.class, src1.symbol,
        src1.open, src2.symbol, src2.
        open
FROM    src1, src2
WHERE   src1.class eq 'STOCK',
        src2.class eq 'STOCK',
        src1.symbol eq 'YHOO',
        src2.symbol eq 'MSFT',
        src1.open < src2.open,
        src2.start_time eq '12-
        Aug-08',
        src2.end_time eq '24-Aug-
         08'
```

Notice that the reserved start_time and end_time attributes can be used to express time constraints in order to query for publications from the past, the future, or both. The sources in the FROM clause specify that two different publications are required to satisfy this query, and are subsequently used to qualify the WHERE constraints. The two publications may come from different publishers and conform to different schema (i.e., advertisements).

The HAVING clause is used to specify constraints across a set of matching publications. The functions $AVG(a_i, N)$, $MAX(a_i, N)$, and $MIN(a_i, N)$ compute the relevant aggregation across attribute a_i in a window of N matching publications. The window may either slide over matching publications or be reset when the HAVING constraints are satisfied. The following subscription returns all publications about YHOO stock quotes in a window of 10 publications whose average price exceeds $20.

Figure 1. Broker network

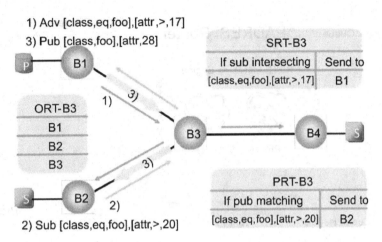

SELECT   class, symbol, price
WHERE   class eq 'STOCK',
         symbol eq 'YHOO'
HAVING   AVG(price, 10) > 20.00

For more information about PSQL, please refer to the technique report (Li, 2008).

Notifications

When a publication matches a subscription at a broker, a *notification* message is generated and further forwarded into the broker network until delivered to subscribers. Notification semantics do not constrain notification results, but transform them. Recall that notifications may include a subset of attributes in matching publications indicated in the SELECT clause in PSQL. Most existing publish/subscribe systems use matching publication messages as notifications whereas PSQL supports projections and aggregations over matching publications. This simplifies the notifications delivered to subscribers and reduces overhead by eliminating unnecessary information.

Broker Network and Broker Architecture

Figure 1 shows a deployed PADRES system consists of a set of brokers connected in an overlay which forms the basis for message routing. Each PADRES broker acts as a content-based router that matches and routes publish/subscribe messages. A broker is only aware of its neighbors (those located within one hop), which information it stores in its *Overlay Routing Tables* (ORT). Clients connect to brokers using various binding interfaces such as Java Remote Method Invocation (RMI) and Java Messaging Service (JMS).

Publishers and subscribers are clients to the overlay. A publisher client must first issue an advertisement before it publishes, and the advertisement is flooded to all brokers in the overlay network. These advertisements are stored at each broker in a Subscription Routing Table (SRT) which is essentially a list of [advertisement, last hop] tuples.

A subscriber may subscribe at any time, and subscriptions are routed based on the information in the SRT. If a subscription intersects an advertisement in the SRT, it is forwarded to the last hop broker the advertisement came from. A subscription is routed hop-by-hop in this way until it reaches the publisher who sent the matching

Figure 2. Router architecture

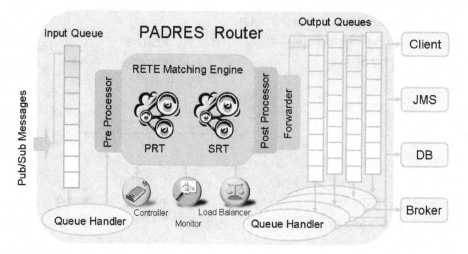

advertisement. Subscriptions are used to construct the Publication Routing Table (PRT). Similar to the SRT, the PRT is a list of [subscription, last hop] tuples, and is used to route publications.

If a publication matches a subscription in the PRT, it is forwarded to the last hop broker of that subscription until it reaches the subscriber that sent the subscription. Figure 1 shows an example PADRES overlay and the SRT and the PRT at one of the brokers. In the figure, in Step 1 an advertisement is published at broker B1. A matching subscription enters through broker B2 in Step 2 and since the subscription overlaps the advertisement at broker B3, it is sent to broker B1. In Step 3 a publication is routed to broker B2 along the path established by the subscription.

Each broker consists of an input queue, a router, and a set of output queues, as shown in Figure 2. A message first goes into the input queue. The router takes the message from the input queue, matches it against existing messages according to the message type, and puts it in the proper output queue(s) which refer to different destination(s). Other components provide other advanced features. For example, the *controller* provides an interface for a system administrator to manipulate a broker (e.g., to shut it down, or to inject a message into it); the *monitor* maintains

statistical information about the broker (e.g., the incoming message rate, the average queueing time and the matching time); the *load balancer* triggers offload algorithms to balance the traffic among brokers when a broker becomes overloaded (e.g., the incoming message rate exceeds a certain threshold); and the *failure detector* triggers the fault-tolerance procedure when a failure is detected in order to reconstruct new forwarding paths for messages and ensure timely delivery of publications in the presence of failures.

PADRES brokers use an efficient Rete-based pattern matching algorithm (Forgy, 1982) to perform publish/subscribe content-based matching. Subscriptions are organized in a Rete network as shown in Figure 3. Each rectangle node in the Rete network corresponds to a predicate and carries out simple conditional tests to match attributes against constant values. Each oval node performs a join between different atomic subscriptions and thus corresponds to composite subscriptions. These oval nodes maintain the partial matching states for composite subscriptions. A path from the root node to a terminal node (a double-lined rectangle) represents a subscription. The Rete matching engine performs efficient content-based matching by reducing or eliminating certain types of redundancy through the use of node sharing.

Figure 3. Rete network

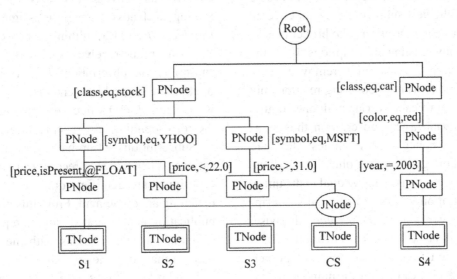

Partial matching states stored in the join nodes allow the matching engine to avoid a complete re-evaluation of all atomic subscriptions each time new publications are inserted into the matching engine. Experiments show that it takes only 4.5 ms to match a publication against 200,000 subscriptions which is nearly 20 times faster than the predicate counting algorithm (Ashayer, 2002). Moreover, the detection time does not increase with the number of subscriptions, but is affected by the number of matched publications. That is, the more publications that match a subscription, the longer it takes the matching engine to process the subscription. This indicates that the Rete approach is suitable for large-scale publish/subscribe systems and can process a large number of publication and subscription messages efficiently. Also, the Rete-based matching engine naturally supports composite subscription evaluation.

Content-Based Routing Protocols

Instead of address-based routing, PADRES uses *content-based routing*, where a publication is routed towards the interested subscribers without knowing where subscribers are and how many of them exist. The content-based address of a subscriber is the set of subscriptions issued by the subscriber. This provides a decoupling between the publishers and subscribers.

PADRES provides many content-based routing optimizations to improve efficiency and robustness of message delivery, including covering-based routing, adaptive content-based routing in cyclic overlays, and routing protocols for composite subscriptions.

Covering and Merging Based Routing

In content-based publish/subscribe systems, subscribers may issue similar subscriptions. The goal of covering-based routing is to guarantee a compact routing table without information loss, thereby avoiding the propagation of redundant messages, and reducing the size of the routing tables and improving the performance of the matching algorithm.

When a broker receives a new subscription from a neighbor, it performs the following steps to determine how to forward it. First, it searches the routing table to determine if the subscription is covered by some existing subscription from the same neighbor. If it is, the new subscription can be safely removed without inserting it into the routing table and, of course, without forwarding it further. If the new subscription is not covered by any existing subscriptions, the broker checks if it covers any existing subscriptions. If so, the covered subscriptions should be removed.

Subscriptions with no covering relations but which have significant overlap with one another can be *merged* into a new subscription, thus creating even more concise routing tables. There are two kinds of mergers: if the publication set of the merged subscription is exactly equal to the union of the publication sets of the original subscriptions, the merger is said to be *perfect*; otherwise, if the merged subscription's publication set is a superset of the union, it is an *imperfect* merger. Imperfect merging can reduce the number of subscriptions but may allow false positives, that is, publications that match the merged subscription but not any of the original subscriptions. These false positives are eventually filtered out in the network, and subscribers will not receive any false positives, but they do contribute to increased message propagations. However, by selectively and strategically employing subscription merging the matching efficiency of the publish/subscribe system can be further improved. For additional information, please refer to (Li, 2005; Li, 2008)

Adaptive Content-Based Routing for General Overlays

The standard content-based routing protocol is based on an acyclic broker overlay network. With only one path between any pair of brokers or clients, content-based routing is greatly simplified. However, an acyclic overlay offers limited flexibility to accommodate changing network conditions, is not robust with respect to broker failures, and introduces complexities for supporting other protocols such as failure recovery.

We propose a TID-based content-based routing protocol (Li, 2008) for cyclic overlays to eliminate the above limitations. In the TID-based routing, each advertisement is assigned a unique *tree identifier* (TID) within the broker network. When a broker receives a subscription from a subscriber, the subscription is bound with the TIDs of its matching advertisements. A subscription with a bound TID value only propagates along the corresponding advertisement tree.

Subscriptions set up paths for routing publications. When a broker receives a publication, it is assigned an identifier equal to the TID of its matching advertisement. From this point on, the publication is propagated along the paths set up by matching subscriptions with the same TID without matching the content of the publication at each broker. This is referred to as *fixed publication routing*.

Alternative paths for publication routing are maintained in PRTs as subscription routing paths with different TIDs and destinations. More alternate paths are available if publishers' advertisement spaces overlap or subscribers are interested in similar publications, which is often the case for many applications with long-tailed workloads. Our approach takes advantage of this and uses multiple paths available at the subscription level. Our *dynamic publication routing* (DPR) algorithm takes advantages of these alternate paths by balancing publication traffic among them, and providing more robust message delivery.

We observe in our experiments that an increase in the publication rate causes the fixed routing approach to suffer worse notification delays. For instance, in Figure 4, when the publication rate is increased to 2400 msg/min, the fixed algorithm becomes overloaded with messages queueing up at brokers along the routing path, whereas the dynamic routing algorithm continues to operate by offloading the high workload across alternate paths. The results suggest that dynamic routing is more stable and capable of handling heavier workloads, especially in a well connected network.

Figure 4. Higher publication rate

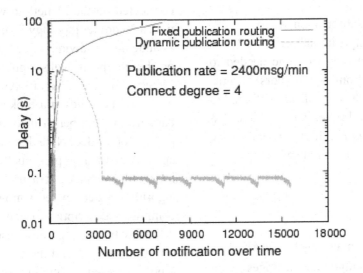

Composite Subscription Routing

Composite events are detected by the broker network in a distributed manner. In *topology-based composite subscription routing* (Li, 2005), a composite subscription is routed as a unit towards potential publishers until it reaches a broker *B* at which the potential data sources are located in different directions in the overlay network. The composite subscription is then split at broker *B*, which is called the *join point broker*. Each component subscription is routed to potential publishers separately. Later, matching publications are routed back to the join point broker for it to detect the composite event. Notice that topology-based routing assumes an acyclic overlay and does not consider dynamic network conditions.

In a general (cyclic) broker overlay, multiple paths exist between subscribers and publishers, and topology-based composite subscription routing does not necessarily result in the most efficient use of network resources. For example, composite event detection would be less costly if the detection is close to publishers with a higher publishing rate, and in a cyclic overlay, more alternative locations for composite event detection may be available.

The overall savings are significant if the imbalance in detecting composite events at different locations is large. PADRES includes a *dynamic composite subscription routing* (DCSR) algorithm (Li, 2008) that selects optimal join point brokers to minimize the network traffic and matching delay while correctly detecting composite events in a cyclic broker overlay. The *DCSR* algorithm determines how a composite subscription should be split and routed based on the cost model discussed below.

A broker routing a composite subscription makes local optimal decisions based on the knowledge available to itself and its neighbors. The cost function captures the use of resources such as memory, CPU, and communication. Suppose a composite subscription *CS* is split at broker *B*. The *total routing cost* (TRC) of *CS* is:

$$TRC_B(CS) = RC_B(CS) + \sum_{i=1}^{n} RC_{B_{N_i}}(CS_{B_{N_i}})$$

and includes the *routing cost* of *CS* at broker *B*, denoted as $RC_b(CS)$, and those neighbors where publications contributing to *CS* may come from, denoted as $RC_{B_{N_i}}(CS_{B_{N_i}})$. $CS_{B_{N_i}}$ denotes the part of *CS* routed to broker B_{N_i}, and may be an atomic or composite subscription.

The cost of a composite subscription *CS* at a broker includes not only the time needed to match publications (from *n* neighbors) against *CS*, but also the time these publications spend in the input queue of the broker, and the time that matching results (to *m* neighbors) spend in the output queues. This cost is modeled as

$$RC_B(CS) = \sum_{i=1}^{n} T_{in} \mid P(CS_{B_{N_i}}) \mid + \sum_{i=1}^{n} T_m \mid P(CS_{B_{N_i}}) \mid + \sum_{i=1}^{m} T_{out_i} \mid P(CS) \mid$$

where T_m is the average matching time at a broker, T_{in} and T_{out_i} are the average time messages spend in the input queue, and output queue to the ith neighbor. $|P(S)|$ is the cardinality of subscription *S*, which is the number of matching publications per unit time. To compute the cost at a neighbor, brokers periodically exchange information such as T_{in} and T_m. This information is incorporated into an M/M/1 queueing model to estimate queueing times at neighbor brokers as a result of the additional traffic attracted by splitting a composite subscription there.

Evaluations of the *DCSR* algorithm were conducted on the PlanetLab wide-area network with a 30 broker topology. The metrics measured include the bandwidth of certain brokers located on the composite subscription routing path. In Figure 5, the solid bars represent the number of outgoing messages at a broker, and the hatched bars are the number of incoming messages that are not forwarded. Note that the sum of the solid and hatched bars represents the total number of incoming messages at a broker. Three routing algorithms are compared: simple routing, in which composite subscriptions are split into atomic parts at the first broker, topology-based composite subscription routing, and the *DCSR* algorithm. The topology-based routing imposes less traffic than simple routing by moving the join point into the network and the *DCSR* algorithm further reduces traffic by moving the join point closer towards congested publishers as indicated by the cost model. In the scenario in Figure 5, compared to simple routing, the *DCSR* algorithm reduces the traffic at Brokers *B*1 by 79.5%, a reduction that is also enjoyed by all brokers downstream of the join point.

Figure 5. Composite subscription traffic

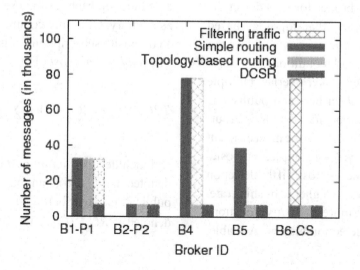

HISTORIC DATA ACCESS

PADRES allows subscribers to access both future and historic data with a single interface as described earlier. The system architecture, shown in Figure 6, consists of a traditional distributed overlay network of brokers and clients. Subscriptions for future publications are routed and handled as usual (Opyrchal, 2000; Carzaniga, 2001). To support historic subscriptions, databases are attached to a subset of brokers as shown in Figure 6. The databases are provisioned to sink a specified subset of publications, and to later respond to queries. The set of possible publications, as determined by the advertisements in the system, is partitioned and these partitions assigned to the databases. A partition may be assigned to multiple databases to achieve replication, and multiple partitions may be assigned to the same database if database consolidation is desired. Partition assignments can be modified at any time, and replicas will synchronize among themselves. The only constraint is that each partition be assigned to at least one database so no publications are lost. Partitioning algorithms as well and partition selection and assignment policies are described in (Li, 2008). Subscriptions can be atomic expressing constraints on single publications or composite expressing correlation constraints over multiple publications. We describe their routing under the extended publish/subscribe model.

Atomic Subscription Routing

When a broker receives an atomic subscription, it checks the start_time and end_time attributes. A future subscription is forwarded to potential publishers using standard publish/subscribe routing (Opyrchal, 2000; Carzaniga, 2001). A hybrid subscription is split into future and historic parts, with the historic subscription routed to potential databases as described next.

For historic subscriptions, a broker determines the set of advertisements that overlap the subscription, and for each partition, selects the database with the minimum routing delay. The subscription is forwarded to only one database per partition to avoid duplicate results. When a database receives a historic subscription, it evaluates it as a database query, and publishes the results as publications to be routed back to the subscriber. Upon receiving an END publication after the final result, the subscriber's host broker unsubscribes the historic subscription. This broker also unsubscribes future subscriptions whose end_time has expired.

Adaptive Routing

Topology-based composite subscription routing (Li, 2005) evaluates correlation constraints in the network where the paths from the publishers to subscriber merge. If a composite subscription correlates a historic data source and a publisher, where the former produces more publications,

Figure 6. Historic data access architecture

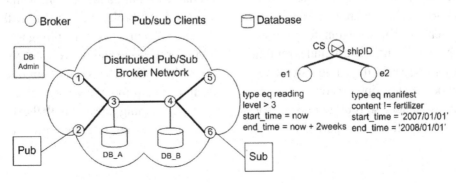

correlation detection would save network traffic if moved closer to the database, thereby filtering potentially unnecessary historic publications earlier in the network. Based on this observation, the *DCSR* algorithm we discussed can be applied here. The WHERE clause constraints of a composite subscription can be represented as a tree where the internal nodes are operators, leaf nodes are atomic subscriptions, and the root node represents the composite subscription. A composite subscription example is represented as the tree in Figure 6. The recursive *DCSR* algorithm (Li, 2008) computes the destination of each node in the tree to determine how to split and route the subscription. The algorithm traverses the tree as follows: if the root of the tree is a leaf, that is, an atomic subscription, the atomic subscription's next hop is assigned to the root. Otherwise, the algorithm processes the left and right children's destination trees separately. If the two children have the same destination, the root node is assigned this destination, and the composite subscription is routed to the next hop as a whole. If the children have different destinations, the algorithm estimates the total routing cost for potential candidate brokers, and the minimum cost destination is assigned to the root. If the root's destination is the current broker, the composite subscription is split here, and the current broker is the join point and performs the composite detection. The algorithm assigns destinations to the tree nodes bottom up.

When network conditions change, join points may no longer be optimal and should be recomputed. A join point broker periodically evaluates the cost model, and upon finding a broker able to perform detection cheaper than itself, initiates a join point movement. The state transfer from the original join point to the new one includes routing path information and partial matching states. Each part of the composite subscription should be routed to the proper destinations so routing information is consistent. Publications that partially match composite subscriptions stored at the join point broker must be delivered to the new join point.

For more detailed description of the historic data access function, please refer to our technique report (Li, 2008).

LOAD BALANCING

In a distributed publish/subscribe system, geographically dispersed brokers may suffer from uneven load distributions due to different population densities, interests, and usage patterns of end-users. A typical scenario is an enterprise-scale deployment consisting of a dozen brokers located at different world-wide branches of an international corporation, where the broker network provides a communication service for hundreds of publishers and thousands of subscribers. It is conceivable that the concentration of business operations and departments, and thus publish/subscribe clients and messages, is orders of magnitudes higher at the corporate headquarters than at the subsidiary locations. Such hotspots at the headquarters can overload the broker there in two ways. First, the broker can be overloaded, if the incoming message rate into the broker exceeds the maximum processing or matching rate of the broker's matching engine. Because the matching rate is inversely proportional to the number of subscriptions in the matching engine, this effect is exacerbated if the number of subscribers is large (Fabret, 2001). Second, overload can also occur if the output transmission rate exceeds the total available output bandwidth. In both cases, input queues at the broker accumulate with messages waiting to be processed, resulting in increasingly higher processing and delivery delays. Worse yet, the broker may crash when it runs out of memory from queueing too many messages.

The matching rate and both the incoming and outgoing message rates determine the load of a

Figure 7. PEER architecture

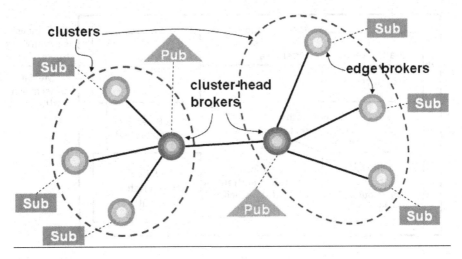

broker. In turn, these factors depend on the number and nature of subscriptions that the broker services. Thus, load balancing is possible by offloading *specific* subscribers from higher loaded to lesser loaded brokers. The PADRES system supports this capability using load estimation methodologies, a load balancing framework, and three offload algorithms (Cheung, 2006).

The load balancing framework consists of the *PADRES Efficient Event Routing* (PEER) architecture, a distributed load exchange protocol called *PADRES Information Exchange* (PIE), and detection and mediation mechanisms at the local and global load balancing tiers. The PEER architecture organizes brokers into a hierarchical structure as shown in Figure 7. Brokers with more than one neighboring broker are referred to as *cluster-head brokers*, while brokers with only one neighbor are referred to as *edge brokers*. A cluster-head broker with its connected set of edge brokers, if any, forms a *cluster*. Publishers are serviced by cluster-head brokers, while subscribers are serviced by edge brokers. Load balancing is possible by moving subscribers among edge brokers of the same or different cluster. With PIE, edge brokers within a cluster exchange load information by publishing and subscribing to PIE messages of a certain cluster ID. For example, a

subscription to PIE messages from cluster C01 is [class, eq, 'LOCAL_PIE'], [cluster, eq, 'C01']. The detector invokes load balancing if it detects overload or the load of the local broker is greater than another broker by a threshold. Load is characterized by three load metrics. First, the input utilization ratio (I_r) captures the broker's input load and is calculated as:

$$I_r = \frac{i_r}{m_r}$$

where i_r is the rate of incoming publications and m_r is the maximum message match rate calculated by taking the inverse of the matching delay. Second, the output utilization ratio captures the output load and is calculated as:

$$O_r = \left(\frac{t_{busy}}{t_{window}} \right)\left(\frac{b_{rx}}{b_{tx}} \right)$$

where t_{window} is the monitoring time window, t_{busy} is the amount of time spent sending messages within t_{window}, b_{rx} represents the messages (in bytes) put into the output queue in time window t_{window}, and b_{tx} represents the messages (in bytes) removed

Figure 8. Components of the load balancer

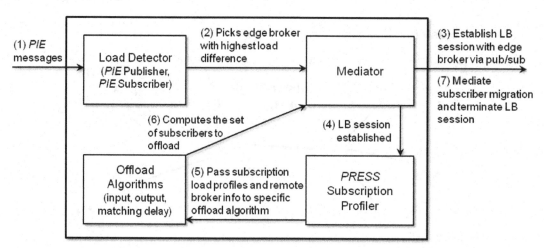

from the output queue and sent successfully in time window t_{window}. A utilization value greater than 1.0 indicates overload. Third, the matching delay captures the average amount of time to match a publication message.

The core of the load estimation is the PADRES *Real-time Event to Subscription Spectrum* (PRESS), which uses an efficient bit vector approach to estimate the input and output publication loads of all subscriptions at the local broker. Together with *locally subscribing* to the load-accepting broker's *covering subscription set*, PRESS can estimate the amount of input and output load introduced at the load-accepting broker for all subscriptions at the offloading broker.

Each of the three offload algorithms are designed to load balance on each load metric of the broker by selecting the appropriate subscribers to offload based on their profiled load characteristics. Simultaneously, the subscriptions that each offload algorithm picks minimize the impact on the other load metrics to avoid instability. For example, the match offload algorithm offloads subscriptions with the minimal traffic, and the output offload algorithm first offloads highest traffic subscriptions that are covered by the load accepting broker's subscription(s.)

This solution inherits all of the most desirable properties that make a load balancing algorithm flexible. PIE contributes to the *distributed* and *dynamic* nature of the load balancing solution by allowing each broker to invoke load balancing whenever necessary. *Adaptiveness* is provided by the three offload algorithms that load balance on a unique performance metric. The local mediator gives *transparency* to the subscribers throughout the offload process. Finally, load estimation with PRESS allows the offload algorithms to account for broker and subscription *heterogeneity*.

The components that make up the load balancing solution, shown in Figure 8, consist of the detector, mediator, load estimation tools, and offload algorithms. The detector detects and initiates a trigger when an overload or load imbalance occurs. The trigger from the detector tells the mediator to establish a load balancing session between the *offloading broker* (broker with the higher load doing the offloading) and the *load-accepting broker* (broker accepting load from the offloading broker). Depending on which performance metric is to be balanced, one of the offload algorithms is invoked on the offloading broker to determine the set of subscribers to delegate to the load-accepting broker based on estimating how much load is reduced and increased at each broker

Figure 9. Internal view of the PADRES broker

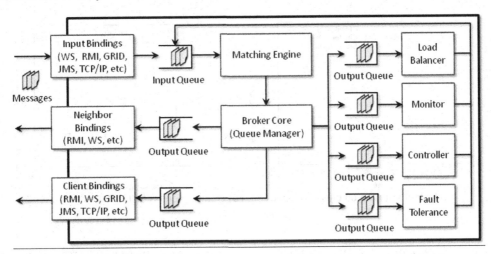

using the load estimation algorithms. Finally, the mediator is invoked to coordinate the migration of subscribers and ends the load balancing session between the two brokers. The load balancing solution is integrated as a stand-alone module in the PADRES broker as shown in Figure 9.

Evaluations of the PADRES load balancing algorithms in the shared wide-area PlanetLab testbed, a dedicated local cluster environment, and a simulator all show that the load balancing algorithms prevent overload by distributing subscribers while simultaneously balancing the three load metrics among edge brokers. The algorithms are effective in both homogeneous and heterogeneous environments and enable the system to scale with added resources. Figure 10(a) shows an experiment with four heterogeneous clusters arranged in a chain with two edge brokers per cluster and having all subscribers join at an edge broker on one end of the chain, namely B1x. As time progresses, subscribers get distributed to other clusters down the chain until the algorithm converges around 3500 s into the experiment. Not shown on this graph is the observation that the subscriptions that sink higher traffic are assigned to brokers with more computing capacity than to brokers with limited capacity. Figure 10(b) shows that the average load of the brokers decreases as

more resources (in the form of clusters) are added. Simultaneously, delivery delay decreases when going from two to four clusters, but increases beyond five clusters due to a longer path length. By adaptively subscribing to load information, the message overhead of the load balancing infrastructure is only 0.2% in the experiments on the cluster test10bed. The results also show that a naive load balancing solution that cannot identify subscription space and load are not only inefficient but can also lead to system instability. The interested reader may consult the full paper for more details about the algorithms and the experiments (Cheung, 2008).

FAULT-TOLERANCE

Fault-tolerance in general refers to the ability of a system to handle the failure of its components and maintain the desired quality of service (QoS) under such conditions. Furthermore, a δ-*fault-tolerant* system operates correctly in presence of up to δ failures. A common class of failures in a distributed system is node *crashes*, in which nodes stop executing instructions, no longer send or receive messages, and lose their internal state.

Figure 10. Experiment results

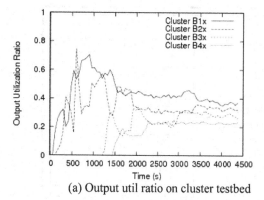

(a) Output util ratio on cluster testbed

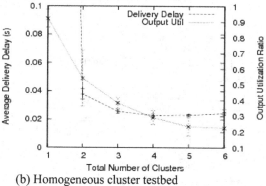

(b) Homogeneous cluster testbed

Failures may be *transient* in which case nodes may *recover* by executing a recovery procedure.

In order to achieve δ-fault-tolerance, PADRES nodes (brokers and clients) transparently collect additional routing information as part of the normal operation of the system and use this information to react to the failure of their neighbors. Two types of information are collected: the broker topology, and the subscription routes. The former allows for increased network connectivity and prevents partitions forming as a result of failures, and the latter is used to decide among alternative routing paths and avoid interruptions in publication delivery.

The remainder of this section presents the system-wide *consistency* properties that correspond to the routing topology and subscription routing state, and describes how this information is used to achieve fault-tolerant routing and recovery.

Consistency

To ensure correct operation of the system (in presence of up to δ failure) the topology and subscription routing information must be kept *consistent* at all times. In our context, consistency is dependant on the desired degree of fault-tolerance of the system, δ, and is thus referred to as δ-*consistency*. The value of δ is chosen by an administrator based on a number of factors including the fan-out of brokers, rate of failures, and average downtimes. To achieve

δ-*consistency* for topology routing information, brokers must know about all peers within a $(\delta+1)$-neighborhood. Distances are measured over the initial acyclic topology which acts as a backbone for the entire system. The δ-consistent topology routing information enhances the connectivity of this acyclic structure by enabling brokers to identify and connect to not only their neighbors, but all nodes within distance $\delta+1$.

On the other hand, δ-*consistency* for subscription routing information is achieved by maintaining references to certain brokers along the subscription propagation paths. These references point to brokers that are up to $\delta+1$ hops closer to the subscriber. More specifically, a broker that is within distance $\delta+1$ of a subscriber stores the subscriber's broker ID as the reference. If it is farther, then the reference points to another broker along the path to the subscriber. This broker is $\delta+1$ hops closer to the subscriber. Figure 11 illustrates a sample network with δ-consistent subscription routing information for two highlighted subscribers.

Fault-Tolerant Forwarding Algorithm

When there are no failures in the system, brokers are connected to their immediate neighbors in the acyclic backbone topology. At the same time, they continuously monitor their communicating

Figure 11. δ-consistent subscription routing information for two subscribers S1 and S2. An arrow from A to B indicates that A holds a reference pointer to B

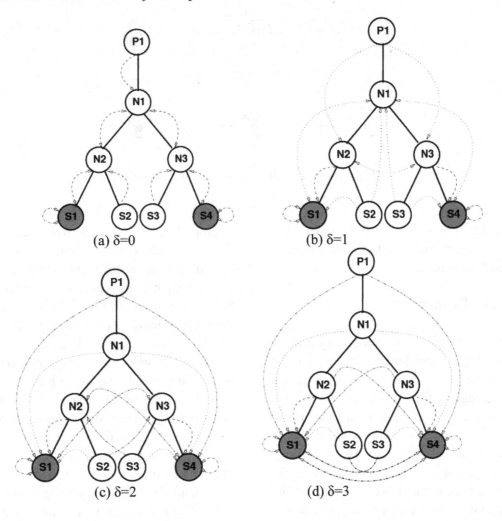

(a) δ=0

(b) δ=1

(c) δ=2

(d) δ=3

peers using a heartbeat based failure detector. It is assumed that the failure detector works perfectly and all broker failures are detected after some time. When a failure is detected, the fault-tolerant forwarding algorithm is triggered at the non-faulty neighbors.

The main objective of fault-tolerant forwarding is to bypass failed neighboring brokers and re-establish the publication flows. For this purpose, having detected a failure, brokers create new communication links to the immediate neighbors of their failed peer, as illustrated in Figure 12. These new brokers, identified using the local topology routing information, may themselves concurrently try to bypass the failed node. Endpoint brokers that establish a new connection (to bypass a failed node) perform an initial handshake and exchange their operational states. Additionally, they exchange the sequence number of the last message tagged by the other endpoint (or "null" if there is no such message). This information is used to determine whether messages previously sent to the failed broker need to be retransmitted. Subsequently, nodes start to (re-)send outstanding messages over a new link if a matching subscriber is reachable through the link. This process maintains the initial

Figure 12. Fault-tolerance forwarding bypasses faulty brokers (filled circles represent final destinations

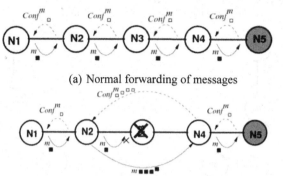

(a) Normal forwarding of messages

(b) Forwarding of messages in the case of a broker failure

arrival order of messages and uses subscription reference pointers to decide to which new peers to forward the messages.

Recovery Procedure

The *recovery procedure* is executed by brokers that have experienced failures in the past and enables them to re-enter the system and participate in message routing. This is in contrast to the fault-tolerance forwarding algorithm that runs on non-faulty brokers, in order to deliver messages in presence of failed brokers. The *recovering* brokers have lost their internal state due to the failure, and have further missed messages (e.g., subscriptions) that were sent during their downtime. Thus, the main objective of the recovery procedure is to restore this lost internal state by establishing a δ-consistent topology and subscription routing information. The recovery procedure involves the following steps: *(i)* identify previous location in the topology, and the nearby brokers; *(ii)* synchronize and receive routing information; *(iii)* participate in message forwarding; *(iv)* end recovery and notify peers.

Recovering brokers can identify their previous location in the topology by accessing their local persistent storage or querying a discovery service that maintains this data. In either case, it is necessary that prior to failures brokers persistently store their topology routing information to disk,

or properly update the discovery service about changes to the topology. The synchronization step involves connecting to the closest non-faulty brokers and requesting updated routing information. Reference pointers in the received subscription routing information is properly manipulated such that the δ-consistency requirements are met.

The synchronization step may involve several nearby brokers and may be lengthened as large volumes of data are transferred or as new failures occur. During this period, new subscription messages may be inserted into the system and the topology tree may undergo further changes. To enable the recovering brokers to keep up with this updated information, they participate in message forwarding in a similar way to fully operational peers. The only exception is that the synchronization points attach additional information determining the destinations of the messages. This is required since the routing information of a recovering broker at this stage may not be complete. Once all the recovery information is transferred from all synchronization points, the recovery is complete and the peers are appropriately notified. From this point onward, the δ-consistent routing properties are established and the recovered broker fully participates in regular message forwarding. More information about the fault-tolerance and recovery procedure is provided in (Sherafat, 2007; Sherafat, 2008).

TOOLS

PADRES includes a number of tools to help manage and administer a large publish/subscribe network. This section presents two of these tools: a monitor that allows a user to visualize and interact with brokers in real time, and a deployment tool that simplifies the provisioning of large broker networks.

Monitor

The PADRES monitor lets a user monitor and control a broker federation. It is implemented as a regular publish/subscribe client and performs all its operations using the standard publish/subscribe interface and messages. Among other benefits, this allows the monitor to be run from anywhere a connection to a broker can be established, and to access any broker in the federation including those that would otherwise be hidden behind a firewall.

Once connected to a broker, the monitor issues a subscription for broker status information that is periodically published by all brokers in the system. This information is used to construct a visual representation of brokers, overlay links, and clients. The display is updated in real time as the monitor continuously discovers and receives updates from brokers.

Figure 13 shows a screenshot of the PADRES monitor connected to a federation of 100 brokers. Nodes in the visualization may be rearranged by manually dragging the nodes around, or various built-in graphing algorithms can decide on the layout automatically. Detailed information of each broker, such as the routing tables, system properties, and various performance metrics can also be viewed. In terms of the control features, the user can pause, resume, and shutdown individual brokers; inject any type of message (including advertisement, unadvertisement, subscription, unsubscription, and publication messages) at any point in the network; and trace and visual-

ize the propagation paths of messages. For more information, please refer to PADRES user guide (Jacobsen, 2004).

PANDA

The *PADRES Automated Node Deployer and Administrator* (PANDA) simplifies the installation, deployment, and management of large broker networks distributed among any number of machines. In addition to starting and terminating processes, PANDA can install and uninstall the required Linux RPM packages, upload and remove PADRES binaries, and even retrieve broker log files from remote machines. As all remote operations are executed via SSH commands, PANDA can manage the deployment of brokers on any machine with SSH access. In addition, PANDA is fully compatible with the PlanetLab wide-area research testbed (PlanetLab, 2006).

PANDA lets a user easily describe complex broker and client networks in a flexible *topology file*. For example, the following ADD command in a topology file indicates that a broker process named *BrokerA* is to be started immediately upon deployment on machine 10.0.1.1 with a set of custom properties, such as a port number of 10000 and an ID of *Alice*.

0.0 ADD BrokerA 10.0.1.1 startbroker.sh -Xms 64 -Xmx 128 -hostname 10.0.1.1 -p 10000 -i Alice

That same process can be scheduled to be terminated at a particular time, say 3500sec, with a REMOVE command as follows:

3500 REMOVE BrokerA 10.0.1.1

All IP addresses referenced in the topology file will be taken into consideration when an install or upload command is issued to install RPMs or upload tarballs. PANDA also supports a unique 2-phase deployment scheme for brokers. Phase

Figure 13. PADRES monitor showing 100 brokers

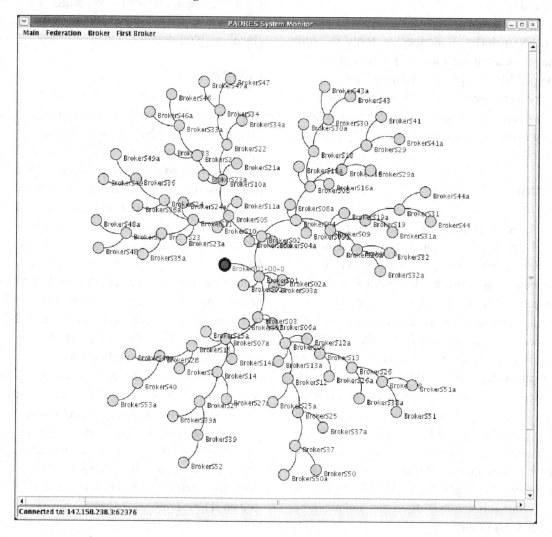

1 includes all broker processes with deployment time of 0.0 where the user wishes the brokers and its overlay links to be fully up and running before deploying any other processes in Phase 2. Phase 2 includes all those processes with deployment times greater than 0. Phase 2 starts only when Phase 1 is complete as detected by PANDA's built-in monitor. For more details and examples of PANDA's topology file, please refer the online PADRES user guide (Jacobsen, 2004).

The PANDA architecture, shown in Figure 14, consists of a Java program that uses helper shell scripts to interact with the remote nodes. The user interacts with PANDA through a text console. Upon loading a topology file or entering a command directly into the console, the input is parsed for correctness and UNIX commands are generated by the CommandGenerator. A Topology Validator validates the input, checking for errors such as duplicate broker IDs. When the user enters the deploy command, the DeploymentCoordinator orchestrates the 2-phase deployment and executes remote UNIX command operations through the ScriptExecutor.

These features of PANDA greatly simplify the management of large broker networks. They can also be used to fully automate any PADRES experiments including starting and stopping brokers and clients at certain times and collecting the experiment log files.

APPLICATIONS

The simple yet powerful publish/subscribe interface supported by PADRES can be applied to a variety of scenarios that run the gamut from simple consumer news filtering to complex enterprise applications.

What follows in this section are illustrations of how the design of sophisticated applications can be simplified by capitalizing on the various features of the publish/subscribe middleware outlined in the preceding sections. Some of these applications exploit properties of the publish/subscribe *model* itself such as the expressiveness of the publish/subscribe language that enables fine-grained event filtering, event correlation and context-awareness, the complex interaction patterns that can be realized such as many to many conversations, the natural decoupling of components that allows for asynchronous and anonymous communication, and the push-based messaging that enables applications to react to events in real-time. Certain scenarios below also take advantage of features of the PADRES publish/subscribe *middleware* including the scalability achieved by a distributed broker architecture, the ability to dynamically load-balance the brokers, and fault-tolerance capabilities that enable the brokers to automatically detect and recover from failures.

To convey the breadth of scenarios to which publish/subscribe can be applied, applications from three domains are presented: *consumer* applications used by individuals for personal productivity or entertainment purposes, *enterprise* applications that are critical to the operation of a business entity, and *infrastructure* services that are used to deploy, monitor, and manage hardware and software infrastructures. These three domains,

Figure 14. PANDA architecture

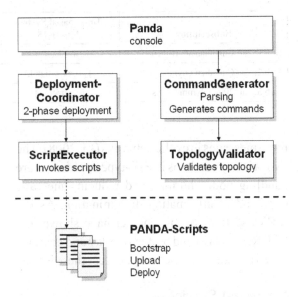

of course, may overlap and are not necessarily mutually exclusive.

Consumer Applications

Interactions in consumer or end-user applications generally follow a client-server model in which a user receives service from a service provider, or a peer-to-peer (P2P) model where a group of users interact in an ad-hoc manner.

The client-server interaction model is very simple to achieve in publish/subscribe systems: the server, such as a newscast service, simply publishes its data, and the clients subscribe to the subset of data they are interested in. Of course, the usability of this type of application depends on the expressiveness with which the clients can express their interests, making an expressive content-based language, such as that provided by PADRES, preferable over simpler publish/subscribe models.

The PADRES content-based publish/subscribe system can also be used to construct a P2P information dissemination service thanks to its distributed broker overlay and its capability to

Table 1. Subscribing using topic-based and content-based publish/subscribe systems

Subscription	Topic-based System (Figure 15)	PADRES Content-based Language
Sub 1: "all Canadian sports news"	T_4, T_5, T_6	[class, eq, 'sports'],[country, eq, 'Canada']
Sub 2: "all Canadian Olympic news"	T_5 (exact match)	[class, eq, 'sports'],[country, eq, 'Canada'],[event, eq, 'Olympics']
Sub 3: "all 2008 Olympic news"	T_2 (super set)	[class, eq, 'sports'],[event, eq, 'Olympics'],[year, eq, '2008']

route messages in cyclic networks (Li, 2008). The two primary concerns in P2P-type application are handling node churning and reducing message overhead in information dissemination. As the earlier sections of this chapter have shown, the PADRES distributed broker overlay and content-based routing can efficiently address these issues.

Newscast Services

Conventional Internet-based newscast services may use a topic-based publish/subscribe system where news items are published under certain topics (for example, "sports" or "local-news") and customers subscribe to one or more of these topics. When the customers want to fine-tune their subscriptions, the topics have to be sub-divided to match their interests. This is handled by creating a hierarchy of topics. For example, when a user wants to subscribe to Canadian Olympic events, a topic hierarchy of "sports → Olympics → Canada" is created as shown in Figure 15.a.

The major issue with a topic hierarchy is its limited expressiveness. Users are constrained to subscribe to only the topics defined by the hierarchy, but constructing a hierarchy to cover all the potential combinations of user interests leads to an explosion of topics and results in poor matching performance and management overhead. Furthermore, when the user interests change, either the topic hierarchy must be restructured, or users must subscribe to broader topics than they are interested. The first solution is impractical, and the second solution results in redundant message overhead

and requires additional processing by the clients to filter out publications that are not of interest.

A content-based publish/subscribe system avoids these issues because client interests are expressed using fine-grained attribute-value tuples. Table 1 and Figure 15 show the difference between topic-based and content-based systems in expressing subscriber interests. Note that the topic-based system can match **Sub 2** exactly with topic T_5, but there is no topic that exactly matches **Sub 3**. Therefore, the client is forced to subscribe to the superset topic T_2 which covers **Sub 3** but also contains unrelated news items. On the other hand, as shown in the table, the results can be filtered more accurately in content-based system by including all the necessary attribute-value tuples in the subscription.

The topic hierarchy also influences the distribution of the matching workload in a distributed system. Consider the network shown in Figure 15 where subscribers S_1 and S_2 connect to broker **B** and subscribe to **Sub 1** and **Sub 2**, respectively. When the topic hierarchy is constructed as shown in Figure 15.a, the publisher matches **Sub 1** to topics T_4, T_5, T_6 and **Sub 2** to topic T_5 and forwards the news items on these topics to broker **B** which forwards them to the respective subscribers. When the topic hierarchy is organized as in Figure 15.b, however, broker **B** need only subscribe to topic T_1, because both **Sub 1** and **Sub 2** are covered by this topic. When the broker receives the news items on T_1, it can immediately forward them to S_1, and forward those news items that match T_2 to S_2. In this way, the matching workload is distributed in the system making the system more scalable.

Figure 15. Topic hierarchy and subscription covering

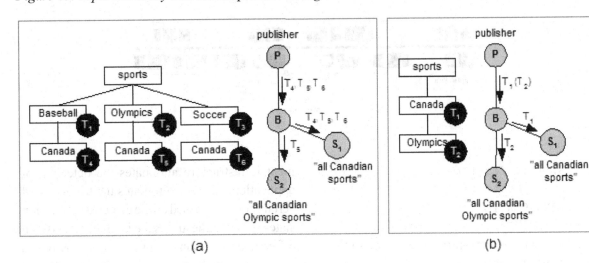

(a) (b)

It is difficult, however, to design a topic hierarchy that effectively distributes the matching workload while simultaneously offering topics that closely correspond to all user interests. This issue does not arise in content-based systems because the language model provides a way of covering subscriptions as described earlier. For example, Table 1 shows that the content-based definition of **Sub 1** covers that of **Sub 2**. Therefore, a content-based publish/subscribe system provides a more scalable design.

Intelligent Vehicular Ad-Hoc Networks

A more sophisticated application of the content-based publish/subscribe paradigm is an intelligent vehicular ad-hoc networks (InVANET). Present day smart car functions involve making decisions based on the data fed from different sensors embedded within a car's infrastructure, such as accident prevention using a proximity radar or air-bag deployment using deceleration sensor.

Automobiles in an InVANET collaborate with one another to construct a distributed sensor that captures the collective knowledge of the individual sensors in each vehicle. The information dissemination in InVANET follows a reactive model where a car detecting a situation triggers an action from another car. A content-based system like PADRES can efficiently implement this event-driven architecture, with each car playing the role of a publisher, subscriber, and content-based router. In this scenario, the events of interest will include accidents, traffic jams, or even the events of cars leaving parking spots.

An example scenario is illustrated in Figure 16. Car *A* is interested in knowing about traffic jams in advance so that it can take an alternate path. It subscribes to a "traffic-jam" event as:

```
[class, eq, 'traffic-jam'],
[location, <, MY_LOC + 10], [dir, eq,
'HW401W']
```

Note that the subscription includes location and directional (HW401W, i.e., Highway 401, West bound) constraints. The location variable MY_LOC is substituted with the current GPS coordinates. The subscription is propagated in the overlay created by the smartcars. When Car *B* detects a traffic jam (perhaps using its internal sensors), it publishes a "traffic-jam" event as:

Figure 16. An example scenario in a InVANET system

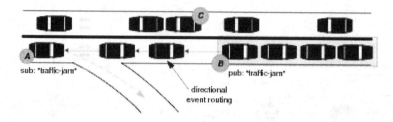

```
[class, 'traffic-jam'],
[location, MY_LOC],
[dir, 'HW401W']
```

This event is reverse-propagated through the overlay until it reaches Car *A*.

A similar publish/subscribe scheme can be used to find a newly available parking spot: when a car leaves a parking spot in a busy downtown area, it can publish the event which is propagated to cars whose driver is interested in finding a parking spot in the vicinity.

An InVANET requires a publish/subscribe middleware that can be implemented over ad-hoc cyclic networks. It should also be noted that subscriptions and publications include location and directional attributes which should be exploited during event routing to reduce message overhead. For example, in Figure 10, the "traffic-jam" event generated by Car *B* need not reach Car *C* that is traveling in the opposite direction. PADRES provides the necessary infrastructure to construct an InVANET, and its matching engine can be extended to support directional and location operators in its subscription language.

Enterprise Applications

The number of applications and users an enterprise manages and supports as well as the amount of data that flows between them grows larger with a growing enterprise. Therefore, enterprises enforce automated service management infrastructures to scale with a growing service base. This manage-ment infrastructure automates the detection of application states; it automates the triggering of certain activities based on the detected application states; and it orchestrates the interaction between different applications and users. These activities require *complex event processing* (CEP) that accepts the different application states as events and process them to detect certain *situations* and invoke relevant actions. A content-based publish/subscribe system, especially a distributed system like PADRES, is the ideal choice for implementing a CEP infrastructure. The applications and users can join the system as clients, situations can be defined by composite event subscriptions, and the interactions between the applications are managed by subscribing to certain events (application states or situations).

Sensor Networks

Sensor networks are created by interconnecting a number of sensors monitoring different parameters at different locations. Sensor networks are commonly used in environment monitoring, traffic control, health care, and battlefield surveillance. Event processing is the primary operation in a sensor network and a content-based publish/subscribe system can simplify this operation. Each sensor can be considered as a publisher that outputs a constant stream of data with a fixed schema (advertisement). The applications that process the sensor data can subscribe to various events from different sensors and produce their own events. For example, a tsunami event can be detected using a composite subscription:

```
([class, eq, 'sesmic'],
 [magnitude, >,3],
 [location, =,$L],
 [time, =,$T]) &&
([class, eq, 'wave'],
 [height, >,10],
 [location, <, $L +5],
 [time, <,$T + 10])
```

It detects an event of a seismic activity of magnitude larger than 3 followed by (within 10min) a wave with a height of more than 10m at a location within 5km from the origin of the seismic event. When this condition is satisfied, a new alert event can be produced as:

```
[class, 'climate-alert'],
[condition, 'tsunami'],
[location, $L], [time, $T]
```

A *radio frequency identification* (RFID) system is a type of sensor network that has already been successfully used in monitoring moving objects. Tracking books in a library, inventory of goods in a store, and automated payments in toll highways are few of the applications of RFID-enabled tags. At present, RFID tags are used mostly to identify the presence (or the lack of presence) of an item at a certain location at a given time. If the RFID readers are networked, the time stamped detection events can be conveyed as publications to a content-based publish/subscribe system that will increase the functionality of the RFID-based systems. For example, a shoplifting event can be detected by subscripting to an appropriate composite event: detecting an event with a certain RFID at the exit sensor without detecting it at a sales counter.

In a sensor network, the event schemata are mostly constant and simple, but the sensors

are distributed and the amount and rate of data produced by them (publications) are often very large. This requires a distributed, fast, and scalable matching and routing infrastructure like PADRES (Petrovic, 2005). In addition, the publish/subscribe middleware used in sensor networks should be self-configuring and fault tolerant, because the sensor networks are sometimes implemented on a mobile network where the lifetimes and the locations of the nodes vary constantly.

Business Process Management

Business process management (BPM) is another important business application where content-based publish/subscribe systems are extremely useful. Business process management organizes a set of enterprise applications and processes in order to facilitate efficient communication among themselves and with clients. One of the key aspects of BPM is *workflow processing*. Figure 17 shows an example workflow of an online retailer.

A workflow describes the interactions between different enterprise applications, processes, and users and includes causal and temporal relationships between applications. Because it follows the model of an event-driven system where the completion of one or many processes activates another, a content-based system can be used to implement it. For example, Figure 17 shows the hypothetical workflow of an online sales application where the availability of an item should be checked and the shipping charge should be calculated before enabling a detailed item view. Therefore, the 'ItemView' module should subscribe to a composite event:

```
([class, eq, 'INVOKE'],
 [service, eq, 'ItemView'],
 [id, eq, $X]) &&
([class, eq, 'RESULT'],
 [service, eq, 'AvailCheck'],
 [id, eq, $X]) &&
```

Figure 17. Example workflow of an online sales application

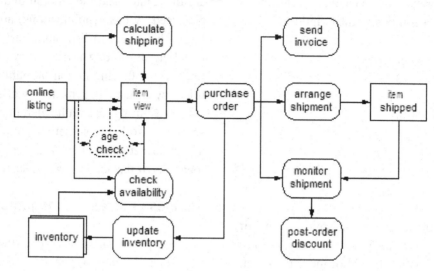

```
([class, eq, 'RESULT'],
 [service, eq, 'CalcShipping'],
 [id, eq, $X])
```

Note that this subscription performs two tasks: the first part of the composite subscription provides the activation command to the 'ItemView' module, but the other two parts restrict the module to be activated only after the results from the relevant 'AvailCheck' and 'CalcShipping' services are received. These events are connected using a variable on the id attribute.

In a workflow, the output of a process can vary depending on the incoming event parameters. For example, in Figure 17, the online retailer might decide to give a post-order discount, if the shipment is delayed more than a specific duration. When a purchase order is placed, a shipment monitor is instantiated as well, which waits for the event of shipment. When the shipment event is received, it will check the purchase agreement and if the shipment failed to match the agreed shipment date, a post-order discount is sent out. This operation can be performed by issuing the following subscription and publication:

```
Sub: ([class, eq, 'INVOKE'],
      [service, eq, 'MonitorShipment'],
      [id, eq, $X],
      [time, =, $T]) &&
     ([class, eq, 'RESULT'],
      [service, eq, 'ItemShipped'],
      [id, eq, $X], [time, >, $T + 10])
Pub: [class, 'INVOKE'],
     [service, 'PostOrderDiscount'],
     [id, 'aaaa']
```

The subscription is used to detect the condition where a post-order discount is to be issued and the publication is used to activate the post-order delivery module. The threshold that triggers a post order discount is 10 days from the order date.

A content-based publish/subscribe system not only efficiently implements a workflow, but it also simplifies reorganizing the workflow when required. For example, in Figure 17, the retailer may decide to invoke a new service to verify the age of the consumers against the approved limit before activating the item view. The modification, shown with the dotted lines in the figure, can be readily implemented by unsubscribing the previous subscription and invoking a new subscription as follows:

```
([class, eq, 'INVOKE'],
 [service, eq, 'ItemView'],
 [id, eq, $X]) &&
([class, eq, 'RESULT'],
 [service, eq, 'AvailCheck'],
 [id, eq, $X]) &&
([class, eq, 'RESULT'],
 [service, eq, 'CalcShipping'],
 [id, eq, $X]) &&
([class, eq, 'RESULT'],
 [service, eq, 'AgeCheck'],
 [id, eq, $X],
 [approve, eq, 'YES'])
```

Business Application Monitoring

Business application monitoring (BAM) is another aspect of BPM. It involves continuously monitoring the performance of applications or processes, producing reports, and triggering actions or notifications when some specific conditions are met.

Again, BAM concerns can be easily realized with publish/subscribe middleware by adding components that subscribe to events generated by various distributed application monitor software agents. Event processing can be used to detect various system conditions and take actions accordingly. The loose coupling properties of publish/subscribe are exploited here to allow BAM components to monitor applications without having to instrument the application they are monitoring. Rather, they simply subscribe to events they are interested in.

Enterprise Service Bus

Service-oriented architectures (SOA) have become a common solution for the problems of enterprise application management. In an SOA, applications are constructed by composing a set of reusable services that are available through standardized interfaces. These applications are themselves exposed as yet another service that can in turn be composed by other applications.

An *enterprise service bus* (ESB) plays a central role in an SOA, mediating the interactions among the services. A publish/subscribe middleware such as PADRES supports the core functionality required of an ESB, and is ideally suited to serve as an ESB in an SOA.

Another important component of an SOA is a *service registry* where services are registered and discovered. In publish/subscribe terminology, the services can be presented as both publishers and subscribers. The historic query capabilities of PADRES can be used to discover services that have registered in the past, while the usual publish/subscribe subscription mechanisms enable applications to continuously monitor for and be notified of new services. The actual execution of a workflow of composed services can then be achieved as described in the BPM discussion above.

Two of the key responsibilities of an ESB are service orchestration and governance. A business orchestration defines the interaction among the services (creating workflow descriptions), whereas governance concerns the management of corporate policies regarding the hosted services and their interactions. An ideal ESB should support an event-driven distributed architecture, an expressive workflow description, a standard-based integration model, flexible data transformation capabilities, and an autonomous and federated environment.

The PADRES middleware provides a highly distributed event-driven architecture that supports many of the required ESB characteristics. Figure 18 shows how PADRES can be used to implement an ESB. The distributed broker network connects applications across a large enterprise, even across the Internet, and the content-based publish/subscribe system provides an event-driven system that supports expressive workflows. Service orchestration is supported by the content-based resource discovery and event-based monitoring and the situation detection mechanism helps enforce governance policies. In addition, the fault tolerant and

Figure 18. Building an enterprise service bus with PADRES

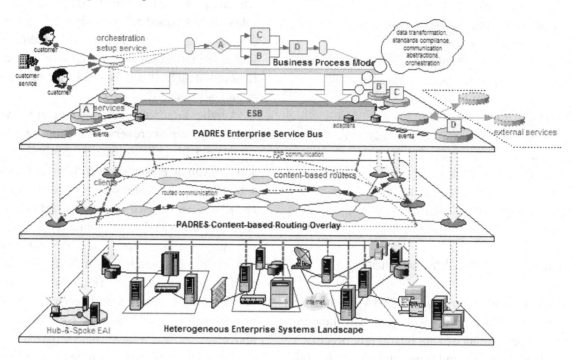

load balancing capabilities of PADRES provide the means to create a robust ESB.

Infrastructure Applications

In an era where applications are increasingly hosted on a distributed infrastructure, composed of components from various partners, used by unknown and possibly hostile users, and yet demand reliable performance, an application developer's concerns no longer end with the development and testing of the product. Effectively monitoring and managing the applications and infrastructure at run-time is necessary to provide an acceptable level of service to users. Owing to its clean decoupling, expressive filtering capabilities and matching predication ability (Liu, 2009), the content-based publish/subscribe paradigm is ideally suited for realizing complex real-time management solutions (Yan, 2009).

Intrusion Detection

One set of infrastructure concerns relates to monitoring a system for malicious attacks. Typically, administrators will specify rules or signatures of attacks they wish to monitor. An attack signature on a Web server may be an excessive request rate for web pages from the same IP address, whereas a credit card fraud may be detected if the same card is used from multiple locations within a short period of time.

For example, the following composite subscription will detect any user attempting to probe a machine for certain open ports.

```
([class, eq, 'TCPOPEN'],
 [dest_port, =, 21],
 [src_ip, eq, $X],
 [dest_ip, eq, $y]) &&
([class, eq, 'TCPOPEN'],
 [dest_port, =, 22],
```

```
[src_ip, eq, $X],
 [dest_ip, eq, $y]) &&
([class, eq, 'TCPOPEN'],
 [dest_port, =, 110],
 [src_ip, eq, $X],
 [dest_ip, eq, $y])
```

To monitor for this signature, a client can issue the above subscription at any time while the system is operational. Likewise, removing the rule is a simple unsubscribe operation by the client and has no effect on the network. The composite subscription above performs in-network filtering so the monitoring client is only notified if the rule is matched, and distributed composite subscription matching algorithms ensure that the rule is detected at the optimal point in the network. For example, if one particular machine in the network is the target of many TCP sessions, the composite subscription detection can automatically move closer to that machine so that events do not have to propagate far into the network before the pattern is matched.

There may be a concern that it is not efficient to require network components to publish events for every conceivable operation such as TCP session state, or link utilization. However, the PADRES system is designed so that events that are of no interest to anyone in the system are immediately dropped and incur no overhead further in the network. For example, if the above subscription is unsubscribed, then TCPOPEN publication would be dropped (assuming there are no other subscriptions interested in these events).

The real power of monitoring a system using the publish/subscribe paradigm is seen when monitoring rules span multiple layers in a system. For example an attack signature may require monitoring both the network infrastructure and application behavior. For example, an attempt to break into a machine and use it as a spam relay may be defined by a pattern of a number of unsuccessful login attempts, followed by a successful one, after which a number of emails are sent.

```
([class, eq, 'LOGIN'],
 [userid, eq, $U],
 [status, eq, 'FAIL'],
 [src_ip, eq, $S]) &&
([class, eq, 'LOGIN'],
 [userid, eq, $U],
 [status, eq, 'FAIL'],
 [src_ip, eq, $S]) &&
([class, eq, 'LOGIN'],
 [userid, eq, $U],
 [status, eq, 'SUCCESS'],
 [src_ip, eq, $S],
 [dest_ip, eq, $D]) &&
([class, eq, 'TCPOPEN'],
 [dest_port, =, 25],
 [src_ip, eq, $D],
 [dest_ip, eq, $X]) &&
([class, eq, 'TCPOPEN'],
 [dest_port, =, 25],
 [src_ip, eq, $D],
 [dest_ip, eq, $Y])
```

In the above subscription, the LOGIN publications are generated by the authentication server (at the application layer), and the network layer components issue the TCPOPEN publications, but the subscription nevertheless is able to retrieve and correlate them in a uniform manner.

Another advantage of using the publish/subscribe model to perform intrusion detection is that the attack signatures are monitored in real-time as the system is running, instead of analyzing log files after the fact. This is important in situations, such as credit card fraud detection where the attack must be managed as soon as possible to prevent further damage.

The ability to correlate events from different layers or applications in a diverse system is also useful in diagnosing the cause of a problem. For example, in an environment where a set of services are deployed on a cluster of machines, an administrator may wish to know which services were invoked shortly before a machine becomes overloaded.

```
([class, eq, 'INVOKE'],
 [service, eq, $A],
 [time, >, $T - 10s]) &&
([class, eq, 'CPULOAD'],
 [machine_id, eq, 'OVERLOAD'],
 [time, =, $T])
```

The above subscription would return all invocations up to 10 seconds before the overload event, and an administrator can use this information to isolate the potential cause of the overload.

Service Level Agreements

System administrators need to be concerned not only with malicious attacks, but with legitimate usage from end users or partners that may affect their applications in unexpected or undesirable ways. To monitor whether their applications are providing (and receiving) the desired performance, it is common for businesses to define service level agreements (SLAs) on large-scale enterprise applications. An SLA defines a contract between a service provider and consumer and precisely outlines how the consumer will use the service, and states the guarantees offered by the provider. Often penalties of not abiding to the terms are also specified in the SLA.

Consider an online retailer running a sales business process shown in Figure 17. Some of the services in this process, such as the shipping services, may be outsourced to other businesses. To provide an acceptable performance to their users, however, the retailer may require certain service guarantees from the shipping services. For example, the retailer may demand that the *Calculate Shipping* service in Figure 17 respond to requests within 0.5 seconds 99.9% of the time within any 24 hour window, and a failure to meet this level of service will result in a loss of payment for that window. Conversely, the SLA may also specify that the retailer will not invoke the

Calculate Shipping service more than 100 times per minute within any one minute window.

Monitoring of such SLAs can be implemented cleanly using a publish/subscribe model (Chau, 2008; Muthusamy, 2008). Suppose the process in Figure 17 is executed in a distributed PADRES execution engine. In this case, invocations of and results from services are represented by publications. The loose coupling of the publish/subscribe paradigm allows the SLA monitoring subsystem to subscribe to these publications without altering the process execution or even stopping the running process. Subscriptions to retrieve the invocation and result publications may look as follows.

```
Sub1: [class, eq, 'INVOKE'],
      [service, eq, 'CalcShipping']
Sub2: [class, eq, 'RESULT'], [ser-
      vice, eq, 'CalcShipping']
```

The monitoring subsystem issuing the above subscription would have to correlate the invocation and result publications and compute the time difference between the two. However, with more complex correlation and aggregation capabilities, it is possible to issue a single subscription that calculates the time difference, and returns the result to the subscriber. For example, the following composite subscription will correlate invocations with their appropriate results and return pairs of publications to the monitoring client. This saves the monitoring client from having to perform the correlation and allows the correlation processing to occur in the network at the optimal point.

```
([class, eq, 'INVOKE'],
 [service, eq, 'CalcShipping'],
 [id, eq, $X]) &&
([class, eq, 'RESULT'],
 [service, eq, 'CalcShipping'],
 [id, eq, $X])
```

Figure 19. SLA monitoring components

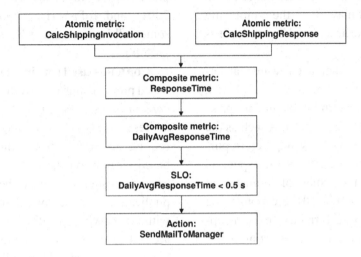

In general, SLAs can be modeled as three types of components: metrics, service level objectives (SLOs), and actions.[4] Metrics measure some phenomenon such as the time when a product is delivered, or the number of times an item is purchased. Sometimes a distinction is made between atomic metrics which measure a property directly, and composite metrics that aggregate the measurements from other metrics. For example, the occurrence of late shipments may be measured by an atomic metric, and the total number of late shipments per day response computed by a composite metric. SLOs are a Boolean expression of some desired state. For example, a desired threshold on the number of late shipments per day may be denoted as *LateShipmentsPerDay < 10*. Finally, actions are descriptions of what should occur when an SLO is violated. Examples of actions include sending an email to a sales manager, generating a publication (that will be processed by another component), and writing to a log file.

Each metric, SLO, and action in an SLA can be mapped to a publish/subscribe client. For example, consider an SLA where an email should be sent to the sales manager if the average daily response time of the *Calculate Shipping* service is greater than 0.5 seconds. Figure 19 shows metrics, SLOs, and actions that realize this SLA. Each component in the figure is modeled as publish/subscribe client, and interactions between them are through publish/subscribe messages. These clients together realize the monitoring of the SLA.

The distributed monitoring architecture coupled with a distributed publish/subscribe system scales well to large SLAs or those that are processing a high volume of publications. Notice that the clients utilize the filtering and in-network aggregation provided by the content-based publish/subscribe model, and being publish/subscribe clients, automatically benefit from the dynamic load-balance and fault-tolerance properties of the system.

An example of in-network filtering is that only those events from the *DailyAvgResponseTime* client in Figure 19 that report a response time greater than 0.5 seconds are delivered to the SLO client. The remaining events are dropped and incur no overhead in the remainder of the network. The *ResponseTime* metric in Figure 19 issues a composite subscription for correlated response and invocation events of the *Calculate Shipping* service. This composite subscription aggregates these pairs of events in the network such that, for example, invocation events with no corresponding response events are not propagated. Furthermore, the load balancing algorithms will ensure that the placement of these clients does not result in load imbalances that may affect the monitoring of the SLA or the execution of the process it is monitoring.

SLAs need not be limited to ensuring that services provided by partners perform as expected, but may also be used internally by an enterprise to monitor various business measures such as the length of time users spend browsing before purchasing an item online, the average time spent on a tech support call, or the number of products that are returned within a month. All these monitoring tasks can be efficiently performed on existing applications without modifying or restarting them.

RELATED WORK

While publish/subscribe was first implemented in centralized client-server systems, current research focuses mainly on distributed versions. The key benefit of distributed publish/subscribe is the natural decoupling of publishers and subscribers. Since the publishers are unconcerned with the potential consumers of their data, and the subscribers are unconcerned with the locations of the potential producers of interesting data, the client interface of the publish/subscribe system is powerful yet simple and intuitive.

Language model: There are several different classes of publish/subscribe systems. *Topic-based* publish/subscribe (Oki, 1993) has a topic associated with each publication which indicates the region of interest for contained data. Clients subscribing to a particular topic would receive all publications with the indicated topic. Topics are similar to the notion of *groups* used in the context of *group communication* (Powell, 1996). *Content-based* publish/subscribe systems add significant functionality by allowing subscribers to specify constraints on the data within a publication. In contrast to the *topic-based* approach, publications are classified according to their content. SIENA (Carzanig, 2001), REBECA (Mühl, 2002), Gry-

phon (Opyrchal, 2000), Le Subscribe (Fabret, 2001), and ToPSS (Liu, 2002; Liu, 2004) are some well-known content-based publish/subscribe prototypes.

Content-based routing: Distributed Content-based publish/subscribe systems typically utilize *content-based routing* in lieu of the standard address-based routing. Messages in content-based routing are routed from source to destination based entirely on the content of the messages. Since publishers and subscribers are decoupled, a publication is routed towards the interested clients without knowing specifically where those clients are and how many such clients exist. Effectively, the content-based *address* of a subscriber is the set of subscriptions it has issued. Two versions of content-based routing are known: simple routing, for example Gryphon (Opyrchal, 2000), and covering-based routing which is discussed in SIENA (Carzanig, 2001), subscription summarization (Triantafillou, 2004) and JEDI (Cugola, 2001). Merging-based routing (Mühl, 2002) is an advanced version of covering based routing. PADRES (Li, 2005; Li, 2008) extends merging-based routing with imperfect merging capabilities that can offer further performance benefits.

General overlays: Most publish/subscribe systems assume an acyclic overlay network. For example, REBECA (Fiege, 2002) explores advanced content-based routing algorithms based on an acyclic broker overlay network, and JEDI (Cugola, 2001) uses a hierarchical overlay for event dispatching. SIENA (Carzaniga, 2001), however, proposes a routing protocol for general overlay networks using reverse path forwarding to detect and discard duplicate messages. In SIENA, any advertisement, subscription or publication message may be duplicated. As well, routing path adaptations to changing network conditions and the implications for composite event detection are not addressed. PADRES (Li, 2008) explores the alternative paths available in a general overlay to provide adaptive and robust message delivery in content-based publish/subscribe systems.

There have been attempts to build content-based publish/subscribe systems over group multicast primitives such as IP multicast (Deering, 1990). To appreciate the challenge in doing so, consider a scenario with N subscribers. In a content-based system, each message may be delivered to any subset of these subscribers, resulting in 2^N "groups". It is infeasible to manage such exponentially increasing numbers of groups, and the algorithms seek to construct a limited number of groups such that the number of groups any given message must be sent to is minimized and the precision of each group is maximized (i.e., minimize the number of group members that are not interested in events sent to that group). This is an NP-complete problem (Adler, 2001), but there have been attempts to develop heuristics to construct such groups (Opyrchal, 2000; Riabov, 2002). To avoid these complexities more recent content-based routing algorithms (Carzanig, 2001; Cugola, 2001) have abandoned the notion of groups and rely on an overlay topology that performs filtering and routing based on message content.

There have been a number of content-based publish/subscribe systems that exploit the properties of distributed hash tables (DHT) to achieve reliability. Hermes (Pietzuch, 2002) builds an acyclic routing overlay over the underlying DHT topology but does not consider alternate publication routing paths as in PADRES. Other approaches (Gupta, 2004; Muthusamy, 2005; Aekaterinidis, 2006; Muthusamy, 2007) construct distributed indexes to perform publish/subscribe matching. PastryStrings (Aekaterinidis, 2006) is a comprehensive infrastructure for supporting rich queries with range and comparison predicates on both numerical and string attributes. It can be applied in a publish/subscribe environment with a broker network implemented using a DHT network. The distinguishing feature of PastryStrings is that it shows how to leverage specific DHT infrastructures to ensure logarithmic message complexity for both publication and subscription

processing. Meghdoot (Gupta, 2004) is a content-based publish/subscribe system build over the CAN DHT. For an application with k attributes, Meghdoot constructs a CAN space of dimension $2k$. Subscriptions are mapped to a point in the CAN space and stored at the responsible node. Publications traverse all regions with possible matching subscriptions. Meghdoot handles routing load by splitting a subscription at a peer to its neighbors. P2P-ToPSS (Muthusamy, 2005), unlike other DHT publish/subscribe systems which focus on large-scale benefits, focuses on small-scale networks. It shows that in small networks (with less than 30 peers) DHTs continue to exhibit good storage load balance of (key,value) pairs, and lookup costs. PADRES, on the other hand, assumes a more traditional dedicated broker network model, one benefit of which is the lack of additional network and computation overhead associated with searching a distributed index to perform publish/subscribe matching. The model in PADRES can achieve lower delivery latencies when there are no failures, but still fall back on alternate path publication routing in case of congestion or failure. Admittedly, the DHT protocols, designed for more hostile network, tend to be more fault-tolerant than the algorithms in this paper which assume a more reliable, dedicated broker network.

Publish/subscribe systems have been developed for even more adverse environments such as mobile ad-hoc networks (MANET). These networks are inherently cyclic but the protocols (Lee, 2000; Petrovic, 2005) require periodic refreshing of state among brokers due to the unreliability of nodes and links, an overhead that is not required by the work in this paper. As well, MANET brokers can exploit wireless broadcast channels to optimize message forwarding. For example, brokers in ODMRP (Lee, 2000) do not maintain forwarding tables, but only record if they lie on the path between sources and sinks in a given group. Brokers simply broadcast messages to their neighbors (discarding duplicates) until the message reaches the destinations. The protocols in PADRES, on the other hand, cannot rely on broadcast transmission and also explicitly attempt to avoid duplicate message delivery. As well, ODMRP does not support the more complex content-based semantics.

Composite Subscriptions: A *composite subscription* correlates publications over time, and describes a complex event pattern. Supporting an expressive subscription language and determining the location of composite event detection in a distributed environment are difficult problems. CEA (Pietzuch, 2004) proposes a Core Composite Event Language to express concurrent event patterns. The CEA language is compiled into automata for distributed event detection supporting regular expression-type patterns. CEA employs polices to ensure that mobile event detectors are located at favorable locations, such as close to event sources. However, CEA's distribution polices do not consider the alternate paths and the dynamic load characteristics of the overlay network.

One of the key challenges in supporting composite subscriptions in a distributed publish/subscribe system is determining how the subscription should be decomposed and where in the network event collection and correlation should occur. While this problem is similar to query plan optimization in distributed DBMS (Özsu, 1999) and distributed stream processing (Kumar, 2006), data in a relation or a stream have a known schema which simplifies matching and routing. Moreover, a database query is evaluated once against existing data, while a subscription is evaluated against publications over time. This may result in different optimization strategies and cost models. In the IFLOW (Kumar, 2006) distributed stream processing engine, a set of operators are installed in the network to process streams. IF-LOW nodes are organized in a cluster hierarchy, with nodes higher in the hierarchy assigned more responsibility, whereas in PADRES (Li, 2005), brokers have equal responsibility.

Load Balancing: Although distributed content-based publish/subscribe systems have been widely studied, load balancing was never directly addressed. The following are various related works that propose load balancing techniques in other publish/subscribe approaches.

Meghdoot (Gupta, 2004) distributes load by replicating or splitting the locally heaviest loaded peer in half to share the responsibility of subscription management or event propagation. Such partitioning and replication schemes are common load balancing techniques used in other DHT-based publish/subscribe systems (Aekaterinidis, 2006; Zhu, 2007). In general, their load sharing algorithm is only invoked upon new peers joining the system and peers are assumed to be homogeneous. (Chen, 2005) proposed a dynamic overlay reconstruction algorithm called *Opportunistic Overlay* that reduces end-to-end delivery delay and also performs load distribution on the CPU utilization as a secondary requirement. Load balancing is triggered only when a client finds another broker that is closer than its home broker. It is possible that subscriber migrations may overload a non-overloaded broker if the load requirements of the migrated subscription exceed the load-accepting broker's processing capacity. Subscription clustering is another technique to achieve load balancing in content-based publish/subscribe systems (Wong, 2000; Riabov, 2002; Riabov, 2003; Casalicchio, 2007). Subscriptions of similar interests are clustered together at different servers to distribute load. However, architecturally, this technique is not applicable to filter-based but only to multicast-based publish/subscribe systems. PADRES differs from the prior three solutions by proposing a distributed load balancing algorithm for non-DHT filter-based publish/subscribe systems that accounts for heterogeneous brokers and subscribers, and distributes load evenly onto all resources in the system without requiring new client joins. As well, a subscriber migration protocol enforces end-user transparency and best-effort delivery to minimize message loss.

Fault-tolerance: Most of the previous work in the fault-tolerant publish/subscribe literature take a best-effort approach and fail to provide strict publication delivery guarantees. Gryphon (Bhola, 2002) is one of the few systems that ensure a similar level of reliability as in PADRES. However, in order to achieve δ-fault-tolerance, the routing information of each Gryphon broker must be replicated on δ+1 other nodes. This design is prone to over provisioning of resources. On the other hand, if a load balancing mechanism is present to improve the node utilization, there are chances that failure of some nodes overwhelms their non-faulty peers (replicas). PADRES improves on these shortcomings by not requiring the assignment of additional replicas. Moreover, the incoming traffic to non-faulty nodes in our system is independent of the number of failures. This implies that in presence of failures, non-faulty peers observe a much lower load increase which is the result of an increase on the number of outgoing messages only. Snoeren et al. (Snoeren, 2001) propose another approach to implement a fault-tolerant P/S system which is based on the construction of several *disjoint* paths between each pair of publishers and subscribers. Publications messages are concurrently forwarded on all disjoint paths enabling the system to tolerate multiple failures. However, this implies that even in non-faulty conditions the publication traffic can be several times the traffic of the system without fault-tolerance support. In many cases, this overhead makes this scheme impractical. On the other hand, this approach has the advantage of minimizing the impact of failures on publication delivery delay, as the delay is equal to the delivery delay of publications propagated on the fastest path.

CONCLUSION

This chapter gave an overview of the PADRES content-based publish/subscribe system. It de-scribed the message format, subscription language, and data model used in the system. Content-based routing was discussed with particular emphasis on how routing is enabled in cyclic overlays. Cyclic overlays provide redundancy in routes between sources and sinks and thus produce alternative paths between them. Therefore, unlike acyclic overlays, cyclic overlays can be more easily exploited to design a system that can tolerate load imbalances, congestion, and broker failures.

In addition to the ability to route around the affected parts of the network, PADRES also implements other efficient load balancing and recovery algorithms to handle load imbalances and broker failures. These techniques were described in details in this chapter.

To exemplify how content-based publish/subscribe can be used in practice, we presented a detailed discussion of example applications that benefit from the content-based nature of the paradigm. These applications can also take advantage of the scalability and robustness of PADRES.

The PADRES code base is released under an open source license (http://padres.msrg.utoronto.ca). The release comprises the PADRES publish/subscribe broker, a client library that allows third party applications to make use of PADRES, a monitoring client, a set of application demonstrations, and the PANDA deployment tool(http://research.msrg.utoronto.ca/Padres/PadresDownload). A user and developer guide is also available.

The PADRES publish/subscribe broker is based on a content-based matching engine that supports the subscription language described in Section 1, including atomic subscriptions, the various forms of historic subscriptions, composite subscriptions with conjunctive and disjunctive operators, the *isPresent* operator, variable bindings, and event correlation with different consumption policies. The PADRES broker was based on the Jess rule engine (Friedman-Hill,2003), not distributed with our release. The released broker is still compatible with the Jess rule engine, which can be used instead of the matching engine distributed in the release. Most of the results reported in our publications are based on the Jess rule engine as the content-based matching and event correlation mechanism for PADRES. All features described in this chapter, except the load balancing and the fault tolerance features, are included in the PADRES open source release.

PADRES is used in several research and development projects. In the eQoSystem project with IBM (Jacobsen, 2006; Muthusamy, 2007; Chau, 2008), PADRES constitutes the enterprise service bus that enables the monitoring and enforcement of SLAs of composite applications and business processes in service oriented architectures. In collaborations with Bell Canada (Jacobsen, 2007), PADRES serves to study enterprise application integration problems pertaining to the integration and execution of business processes across existing integration hubs. In collaborations with CA and Sun Microsystems, PADRES is used to explore the event-based management of business processes and business activity monitoring (Li, 2007). In collaborations with the Chinese Academy of Sciences, PADRES is used for service selection (Hu, 2008) and for resource and service discovery in computational Grids (Yan, 2009).

ACKNOWLEDGMENT

The PADRES research project was sponsored by CA, Inc., Sun Microsystems, the Ontario Centers of Excellence, the Canada Foundation for Innovation, the Ontario Innovation Trust, the Ontario Early Researcher Award, and the Natural Sciences and Engineering Research Council of Canada. The completion of the research described in this chapter was also made possible in part thanks to the support through Bell Canada's Bell University Laboratories R&D program, the IBM's Center for Advanced Studies and various IBM Faculty Awards. The authors would like to thank Serge Mankovskii, CA, Inc. for valuable input to the research presented in this chapter. The authors would also like to acknowledge the contributions of other past and present members of the PADRES research team for their contributions to the project. This includes Chen Chen, Amer Farroukh, Eli Fidler, Gerald Chen, Ferdous Jewel, Patrick Lee, Jian Li, David Matheson, Pengcheng Wan, Alex Wun, Shuang Hou, Songlin Hu, Naweed Tajuddin, and Young Yoon.

REFERENCES

Adler, M., Ge, Z., Kurose, J., Towsley, D., & Zabele, S. (2001). Channelization Problem in Large Scale Data Dissemination. In *Proceedings of the Ninth International Conference on Network Protocols*(pp.100-110). Washington, DC: IEEE Computer Society.

Aekaterinidis, I., & Triantafillou, P. (2006). *PastryStrings: A comprehensive content based publish/subscribe DHT network*. New York: Springer.

Andrews, T. (2003). *Business Process Execution Language for Web Services*. Retrieved Oct. 31 2006, from http://www.ibm.com/developerworks/library/specification/ws-bpel/

Ashayer, G., Leung, H., & Jacobsen, H.-A. (2002). Predicate Matching and Subscription Matching in publish/subscribe Systems. In *Proceedings of the 22nd International Conference on Distributed Computing Systems*(pp. 539 - 548). Washington, DC: IEEE Computer Society.

Bhola, S., Strom, R. E., Bagchi, S., Zhao, Y., & Auerbach, J. S. (2002). Exactly-once Delivery in a Content-based Publish-Subscribe System. In *Proceedings of the 2002 International Conference on Dependable Systems and Networks,* (pp 7-16). Washington, DC: IEEE Computer Society.

Birman, K. P., Hayden, M., Ozkasap, O., Xiao, Z., Budiu, M., & Minsky, Y. (1999). Bimodal multicast. *ACM Transactions on Computer Systems, 17*(2), 41–88. doi:10.1145/312203.312207

Bittner, S., & Hinze, A. (2007). The arbitrary Boolean publish/subscribe model: making the case. In *Proceedings of the 2007 inaugural international conference on Distributed event-based systems,* (pp 226 - 237). New York: ACM.

Carzaniga, A., Rosenblum, D. S., & Wolf, A. L. (2001). Design and Evaluation of a Wide-Area Event Notification Service. *ACM Transactions on Computer Systems, 19*(3), 332–383. doi:10.1145/380749.380767

Carzaniga, A., & Wolf, A. L. (2003). Forwarding in a Content-Based Network. In *Proceedings of the 2003 conference on Applications, technologies, architectures, and protocols for computer communications* (pp 163-174). New York: ACM.

Casalicchio, E., & Morabito, F. (2007). Distributed subscriptions clustering with limited knowledge sharing for content-based publish/subscribe systems. In Proceedings of Network Computing and Applications, (pp 105-112). Cambridge, MA.

Castro, M., Druschel, P., Kermarrec, A. M., & Rowstron, A. (2002). SCRIBE: A large-scale and decentralized application-level multicast infrastructure. *IEEE Journal on Selected Areas in Communications, 20*(8), 1489–1499. doi:10.1109/JSAC.2002.803069

Chau, T., Muthusamy, V., Jacobsen, H. A., Litani, E., Chan, A., & Coulthard, P. (2008). Automating SLA modeling. In *Proceedings of the 2008 conference of the Centre for Advanced Studies on Collaborative Research,* Richmond Hill, Canada.

Chen, Y., & Schwan, K. (2005). Opportunistic Overlays: Efficient Content Delivery in Mobile Ad Hoc Networks. In *Proceedings of the 6th ACM/IFIP/USENIX International Middleware Conference* (pp 354-374). New York: Springer.

Cheung, A., & Jacobsen, H.-A. (2006). Dynamic Load Balancing in Distributed Content-based Publish/Subscribe. In *Proceedings of the 7th ACM/IFIP/USENIX International Middleware Conference* (pp 249-269). New York: Springer.

Cheung, A., & Jacobsen, H.-A. (2008). *Efficient Load Distribution in Publish/Subscribe (Technical report).* Toronto, Canada: Middleware Systems Research Group, University of Toronto.

Cugola, G., Nitto, E. D., & Fuggetta, A. (2001). The JEDI event-based infrastructure and its application to the development of the OPSS WFMS. [Piscataway, NJ: IEEE Press.]. *IEEE Transactions on Software Engineering, 27*(9), 827–850. doi:10.1109/32.950318

Deering, S., & Cheriton, D. R. (1990). Multicast routing in datagram internetworks and extended LANs. *ACM Transactions on Computer Systems, 8*(2), 85–111. doi:10.1145/78952.78953

Eugster, P. T., Felber, P. A., Guerraoui, R., & Kermarrec, A. M. (2003). The many faces of publish/subscribe. *ACM Computing Surveys, 35*(2), 114–131. doi:10.1145/857076.857078

Fabret, F., Jacobsen, H.-A., Llirbat, F., Pereira, J., Ross, K. A., & Shasha, D. (2001). Filtering algorithms and implementation for very fast publish/subscribe systems. In *Proceedings of the 2001 ACM SIGMOD international conference on management of data* (pp 115-126), New York: ACM.

Fawcett, T., & Provost, F. (n.d.). Activity monitoring: Noticing interesting changes in behavior. In *Proceedings of the fifth ACM SIGKDD international conference on Knowledge discovery and data mining* (pp 53-62), New York: ACM.

Fiege, L., Mezini, M., Mühl, G., & Buchmann, A. P. (2002). Engineering Event-Based Systems with Scopes. In *Proceedings of the 16th European Conference on Object-Oriented Programming* (pp 309-333), Berlin, Germany: Springer.

Forgy, C. L. (1982). Rete: A Fast Algorithm for the Many Pattern/Many Object Pattern Match Problem. *Artificial Intelligence, 19*(1), 17–37. doi:10.1016/0004-3702(82)90020-0

Friedman-Hill, E. J. (2003). *Jess, The Rule Engine for the Java Platform*. Retrieved from http://herzberg.ca.sandia.gov/jess/

Gupta, A., Sahin, O. D., Agrawal, D., & Abbadi, A. E. (2004). Meghdoot: Content-Based publish/subscribe over P2P Networks. In *Proceedings of the 5th ACM/IFIP/USENIX International Middleware Conference* (pp 254-273), New York: Springer.

Hu, S., Muthusamy, V., Li, G., & Jacobsen, H.-A. (2008). Distributed Automatic Service Composition in Large-Scale Systems. In *Proceedings of the second international conference on Distributed event-based systems* (pp 233-244), New York: ACM.

IBM. (2003). *Web Service Level Agreements (WSLA) Project*. Retrieved July 12th, 2007, from http://www.research.ibm.com/wsla/

Jacobsen, H.-A. (2004). *PADRES User Guide*. Retrieved July 19, 2006, from http://research.msrg.utoronto.ca/Padres/UserGuide

Jacobsen, H.-A. (2006). *eQoSystem*. http://research.msrg.utoronto.ca/Eqosystem

Jacobsen, H.-A. (2007). *Enterprise Application Integration*. http://research.msrg.utoronto.ca/EAI/

Koenig, I. (2007). Event Processing as a Core Capability of Your Content Distribution Fabric. In *Proceedings of the Gartner Event Processing Summit,* Orlando, FL.

Kumar, V., & Cai, Z. (2006). Implementing Diverse Messaging Models with Self-Managing Properties using IFLOW. *IEEE International Conference on Autonomic Computing* (pp 243-252). IEEE Computer Society. Washington, DC.

Lee, S., Su, W., Hsu, J., Gerla, M., & Bagrodia, R. (2000). A performance comparison study of ad hoc wireless multicast protocols. In *Proceedings of 9th Annual Joint Conference of the IEEE Computer and Communications Societies* (pp 565-574), IEEE Computer Society. Washington, DC.

Li, G., Cheung, A., Hou, S., Hu, S., Muthusamy, V., Sherafat, R., et al. (2007). Historic data access in publish/subscribe. In *Proceedings of the 2007 inaugural international conference on Distributed event-based systems* (pp 80-84), Toronto, Canada.

Li, G., Hou, S., & Jacobsen, H.-A. (2005). A Unified Approach to Routing, Covering and Merging in Publish/Subscribe Systems based on Modified Binary Decision Diagrams. In *Proceedings of the 25th IEEE International Conference on Distributed Computing Systems* (pp 447-457), Columbus, OH.

Li, G., Hou, S., & Jacobsen, H.-A. (2008). Routing of XML and XPath Queries in Data Dissemination Networks. In *Proceedings of the 28th IEEE International Conference on Distributed Computing Systems* (pp 627-638), Beijing, China.

Li, G., & Jacobsen, H.-A. (2005). Composite Subscriptions in Content-Based publish/subscribe Systems. In *Proceedings of the 6th ACM/IFIP/USENIX International Middleware Conference* (pp 249-269), Berlin, Germany: Springer.

Li, G., Muthusamy, V., & Jacobsen, H.-A. (2007). *NIÑOS: A Distributed Service Oriented Architecture for Business Process Execution. (Technical report)*. Tortonto, Canada: Middleware Systems Research Group, University of Toronto.

Li, G., Muthusamy, V., & Jacobsen, H.-A. (2008). Adpative content-based routing in general overlay topologies. In *Proceedings of the 9th ACM/IFIP/USENIX International Middleware Conference* (pp 249-269), Berlin, Germany: Springer.

Li, G., Muthusamy, V., & Jacobsen, H.-A. (2008). *Subscribing to the past in content-based publish/subscribe. (Technical report)*. Toronto, Canada: Middleware Systems Research Group.

Liu, H., & Jacobsen, H. A. (2002). A-ToPSS: A Publish/Subscribe System Supporting Approximate Matching. In *Proceedings of 28th International Conference on Very Large Data Bases* (pp 1107-1110), Hong Kong, China.

Liu, H., & Jacobsen, H. A. (2004). Modeling uncertainties in publish/subscribe systems. In *Proceedings of the 20th International conference on Data Engineering* (pp 510-522), Boston, MA.

Liu, H., Muthusamy, V., & Jacobsen, H. A. (2009). *Predictive Publish/Subscribe Matching. (Technical report)*. Toronto, Canada: Middleware Systems Research Group, University of Toronto.

Liu, H., Ramasubramanian, V., & Sirer, E. G. (2005). Client behavior and feed characteristics of RSS, a publish-subscribe system for web micronews. In *Proceedings of the 5th ACM SIGCOMM conference on Internet Measurement* (pp 3-3), Berkeley, CA: USENIX Association.

Mühl, G. (2002). *Large-scale content-based publish/subscribe systems.* Unpublished doctoral dissertation, Darmstadt, Germany: Darmstadt University of Technology.

Mukherjee, B., Heberlein, L. T., & Levitt, K. N. (1994). Network intrusion detection. *IEEE Network, 8*(3), 26–41. doi:10.1109/65.283931

Muthusamy, V., & Jacobsen, H.-A. (2005). Small-scale Peer-to-peer Publish/Subscribe. *Proceedings of the MobiQuitous Conference* (pp 109-119), New York: ACM.

Muthusamy, V., & Jacobsen, H.-A. (2007). *Infrastructure-less Content-Based Pub. (Technical report)*. Toronto, Canada: Middleware Systems Research Group, University of Toronto.

Muthusamy, V., & Jacobsen, H.-A. (2008). SLA-driven distributed application development. In *Proceedings of the 3rd Workshop on Middleare for Service Oriented Computing* (pp 31-36), Leuven, Belgium.

Muthusamy, V., Jacobsen, H.-A., Coulthard, P., Chan, A., Waterhouse, J., & Litani, E. (2007). SLA-Driven Business Process Management in SOA. In *Proceedings of the 2007 conference of the center for advanced studies on Collaborative research* (pp 264-267), Ontario, Canada.

Nayate, A., Dahlin, M., & Iyengar, A. (2004). Transparent information dissemination. In *Proceedings of the 5th ACM/IFIP/USENIX International Middleware Conference* (pp 212 - 231), Berling, Germany: Springer.

Oki, B., Pfluegl, M., Siegel, A., & Skeen, D. (1993). The Information Bus: an architecture for extensible distributed systems. In *Proceedings of the fourteenth ACM symposium on Operating systems principles* (pp 58-68). New York.

Opyrchal, L., Astley, M., Auerbach, J., Banavar, G., Strom, R., & Sturman, D. (2000). Exploiting IP multicast in content-based publish-subscribe systems. *IFIP/ACM International Conference on Distributed systems platforms,* (pp 185-207), New York: Springer.

Ostrowski, K., & Birman, K. (2006). Extensible Web Services Architecture for Notification in Large-Scale Systems. *Proceedings of the IEEE International Conference on Web Services* (pp 383-392), Washington, DC: IEEE Computer Society.

Özsu, M. T., & Valduriez, P. (1999). *Principles of Distributed Database Systems*. Upper Saddle River, NJ: Prentice Hall.

Petrovic, M., Muthusamy, V., & Jacobsen, H.-A. (2005). Content-based routing in mobile ad hoc networks. In *Proceedings of the Second Annual International Conference on Mobile and Ubiquitous Systems: Networking and Services* (pp 45-55), San Diego, CA.

Pietzuch, P. R., & Bacon, J. (2002). Hermes: A Distributed Event-Based Middleware Architecture. In *Proceedings of the 22nd International Conference on Distributed Computing Systems*, (pp 611-618), Washington, DC: IEEE Computer Society.

Pietzuch, P. R., Shand, B., & Bacon, J. (2004). *Composite Event Detection as a Generic Middleware Extension. IEEE Network Magazine, Special Issue on Middleware Technologies for Future Communication Networks* (pp. 44–55). Washington, DC: IEEE Computer Society.

PlanetLab. (2006). *PlanetLab*. Retrieved from http://www.planet-lab.org/

Powell, D. (1996). Group communication. [New York: ACM.]. *Communications of the ACM, 39*(4), 50–53. doi:10.1145/227210.227225

Riabov, A., Liu, Z., Wolf, J. L., Yu, P. S., & Zhang, L. (2002). Clustering algorithms for content-based publication-subscription systems. In *Proceedings of the 22nd International Conference on Distributed Computing Systems* (pp 133-142), Washington, DC: IEEE Computer Society.

Riabov, A., Liu, Z., Wolf, J. L., Yu, P. S., & Zhang, L. (2003). New Algorithms for Content-Based Publication-Subscription Systems. In *Proceedings of the 23nd International Conference on Distributed Computing Systems* (pp 678- 686), Washington, DC: IEEE Computer Society.

Rose, I., Murty, R., Pietzuch, P., Ledlie, J., Roussopoulos, M., & Welsh, M. (2007). Cobra: Content-based Filtering and Aggregation of Blogs and RSS Feeds. In *Proceedings of the 4th USENIX Symposium on Networked Systems Design & Implementation* (pp 231-245), Cambridge, MA.

Schuler, C., Schuldt, H., & Schek, H. J. (2001). Supporting Reliable Transactional Business Processes by publish/subscribe Techniques. In *Proceedings of the Second International Workshop on Technologies for E-Services* (pp 118-131), London, UK: Springer-Verlag.

Sherafat Kazemzadeh, R., & Jacobsen, H.-A. (2007). *Fault-Tolerant Publish/Subscribe systems. (Technical report)*. Toronto, Canada: Middleware Systems Research Group, University of Toronto.

Sherafat Kazemzadeh, R., & Jacobsen, H.-A. (2008). *Highly Available Distributed Publish/Subscribe Systems. (Technical report)*. Toronto, Canada: Middleware Systems Research Group, University of Toronto.

Snoeren, A. C., Conley, K., & Gifford, D. K. (2001). Mesh-based content routing using XML. *ACM SIGOPS Operating Systems Review., 35*(5), 160–173. doi:10.1145/502059.502050

Tock, Y., Naaman, N., Harpaz, A., & Gershinsky, G. (2005). Hierarchical Clustering of Message Flows in a Multicast Data Dissemination System. In *Proceedings of the* 17th *IASTED International Conference on Parallel and Distributed Computing and Systems* (pp 320-326), Calgary, Alberta: ACTA Press.

Triantafillou, P., & Economides, A. Subscription Summarization: A New Paradigm for Efficient publish/subscribe Systems. In *Proceedings of the 24nd International Conference on Distributed Computing Systems,* (pp 562-571), Washington, DC: IEEE Computer Society.

Wong, T., Katz, R. H., & McCanne, S. (2000). An evaluation of preference clustering in large-scale multicast applications. In *Proceedings of the conference on computer communications* (pp 451-460), Washington, DC: IEEE Computer Society.

Yan, W., Hu, S., Muthusamy, V., Jacobsen, H.-A., & Zha, L. (2009). Efficient event-based resource discovery. In *Proceedings of the 2009 inaugural international conference on Distributed event-based systems*, Nashville, TN.

Zhu, Y., & Hu, Y. (2007). Ferry: A P2P-Based Architecture for Content-Based publish/subscribe Services. *IEEE Transactions on Parallel and Distributed Systems*, *18*(5), 672–685. doi:10.1109/TPDS.2007.1012

ENDNOTES

[1]   The project name PADRES is an acronym that was initially comprised of letters (mostly first letters of first names) of the initial group of researchers working on the project. Over time, the acronym was also synonymously used as name, simply written Padres. Both forms are correct. Also, various re-interpretations of the acronym have been published, such as *Publish/subscribe Applied to Distributed REsource Scheduling*, *PAdres is Distributed REsource Scheduling*, etc.

[2]   The PADRES language is nevertheless fully content-based and supports a rich predicate language.

[3]   Operator 'eq' is used for String type values and '=' is used for Integer and float type values.

[4]   These terms are borrowed from the Web Service Level Agreements (WSLA) specification.

Chapter 9
TinyDDS:
An Interoperable and Configurable Publish/Subscribe Middleware for Wireless Sensor Networks

Pruet Boonma
University of Massachusetts, USA

Junichi Suzuki
University of Massachusetts, USA

ABSTRACT

Due to stringent constraints in memory footprint, processing efficiency and power consumption, traditional wireless sensor networks (WSNs) face two key issues: (1) a lack of interoperability with access networks and (2) a lack of flexibility to customize non-functional properties such as event filtering, data aggregation and routing. In order to address these issues, this chapter investigates interoperable publish/subscribe middleware for WSNs. The proposed middleware, called TinyDDS, enables the interoperability between WSNs and access networks by providing programming language interoperability and protocol interoperability based on the standard Data Distribution Service (DDS) specification. Moreover, TinyDDS provides a pluggable framework that allows WSN applications to have fine-grained control over application-level and middleware-level non-functional properties. Simulation and empirical evaluation results demonstrate that TinyDDS is lightweight and efficient on the TinyOS and SunSPOT platforms. The results also show that TinyDDS simplifies the development of publish/subscribe WSN applications.

INTRODUCTION

Wireless sensor networks (WSNs) have been used to detect events and/or collect data in various domains such as environmental observation, structural health monitoring, human health monitoring, inventory tracking, home/office au-tomation and military surveillance. Due to their deeply-embedded pervasive nature, WSNs have a potential to revolutionize the way that humans understand and construct complex natural/physical systems (Estrin et al., 1999).

A WSN application requires per-node embedded software that imposes stringent constraints in memory footprint, processing efficiency and power consumption. In order to satisfy these

DOI: 10.4018/978-1-60566-697-6.ch009

constraints, traditional WSN applications often result in *vertically integrated* and *tightly coupled* designs. Vertically integrated designs make WSN applications less interoperable. For example, most of traditional WSNs lack interoperability with access networks, which allow human users to connect to WSNs and perform information retrieval such as data collection and event detection (Henricksen & Robinson, 2006; Romer et al., 2002; Hadim & Mohamed, 2006). Despite the interoperability can foster the practicality and production deployment of WSNs, they have been investigated and designed separately from access networks. As a result, it is often ad-hoc, expensive and error-prone to build a gateway node, which is responsible for protocol bridging and data conversion between WSNs and access networks. Currently, gateways need to be rebuilt from scratch when WSNs and access networks use different programming languages and protocols.

Tightly coupled designs make WSN applications less flexible. In WSN applications, it is hard to flexibly introduce, reuse, customize and replace various non-functional properties such as event correlation, event filtering, data aggregation and routing policies. Currently, changes in non-functional properties require substantial re-designs and re-programming of WSN applications. As a result, the productivity of WSN application development remains low, and the cost of application maintenance remains high.

In order to address the aforementioned interoperability and flexibility issues, this chapter investigates interoperable publish/subscribe communication with TinyDDS, which is open-source[1], standards-based and configurable middleware for WSNs. It is designed and implemented generic enough to aid in developing a wide range of event detection and data collection applications. Compliant with Object Management Group (OMG)'s standard Data Distribution Service (DDS) specification (Object Management Group, 2007), TinyDDS provides two types of interoperability: *programming language interoperability* and *protocol interoperability*.

Programming language interoperability is the ability of TinyDDS to interoperate applications written in different programming languages. TinyDDS implements a set of standard DDS APIs in nesC[2] (Gay et al., 2003) and Java Micro Edition (Simon & Cifuentes, 2005) by providing mappings of the OMG Interface Definition Language (IDL) (Object Management Group, 2007) to the two languages. TinyDDS' nesC version operates on the TinyOS platform (Levis, et al., 2005), and its Java version operates on the SunSPOT platform (Simon & Cifuentes, 2005). This allows different applications to use different languages with the same DDS APIs. For example, an access network application (or end-user application) may be implemented with Java or JavaScript, while a WSN application may be implemented with nesC or Java. Application developers do not have to learn/use different APIs for different applications. This can significantly improve their productivity in application development.

Protocol interoperability is the ability of TinyDDS to interoperate WSNs and access networks built with different MAC (L2), routing (L3) and transport (L4) protocols. TinyDDS implements a session (L5) protocol, called TinyGIOP, which is a subset of the OMG General Inter-ORB Protocol (GIOP) (Object Management Group, 2007). Similar to GIOP, TinyGIOP is independent from underlying L2 to L4 protocols. It transmits data formatted with TinyCDR, which is a subset of the OMG Common Data Representation (CDR) (Object Management Group, 2007). CDR defines the standard binary representations of IDL data types. Taking advantage of TinyGIOP and TinyCDR, TinyDDS allows gateway nodes to be reusable to bridge various WSNs and access networks even if the two networks use different L2, L3 and L4 protocols. This way, it is intended to reduce the costs (time and labor) to build and maintain gateways. TinyDDS is the first DDS implementation for the TinyOS and SunSPOT platforms.

TinyDDS addresses the flexibility issue described earlier by providing a pluggable framework that decouples various non-functional properties from WSN applications. The framework allows WSN applications to flexibly reuse and configure non-functional properties according to their requirements. For example, an event detection application may require in-network event correlation and filtering as its non-functional properties in order to eliminate false positive sensor data in the network. A data collection application may require data aggregation as its non-functional property in order to reduce traffic volume and expand the network's lifetime. Without breaking the generic architecture of TinyDDS, its pluggable framework allows WSN applications to have fine-grained control over non-functional properties and specialize in their own requirements. Currently, TinyDDS supports two types of non-functional properties: *application-level* and *middleware-level* non-functional properties. TinyDDS is the first middleware for WSN applications to flexibly configure the two types of non-functional properties.

This chapter describes the design, implementation and performance of TinyDDS. It discusses the layered architecture of TinyDDS, followed by details of each layer, application development process with nesC and Java, and a pluggable framework for non-functional properties. This chapter also evaluates TinyDDS' performance through blackbox and whitebox measurements in simulation and empirical experiments. TinyDDS is lightweight and efficient, and simplifies the development of publish/subscribe WSN applications.

BACKGROUND

This section overviews the publish/subscribe communication scheme and describes the standard DDS specification.

Publish/Subscribe Communication in WSNs

The publish/subscribe (pub/sub) communication scheme (Banavar et al., 1999; Eugster et al., 2003) is expected to significantly aid in developing and maintaining WSN applications by decoupling space and time among event source nodes (publishers) and sink nodes (subscribers) (Hadim & Mohamed, 2006; Wang et al., 2008; Henricksen & Robinson, 2006). In the pub/sub scheme, a subscriber has the ability to express its interest in an event or a pattern of events in order to be notified subsequently. Each interest is subscribed to a publisher(s), and the publisher(s) notifies an event to a subscriber(s) when the event matches a subscribed interest. Publishers do not need to know the number and locations of subscribers, and vice versa. Thus, publishers indirectly publish events to subscribers, and subscribers indirectly subscribe their interests to publishers. Moreover, publishers do not need to know the availability of subscribers, and vice versa. For example, subscribers may be active, sleeping or dead due to a lack of battery when a publisher publishes an event to them. Event subscription and publication are asynchronously transmitted among publishers and subscribers. By decoupling publishers and subscribers in space and time, the pub/sub scheme can improve scalability in terms of network size and traffic volume.

OMG DDS Specification

DSS is a specification that OMG standardizes for pub/sub middleware. It provides standard interfaces for event subscription and publication in Interface Definition Language (IDL). TinyDDS implements them with nesC and Java for the TinyOS and SunSPOT platforms, respectively. DDS consists of two layers: a lower-level fundamental layer called Data-Centric Publish-Subscribe (DCPS) and a higher-level optional layer called Data Local Reconstruction Layer (DLRL). DCPS defines a set of interfaces for event subscription and publication. Using the interfaces, each event is defined with an associated *topic*. The interfaces also allow applications to declare their intents to become publishers and subscribers and transmit event subscriptions/publications between publishers and subscribers. DLRL automatically obtains events from a remote publisher and allows a subscriber to access the events as if they were locally available. Currently, TinyDDS implements DCPS only to minimize its memory footprint and processing overhead.

Figure 1 and Figure 2 show key components in DDS. Figure 3 and Figure 4 illustrate how these DDS components are used in the subscription and publication processes, respectively. Each node operates a single instance of DomainPartipant for each domain. A domain is a context to which a

Figure 1. DDS architecture

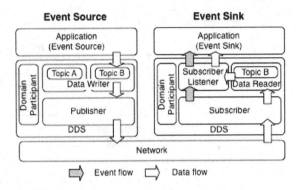

DDS application is associated. A DomainPartipant maintains references to all objects associated to the same domain.

When an event-sink application subscribes to an event(s), it instantiates Subscriber with the local DomainParticipant (Figure 3). Then, it creates an

Figure 2. Standard DDS interfaces

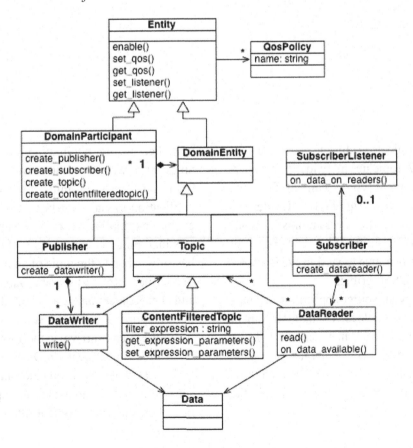

Figure 3. Subscription process

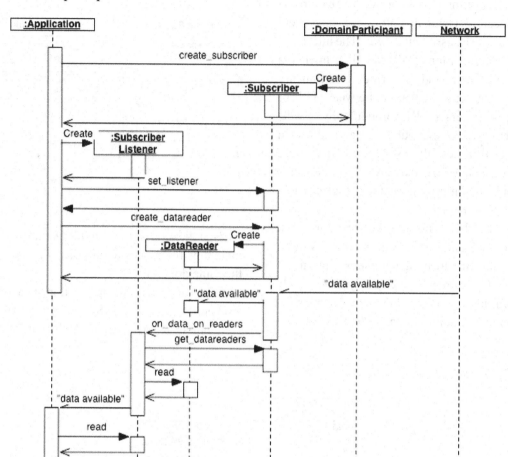

instance(s) of Topic according to the event(s) it is interested in. A topic uniquely identifies a particular event's content type or context. For each topic, the application instantiates DataReader and SubscriberListener as the access points for reading event data in the future (Figure 3). An event subscription is transmitted toward an event-source application(s) via Subscriber.

Similarly, an event-source application instantiates Publisher with the local DomainParticipant (Figure 4). It creates an instance(s) of Topic according to the event(s) it generates. For each topic, the application instantiates DataWriter as the access point for writing out event data in the future (Figure 4).

When an event-source application generates an event, it writes out the event to a DataWriter. Then, the event is transmitted toward an event-sink application(s) via Publisher. At a node where an event-sink application runs, a Subscriber monitors incoming event messages. If the application has subscribed to the topic of an incoming event, Subscriber informs the local SubscriberListener and DataReader that are associated with the event topic (Figure 1). Then, the SubscriberListener informs the event's arrival to the application, which in turn reads the event via DataReader.

Instead of receiving all incoming events of the subscribed topics, an event-sink application can filter them out with a ContentFilteredTopic, which

Figure 4. Publication process

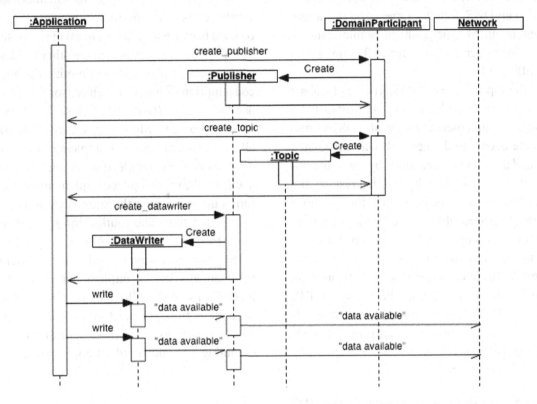

TINYDDS ARCHITECTURE

derives Topic (Figure 2). ContentFilteredTopic is used to specify a subscription interest in the events whose contents satisfy certain criteria. For example, an event-sink application can specify an interest in the events whose topic is Temperature and whose contents is in between 100 and 150 degrees by defining "Temperature > 100 AND Temperature < 150" as criteria in a ContentFilteredTopic. In DDS, event filtering expression is described with a subset of SQL syntax.

Besides a set of standard DDS interfaces, the DDS specification defines no algorithms/protocols for event publication and subscription; they are left to DDS implementations. TinyDDS implements a subscription protocol and several publication protocols as subsequent sections discuss.

In Figure 5, TinyDDS architecture running in each sensor nodes is shown in the figure, with labeled TinyDDS. Currently, there are two implementations of TinyDDS, one for TinyOS platform, for example, Mica Z, Mica 2 or iMotes2 sensor nodes. The other is for Sun Microsystem's SunSPOT platform. The figure shows the architecture of TinyDDS for TinyOS running on TinyOS, i.e., inside a Mica Z mote, on the left hand side and the TinyDDS for SunSPOT platform on the right hand side. TinyDDS running on TinyOS-based sensor nodes is implemented in nesC programming language. On the other hand, TinyDDS running on SunSPOT platform is implemented on Java programming language and operates on top of Squawk Virtual Machine inside SunSPOT sensor nodes. With respect to TCP/IP reference model, TinyDDS operates in transport layer and work on top of any network layer (L3) implementation. TinyDDS follows Layer design pattern (Buschnmann, Meunier, Rohnert, Sommerland, & Stal, 1996) by separating different functionalities into different layers.

At the top layer, TinyDDS provides a subset of DDS interfaces to be used by applications. An application implemented on top of DDS can disseminate events to the network with associated topic and the events are captured by any subscribers, i.e., base station, who has interest on the topic of the events. The implementation of those interfaces, as described in the previous section, operates on top of an overlay network for event routing. Different routing protocols can be used to implement the overlay network by implementing in the Overlay Event Routing Protocols (OERP) layer. This OERP layer allows application developer to choose appropriate routing protocol to suit their requirements and constraints. For example,

in sensor network with very limited memory space sensor nodes, spanning-tree routing protocol may be used because it needs minimal memory space to maintain routing table. On the other hand, sensor network, which try to minimize the energy consumption of memory rich sensor nodes, may use DHT-based (Distributed Hash Table based) routing protocol. Moreover, OERP layer hides all event routing protocol implementation from developers. For example, if spanning-tree is used in OERP, the spanning-tree implementation wills forms the tree where the subscriber node is the root of the tree. The routing information, e.g., tree structure in spanning-tree, is performed when sensor nodes are started up and maintained automatically by the implementation in OERP layer. By using this OERP layer, TinyDDS frees developers from the need to manually maintain the event routing between nodes and the limitation of routing algorithm used in network layer, which

Figure 5. Architectural components in TinyDDS

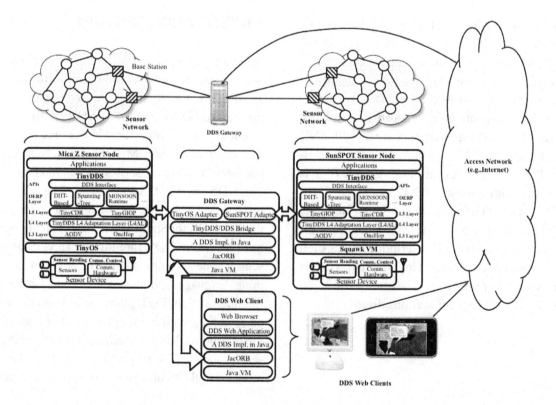

generally depends on sensor node platform. The routing protocol in OERP layer utilizes low-level network layer implementation through a L5 layer called TinyGIOP. TinyGIOP encapsulates data into transportation messages and interacts with the DDS Gateway for exchanging data with DDS applications. Only the nodes, i.e., base station, that are physically connected to the DDS gateway through serial interface can exchange data with the DDS gateway. Beside, another component in this layer called TinyCDR provides an interchangeable data format, which allows different implementations of TinyDDS or DDS exchange data. For transmitting/receiving data to/from the other sensor nodes in the WSN, TinyGIOP utilizes a transport layer interface called TinyDDS L4 Adaption Layer (L4AL). L4AL allows TinyOS to operates with any network (L3) and MAC layer (L2) protocol, such as AODV and Zigbee (IEEE 802.15.4) respectively.

DDS Interfaces

In the top layer, TinyDDS provides an API for application developers. This API provides a subset of DDS for creating topics, subscribe to events of topics and publish events for particular topics. For each function, the implementation is provided so application developers do not need to implement that functionality themselves. The implementation for the DDS interfaces is written in nesC and Java programming language and optimized for small sensor nodes platform such as TinyOS and SunSPOT, respectively.

Listing 1. An Example TinyDDS Application with nesC

```
1   typedef struct {
2       cdr_short temperature;
3       cdr_long time;
4   } TempData_t;
5   Publisher_t publisher;
6   Topic_t topic;
7   DataWriter_t data_writer;
8   TempData_t temp_data;
9   command result_t StdControl.start() {
10      publisher = call DomainParticipant.
        create_publisher();
11      topic = call DomainParticipant.
        create_topic("TempSensor");
12      data_writer = call Publisher.create_
        datawriter(publisher, topic);
13      temp_data.temperature = TempSensor.
        read();
14      temp_data.time = call Time.getLow32();
15      call DataWriter.write(data_writer,
        serialize(data), sizeof(TempData_t));
16  }
```

Listing 2. An Example TinyDDS Application with Java

```
1   class TempData extends Data {
2       public short temperature;
3       public int time;
4   }
5   public class Application {
6       Publisher publisher;
7       Topic topic;
8       DataWriter dataWriter;
9       TempData tempData;
10      DomainParticipant domainParticipant;
11      public Application() {
12          domainParticipant = new
            DomainParticipant();
13          publisher = domainParticipant.
            create_publisher();
14          topic = domainParticipant.
            create_topic("TempSensor");
15          dataWriter = publisher.
            create_datawriter(topic);
16          tempData = new TempData();
17          tempData.temperature =
            TempSensor.read();
18          tempData.time = (new Date()).
            getTime();
19          dataWriter.write(data.
            marshall());
20      }
21  }
```

Listing 1 and Listing 2 show an example of an event source application implemented on top of TinyDDS using nesC and Java, respectively. In Listing 1, a user-defined data type is defined at line 1-4. Then, at line 10, a *Publisher* is created. Line 11-12, a *DataWriter* is created associate with topic "TempSemsor". At line 13 and 14, a sensor reading is captured from temperature sensor and also the current time is read from a local clock, both data are stored in a variable of the user-define data type. Finally, at line 15, the data in user-defined data type is serialized into byte stream and published through *DataWriter* interface. Listing 2 shows the same application implemented in Java. Because both applications are developed based on the same API, i.e., TinyDDS, the applications are very similar. Thus, application developer can easily port an application from one platform to the other platform.

Overlay Event Routing Protocols

The OERP layer provides an overlay network over sensor network's physical ad-hoc networks. The overlay network is used for transporting published events, i.e. sensor data, to all nodes that subscribes to the events. The published event is routed to each subscriber according to the event routing protocols deployed within OERP layer. Application developers can specify the deployed routing protocols to suit their need. The OERP layer encapsulates the overlay network algorithm and implementation from DDS interfaces and the lower level physical network. Routing protocols in OERP layer work with lower-level network protocol through the L4AL. In the other words, routing protocols can be seen as a non-functional property of TinyDDS, which can be deployed to meet application developers' non-functional requirements. For example, application developers who want to reduce the price and size of sensor nodes by using small-memory sensor nodes may choose to use spanning-tree routing protocol, which will use very small memory space. The

routing protocols used in this OERP framework are developed by library developers and can be used in any TinyDDS-based applications in the same platform.

Example of event routing protocols currently implemented for OERP layer are spanning-tree based, DHT-based and MONSOON. When spanning-tree based event routing protocol is used in OERP layer, subscriber nodes flood the subscription messages, i.e., the topic of interest, through their neighbors. The subscription messages have a counter that is increased hop-by-hop. Each node keeps the value of the counter of each incoming message, and its topic, and also the neighboring node who has lower counter value. Then, when a sensor node collect an event matched with the topic interested by a subscriber node, the event message is forwarded from the collecting node to a neighboring node who has lower counter of that topic. The event message is forwarded toward the subscriber node by climbing up the gradient, i.e., the counter value.

DHT-based event routing relies on unique information belongs to each sensor node such as sensor node ID in MicaZ or MAC address in SunSpot. Then, by using a hash function, a topic can be mapped to a sensor node ID or MAC address. This node can be called hashed node. Thus, when a sensor node subscribes to a topic, the subscription and the node information is sent to the hashed node that is associated with that topic. Similarly, when an event is published, the event is sent to the hashed node associated with the topic of the event. Consequently, the event is forwarded to subscribed node using subscribing information maintained by the hashed node.

In MONSOON, a biologically-inspired adaptation routing mechanism, event is delivered by *software agent*, from publishers to subscribers (Boonma & Suzuki, 2008a). In contrast with spanning-tree and DHT-based routing protocol, MONSOON employs a constrained-based evolutionary multiobjective optimization algorithm which allows agents to adapt to changed network condition and also self-healing against partial network failure.

TinyCDR

The Common Data Representation (CDR) (Object Management Group, 2007) is the format for exchanged data in DDS. CDR is the format for exchanging data in DDS standardized by OMG. CDR enables different parties, i.e., sensor nodes and client applications, which utilizes different programming languages, such as nesC or Java, to be able to exchange data. CDR defines standard data type with specific size and endian, which have to be followed by each party in order to guarantee seamlessly data exchanging. TinyCDR is a subset of CDR, which allows TinyDDS applications to directly exchange data with DDS applications and TinyDDS in different platform, such as between desktop application using DDS and sensor node application running in Mica Z and SunSPOT. Table 1 shows the mapping of primitive data type between CDR version 1.3, TinyCDR for nesC and for Java programming language. Listing 1 and Listing 2 show how the data type defined in TinyCDR for nesC and for Java, respectively, is used in application.

In the table, TinyCDR does not support *wchar* type (wide character, i.e., Unicode characters) because it is not necessary in WSN environment. Also, TinyCDR for Java does not support *unsigned long long* type because there is no Java primitive type that has the same storage size for this data type. Beside primitive data types, TinyCDR also supports CDR constructed types such as *struct*, *union* and *array*. TinyCDR serializes constructed data structure into an octet stream, which is compatible with CDR octet stream. Therefore, TinyDDS applications can exchange data formatted in TinyCDR directly with DDS applications using CDR data format.

Table 1. CDR mappings to nesC and Java

CDR Type	TinyCDR Type in nesC	TinyCDR Type in Java
char	cdr_char	byte
wchar	N/A	N/A
octet	cdr_octet	byte
short	cdr_short	short
unsigned short	cdr_ushort	int
long	cdr_long	int
unsigned long	cdr_ulong	long
long long	cdr_longlong	long
unsigned long long	cdr_ulonglong	N/A
float	cdr_float	float
double	cdr_double	double
long double	cdr_longdouble	double
boolean	cdr_boolean	boolean

TinyGIOP

TinyGIOP defines message format use for exchanging between TinyDDS/DDS applications, based on General Inter-ORB Protocol (GIOP) version 1.3 (Object Management Group, 2007). GIOP is an abstract protocol for communicating between object request brokers (ORBs). There are several concrete implementation based on GIOP such as Internet Inter-ORB Protocol (IIOP), an implementation of GIOP over TCP/IP, and HyperText Inter-ORB Protocol (HTIOP), an implementation of GIOP over HTTP. GIOP consists of three components; CDR, Interoperable Object Reference (IOR), and a set of message types. In TinyDDS, the CDR part of GIOP is addressed by TinyCDR while the message type is addressed by TinyGIOP. Given the limited resources of sensor nodes, IOR is not supported by TinyDDS.

TinyGIOP supports three message formats; *Request*, *Reply* and *CancelRequest*. When a TinyDDS application wants to communicate with the other TinyDDS application, for example, for subscribing to a topic, it sends out *Request* message. The message will be serialized and passed to lower level, i.e. OERP, or to DDS Gateway for delivering to DDS applications. *Reply* message is used for answering the request, e.g., when a TinyDDS application publish an event subscribed by another TinyDDS application, the publisher sends out *Reply* message to the subscriber. *CancelRequest* is used for withdraw request sending out earlier. Contrast with GIOP, TinyGIOP does not support object location message formats because there is no notion of object in TinyDDS.

TinyDDS L4 Adaptation Layer

To access to low level physical network, TinyGIOP make use of low-level physical network through a network abstract layer called TinyDDS L4 Adaptation Layer (L4AL). This L4AL utilize Bridge design pattern to separate the real low-level physical network implementation from the higher-level overlay network. Thus, TinyDDS can be portable among different sensor node hardware, which utilize TinyOS, such as Mica Z, Mica 2 or iMotes2. In particular, L4AL provides an interface to access physical network functions such as how to get the list of neighboring nodes, how to get the link quality to each neighboring node and also how to send/receive data to/from particular nodes in the network. These functions are used by the TinyGIOP and above OERP routing protocol layer and implemented by the Network Layer implementation. Internally, L4AL contains a set of tables that maintains the information of network, such as neighbor list and link quality, and a set of event queues. There are two types of event queues, incoming queues and outgoing queues. The events submitted from OERP for sending out to physical network is put to the end of outgoing queue while the events collected from physical network are put to the end of incoming queue, waiting to be processed by the routing protocol in OERP.

Application Development with TinyDDS

Figure 6 shows the development model of a TinyDDS application on TinyOS platform. There are three main elements of the development model, *TinyDDS middleware*, *TinyDDS Library* and the *application*. The TinyDDS middleware comprises of two parts, the DDS interfaces definition and the TinyDDS implementation of the interfaces. The DDS interfaces definition is directly generated from the *dds.idl*, which is the official DDS interfaces definition in IDL format from OMG. The *dds.idl* is first converted into XML format. Then, *IDL2nesc* converts the DDS interfaces definition from XML format to *TinyDDS interfaces* and *Application Configuration*. The Application Configuration follows Facade design pattern (Gamma, Helm, Johnson, & Vlissides, 1995) and describes how to connect each interfaces and implementation together. IDL2nesc also uses an *Application Specification*, written in XML, in order to generate appropriate Application Configuration. In particular, Application Configuration consults Application Specification how to connect comments together. Thus, developers can customize TinyDDS by adjusting the configuration in the Application Specification. For example, Application Specification specifies which routing protocol will be used in OERP layer, then Application Configuration connects the implementation of the routing protocol into OERP interface. In particular, the event routing protocols implemented for OERP layer needs to implement a specific nesC interface; therefore, they can be integrated with DDS implementation on the upper layer and L4AL component on the lower layer.

The second element is the *TinyDDS Library*. TinyDDS Library consists of two non-function properties implementation, namely, application-level and middleware-level non-functional properties. The application-level non-function properties provide a set of services, which can

Figure 6. Application development model on TinyOS

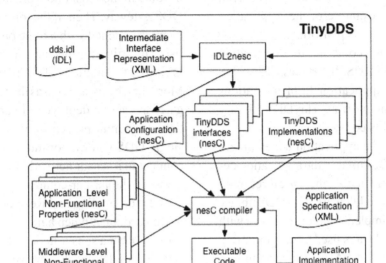

be used by application, such as data aggregation and event detection. The middleware-level non-function properties provide the services inside the middleware, for example, routing protocols in OERP layer. Library developer develops this functionality in the TinyDDS Library and the TinyDDS Library can be used in any application implemented on the same platform.

Listing 3. Application Specification

```
1   <?xml version="1.0" encoding="ISO-8859-
    1" ?>
2   <configuration platform= "micaz" thread-
    ing= "per-event">
3       <includes>
4           <header name="BaseUART" />
5           <header name="DDS_utils" />
6           <component name="DDS_
            DataAggregrator" />
7           <component name="LedsC" />
8           <component name=
            "SpanningTree" />
9       </includes>
10      <implementations>
11          <implementation com-
            ponent= "SpanningTree"
            interface="OERP" />
12      </implementations>
13      <connections>
14          <connection from="Main.
            StdControl" to="*" />
15          <connection from=
            "Application.DG" to ="DDS_
            DataAggragation" />
16          <connection from="Application.
            Leds" to="LedsC" />
17      </connections>
18  </configuration>
```

The third element is the application. Every application implemented on TinyDDS consists of two parts, the *Application Specification* which is used by the IDL2nesc compiler and the *Application Implementation*. The Application Implementation is developed by application developer and performs a certain task such as data collection and event detection. The Application Specification describes the overview of the application, for example, what is the target platform, or which routing protocol will be used in OERP layer. Listing 3 shows an example of the Application Specification for TinyOS platform.

The nesC compiler combines the Application Configuration, TinyDDS Interfaces, TinyDDS Implementations, Application Implementations and the implementation from TinyDDS Library into target executable code.

Figure 7 shows the application development model in SunSPOT. Similar to the application development model on TinyOS, there are three main elements, TinyDDS middleware, TinyDDS library and the application implementation. Contrast with the TinyOS model, Java is used as programming language instead of nesC. Also, a Java compiler, i.e., javac, is used for combining all parts into a Java byte code suitable for deploying in a SunSPOT sensor node. Also, Sun Microsystem's IDLJ is used for converting DDS's IDL to Java interfaces and the main class generator program is used for processing Application Specification and generating the main class used by Java compiler.

DDS Gateway

Figure 5 shows that TinyDDS uses TinyGIOP to communicate with the DDS gateway in order to exchange data with DDS applications. The DDS gateway is a Java application that interacts with TinyDDS running in sensor nodes through serial port using TinyOS' serial adapter and SunSPOT serial adapter (i.e., Host API) Java class. The DDS gateway uses JacORB (Brose, 1997) and a Java implementation of DDS (Allaoui, Yehdih, & Donsez, 2005) to communicate with another DDS application. A TinyDDS/DDS bridge operates on top of DDS implementation and communicates with TinyGIOP to exchange data between TinyDDS and DDS. When a DDS client subscribes to a topic, the subscription information is distributed on the access network over DDS. As a consequence, DDS gateway can intercept the subscription information and store it in a subscription list. Thus, when a sensor node publishes an event, the event is dis-

tributed in the sensor network and intercepted by DDS gateway. Then, if the topic of the published event is matched with a topic on the subscription list. This filtering mechanism in DDS gateway is similar to traditional TCP/IP bridge operation. Moreover, by using DDS as the backend, the DDS gateway can be deployed in variable configurations, e.g., multiple sensor networks connect to a single DDS gateway or multiple DDS gateways connect to a sensor network. Each DDS gateway supports thread-per-connection, thread pooling, connection pooling and Reactor.

In particular, when a message is pushed from TinyGIOP in a sensor node to DDS gateway, the TinyDDS/DDS bridge translate message into DDS format, i.e., encapsulate with GIOP header, and send out to DDS network. This is called **downstream** message transmissions because the message is sent from source (sensor nodes) to sink (client applications). On the other hand, when the DDS gateway receives messages destine to the sensor network from the DDS network, TinyDDS/DDS bridge translates the message into TinyDDS format, i.e., encapsulate with TinyGIOP header, and injects the message into sensor nodes through serial interface. This is called **upstream** message transmission. Also, the message from one TinyDDS will be passed to another TinyDDS, for example, TinyDDS on a TinyOS nodes can pass message through its TinyGIOP to DDS Gateway, which will also pass the message to TinyDDS on the SunSPOT as well.

To manage the flow control from a WSN to an access network, TinyDDS utilizes the hop-by-hop flow control mechanisms available at the MAC layer of TinyOS and SunSPOT. In particular, the flow control mechanisms at the MAC layer allow a sensor node to send out a packet only when it receives acknowledgment from the destination node; there fore, the source node cannot overwhelm the destination node. However, TinyDDS does not consider end-to-end flow control. For the flow control from an access network to a WSN, TinyDDS uses TCP's end-to-end flow control.

Figure 7. Application development model in SunSPOT

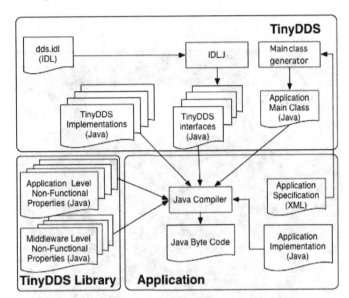

DDS Web Clients

On the lower part of the Figure 5, a DDS web client is shown connected to DDS gateway. Currently, a DDS web client is implemented as a Java application provides HTTP service to any web browser. The DDS web client is able to operate on ordinary desktop computers or mobile devices such as Apple's iPhone. By using JacORB, the DDS web client can communicate with DDS gateway in order to subscribe to data published from sensor networks and show the result on the Google map. Figure 8 and Figure 9 show examples of web interface running on a desktop computer and an iPhone respectively. In the Figures, the small dots show location of each sensor node, bright dots represent sensor nodes, which report data.

NON-FUNCTIONAL PROPERTIES OF TINYDDS

To ease the application development on sensor nodes, TinyDDS provides non-functional properties both on application and middleware level collectively as a library, called TinyDDS library in the Figure 6 and Figure 7. The application-level non-functional properties accelerate the application development process by providing frequently used non-functional properties such as data aggregation and event detection. Thus, application developers can focus more on their application functionality, e.g. how to interpret and process data and event. Moreover, utilizing non-functional properties can reduce the application complexity and thus improve the maintainability. On the other hand, middleware-level non-functional properties allows application developers to adjust the behavior of the middleware to suit their need and constraints, i.e., choosing event routing protocol which suite the application or specify the QoS of each middleware components. In addition, TinyDDS library is designed to be portable and can be used by many TinyDDS based application. Therefore, by using both application and middleware-level non-functional properties, application developers can gain better reusability, maintainability, composability and performance.

Figure 8. A DDS web client on a desktop computer

Figure 9. A DDS web client on an iPhone

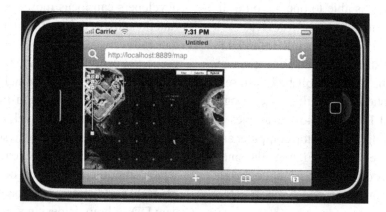

Application-Level Non-Functional Properties

In the application level, non-functional properties in TinyDDS help application developers to rapidly develop their applications. Application developers can use application-level middleware services such as data aggregation instead of subscribing/publishing directly to TinyDDS middleware. Data aggregation collects and process data from sensor network and provides processed data to application. Processing operators supports by data aggregation are, for example, summation, average, maximum, and minimum. The data aggregation component is pluggable, application developers can include this in their application using Application Specification (see section *Application Development with TinyDDS*). In addition, library developers can develop any non-functional components, such as network security and persistence storage, which are reusable and pluggable to all TinyDDS applications.

Listing 4 shows a fragment of an application using data aggregation in nesC. Instead of using DataReader for collecting data (see Figure 2), this application uses DataAggregrator. Then, when new data is available, the TinyDDS middleware informs the application using data_available event and provide aggregated data, in this case, average temperature sensor, to the application. Listing 5 shows the same application implemented in Java.

From the listing, Java implementation of TinyDDS adopts Observer Design Pattern. So, when an event is collected, SubscriberListener informs the application by calling eventNotified. Then, the application can collect aggregated data through an instance of DataAggragation class.

Listing 4. An Example of nesC Application Using Data Aggregation

```
1   Subscriber_t subscriber;
2   Topic_t ts_topic;
3   DataAggregator_t data_aggregrator;
4   SubscriberListener_t listener;
5   command result_t StdControl.start() {
6       subscriber = call DomainParticipant.
        create_subscriber();
7       ts_topic = call DomainParticipant.
        create_topic("TempSensor");
8       listener = call SubscriberListener.
        create(ts_topic);
9       call Subscriber.set_listener(listener);
10      data_aggregrator = call Subscriber.
        create_data_aggregrator(ts_topic,
        AVERAGE, listener);
```

```
11  }
12  event ReturnCode_t SubscriberListener.
    data_available(Topic_t topic) {
13      Data data;
14      if (topic == ts_topic) {
15          data = call DataAggregator.
            read(data_aggregrator);
16          // processing aggregated data..
17      }
18  }
```

Listing 5. An Example of Java Application Using Data Aggregation

```
1   public class Application implement Observer {
2       Subscriber subscriber;
3       Topic topic;
4       DataAggregator dataAggregrator;
5       SubscriberListener listener;
6       DomainParticipant domainParticipant;
7       public Application() {
8           domainParticipant = new
            DomainParticipant();
9           subscriber = domainParticipant.
            create_subscriber();
10          topic = domainParticipant.
            create_topic("TempSensor");
11          listener = new SubscriberListener
            (topic, this);
12          subscriber.set_listener(listener);
13          data_aggregrator = subscriber.
            create_data_aggregrator(topic,
            AVERAGE, listener);
14      }
15      public void eventNotified(Object orig,
        Object event) {
16          TempData data;
17          if(orig == listener) {
18              data = data_aggregrator.read();
19          }
20      }
21  }
```

Middleware-Level Non-Functional Properties

TinyDDS supports three non-functional properties in the middleware level, the pluggable routing protocols in OERP layers, concurrency, and the QoS policy. Application developers can specify the concurrency model used in TinyDDS. In particular, TinyDDS supports three concurrency model, thread-per-event, thread-per-event-topic and Reactor (Pyarali, Harrison, Schmidt, & Jordan, 1997). In the thread-per-event model, TinyDDS creates a thread, e.g., a Task in TinyOS or a thread in Java, for each event submitted from application or collected from the network. This model gives same priority for each event. In the other words, TinyDDS has only two event queues, one for outgoing events and the other for incoming events, both of them working in first-come-first-serve fashion. The thread-per-event-topic allows application developer to specify different priority for each event topic. In particular, TinyDDS creates two message queues for each event topic, one for incoming events and the other for outgoing events. Then, TinyDDS processes the event queues based on priority of each event topic; for example, TinyDDS publishes events with high priority topic more frequent than one with low priority topic. The concurrency model and its working parameters, e.g. topic priority, can be specified in Application Specification (see Listing 3). For the Reactor model, it is mapped to single-threaded mode in TinyOS and SunSPOT.

For the QoS policy, TinyDDS utilizes some of the QoS model of DDS, such as latency budget and reliability. Latency budget QoS policy specifies the maximum accepted latency from the time the event is published until the event is available to the destination subscribers. Listing 6 and Listing 7 show fragment of an application using latency budget QoS policy in nesC and Java, respectively. From the listing, the application creates a latency budget QoS policy with 100 ms constraint. Then, the QoS policy is applied to the Topic instance, which imply that the data in this topic should be delivered with less than 100 ms latency. TinyDDS uses mechanisms in L4AL to satisfy the QoS policies. For example, to satisfy the latency budget QoS policy, TinyDDS rearranges the order of event in the event queues such that the event with has a high chance to break the QoS policy, e.g., event's actual latency is already very close to the desired latency, will be published earlier than the event which has the less chance to break the QoS policy, e.g., event's actual latency is very far from the desired latency. The reliability QoS policy indicates the level of data transmission reliability provides by TinyDDS. In particular, TinyDDS supports two reliability model, RELIABLE and BEST_EFFORT. When reliability QoS policy is set to RELIABLE, TinyDDS attempts to deliver all events. The missed events are retransmitted until the number of transmission is greater than a threshold or the transmission is success. On the other hand, when reliability QoS policy is set to BEST_EFFORT, TinyDDS sends out each event only once and relies on MAC layer for succeeding the transmission.

Listing 6. Latency Budget QoS Parameters Settings in nesC.

```
1     Publisher_t publisher;
2     Topic_t topic;
3     DataWriter_t data_writer;
4     QosPolicy_t qos;
5     Data data;
6     command result_t StdControl.start() {
7         publisher = call DomainParticipant.
          create_publisher();
8         qos = call QosPolicy.
          create_latency_budget_qos(100);
9         topic = call DomainParticipant.
          create_topic("TempSensor");
10        call Topic.set_qos(topic, qos);
11        data_writer = call Publisher.cre-
          ate_datawriter(publisher, topic);
12    // Get sensor reading, and put to data variable
```

```
13      call  DataWriter.write(data_writer,
        data);
14  }
```

Listing 7. Latency Budget QoS Parameters Settings in Java.

```
1   public class Application {
2       Publisher publisher;
3       Topic topic;
4       DataWriter dataWriter;
5       TempData tempData;
6       DomainParticipant domainParticipant;
7       QosPolicy qos;
8       public Application() {
9           domainParticipant = new
            DomainParticipant();
10          publisher = domainParticipant.
            create_publisher();
11          qos = QosPolicy.create_latency_
            budget_qos(100);
12          topic = domainParticipant.
            create_topic("TempSensor");
13          topic.set_qos(qos);
14          dataWriter = publisher.
            create_datawriter(topic);
15          tempData = new TempData();
16          tempData.temperature =
            TempSensor.read();
17          tempData.time = (new Date()).
            getTime();
18          dataWriter.write(data.
            marshall());
19      }
20  }
```

EVALUATION

This section evaluates TinyDDS through simulations and empirical experiments. In all simulations and experiments, each node runs an application that accepts a one-time event subscription from a subscriber and publishes subsequent events to the subscriber. It is configured with an application

specification shown in Listing 3. The duration of each simulation/experiment is 120 seconds, and 25 nodes transmit an event to the base station (i.e., subscriber) every 2 seconds. The TinyOS-based implementation of TinyDDS is evaluated with 25 nodes simulated on the PowerTOSSIM simulator with the MICA 2 power consumption model (Shnayder, Hempstead, Chen, Allen, & Welsh, 2004). The SunSPOT-based implementation of TinyDDS is evaluated with 20 nodes simulated on the Solarium emulator (Goldman, 2008) and 5 nodes deployed on real/physical SunSPOT platforms. Figure 10 shows a screenshot of the Solarium emulator. 20 light gray SunSPOT icons represent simulated nodes, and 5 dark gray SunSPOT icons represent real nodes.

On TinyOS, TinyDDS is compared with Surge, which is a simple data collection application bundled in TinyOS. For a purpose of performance comparison, TinyDDS uses a spanning tree-based protocol as its OERP as Surge does. On SunSPOT, a Surge-like data collection application is implemented to for the comparison with TinyDDS. (It uses a spanning tree-based protocol too.)

Per-Packet Header Overhead

In order to evaluate how much data TinyDDS requires to transmit an event, Table 2 shows the header overhead at each layer of TinyDDS. Packet size is limited at 48 bytes on TinyOS. Given this limit, TinyDDS can transmit event data of 16 bytes per packet; packet overhead is 32 bytes (66%) per packet. Given 16 bytes, TinyDDS can transmit a sequence/array of two unsigned long long values. (Unsigned long long is the largest fixed-length IDL type; it occupies 8 bytes.) Also, It can contain 16 of the shortest fixed-length IDL data such as char, octet and Boolean per packet. Surge's packet overhead is approximately 11 bytes (40%) per packet. Packet size is limited at 64 bytes on SunSPOT; packet overhead is 52 bytes (82%) per packet. TinyDDS can transmit event data of 12 bytes per packet. In a Serge-like

Figure 10. A screenshot of the Solarium Emulator showing 25 simulated and real nodes

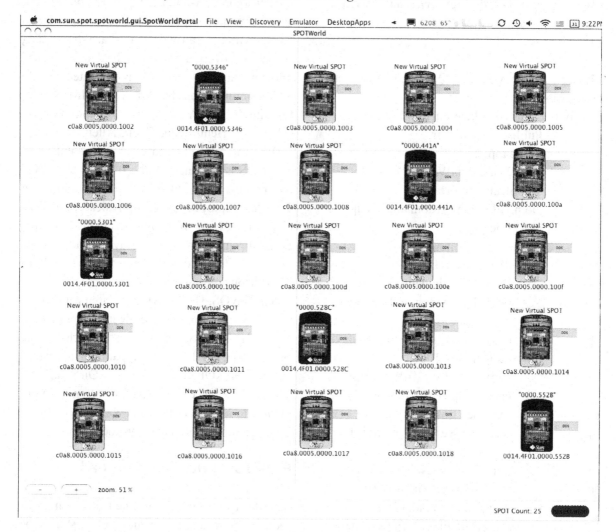

data collection application, packet overhead is 32 bytes (66%) per packet.

TinyDDS incurs larger header overhead because it encapsulates more information in each packet; for example, a timestamp from DDS, routing information from OERP and session information from TinyGIOP. With these extra information, TinyDDS can perform advanced functionalities that simple data collection applications do not have. For example, TinyDDS can use timestamp information to calculate the delay in an event publication and prioritize packets hop by hop. Given this consideration, the authors of the chapter believe that the measured header overhead is acceptable and small enough for a number of WSN applications.

Memory Footprint

Table 3 shows the memory footprint of TinyDDS on TinyOS and SunSPOT. Without running applications on TinyOS, TinyDDS consumes 36,132 bytes of Flash memory and 3,360 bytes of RAM. Including an application, memory footprint slightly increases to 36,246 bytes in Flash memory and 3,392 bytes of RAM. Surge consumes 34,430 byte

Table 2. Per-packet header overhead of TinyDDS on TinyOS and SunSPOT

Size (Bytes)	TinyOS	SunSPOT
Event Data	16	12
DDS Overhead	10	21
OERP (Spanning Tree) Overhead	4	4
L5 Overhead	8	12
L4 Overhead	6	8
L3 (OneHop) Overhead	4	8
Total Overhead	32	52
Total Size	48	64

of Flash memory and 1,929 byte of RAM. The difference of memory footprint between TinyDDS and Surge is very small on TinyOS; less than 2 KB on both Flash memory and RAM.

Without running applications on SunSPOT, TinyDDS consumes 35,832 bytes of Flash memory and 100,268 bytes of RAM. RAM consumption is higher on SunSPOT than TinyOS because SunSPOT requires to copy every executable code, including the Squawk VM and applications, and execute them in RAM. Including an application, TinyDDS consumes 37,285 bytes of Flash memory and 104,404 bytes of RAM. A Surge-like data collection application consumes about 29,519 bytes of Flash memory and 82,739 bytes of RAM. The difference of memory footprint between TinyDDS and a Surge-like application is small enough on SunSPOT. TinyDDS is implemented lightweight, and it can operate in resource-limited nodes such as MICA2. Figure 11 shows the breakdown of memory footprint in each layer in TinyDDS.

Processing and Power Efficiency

When TinyDDS is deployed on TinyOS, the average total latency is 0.79 second to route an event between source and destination nodes in a single hop. Of this total latency, TinyDDS spends 0.08 second; 0.03 second on a source node and 0.05 second on a destination node. This means TinyDDS occupies approximately 10% of the total latency.

When TinyDDS is deployed on SunSPOT, the average total latency is 0.675 second to route an event between source and destination nodes in a single hop. Of this total latency, TinyDDS spends 0.0247 second; 0.0128 second on a source node and 0.0119 second on a destination node. Table 4 shows the breakdown of the latency on both source and destination nodes. TinyDDS' latency occupies approximately 10% of the total latency. When a Serge-like application is used for routing event, the average total latency is 0.58 second to route an event in a single hop. Thus, latency of event routing in TinyDDS is about 10% higher than in Surge. This difference is small enough.

Table 5 shows the average and standard deviation (SD) of power consumption from 25 simulated Mica 2 nodes using TinyDDS on TinyOS platform and from five actual SunSPOT nodes and 20 emulated nodes running TinyDDS. Without a subscriber, TinyDDS transmits no data; thus, its power consumption remains small. In contrast, Surge always transmits data to the base station; it consumes much more power than TinyDDS. With a subscriber, TinyDDS consumes a comparable amount of power compared with Surge. In SunSPOT, without a subscriber, TinyDDS transmits no data; thus, its power consumption remains small. In contrast, a simple data collection application always transmits data to the base station; it consumes much more power than TinyDDS. With a subscriber, TinyDDS consumes a comparable amount of power compared with the simple data collection application. TinyDDS is implemented power efficient.

Size of Application Source Code

With the TinyOS-based implementation of TinyDDS, only 60 lines of nesC code are required

Table 3. Memory footprint of TinyDDS on TinyOS and SunSPOT

Memory Footprint (Bytes)	TinyOS		SunSPOT	
	Flash memory	RAM	Flash memory	RAM
Application	1440	34	1453	4136
DDS Implementation	1784	2240	10388	404
OERP (Spanning Tree)	4644	172	4907	3364
L5	5452	175	1352	19800
L4	1827	150	1457	11892
L3 (OneHop)	3905	205	5693	16248
TinyOS/ Squawk VM	18520	418	12035	48560

Figure 11. Memory footprint of components on TinyOS and SunSPOT platform

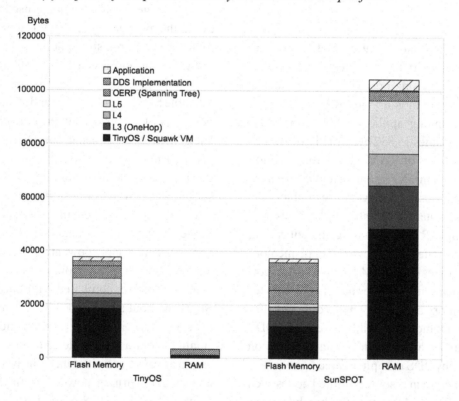

to implement the same application as Surge. Surge is implemented with 300 lines of nesC code. Moreover, it takes less than 3 seconds for idl2nesc to nesC source code. With the SunSPOT-based implementation of TinyDDS, 80 lines of Java code are required for implementing a Surge-like application. On the other hand, without TinyDDS, more than 350 lines of Java code are required to implement the same application from scratch. It takes less than 10 seconds for idl2j to generate Java code. These results demonstrate that TinyDDS effectively simplifies the development of WSN applications.

Table 4. Latency of each layer on SunSPOT

	Source Node (msec)	Destination Node (msec)
DDS	0.1	0.3
OERP (Spanning-Tree)	0.5	0.75
L5	5.5	10.15
L4	4.3	0.3
L3	2.4	0.4
Total	12.8	11.9

RELATED WORK

This chapter describes a set of extensions to the authors' prior work (Boonma & Suzuki, 2008b; Boonma & Suzuki, 2009b). While the prior work studied TinyDDS on TinyOS, this chapter investigates it on both TinyOS and SunSPOT. Moreover, this chapter reveals TinyDDS' performance implications in more detail. An event routing protocol of TinyDDS, called MONSOON, is investigated in (Boonma & Suzuki, 2008a; Boonma & Suzuki, 2009a), while it is out of scope of this chapter.

There exist several pub/sub middleware for WSNs. TinyCubes, Mires and Runes are similar to TinyDDS in that they implement pub/sub communication on TinyOS and provide reconfigurable middleware services to customize application-level non-functional properties (Marrón et al., 2005; Souto, et al., 2005; Costa et al., 2007; Nam et al., 2008). However, unlike TinyDDS, they do not consider middleware-level non-functional properties and interoperability between WSNs and access networks. SMC is pub/sub middleware for body-area sensor networks (Keoh, et al., 2007). It is assumed to operate on a Java VM atop powerful Linux node; memory footprint and power efficiency are not important issues in SMC. In contrast, TinyDDS assumes resource-limited nodes as its target platforms. SMC does not consider middleware-level non-functional properties, programming language interoperability and protocol interoperability as TinyDDS does. DSWare is pub/sub middleware that makes middleware-level non-functional properties (e.g., data storage and caching policies) configurable for WSN applications (Li et al., 2004). However, it does not consider application-level non-functional properties, programming language interoperability and protocol interoperability as TinyDDS does.

Schönherr et al. (2008) propose a clustered event routing protocol for WSNs. It allows nodes to form clusters and transmit event data to destinations (subscribers) via cluster head nodes. Costa and Picco (2005) propose a semi-probabilistic routing protocol in which event data are randomly broadcasted when intermediate nodes do not know subscribers. TinyDDS does not focus on particular event routing protocols. Instead, it focuses on its generic and pluggable framework to support a wide range of event routing protocols.

Li et al. (2004), Yoneki and Bacon (2005) and Sivaharan et al. (2005) propose event subscription languages for the pub/sub scheme in WSNs. Yoneki and Bacon (2005) extend a traditional subscription language by introducing the expressiveness for spatial and temporal concerns. Li et

Table 5. Power consumption in Mica 2 (TinyOS) and SunSPOT

Power Consumption (mW)		Mica 2 (TinyOS)		SunSPOT	
		TinyDDS	Surge	TinyDDS	A Data Coll. App.
Without a Subscriber	Average	189.59	3743.0	295.2	5544.4
	SD	61.24	76.03	89.5	76.03
With a Subscriber	Average	3900.9	3924.97	5323.75	5601.3
	SD	52.55	76.03	83.4	75.4

al. (2004) investigates an SQL-like subscription language that supports the notion of confidence in subscription and event correlation. Sivaharan et al. (2005) propose an extensible subscription language, called Filter Expression Language (FEL), to define topic, content and context filtering policies for published events. In contrast, TinyDDS currently does not focus on subscription semantics but reuse the standard subscription language in the DDS specification.

Several research efforts have focused on the interoperability between WSNs and access networks. Girod et al. (2004), Pietzuch et al. (2004), Spiess et al. (2006) and Hunkeler et al. (2008) propose interoperable middleware/frameworks. Adam et al. (2004), Shu et al. (2006) and Schott et al. (2007) investigate TCP/IP-based and other protocol stacks. These work provide interoperability by unifiying low-level (L2 to L4) protocols between WSNs and access networks. In contrast, TinyDDS provides interoperability by introducing an interoperable session (L5) protocol over heterogeneous L2 to L4 protocols. Given the fact that heterogenity have been increasing in WSN platforms, TinyDDS retains the heterogenity at L2 to L4 rather than unifying them. Moreover, all of these existing work do not provide programming language interoperability. They also do not consider the pub/sub communication scheme as a networking/programming abstraction.

CONCLUSION

TinyDDS is pub/sub middleware that allows applications to interoperate regardless of programming languages and protocols across the boundary of WSNs and access networks. Moreover, it allows WSN applications to have fine-grained control over application-level and middleware-level non-functional properties and flexibly specialize in their own requirements. Evaluation results demonstrate that TinyDDS is lightweight and efficient. They also show that TinyDDS simplifies the development of publish/subscribe WSN applications.

Several future extensions are planned for TinyDDS. One of them is to investigate an end-to-end flow control mechanism between WSNs and access networks. Another is to evaluate various I/O and resource management mechanisms in scalability measurements with varying network sizes and traffic patterns. Particularly, the authors of the chapter are interested in studying staged event-driven architecture (SEDA) (Welsh, Culler, & Brewer, 2001) on a gateway node and the Proactor design pattern (Pyarali, Harrison, Schmidt, & Jordan, 1997) on individual sensor nodes.

REFERENCES

Adam, D., Thiemo, V., & Juan, A. (2004). Making TCP/IP Viable for Wireless Sensor Networks. *European Workshop on Wireless Sensor Networks* (p. 4). Berlin, Germany: Springer.

Allaoui, F., Yehdih, A., & Donsez, D. (2005, August 20). *Open-source Java-based DDS (Data Distribution Service) Implementation*. Retrieved August 25, 2008, from Open-source Java-based OMG DDS Implementation: http://www-adele.imag.fr/users/Didier.Donsez/dev/dds/readme.html

Banavar, G., Candra, T. D., Strom, R. E., & Sturman, D. C. (1999). A Case for Message Oriented Middleware. *International Symposium on Distributed Computing* (pp. 1-18). Bratislava, Slovak Republic: Springer.

Boonma, P., & Suzuki, J. (2008). Exploring Self-star Properties in Cognitive Sensor Networking. *International Symposium on Performance Evaluation of Computer and Telecommunication Systems*. Edinburgh, Scottland: IEEE/SCS.

Boonma, P., & Suzuki, J. (2008). Middleware Support for Pluggable Non-Functional Properties in Wireless Sensor Networks. *International Workshop on Methodologies for Non-functional Properties in Services Computing* (pp. 360-367). Honolulu, HI: IEEE.

Boonma, P., & Suzuki, J. (2009). Self-Configurable Publish/Subscribe Middleware for Wireless Sensor Networks. *International Workshop on Personalized Networks.* Las Vegas, NV: IEEE.

Boonma, P., & Suzuki, J. (2009). Toward Interoperable Publish/Subscribe Communication between Wireless Sensor Networks and Access Networks. *International Workshop on Information Retrieval in Sensor Networks.* Las Vegas, NV: IEEE.

Brose, G. (1997). JacORB: Implementation and Design of a Java ORB. *International Working Conference on Distributed Aplications and Interoperable Systems* (pp. 143-154). Cottbus, Germany: Chapman & Hall.

Buschnmann, F., Meunier, R., Rohnert, H., Sommerland, P., & Stal, M. (1996). *Pattern-Oriented Software Architecture - A: System of Patterns.* New York: Wiley and Sons.

Costa, P., Coulson, G., Mascolo, C., Mottola, L., Picco, G. P., & Zachariadis, S. (2007). A Reconfigurable Component-based Middleware for Networked Embedded Systems. *Springer Journal of Wireless Information Networks, 14*(2), 149–162. doi:10.1007/s10776-007-0057-2

Costa, P., & Picco, G. P. (2005). Publish-Subscribe on Sensor Networks: A Semi-Probabilistic Approach. *International Conference on Mobile Adhoc and Sensor Systems* (p. 332). Washington, DC: IEEE Press.

Estrin, D., Govindan, R., Heidemann, J., & Kumar, S. (1999). Next Century Challenges: Scalable Coordination in Sensor Networks. *International Conference on Mobile Computing and Networks* (pp. 263-270). Seattle, WA: ACM.

Eugster, P. T., Felber, P. A., Guerraoui, R., & Kermarrec, A.-M. (2003). The Many Faces of Publish/Subscribe. *ACM Computing Surveys, 35*(2), 114–131. doi:10.1145/857076.857078

Gamma, E., Helm, R., Johnson, R., & Vlissides, J. (1995). *Design Patterns: Elements of Reusable Object-Oriented Software.* Reading, MA: Addison-Wesley Professional.

Gay, D., Levis, P., Behren, R. v., Welsh, M., Brewer, E., & Culler, D. (2003). *The nesC Language: A Holistic Approach to Networked Embedded Systems. Programming Language Design and Implementation* (pp. 1–11). San Diego, CA: ACM.

Girod, L., Elson, J., Cerpa, A., Stathopoulos, T., Ramanathan, N., & Estrin, D. (2004). Emstar: A Software Environment for Developing and Deploying Wireless Sensor Network. *USENIX Technical Conference* (pp. 24-38). Boston, MA: USENIX.

Goldman, R. (2008, June 1). *Using the SPOT Emulator in Solarium.* Retrieved December 24, 2008, from SunSPOTWorld: http://www.sunspotworld.com/docs/Blue/SunSPOT-Emulator.pdf

Hadim, S., & Mohamed, N. (2006). Middleware Challenges and Approaches for Wireless Sensor Networks. *IEEE Distributed Systems Online, 7*(3), 1. doi:10.1109/MDSO.2006.19

Hadim, S., & Mohamed, N. (2006). Middleware for Wireless Sensor Networks: A Survey. *International Conference on Communication System Software and Middleware,* (pp. 1-7). New Delhi, India.

Henricksen, K., & Robinson, R. (2006). A Survey of Middleware for Sensor Networks: State-of-the-art and Future Directions. *International Workshop on Middleware for Sensor Networks* (pp. 60-65). Melbourne, Australia: ACM.

Hunkeler, U., Truong, H. L., & Stanford-Clark, A. (2008). *MQTT-S—A Publish/Subscribe Protocol for Wireless Sensor Networks. Communication Systems Software and Middleware and Workshops* (pp. 791–798). Dublin, Ireland: ICST.

Keoh, S. L., Dulay, N., Lupu, E., Twidle, K., Schaeffer-Filho, A. E., Sloman, M., et al. (2007). Self-Managed Cell: A Middleware for Managing Body-Sensor Networks. *International Conference on Mobile and Ubiquitous Systems* (pp. 1-5). Philadelphia, PA: IEEE.

Levis, P., Madden, S., Polastre, J., Szewczyk, R., Whitehouse, K., Woo, A., et al. (2005). TinyOS: An Operating System for Sensor Networks. In W. Weber, J. Rabaey, & E. Aarts, Ambient Intelligence (pp. 115-148). Berlin, Germany: Springer.

Li, S., Lin, Y., Son, S. H., Stankovic, J. A., & Wei, Y. (2004). Event Detection Services Using Data Service Middleware in Distributed Sensor Networks. *Springer Telecommunication Systems, 26*(2-4), 351–368. doi:10.1023/B:TELS.0000029046.79337.8f

Marchiori, A., & Han, Q. (2008). A Foundation for Interoperable Sensor Networks with Internet Bridging. *Workshop on Embedded Networked Sensor.* Charlottesville, VA: ACM.

Marrón, P. J., Lachenmann, A., Minder, D., Gauger, M., Saukh, O., & Rothermel, K. (2005). Management and Configuration Issues for Sensor Networks. *Wiley International Journal of Network Management, 15*(4), 235–253. doi:10.1002/nem.571

Nam, C.-S., Jeong, H.-J., & Shin, D.-R. (2008). Design and Implementation of the Publish/Subscribe Middleware for Wireless Sensor Networks. *International Conference Networked Computing and Advanced Information Management* (pp. 270-273). Gyeongju, South Korea: IEEE.

Object Management Group. (2007). *Common Object Request Broker Architecture (CORBA) Specification, Version 3.1; Part 2: CORBA Interoperability.* Needham, MA: OMG.

Object Management Group. (2007). *Data Distribution Service (DDS) for real-time systems, v1.2.* Needham, MA: OMG.

Pietzuch, P., Ledlie, J., Shneidman, P., Roussopoulos, M., Welsh, M., & Seltzer, M. (2004). Network-Aware Operator Placement for Stream-Processing Systems. *International Conference on Data Engineering* (p. 49). Atlanta, GA: IEEE.

Pyarali, I., Harrison, T., Schmidt, D. C., & Jordan, T. D. (1997). Proactor--An Object Behavioral Pattern for Demultiplexing and Dispatching Handlers for Asynchronous Events. *Pattern Languages of Programming Conference.* Monticello, IL: Washington University.

Romer, K., Kasten, O., & Mattern, F. (2002). Middleware Challenges for Wireless Sensor Networks. *ACM Mobile Computing and Communications Review, 6*(4), 59–61. doi:10.1145/643550.643556

Schönherr, J. H., Parzyjegla, H., & Mühl, G. (2008). Clustered Publish/Subscribe in Wireless Actuator and Sensor Networks. *International Workshop on Middleware for Pervasive and Ad-Hoc Computing* (pp. 60-65). Leuven, Belgium: ACM.

Schott, W., Gluhak, A., Presser, M., Hunkeler, U., & Tafazolli, R. (2007). e-SENSE Protocol Stack Architecture for Wireless Sensor Networks. Mobile and Wireless Communications Summit (pp. 1-5). Budapest, Hungary: IST.

Shnayder, V., Hempstead, M., Chen, B.-r., Allen, G., & Welsh, M. (2004). Simulating the Power Consumption of Large-Scale Sensor Network Applications. *International Conference on Embedded Networked Sensor Systems* (pp. 188-200). Baltimore, MD: ACM.

Shu, L., Wang, J., Xu, H., Jinsung, C., & Sungyoung, L. (2006). Connecting Sensor Networks with TCP/IP Network. *International Workshop on Sensor Networks* (pp. 330-334). Harbin, China: Springer.

Simon, D., & Cifuentes, C. (2005). The Squawk Virtual Machine: Java™ on the Bare Metal. *Conference on Object Oriented Programming Systems Languages and Applications* (pp. 150-151). San Diego, CA: ACM.

Sivaharan, T., Blair, G., & Coulson, G. (2005). *GREEN: A Configurable and Re-configurable Publish-Subscribe Middleware for Pervasive Computing. On the Move to Meaningful Internet Systems* (pp. 732–749). Agia Napa, Cyprus: Springer.

Souto, E., Guimarães, G., Vasconcelos, G., Vieira, M., Nelson, R., & Ferraz, C. (2005). Mires: A Publish/Subscribe Middleware for Sensor Networks. *Springer Personal Ubiquitous Computing*, *10*(1), 37–44. doi:10.1007/s00779-005-0038-3

Spiess, P., Vogt, H., & Jütting, J. (2006). *Integrating Sensor Networks With Business Processes. Real-World Sensor Networks Workshop.* Uppsala, Sweden: ACM.

Wang, M.-M., Cao, J.-N., Li, J., & Dasi, S. K. (2008). Middleware for Wireless Sensor Networks: A Survey. *Springer Journal of Computer Science*, *23*(3), 305–326.

Welsh, M., Culler, D., & Brewer, E. (2001). SEDA: An Architecture for Well-Conditioned, Scalable Internet Services. *Symposium on Operating Systems Principles* (pp. 230-243). Banff, Canada: ACM.

Yoneki, E., & Bacon, J. (2005). Unified Semantics for Event Correlation over Time and Space in Hybrid Network Environments. *International Conference on Cooperative Information Systems* (pp. 366-384). Agia Napa, Cyprus: IFIP.

ENDNOTES

[1]   TinyDDS is available at dssg.cs.umb.edu.

[2]   nesC is a dialect of the C language for WSNs.

Chapter 10
Knowledge–Based Networking

John Keeney
Trinity College Dublin, Ireland

Dominic Jones
Trinity College Dublin, Ireland

Song Guo
Trinity College Dublin, Ireland

David Lewis
Trinity College Dublin, Ireland

Declan O'Sullivan
Trinity College Dublin, Ireland

ABSTRACT

Knowledge-Based Networking, which is built on-top of Content-based Networking (CBN), involves the forwarding of events across a network of brokers based on subscription filters applied to some semantics of the data and associated metadata of the events contents. Knowledge-based Networks (KBN) therefore support the efficient filtered dissemination of semantically enriched knowledge over a large, loosely coupled network of distributed heterogeneous agents. This is achieved by incorporating ontological semantics into event messages, allowing subscribers to define semantic filters, and providing a subscription brokering and routing mechanism. The KBN used for this work provides ontological concepts as an additional message attribute type, onto which subsumption relationships, equivalence, type queries and arbitrary ontological relationships can be applied. It also provides a bag type to be used that supports bags equivalence, sub-bag and super-bag relationships to be used in subscription filters, composed with traditional CBN subscription operators or the ontological operators. When combined with the benefits of Content–based Networking, this allows subscribers to easily express meaningful subscription interests and receive results in a more expressive and flexible distributed event system than heretofore. Within this chapter the detailed analysis of ontological operators and their application to a publish/subscribe (pub/sub) domain will be fully explored and evaluated.

DOI: 10.4018/978-1-60566-697-6.ch010

INTRODUCTION

Content-based networks (CBN), such as (Carzaniga et. al., 2001; Peitzuch et. al., 2002; Segal et. al., 2000), formed around the necessity to match a varying subscriber base to that of a network's publication creators. However, open standards for CBNs have been slow to emerge due to the difficulty in reaching a general compromise between the expressiveness of event attribute types and subscription filters, exasperated by the need to both match these efficiently at CBN nodes and to efficiently maintain routing tables. The need for efficient network utilisation requires that notifications are routed towards nodes and subscribers that are interested in a particular message, using a routing table composed of subscription filters, rather than flooding the network in the search for all interested possible parties. This is usually combined with a mechanism to exploit multicast efficiencies made possible by aggregating/covering subscriptions with ones that match the same or wider range of messages. For example, in the Siena CBN (Carzaniga et. al., 2001) subscription covering is achieved by restricting attribute types and subscription filters to simple number, string and boolean types using a set of transitive operators to filter across them (i.e. greater/less than, super/sub string etc.). The underlying routing structure on which messages pass allows for a message inserted on one side of the network to propagate across the network based on positive matches to the filters (subscriptions) until every client, and only those interested in the message, have been delivered the message.

CBN Subscriptions are specified as a set of filtering constraints constructed using a filter operator and some values used to compare against the contents of any incoming notification. The range of these operators determines the type of pub/sub network in which the message is being sent. Mühl et. al. (Mühl et. al., 2006) describe this content-based matching as a set of "Filters which are evaluated against the whole contents of notifications". Within this work, notifications can be thought of as publications. Publications are only forwarded to a user when the contents of their subscriptions' filtering constraints matches a subset of the message's contents. This allows for a more flexible message format and, in comparison to topic-based networks, allows an even looser coupling between publishers and subscribers.

As an example, in the Siena CBN system (Carzaniga et. al., 2001) a notification is seen a set of typed attributes. Each attribute is comprised of a name, a type and a value. The existing version of Siena supports the following types: String, Long, Integer, Double and Boolean. A Siena subscription is a conjunction of filtering constraints, where constraints are comprised of the attribute name, an operator, and a value. A subscription matches a notification if the notification matches all the filtering constraints of the subscription's filter. The notification is then delivered to all of the clients that submitted those subscriptions that match against that publication. Carzaniga (Carzaniga et. al., 2008) defined three basic types of Siena topology: hierarchical client/server, acyclic peer-to-peer, and general peer-to-peer. All topologies provide the same functionality, however they differ in non-functional features, like time complexity, scalability and fault tolerance.

The work within this chapter focuses on Knowledge-Based Networks (KBNs) (Lewis et. al., 2006; Keeney et. al., 2008a; Keeney et. al., 2008b) rather than CBNs. It is important to note that Knowledge-based networks are defined, within this chapter, as publish/subscribe networks in which the semantics of the message play an important part in the matching of publications to subscriptions. The particular flavour of KBN investigated in this work is an extension of the Java Siena CBN middleware (Carzaniga et. al., 2001). The specific Siena-based KBN implementation introduces two extensions to the existing type set and the set of filter operators already provided by the hierarchical Java version (Carzaniga et. al., 2008). These extensions add support for bags of values and the use of ontological types and operators. Specifically, the KBN implementation supports all Siena types in addition and union with ontological classes, properties and individuals (owl, 2004). In addition to the standard Siena operators, operators have been added to exploit ontological type checking, subsumption, equivalence, disjointedness, and the use of arbitrary ontological object properties to capture relationships between individuals. The second extension examined in this chapter adds support for bags (unsorted mulisets) of values to be used in subscription filter constraints. Most CBN implementations only allow a single comparison value to be specified in each filter. Hence multiple constraints in a single subscription are usually combined as a conjunction, this greatly restricts the expressiveness of any single subscription filter. Rather than extending the subscription mechanism to support a disjunction of filtering constraints, which would greatly affect the ability to aggregate filters for efficient routing tables, our extension instead supports a disjunction of values within a single filtering constraint, with multiple constraints being combined as a conjunction as before.

Many of the additional capabilities of a KBN over a CBN could be achieved with complicated and unwieldy CBN subscriptions, which become unmanageable within both the network and with regard to the human creating the subscription. In addition, by extending the expressiveness of the subscription language we risk adversely affecting the potential for subscription aggregation in the broker network. However we have shown that more expressive subscriptions lead to more accurate and more concise subscriptions (Keeney et. al., 2007a), thereby improving performance and making a KBN more flexible and applicable in a wide range of use-cases.

RELATED WORK

The use of ontological information in publish/subscribe systems is still on the edge of wide-scale deployment. Exploratory research and a detailed review of the State of the Art shows a gradual shift towards an understanding of both the importance and power of semantics when combined with pub/sub networks. In this section we discuss seven influential systems showing similarities to the KBN implementation discussed here. For convenience, a comparison table is provided at the end of this discussion.

S-ToPSS (Petrovic et. al., 2003; Burcea et. al., 2003) offers a content-based network that is targeted at making an existing centralized syntactic matching algorithm semantically-aware whilst keeping the efficiency of current event matching techniques. S-ToPSS proposes a two-level distributed semantic pub/sub system. The top-level brokers have only high level descriptions of ontologies gained from the lower level routers. The lower level brokers maintain their own ontology for communication, heterogeneous ontological information for each application is distributed between multiple routers. These low level brokers advertise more general descriptions of the ontologies they hold to higher level brokers. Petrovic (Petrovic et. al., 2003) describes this additional semantic matching mechanism in an employer-employee scenario. The employer is looking for a candidate from a "certain university", "with a PhD degree", "with at least 4 years work experience." This is matched using the semantically enhanced pub/sub system, S-ToPSS, in which an employee who has a PhD, from a particular school, with at least four years work experience is matched with the prospective employer. The match is only made because the semantic system is aware that "school" and "university" have the same semantic meaning, particularly in the North American Educational System. S-ToPSS manipulates subscriptions as they arrive to add semantic synonyms and super-class concepts to the subscription (with support for more general mapping functions). In addition, when events enter the broker, new synthetic events are created depending on the semantics contained in the event. This way the semantic stage of subscription merging and publication matching is performed first, outside of the standard matching engine; unlike within a KBN. We argue that the KBN solution discussed in this paper is more general, flexible and expressive than S-ToPSS. More significantly, however, no report of an implementation or evaluation of this proposal has yet emerged.

An ontological pub/sub system called **Ontology-based Pub/Sub (OPS)** is presented in (Wang et. al., 2004), and shares our motivations to improve the expressiveness of events and subscriptions within the system. Wang's work is achieved using RDF and DAML+OIL techniques to semantically describe events and subscriptions. Both of which are represented as RDF graphs and graph patterns respectively. The OPS system is shown to match events and subscriptions both semantically and syntactically. Central to the OPS matching algorithm is an efficient and scalable index structure based on the complete set of possible statement patterns (decomposed by RDF graph patterns) which are used as the basic unit of matching using AND-OR trees as matching trees, which subsequently avoids the backtracking of the RDF graphs. One criticism of the OPS system is that it does not include/provide the ability to perform generic content-based subscriptions. It is important to note at this point that the KBN documented within this chapter not only provides the semantic extensions documented, but couples these with the full set of Siena CBN operators, thereby allowing semantic and syntactic operators to coexist, increasing both the expressiveness and flexibility of the KBN.

The **Ontology-based Pub/Sub System (OBPS)** (Skovronski et. al., 2006) is another ontologically based publish/subscribe system.

This system expresses event models by using the XML language: notifications are composed of an XML message where the tags are the names of the ontological classes or property within the ontology and the root node of the notification must contain the name of the ontology to which the notification is destined. To assist in the matching of events to the interested subscriptions, each publisher has its own publisher agent which is responsible for processing published events within the routing broker. Each agent inside the router maintains their own ontological model that stays in scope as long as the publisher continues publishing messages. The interested subscribers register a topic which defines the ontology in which they are interested, guaranteeing that all notifications for that topic are routed to all interested subscribers. The system uses the SPARQL query language as its subscription language, allowing the subscribers to easily parse the returned notification message based on their knowledge of which XML tags will be within the notification. Again like OPS, this system cannot perform generic content-based subscriptions. All the processing of messages and query analysis is done on the client-side. In addition, the substantial overhead introduced by SPARQL and the lack of mechanisms to aggregate subscriptions and construct network overlays means that the overhead and scalability of this system is uncertain.

Continuous Querying Syndication System (CQS) (Christian et. al., 2007) targets the formalisation of a syndication architecture that utilizes web ontologies and logic-based reasoning for a large volume of selective content dissemination on the web. The aim of this is towards an expressive syndication system, where each of its subscriptions is comprised of a conjunctive ABox (ontology instance) query; while the publication is defined to be composed of a set of ABox assertions. Intuitively, CQS offers only one syndication broker which maintains a local knowledge base in which newly published information is integrated.

To match newly published information with subscription requests in an efficient and practical manner, the events are matched to subscribers by employing a composition matching algorithm as follows: "information matches" refers to the instances that are stored in the local knowledge base bound to the variables of a continuous query representing a subscription. Hence the result returned to the interested subscriber is actually the query answer in the local knowledge base rather than the matched publications. Whilst "publication matches" refers to the collection of publications satisfying subscribers' subscriptions, the broker delivers the minimal sets of publications to the interested subscribers.

Elvin with Ontology (ElvinO) is a proof-of-concept system implemented by the authors of this chapter to investigate the feasibility of knowledge based networking (Keeney et. al., 2006b). The system uses the closed-source Elvin CBN (Segal et. al., 2000) as an underlying subscription matching mechanism. In this system ontological synonyms were added to publications at the edge of the network in a generic mapping gateway, allowing subscribers to register subscriptions according to their own ontology, and have the matching publications delivered to them, where the publication may have contained information defined in a different (mapped) ontology. Alongside the semantic subscriptions, this system maintains the very expressive Elvin subscription language thereby supporting semantic and content based subscriptions. Additionally, the Elvin subscription language is more expressive than that provided by Siena, in particular supporting logical combinatorial operators, thereby additionally supporting combinations and disjunctions of constraints. The closed-source nature of the Elvin implementation is also prohibitive in terms of enhancing the content based routing algorithms to include more complex datatypes, such as OWL classes, instances or properties inside the broker rather than at the edge of the network.

Semantic Pub/Sub with Super-Peers (SPS-SP) (Chirita et. al., 2004) is an RDF-based system for managing arbitrary digital resources. Subscription queries are expressed by using a typed first-order RDF query language called "L". A subscription is a conjunction of RDF triples, where an RDF triple is comprised of a subject (s), predicate (p), and an object (o). The predicate can be the operator of ">" for integer type or "⊇" meaning "contains" for string type. The publication is a pair (T, I), where T is a set of ground atomic formulas of L of the form of an RDF Triple (s, p, o) with the same constant s, and I being a client identifier. A publication PUB matches a subscription SUB if the ans(SUB, T) ≠ ∅; where the ans(SUB, T) denotes the answer set of SUB when it is evaluated over T. Publications and subscriptions are matched at super-peers and appropriate subscribers notified. The advertisement of messages is also used to help super-peers route information efficiently. The SPS-SP system is comprised of peers with the role of publisher or subscriber and a super-peer that offers a more powerful capability than a typical peer. Super-peers are organized within the HyperCup (Schlosser et. al., 2002) semantic topology which takes care of processing publications, advertisement and subscriptions using P2P semantic queries. The HyperCup algorithm is capable of organizing super-peers into a binary hyper cube that is a type of Cayley graphs. The advantage of HyperCup is that it limits the path between any two super-peers to $\log_2 N$, which enables efficient and non-redundant information broadcasting. The SPS-SP system utilizes a selective advertisement forwarding algorithm to selectively broadcast advertisement messages to super-peers. Therefore, the peers that submit subscriptions can exploit the advertised information to calculate which peers contains information that should match their subscriptions. Again, like several of the systems discussed, this system does not include capability to perform generic content-based subscriptions.

Other Relevant Work

Semantic web technologies aid a user in the continued search for relevant information. Grossnickle in (Grossnickle et. al., 2005) show how the introduction of semantically rich meta-data to a rapidly changing source of RSS news feeds increases the ease in which searched information is delivered to a user. This is supported by Borsje et al. in (Borsje et. al., 2008) using a localised SPARQL query. This query could be represented in pub/sub terms, as a user's subscription. Knowledge-distribution systems allow semantically marked-up data to be delivered to all interested parties in both a timely and efficient manner. This is achieved by embedding ontological mark-up in the feeds. However, with the substantial overhead of using SPARQL as the query mechanism and the lack of subscription aggregation techniques, the overhead and scalability of this system is unclear.

Baldoni (Baldoni et. al., 2007) supports the rationale behind the introduction of knowledge representation through the use of tagging each publication within the network with a topic. This can be likened to assigning a subject line to an email, except in this case the message is instead assigned a topic tag. The subscription table is constructed using a list of couples <t,i> where t is the topic a node is subscribed to and i is the corresponding topic overlay identifier. Upon retrieving a new subscription for a topic, the subscription management component adds an entry for the topic to the subscription table and then passes the task of connecting the corresponding overlay networks. In the tagging of messages it is easy to see that this increased level of descriptiveness is beneficial in the task of matching future publications to subscriptions.

Li (Li et. al., 2004) presents a semantic-based system based on a centralised pub/sub bus implementation. Its application is limited to enterprise scale deployment, not fully offering true CBN capabilities. Li offers approaches to the enhancement of subscriptions and events within the proposed semantically enhanced system. Li does not, however, address the issues involved in introducing ontology reasoning within a distributed event dissemination network and does not offer an implementation or evaluation of the proposed system.

The pub-sub middleware presented by Cilia (Cilia et. al., 2003) allows semantics to be used in the publications and subscriptions. However, these semantics are used on the edge of the network in a manner similar to that presented in number of systems above.

Semantic techniques are also being used by several systems for the retrieval of distributed ontologically encoded knowledge in P2P DHT systems (Tempich et. al., 2004; Cai et. al., 2004; Loser et. al., 2003). These systems, such as the one presented in (Borsje et. al., 2008), focus on query-response communication as opposed to the pub/sub model presented within this chapter.

As shown by Meier (Meier et. al., 2005) and Eugster (Eugster et. al., 2003) there are various type- and topic-based/subject-based distributed event-based systems whose popularity has both increased and subsequently decreased over time. Several of these systems use hierarchical addressing in the organisation of topics using a containment relationship in a manner that is loosely similar to the taxonomical class hierarchy defined in an ontology. The mechanisms addressed by these systems allows the use of the notion of sub- and super-class subscriptions with regard to the topic hierarchies, which in this respect allow a simple comparison to KBN subsumption operators. However the subscriptions using topic-based hierarchical addressing only operate in an "up/down" subscription creation algorithm. With the use of equivalence and disjoint relationships between classes, a KBN allows a multidimensional type tree to be created in which sub-classes not only point to classes further down the class hierarchy, but can point to other classes in other parts of the ontology, a unique addition to the traditional topic-based subscription.

Table 1.

	Event Model	Subscription Language		Overlay Type	Routing Scheme		Event Matching
System	**Data Expression:***The format of knowledge within the network.*	**Subscription Expression:***How user subscriptions are formed.*	**Subscription Operators:***The operators used in the user subscriptions.*	*The overlay network design and deployment*	**Sub Forwarding:***The method in which a broker forwards subscriptions to other nodes in the network*	**Event Forwarding:** *How the pubs are forwarded to interested subs*	**Pub to Sub Matching:***How incoming pubs are matched to stored subscriptions.*
KBN	***Basic Format:*** [attribute_ name, value] ***Non-ontological_value:*** Integer, Long, Double, Boolean, String ***Ontological_value:*** OWL Classes, Properties, and Instances.	***Basic Format:*** [attribute, operator, value] ***Note:*** Similar to Siena	***Non-ontological:*** $>$, $<$, $=$, $\leq$, $\geq$, etc. ***OWL operators:*** equivalent, subsumes, subsumed by, instance_of, object property ***Bag operators:*** (e.g., subbag, superbag)	***Hierarchical structure Semantic-based cluster***	***Sub Aggregation method:*** aggregated by subscription covering. ***Forwarding method:*** The most general subscriptions are hierarchically sent to master servers.	***Reverse path forwarding:*** follows the reverse path of subs.	***Matching tree:*** Event iterates through sub tree using breadth first search.
S-ToPSS	***Basic Format:*** [attribute_ name, value] ***Non-ontological_value:*** Integer, Long, Double, Boolean, String ***Ontological_value:*** uses three extra matching algorithms to make a non-ontological event semantically.	***Basic Format:*** [attribute, operator, value] ***Note:*** Similar to ToPSS	***Crisp:*** $>$, $<$, $=$, $\leq$, $\geq$, etc. ***Approximate:*** (e.g., $\sim=$ modelling approximate equality) ***Probability:*** (e.g., modelling a random event)	***Centralized:*** The system is deployed as an information dissemination service	***None:*** it is a centralized, therefore no routing	***None:*** it is a centralized, therefore no publication forwarding	***Matching tree:*** Event iterates through sub tree using breadth first search. ***For semantic matching:*** Existing matching assisted with: -Synonyms matching -Concept matching -Hierarchy matching
OPS	***Basic Format:*** [attribute_ name, value] ***Non-ontological_value:*** Integer, Long, Double, Boolean, String ***Ontological:*** RDF graph (concepts within Events form an ontology)	***Basic Format:*** RDF graph pattern (subject, object, meta-statement, [filter_ func(object)])	*Note:* When the object is a variable and its type is literal, it includes *filter_ func(object)* to refine value *filter:* $>$, $<$, $=$, $\leq$, $\geq$, etc	***Hierarchical structure Peer-to-peer***	***Aggregation method:*** Aggregation by graph merging. ***Forwarding method:*** the most general subscriptions are sent to parent servers.	***Reverse path forwarding:*** Follows the reverse path of subs.	***AND-OR matching tree:*** Node and arc in sub graph must be mapped to node and arc in the event graph.

continued on the following page

Table 1. continued

	Event Model	Subscription Language		Overlay Type	Routing Scheme		Event Matching
OBPS	***Basic Format:*** [attribute_ name, value] ***Non-ontological_value:*** Integer, Long, Double, Boolean, String ***Ontological_value:*** XML-based (Tags represent the name of classes & properties of an ontology model)	***Basic Format:*** SPARQL query.	***No operators:*** subscription is represented by SPARQL query.	***Centralized:*** a single router	***None:*** it is a centralized, therefore no routing	***None:*** it is a centralized, therefore no publication forwarding	***Ontological topic based matching using a SPARQL engine.***
CQS	***Basic Format:*** (*a, t, p*) where *a* is a set of DL ABox assertions, *t* is time unit and *p* is the client identifier	***Basic Format:*** (*Q, t*) where Q is a Continuous conjunctive ABox query with respect to a DL KB and t is time unit	***No operators:*** Subscription is represented by ABox query.	***Centralized:*** a single syndication broker	***None:*** it is a centralized, therefore no routing	***None:*** it is a centralized, therefore no publication forwarding	***Composition matching algorithm:*** -Information matches. -Publication matches. -Composite matches.
Elvin(O)	***Basic Format:*** [attribute_ name, value] ***Non-ontological_value:*** Integer, Long, Double, Boolean, String.	***Basic Format:*** [Attribute, operator, value]	***Non-ontological:*** >, <, =, etc. ***Logic:*** ||, &&, !, etc.	***Clustering:*** local area server replication for fault tolerance ***Federation:*** static wide area hierarchical subscription space partitioning	***Forwarding method:*** Subscription kept locally, possibly broadcast within clusters (limited information available)	***Hybrid:*** Between federations publications are routed according to subscription space partitioning, possibly broadcast within clusters (limited information available)	***Non-ontological:*** unknown, internal to Elvin ***String based label Ontological matching:*** OWL labels and synonyms are merged into/from string labels at the edge of network.
SPS-P2P	***Basic Format:*** (*T, I*) where *T* is a set of *t*(*s, p, o*) and *I* is a client identifier ***Ontological_value:*** RDF Event Model	***Basic Format:*** (*s, p, o*) where *s* is a variable, *p* is constant, *o* is distinct variable ***Note:*** Datalog-inspired RDF Query	***Binary:*** >, <, =, ≤, ≥, etc. ***Subsumes:*** ⊒	***Peer-to-Peer Note:*** Superpeer arranged in HyperCup topology.	***Aggregation method:*** Aggregation by subscription covering (similar to SIENA). ***Forwarding method:*** the most general subscriptions are broadcasted to all neighbours	***Reverse path forwarding:*** Follows the reverse path of subs (similar to SIENA).	***Matching tree:*** The event iterates through the subscription tree by using breadth first search algorithm (similar to SIENA).

XML-based systems (Diao et. al., 2003; Gupta et. al., 2003; Chan et. al., 2006) publish events as XML documents where XPath (Clark et. al., 1999) queries allow expressive and flexible subscriptions, and arbitrarily complex queries to be applied to a DOM tree derived from the published XML document. The message is then forwarded only when a match is found. XML-based systems provide an increased level of expressiveness in comparison to topic-based systems. However, the message architecture is exclusively based on tree patterns and thus is less expressive and flexible than KBN messages (See Table 1).

THE SEMANTIC EXTENSION

Knowledge-based Networking, an extension to Content-based Networking, involves the routing of events across a network, based not just on the values of the event contents but also on some semantics of the data and associated meta data contained in the event. We have developed a model for the filtered dissemination of semantically enriched knowledge over a large loosely coupled CBN of distributed heterogeneous agents. We call such a semantic-based CBN a Knowledge-Based Network (KBN). In (Lynch et. al., 2006; Lewis et. al., 2006; Keeney et. al., 2007a; Keeney et. al., 2008b) a KBN implementation is presented that extends Siena by providing three additional ontological base types: properties, concepts/classes and individuals/instances, as described in ontologies originating from the semantic web community. It also supports subsumptive subscription operators, i.e. sub-class/property (MORESPEC) i.e. *more specific*, super-class/property (LESSSPEC) i.e. *less specific*, and semantic equivalence (EQUIV). For example, as seen in the Wine ontology (w3c, 2003), the ontological type "wine" is less specific than (subsumes) the type "white wine", as "white wine" is more specific than "wine" since "wine" is a superclass of "white wine". Producers of knowledge express the semantics of their

available information based on the ontological representation of that information. Consumers express subscriptions upon that information as simple semantic queries. If an event consumer was interested in receiving events about some ontological entity type E, classes equivalent to E, or entities more specific than E, this can be easily achieved by creating a filtering constraint such that the entity type described in a field x of the message is subsumed by E, i.e., (x MORESPEC E). E.g. A subscriber can subscribe to all KBN messages that contain an attribute whose value is a concept more/less specific than the named concept in the subscription. This approach provides loose semantic coupling between applications, which is vital as a new wave of applications increasingly rely on using the application information, context and services offered by existing heterogeneous distributed applications. To achieve this, each KBN router holds a copy of a shared OWL ontology (owl, 2004), within which each ontological class, property and individual used is described and reasoned upon.

The new ontological types, namely classes, individuals, and properties, are first class KBN types and can be used in any KBN subscription or notification, along-side the standard CBN (Siena) types and operators. This allows messages to be matched to subscriptions based on extensible type information, which can effectively represent metadata for the message without having to maintain an ever-growing set of universal attribute names. Instead, a simple set of shared attribute names can be used for a concept type, which uses values from a taxonomy that is maintained, distributed and reasoned over at run-time using existing standardised ontology techniques. In addition to the equivalent class, equivalent property, equivalent individual (EQUIV) operator; subclass, subproperty (MORESPEC) operator; and superclass and superproperty (LESSSPEC) operators we also discuss the following operators: ISA, IS_NOT_A, ONTPROP, and NOT_EQUIV.

The ISA operator is used to match an ontological individual/instance against its ontological types/classes. If an individual I is defined as being of an instance of a certain type C then the ISA operator will match the individual I to class C, all classes equivalent to C, and all superclasses of C. For example if the person "John" is represented in an ontology as an individual of the class "person", and the class "person" is defined to be equivalent to the class "human", and the class "human" is a subclass of the class "mammal", then the individual "John" is related by the ISA relationship to the classes "person", "human", and "mammal". So the subscription filter (x ISA "human"), where x is an ontological individual, would match a notification that contains the named attribute (x: "John"). This kind of a subscription filter was not previously possible since the EQUIV, MORESPEC, and LESSSPEC operators could only compare classes with classes, properties with properties, and individuals with individuals.

The IS_NOT_A operator is again used to compare an ontological individual/instance against its ontological types. If an individual I is defined as being of an instance of a certain type C then the IS_NOT_A operator will match the individual I to all classes except class C, all classes equivalent to C, and all superclasses of C. Based on the example above the individual "John" is a "human" and is a "mammal", but is not a "cow" where "cow" is a subclass of "mammal". So the subscription filter (x IS_NOT_A "cow") would match a notification that contains the named attribute (x: "John").

The ONTPROP operator is used to match ontological individuals against each other using any ontological object property. Ontological object properties define named relationships between individuals of two classes. For example the object property "eats" might be defined between individuals of type "animal" and individuals of type "food", so the individual called "Colleen" of type "cow" (where "cow" is a subclass of type "animal") would be related by this "eats" relationship to an individual called "grass" of type "food". So the filter (y ONTPROP $_{\text{"eats"}}$ "grass"), where y is an ontological individual, would match a notification that contains the named attribute (y: "Colleen").

Siena does not support a generic NOT (!) operator modifier, and so the NOT_EQUIV operator was added for completeness. This operator is the opposite of the ontological EQUIV operator discussed above. It is used to compare classes with classes, properties with properties, and individuals with individuals.

THE BAG EXTENSION

According to Weisstein (Weisstein, 2002), a bag (also called multiset) is a set-like object in which order is ignored, but multiplicity is explicitly significant. Therefore, bags {1, 2, 3} and {2, 1, 3} are equivalent, but {1, 1, 2, 3} and {1, 2, 3} differ. A bag differs from a set in that each member has a multiplicity indicating how many times it is a member.

As presented by Roblek (Roblek, 2006) a bag value can contain any valid Siena/KBN values, including other bag values. A bag is not allowed to contain itself, either directly or indirectly via other bags. Elements of a bag do not need to be of a uniform type. In the extension presented here bags are first order members of the Siena/KBN type set. They can appear in notifications as well as in subscription filters like any other Siena/KBN type. Siena advertisements, which are part of the theoretical Siena model, are not supported in the hierarchical version of Siena but, should they be, bag type should work seamlessly with them.

Therefore, some examples bags are: {3, 345, 27, 35, 3476, 0, 27, 27}, {"Ljubljana", 2, "Ljubljana", 3.14159}, {"Ljubljana", "Vienna", "Amsterdam", "Dublin"}. Since a set is a bag where all elements have a cardinality of one, this extension also implicitly supports sets. The bag extension adds simple binary bag operators and composite binary bags operators.

Simple Bag Subscription Operator

The simple operators support the three well-known binary bag relations: equal, subbag, and superbag. Two bags A and B are equal (A=B), if the number of occurrences of each element in A or B is the same in each bag. For example the bags {'b', 'o', 'o', 'k'} and {'b', 'o', 'k', 'o'} are equal bags but {'b', 'o', 'o', 'k'} and {'b', 'o', 'k'} are not equal bags.

For the subbag relationship, we can define A to be a subbag of B, $A \subseteq B$, if the number of occurrences of each element χ in A is less than or equal to the number of occurrences of χ in B. For example {'b', 'o', 'k'} is a subbay ($\subseteq$) of {'b', 'o', 'o', 'k'} but {'b', 'o', 'w'} is not a subbag of {'b', 'o', 'o' 'k'}. It follows from the definition of subbag that bags A and B are equal bags, if and only if $A \subseteq B$ and $B \subseteq A$. Another way to describe the subbag relationship is to say that if $A \subseteq B$ then bag B includes or contains all of the elements in A. The superbag relationship is the inverse of the subbag relationship. If B is a subbag of A, $B \subseteq A$, then A is a superbag of B, $A \supseteq B$. Also note, a bag can contain other bags. For example {{'g', 'o', 'o', 'd'}, {'b', 'o', 'o', 'k'}} $\supseteq$ {{'b', 'k', 'o', 'o'}}. All three simple bag relations, namely equal, subbag, and superbag, are transitive and reflexive. The transitivity and reflexivity of simple bag relations follows from the transitivity and reflexivity of numerical *equal* and *less than or equal* relations that were used in the definition of the simple bag relations.

The main advantage of the bag type and bag operators lies in the ability to define much more expressive and flexible subscriptions. For example, without the use of bags, multiple filter constraints in a single subscription filter are joined by conjunction (using the boolean AND operator). A disjunction of constraints (using the boolean OR operator) could only be specified using multiple distinct subscriptions. Using bags and the subbag operator the subscriber can create a filter where matching values should be drawn from a bag of possible values for example. A particularly useful example of this is in Keyword matching.

Composite Bag Subscription Operator

While the simple bag operators defined above prove useful for a number of case studies, they are restricted to comparing the sizes and structures of bags rather than performing more useful comparisons over the *elements contained* in the bags. Using the simple bag operators, bags are only compared based on the presence of *matching* elements in each bag. For example, a subscriber may be interested in a bag of strings that is larger than that specified in her subscription (superbag), but the strings used in her subscription bag should be substrings of those in matching publications rather than equal strings. The simple bag operators can also be combined with other Siena or KBN operators to produce composite bag operators. The composite bag relation is also a binary relation over bags, but is composed of (i) a simple binary bag relation over the bags and (ii) a sub-relation over the bags' elements. In this way two bags can be compared not just on the presence of *matching* elements but rather by using more expressive operators to compare the elements that make up each bag.

Suppose Φ is a simple binary relation over bags (Φ is a simple bag operator), and λ is an arbitrary binary relation (λ is *any* non-bag subscription operator). Bag P is Φ-related to bag Q when sub-relation λ is applied between the elements of P and Q, written as $P \, \Phi_\lambda \, Q$, if and only if there exist some sequences X and Y, (X is some ordered list of elements from P, and Y is some ordered list of the elements from Q), so that all of the following statements are true:

1. P is Φ-related to $\tau(X)$, where $\tau(X)$ denotes the bag of all elements in sequence X. bag P is Φ-related to X when X is expressed as a bag

2. $\tau(Y)$ is Φ-related to Q, where $\tau(Y)$ denotes the bag of all elements in sequence Y. when Y is expressed as a bag, that bag is Φ-related to Q

3. $|X| = |Y|$ sequences X and Y have the same number of elements 4. $\forall$ i $\in$ $\mathbb{N}$ (natural number), i $<|X|$, X_i is λ-related to Y_i. for (int i$=0$; i$<|X|$; i++) X.elemantAt(i) is λ-related to Y.elemantAt(i)

We call relation Φ the primary relation of the composite bag relation, and relation λ the sub-relation of the composite bag relation. So for bags P and Q, simple bag relation Φ, and any relation λ, P Φ_λ Q means that P is Φ-related to bag Q when sub-relation λ is applied. If the composite Φ_λ relation is being used as a subscription operator we call Φ the primary bag operator of the λ the suboperator.

The bag of integers $\{1, 1, 2, 3, 4\}$ is a superbag of $\{2, 4, 3\}$ using the default "equals" ($=$) sub-relation, i.e. $\{1, 1, 2, 3, 4\} \supseteq_= \{2, 4, 3\}$ *(for every element in the second bag, there exists an element in the first bag that is equal to the element, with no reused elements in either bag)*.

The bag of integers $\{1, 2, 3\}$ is an equal-bag of $\{2, 3, 4\}$ using the "less than" ($<$) sub-relation *(for every element in the second bag, there exists an element in the first bag that is less than the element, with no unused or reused elements in either bag)*, so $\{1, 2, 3\} =_< \{2, 3, 4\}$.

The bag of Strings $\{$"ood", "boo"$\}$ is an sub-bag of $\{$"a", "good", "book"$\}$ using the "substring" (substr) sub-relation *(for every element in the first bag, there exists an element in the second bag such that the element in the first bag is a substring of the element in the second bag, with no reused elements in either bag)*, so $\{$"ood", "boo"$\} \subseteq_{substr} \{$"a", "good", "book"$\}$.

Note, the simple bag operators defined in the previous sections can be defined as composite bag operators using the default "equals" ($=$) sub-relation. More generally, any simple bag operator Φ is equivalent to the $\Phi_=$ composite bag relationship.

Where Φ be a simple binary relation over bags, and λ a binary relation over bag elements. (i.e. Φ_λ, Φ is the primary bag operator, λ is the sub operator). If λ is transitive, then Φ_λ is also transitive. If λ is reflexive, then Φ_λ is also reflexive. Another interesting observation is that if A $\subseteq_>$ B then B $\supseteq_{\leq}$ A. More generally, if Φ is a simple binary relation over bags, λ is a binary relation over bag elements, Φ^{-1} is the inverse relation of Φ, and λ^{-1} is the inverse relation of λ, then PΦ_λQ exactly when Q Φ^{-1}_λ-1 P. This observation only holds where inverse subscription operators exist. For example, some operators have no inverse (e.g. substring, some ontological properties), and as mentioned, Siena and the presented KBN implementation does not support a generic boolean NOT (!) operator modifier.

Of further note, a composite bag operator may have another composite bag operator as its sub operator. For example given:

$$V = \{\{\}, \{0, 0\}, \{1, 2, 3, 4\}\}$$

$$W = \{\{8\}, \{0\}, \{1, 1, 1\}, \{2, 3, 4, 5, 6\}\}$$

then

$$V \subseteq_{(\subseteq<)} W$$

since

$$\{\} \subseteq_< \{8\}, \{0, 0\} \subseteq_< \{1, 1, 1\},$$

$\{1, 2, 3, 4\} \subseteq_< \{2, 3, 4, 5, 6\}$, and bag $\{0\}$ in W is unused.

Compared to simple bag relations, composite bag relations make looser comparisons of bags possible. They allow for "inexact" matches that would not be possible should we use only simple bag relations. Several content-based publish/subscribe systems support disjunction type subscriptions and the use of sets in the formation and matching of subscriptions. However, the Bag extension presented within this chapter (and combined with ontological subscription formation) allows Siena to now support such queries. However we argue that the composite bag operators supported by our KBN surpasses this with the ability to define even more expressive and flexible subscriptions, especially when the composite bag operators (unique to this work) are used and particularly when used together with the ontological types and operators.

Performance of Subscription Tree Merge and Lookup using Simple Bag Operators

Suppose there exists some bags B and C. Without loss of generality we can assume that $|B| \leq |C|$. The simple bag operator comparison has the best time complexity $O(|B|)$ and the worst time complexity $O(|C|^2)$. If bags are pre-ordered it is possible to use a better algorithm that has the worst time complexity of $O(|C|)$, assuming the bag elements can be totally ordered. Since sorting a list of n element has an average complexity of $O(n\log n)$, with both bags requiring pre-sorting, we did not opt to explicitly pre-sort each bag before attempting bag comparison. This decision should have a relatively small performance affect unless the bags being compared are particularly large. In the current KBN implementation bags are either pre-ordered or not depending on the order in which values are encoded into subscriptions or publications by the subscribing/publishing application.

Performance of Subscription Tree Merge and Lookup using Composite Bag Operators

With the brute force algorithm used to compare bags with the composite bag operator, in the most optimistic case the algorithm finds matching elements immediately. In this case the time complexity of the algorithm is $O(|B|)$ for some bags B and C where $|B| \leq |C|$.

In the most pessimistic case the algorithm must fully match all possible arrangements of the elements of the smaller bag with the elements of the bigger set. In this case the time complexity is:

$$O\left(\frac{|C|!}{|C|-|B|!}\right) < O\left(|C|^{|B|}\right)$$

We can see from the formula above that the time complexity depends on the size of the smaller bag because the size of the smaller bag appears in the exponent of the time complexity. It is evident that the algorithm for composite bag operators implemented in the scope of this research is truly useful only for bag comparisons where at least one of the bags is small.

It is possible to develop much more effective comparison algorithms for certain specialized composite bag operators. For example, as discussed above it is possible to exploit the ordering of the elements to improve the algorithm. For composite bag operators over bags containing numbers, where the sub-operator is one of $<, \leq, >, \geq$ it is possible to develop a very effective comparison algorithm if bags are pre-ordered. This is because the set of all integers are a totally ordered set with regard to $\leq$, or $\geq$. Such an algorithm would have a very low time complexity of $O(\max(|B|,|C|))$. Unfortunately not all sets are totally ordered, or can be totally ordered. For example, ontological instances cannot be naturally ordered, ontological concepts form only a partially ordered set with regard to subsumption. An effective algorithm for composite bag operators over partially ordered sets could be a subject of the future research. Alternatively, there might exist some constraints that narrow the set of all possible bags that are to be published in notifications or subscription filters, to some subset, for which good performance of the composite bag operator could be guaranteed.

IMPLEMENTING A KBN

The two extensions discussed in this paper have been fully implemented and tested, and a deployable Java based KBN implementation is available. Though our initial measurements described in (Keeney et. al., 2006b) used the Elvin CBN, this was a centralised system and our scalability goals required us to consider a decentralised CBN scheme. For the design and implementation presented here we opted to build upon the Siena CBN (Carzaniga et. al., 2001). This was mainly due to source code availability and an abundance of associated technical reports and papers, and in addition, its focus on expressiveness in a wide-area distributed environment. The approach taken was to extend the CBN to use ontological reasoning using Jena (Carroll et. al., 2004) for ad hoc matching and filtering inside the network. The open-source nature of Siena also allows us to migrate the semantic subscription and message matching from the endpoint network nodes to within the network itself.

Previous works by the authors have shown that semantic types and operators can be supported by incorporating an ontological knowledge base and an ontological reasoner into each Siena KBN router/broker (Lynch et. al., 2006; Lewis et. al., 2006; Keeney et. al., 2007a; Keeney et. al., 2008b). Each KBN router/broker holds a copy of a shared OWL ontology, within which each ontological class, property and individual used is described and reasoned upon. To achieve this there have been significant additions to the codebase to support the new ontological types and operators. However, it must be noted that the operation of the described KBN as a non-semantic Siena CBN has not been compromised in any way. The semantic extensions

(and bag extensions) are provided to supplement the Siena CBN system.

Previous works by the authors have also shown that the loading of new ontologies into a reasoner embedded in a KBN node is computationally expensive due to load-time inference (Lewis et. al., 2006). Therefore the frequency of changes to the ontological base of a given KBN must be minimised since changes need to be distributed to each of the nodes in the network. Secondly, ontological reasoning is memory intensive and memory usage is proportional to the number of concepts and relationships loaded into the reasoner so reasoning latency can be controlled by limiting this number in any given KBN node. However, once loaded and reasoned over, the querying of such an ontological base is relatively efficient with performance relative to size of the ontological base (Lewis et. al., 2006).

A large number of ontology reasoners are available, including: KAON2 (Motik et. al., 2006), Pellet (Parsia et. al., 2004), Racer (Haarslav et. al., 2001), FaCT (Tsarkov et. al., 2005), and F-OWL (Zou et. al., 2004). Any choice of a reasoner must be based on examination of performance evaluations in the literature, such as (Pan, 2005; Guo et. al., 2004; Pellet Performance, 2003; Motik et. al., 2006), as well as with separate benchmarking. These evaluations must also be compared to the performance characteristics of domain specific reasoners, or existing reasoners cut-down to give reduced but sufficient results in return for enhanced performance. This trade-off around reasoning performance versus expressiveness and accuracy of the model, after the inference cycle, is of particular importance where such reasoning may be required for efficient and correct routing in the network. The differing performance characteristics of different reasoners under different conditions, such as, the impact of the ratio of concepts to relationships of subsumption relationships to user defined predicates, must also be evaluated. The performance of different reasoners, and the reasoning load, will also change in a non-linear fashion depending on the size and expressiveness of the ontologies used and the level of ontology language used (e.g. OWL-Lite vs. OWL-DL) (Pan, 2005; GuoY et. al., 2004; GuoY et. al., 2005; Pellet Performance, 2003; Motik et. al., 2006). Of particular importance is the amount of reasoning that can be performed at ontology load time versus when the first or subsequent queries are submitted to the ontology. While the load time overhead of incorporating an ontology and a reasoner at every broker is substantial, we have shown that the runtime overhead of querying the ontology is minimal when performing subscription merging or publication/subscription matching (Lewis et. al., 2006). We have also begun working towards a fully composable KBN reasoner which minimises the reasoning overhead and memory footprint of the reasoning depending on the ontologies to be reasoned and application requirements.

One of the crucial tasks in the development of the Siena bag extension was the implementation of the algorithm for the comparison of bags by the composite bag operator. The algorithm used is a very simple brute force algorithm that clearly illustrates the performance of the composite bag operator. It simply checks all possible arrangements of one bag with another until a matching arrangement is found, or else it returns unsuccessful. As discussed above this is not ideal. An optimisation of this algorithm would be to sort the bags (at least partially) before applying the bag comparison to the bag. This has no effect on the contents of the bags since bags are defined to be unordered, but depending on the size of the bags it can optimise the matching of bags.

MOTIVATIONAL CASE STUDIES

The previously introduced KBN semantic operators support the ontological comparison and selection of ontological classes, instances and properties with regard to one another. Additionally, reasoning and inference allows ontological

rules and restrictions placed within the ontology by the author, to be represented within the reasoned model. Both of these support the central and pivotal role in which the ontology takes with regard to KBN operation. The ontology can be seen as the defining structure on which semantic publications and subscriptions are formed, and the use of ontologies differentiates KBN from CBN technologies, where a KBN is an extension to a CBN implementation. The application domain in which a KBN operates dictates and defines how the ontology is formed, all of which is central to a KBN's applicability to the chosen application domain. Here we introduce some of the factors which illustrate the technical considerations required when evaluating the applicability of a KBN in certain types of application domains. We then introduce application domains that may benefit from a KBN deployment.

Based on a synthetic benchmark for evaluating a KBN (Keeney et. al., 2006a), a number of factors need to be established to analyse the applicability of a KBN deployment in an application domain. These include:

1. **Publication to Subscription ratio:** Subscriptions are stored in a hierarchical manner in the routers subscription tables, where subscription tables are paired with interested subscribers. Once a publication has been received at a broker/router then all subscriptions within that router must be searched for any possible matches. Therefore the performance of a knowledge-based (or content-based) network is highly dependent on the size and organisation of subscription tables in its brokers. The Siena subscription merging/matching algorithm, extended in the presented KBN implementation, is focused on the efficient partial ordering of subscriptions to maximise subscription coverage to minimise the number of subscriptions which must be checked against each publication. This creates a large overhead when adding

or removing subscriptions, but is optimised for scenarios operating in a publication heavy manner. Therefore a high publication to subscription ratio would be of benefit for the presented KBN implementation. Other KBN implementations that do not expend as much effort optimising their subscription table would perform better where there are many more subscriptions than publications.

2. **The Steady source of "live" Publication Data:** Having established that a high publication rate combined with a small subscription table is the preferred operational conditions for a KBN based on the Siena CBN, the next consideration surrounds the source of the publications themselves. Publications need to be sourced from data which is being rapidly updated. Looking to a generic stock market example it is easy to see that the data relating to a stock's price changes per-second whilst the markets are open (continuously with regard to currency). This level of change is important within a KBN and indeed any event-based system. Most pub/sub systems, including Siena, operate in a non-caching manner, i.e. once a publication has been checked against a routers subscription table it will not be checked again, and if a client was not subscribed to receive that publication there is no possibility of receiving it until it occurs again, if it ever does occur again. This non-caching operation prescribes that publication can occur often and change rapidly.

3. **Subscription Churn and un-subscribes:** Within a CBN/KBN, subscriptions can (at the users request) be followed by un-subscriptions. This removes all traces of the subscription from all subscription tables across the broker network. Using the Stock quote analogy it is easy to see that as a stock broker decides to purchase a specific stock a subscription is made for that stock, once

trading is complete, an unsubscribe removes their initial subscription. This "churn" forces the brokers to apply cleanup mechanisms to their subscription/routing tables to remove stale information which speeds-up the subscription matching mechanism as fewer subscriptions are searched for a possible match to incoming publications. This mechanism is however dependent upon a source of subscribing clients with dynamic and shifting subscription interests. In the Siena CBN subscriptions are aggregated together so that only the most general of subscriptions are routed between brokers, thereby minimising the size of each brokers subscription tables. This means however that when a general subscription is unsubscribed it may have a serious knock-on effect throughout the network as subscriptions that were previously covered by that subscription now need to be sent to neighbouring brokers. In scenarios with high subscription rates and substantial subscription churn this can seriously undermine the operation of the broker network as subscription table optimisation messages propagate. In such scenarios a KBN implementation that is not based on the Siena CBN (with regard to subscription propagation) would perform better.

4. **Full Utilisation of Ontological operators:** to be able to fully demonstrate and exploit the range of ontological operators it is important to understand the types to which those operators can be applied. Typically CBN operators, as previously outlined, operate across Strings, Integers, Doubles, Floats, Booleans and Byte arrays. The introduction of the ontological operators and types extends the operator set and the type set over which publications and subscriptions can be formed. This additional semantic descriptiveness is only as useful as the percentage of messages within the network that use the

new semantically rich message format. To demonstrate this it is necessary for messages to be formed from a rich semantic knowledge base, defined in an ontology, which can be used throughout the network and which matches the domain in which the network is operating. Full utilisation of ontological operators will only become viable once in a domain which supports full ontological messages. An important characteristic of the presented KBN implementation is that the new types and operators add no additional runtime overhead if they are unused. In such cases the KBN implementation performs identically to the Siena implementation from which it was extended.

5. **Availability of usage statistics:** in a hypothetical world, the chosen application domain would be associated with a publically accessible set of usage statistics which would be able to be used for simulation and evaluation of the chosen KBN implementation. Simulations will generally occur in a concentrated time frame, e.g., simulating up to a year's worth of usage in a period of several hours or days. In searching for the ideal KBN application domain it must also provide a set of data which will allow the various rates within the network to be set based on documentable data. This data is often difficult to obtain but can, with extended research, be found or estimated. Such data can then be used to demonstrate the feasibility and utility of a KBN and allow similar systems to be evaluated, compared, and improved in an objective and scientific manner.

Decentralised Semantic Service Discovery

A service-orientated architecture provides the opportunity to compose services from a number of elementary services operated by various or-

ganisations across various sites. The discovery and availability of these services are currently limited to centralised registries and compositions. The typical web service scenario consists of three parts: the first being the service provider which creates and publishes the web service, the second part involves the service brokers which maintain and manage a registry of published services and aid in their discovery, and finally the service consumers, which search the service brokers' registries for a service to fulfil their requirements. Most importantly, service discovery is reliant upon the searching of centralised repositories recording the various offerings provided by various services.

With the use of Knowledge-based Networking the dynamic and decentralised discovery of semantic services has been developed and implemented, as documented by Roblek (Roblek, 2006). The Decentralised Semantic Service Discovery system exploits ontological and syntactic descriptions of Web services using OWL-S (Martin et. al., 2004) in order to provide an effective, efficient and distributed rendezvous of loosely coupled service providers and consumers. Participating services are described based on required capabilities in terms of inputs, outputs, preconditions and effects. Central to the detailed process of composing a composite service involves the use of constituent services where the outputs of one service correspond to the inputs of the next service, and the preconditions required by each service can be satisfied by the resulting effects of the previous service in the workflow. This combines to provide a composite service which performs the tasks required given the available inputs and provides the required outputs so together the constituent services achieve the goal of the composite service. The entire process of discovering and orchestrating services is explained in detail by Roblek (Roblek, 2006). However here we focus on the matching of semantic service inputs and outputs, as explained below.

In this application domain, distributed services can announce their presence with a KBN notification. The core of this notification includes the description of the service inputs, as a bag of semantic classes, and the service outputs also represented as a bag of semantic classes. For each required service, a KBN subscription is created. The subscription uses the previously discussed composite bag operators to search for and discover compatible services or sub-services. With the routing of semantically enhanced messages being central in each KBN broker this allows for the routing of service notifications from service providers to consumers. If the bag of inputs required by an available service is a sub-bag of the bag of available inputs when the superclass suboperator is applied ($AdvertisedServiceInputs \subseteq_{LESSSPEC} AvailableInputs$), and the bag of outputs from an available service is a super-bag of the bag of the required outputs when the subclass operator is applied ($AdvertisedServiceOutputs \supseteq_{MORESPEC} RequiredOutputs$), then the available service's interface is appropriate. If this is encoded as a KBN subscription, allowing for any additional semantic or content-based subscription filters to be also included in the subscription, then the KBN can act as a decentralised service discovery platform. For more details on how the other aspects of discovering, orchestrating and choreographing services is achieved, please refer to (Roblek, 2006).

Recently a similar approach was taken with the PADRES pub/sub system for the decentralised discovery and composition of services (Songlin et. al., 2008). This system maps service interface specifications to content-based pub/sub messages and is based on type-matching outputs of one service to the inputs of the next. This is, however, restricted to non-semantic services and so takes no account of the behaviour or semantics of the services being composed or the semantics of their inputs or outputs, and so the end-to-end behaviour or semantics of the composed process cannot be clearly established.

Distributed Correlation of Faults in a Managed Network

Increasingly there is a demand for more scalable fault management schemes to cope with the ever increasing growth and complexity of modern networks. However, traditional fault management approaches typically involve rigid and inflexible hierarchical manager/agent topologies and rely upon significant human analysis and intervention, both of which exhibit difficulties as scale and complexity increases. Our distributed correlation scheme, designed and implemented by Tai (Tai, 2007; Tai et. al., 2008) distributes correlation tasks amongst an entire network of fault agents, where each agent takes a role in part of the correlation. These distributed agents are arranged so that low level correlators provide sub correlation results for higher level correlation agents, and the whole correlation task for the managed network can then be performed hierarchically.

Event information, correlation rules and the event correlation graphs are all represented in this scheme as ontologies. The use of an ontological representation not only enables these elements to be easily changed, but also (through reasoning) provides an opportunity for self configuration of the fault correlation system itself to be achieved automatically in reaction to context changes. We have also published numerous works describing the benefits of using semantic mark-up in the area of network fault management (Keeney et. al., 2006b; Keeney et. al., 2007a; Lewis et. al., 2005)

In one part of the work described by Tai (Tai et. al., 2008; Tai, 2007) we arranged high-level and low level events in a hierarchical manner according to a "caused_by" relationship, where low level events cause high-level events. This was then codified using the ontological subclass/superclass relationship. A correlation agent would then subscribe to all events at a certain level or all events that could cause a specific event using the semantic MORESPEC operator. If the agent was interested in a combination of events then it can subscribe to a flexible bag of causing events that may have occurred together. Once an agent discovers or calculates a correlation it announces this as a higher-level event using a KBN publication. (By including a bag of information about what triggered this correlation a top level agent could then perform root-cause analysis of what caused a top-level fault!).

However, we found that this mapping of caused-by relationships may not be easily mapable to a subclass/superclass relationship, and the use of this relationship to codify a "caused-by" relationship was breaking the semantics of the concept hierarchy. It was this, combined with several other factors, that prompted us to develop the generic ONTPROPERTY operator, where the "caused-by" relationship and similar relationships could be codified directly as ontological object properties without rearranging the natural hierarchy of event types. Therefore agent subscription(s) would then match interesting events according to this causes/caused-by ontological property, (*Fault-Instance* ONTPROPERTY$_{CAUSEDBY}$*SubFault*). This could then be easily expanded to make use of the bag extension.

There is neither tight coupling between the network of managed elements nor specific correlation agents due to the usage of the semantic publish/subscribe middleware. All events, including raw fault events are pushed into the fault correlation network, but if no agent is interested in that event then the event is quenched immediately. If an agent is interested in the event then it is routed to that agent. If there are no events in the network the correlation agent takes up minimal resources. In addition, a failure in one specific correlation agent can not disable the whole fault management system, as another correlation agent can assume the correlation task of the failed agent by adjusting its subscription.

News Feed Distribution and Subscription

Modern web users are increasingly interested in being delivered in a timely and efficient manner, content which is being freshly created and posted online as opposed to the user locating existing knowledge from a multitude of sources. A prime example of this being the subscription to a blog as a source of information as opposed to a reading a leading encyclopaedia. Additionally the interval in-between which information on the internet is posted is becoming more and more important with regard to the content of the posting. Only the most up-to-date and relevant information is required by the modern user, e.g. blog postings rapidly fade in importance as time passes. The internet community has responded to this with the wide-spread adoption of RSS, which allows subscribers to be notified immediately of changes in a publisher's content. News-based RSS feeds have emerged, with blog subscriptions, as a cornerstone of the Web 2.0 movement. This system still relies on users actively seeking feeds which they are interested in, or being delivered information on feeds they have already subscribed to. RSS does not "push" feeds to users, users search and subscribe to feeds which match their interests. Feeds aggregators aid users in the search for information, but offer only rudimentary searches on categorised and crawled feeds.

The pub/sub model provides good suitability with regard to an application such as RSS. The examples in use typically are limited to a number of static events and event types, and for this reason semantic web techniques have yet to be fully deployed in the efficient distribution of events. The work presented in (Keeney et. al., 2008b) introduces a real-world study of the distribution and subscription of multiple consumers to podcast feeds, utilising the Apple podcast schema, which adds XML tags to RSS feeds and news postings. It was found that podcast feeds and their individual postings are marked up with a small amount of metadata and that this data could be extended, naturally, to utilise the semantic extensions offered through the use of knowledge-based networking. This hidden metadata includes information such as classification, categorisation and keywords associated with the podcast (which would form the publication). The categorisation of the feeds was found to be usually drawn from a taxonomy of categories, which could be easily ported to and extended into the structure of a rich if somewhat flat ontology.

Having encoded the scope of the podcasts categorisation ontologically, it became possible to encode the publication of a new podcast in the form of a KBN notification which contained authorship and ownership information, the URL of the actual podcast post, an attribute containing a bag of ontological classes as subject categories and a bag of keywords. These publications were passed to the KBN brokers in line with typical RSS usage statistics collected and documented in (Keeney et. al., 2008b). Having established the source of publications, consumers of events could then receive events based on a KBN subscription. Among other filters the subscriber would specify a bag of zero or more required keywords and a bag of zero or more categories. Firstly the bag of keywords in the event notification were formed around a simple superbag of the keywords in the subscription ($EventKeywords \supseteq RequiredKeywords$). Subsequently if the subscription contained a bag of *required* categories then the bag of categories in the event notification should be a *superbag* of the bag of categories requested, or equivalent. If the subscription contained a bag of *suggested* categories then the bag of categories in the event notification should be a *subbag* of the bag of categories requested or equivalent. Since the categories were arranged taxonomically, the subscription should match equivalent categories and their sub-categories, so the subscription used the MORESPEC suboperator ($EventCategories$ $\supseteq_{MORESPEC}RequiredCategories$) or ($EventCategories \subseteq_{MORESPEC}SuggestedCategories$). For more

details on this scenario, and a detailed evaluation of the performance of the KBN in this usage scenario, refer to (Keeney et. al., 2008b).

Similar approaches for RSS feed aggregation have also been proposed, both semantic based, e.g. S-ToPSS (Petrovic et. al., 2005) and non-semantic based, e.g. (Roitman et al, 2008). Both systems attempt to maintain the interests (profile) of a user as a set of shifting subscriptions, whereby RSS feeds are polled, aggregated and presented to the user according to their interests and/or semantics. Both works further strengthen the argument behind the need for a knowledge-discovery mechanism, as the size of searched content increases and the need for a dynamic mechanism for the management of knowledge further becomes a necessity.

Context Distribution

Pervasive computing promises to make available a vast volume of context messages from environmental sensors embedded in the fabric of everyday life, reporting on user location, sound levels and temperature changes, to name but a few. Any scalable context delivery system must ensure therefore the accurate delivery of context events to the consumers that require them. However, the wide range of sensors and sensed information, and the mobility of consuming clients, will present a level of heterogeneity that prevents consumers accurately forming queries to match possibly unknown forms of relevant context events. As context-aware systems become more widespread and more mobile there is an increasing need for a common distributed event platform for gathering context information and delivering to context-aware applications. However, most pub/sub systems require agreements on message types between the developers of producer and consumer applications. This places severe restrictions on the heterogeneity and dynamism of client applications. Here we see an ideal potential application of Knowledge-based Networks for the filtered dissemination of context over a large loosely coupled network of distributed heterogeneous agents, while removing the need to bind explicitly to all of the potential sources of that context. The likely heterogeneity across the body of context information can be addressed using runtime reasoning over ontology-based context models. A KBN based solely on semantically enhanced messages, and corresponding expressive and flexible queries, is far more flexible, open and reusable to new applications. For this reason we foresee the application of, and have already applied, KBNs in numerous context-aware scenarios (Keeney et. al., 2006a; Keeney et. al., 2006b; Keeney et. al., 2007a).

Discussion of Case Studies

As described in (Raiciu et. al., 2006) no single implementation or configuration of a content-based (or knowledge-based) pub/sub system will perform well for all application scenarios. Different scenarios require different configurations depending on a number of factors: subscription language expressiveness, publisher bandwidth, latency tolerances, clustering capabilities, the number of subscribers and publishers, the frequency of publications and subscriptions etc.. When extended to knowledge-based networking the main additional factors centre on the amount, complexity, expressiveness and spread of semantics across the network of brokers, publishers and subscribers, just as the addition of semantics marks a KBN different from a CBN. A full list of the content-based and knowledge-based factors that characterise a KBN deployment is presented in (Keeney et. al., 2006a).

Just as a single CBN implementation or configuration will struggle to perform optimally for all applications, the KBN implementation presented is not intended for all applications and deployments since it inherits the characteristics of the Siena CBN upon which it is implemented. The motivational case studies presented are intended to illustrate the contributions and relative

advantages of exploiting semantics deep in the pub/sub network. It is envisioned that as more knowledge-based networking implementations (in addition to those surveyed earlier) begin to appear, this will provide a more diverse set of systems to cover such a diverse set of applications and deployments.

The performance and appropriateness of a KBN deployment for decentralised semantic service discovery is very similar to that of a CBN for non-semantic web services. Semantic web services are generally well marked up in terms of the semantics of the behaviour of the service, the semantics of its inputs, outputs, preconditions and effects (Martin et. al., 2004). A KBN provides a mechanism for service consumers to form expressive queries for services, not just in terms of semantics but also in terms of non-semantic metadata describing the services. It is likely that the semantic operators used in such subscriptions will be relatively restricted to the ontological concept subsumption operators rather than instance or property operators. As event notifications are generally not cached in a pub/sub system, the lack of a registry means that publishers must regularly announce themselves, where the frequency of these announcements is dependent on the requirements of the consumers to find services in a timely manner. This means that the number of publications grows quickly, and so publications that do not match a subscription must be quenched as close to the source as possible. This necessitates that subscriptions are widely distributed in the network. At any time some popular services will likely have many subscribers while most will likely have no subscribers. Subscriptions will generally be quite short lived as the service consumer will remove their subscription after they have found an appropriate service, so the KBN configuration chosen must be tolerant of substantial subscription churn.

The performance and appropriateness of a KBN deployment for distributed fault correlation in network management depends largely on the size of the network being managed and the services it supports. When faults are signalled as publications, it is necessary that the pub/sub mechanism can support huge numbers of events when an "event storm" occurs with possibly hundreds of events per second in irregular bursts. In this scenario there will be relatively few subscribers, but the total number of subscriptions will depend on the faults that the network administrator anticipates and for which correlation tasks have been encoded as subscriptions. It is still expected that the number of publications (in bursts) will greatly outnumber the number of subscriptions. It is also expected that the subscriptions will be long-lived, un-subscriptions will be rare and so subscription churn will be low. The advantage of a KBN over other types of pub/sub mechanisms is directly related to the use of semantics. While low-level events from network elements will have relatively little semantics (aside from statically encoded metadata), the higher-level events will become increasingly semantically rich as correlation tasks are performed. Therefore, only rudimentary operators are used at the lowest levels, but with more expressive operators used higher up to exploit the flexibility of the subscription mechanism for use in the complicated correlation process.

In a news feed distribution scenario scalability becomes the major concern. As users of the web are increasingly interested in tracking the appearance of new postings rather than locating existing knowledge the use of a pub/sub mechanism is seen as ideal for this scenario. Some feeds have large numbers of subscribers, where many feeds have very few subscribers. Based on information from (Keeney et. al., 2008b) subscriptions are generally long-lived, with considerable semantic drift in new subscriptions but subscription churn is small. The number of publishers are moderately large, but the number of publications each publisher produces is relatively few. In contrast there are more subscribers than publishers so it is necessary to efficiently aggregate, and perhaps cluster subscriptions across the network. Using a CBN, user subscriptions are limited to simple syntactic

matches (typically integers, strings and Booleans) but with the popularity of semantic-based metadata within the Web 2.0 community more expressive and flexible semantic subscriptions are required. A problem arises from the vast and shifting semantic knowledge-base required for this scenario. It is necessary that any large scale KBN implementation can partition or semantically cluster the network to localise semantic drift and handle the large semantic spread across the network and in the subscription base.

The context distribution domain is much more difficult to quantify. The number of subscribers and publisher, the number of messages, subscription churns, the use of semantics, the expressiveness of the subscriptions etc. are all impossible to quantify and are completely application dependent. For example, large deployments of sensors will likely produce a vast number of publications, but with relatively little semantics in each. There will probably be few subscribers interested in such low-level information. In contrast, context aggregators, application level context producers, sophisticated sensors, and environment controllers will publish few events but they will be contain rich semantics and may be of interest to a larger set of subscribers. There will be little semantic drift as the semantics of the context will not change, even if the context values may charge rapidly. Subscriptions for low level events will probably be long lived, but high level semantic subscription may be shorter lived as the subscriber discovers the context they were interested in, but overall there will be little subscription churn and subscription patterns will be relatively predictable.

Additional Proposed Scenarios

Academic Calls for Papers are unique in their global interest combined with regular churn of both subscribers and publishers, in the form of both conference organisers and delegates. Semantically rich messages in the form of contribution calls, can focus mainly on the semantically rich message format as used within a KBN as opposed to static type comparison used within most Content-based Networks. The relationship between areas in computer science, the conference location, the types of papers being solicited and the dates and duration of the conference can all be represented within a rich and expressive CFP ontology, from which both publication and subscription can be formed.

Commodities and Currencies provide two interesting and popular areas in which KBNs could successfully operate, with the demand for timely and accurate information requiring such a semantically enhanced pub/sub system. Commodities can be classified as products such as Heating Oil or Light Crude, Silver or Copper and Live Cattle or Lean Hogs, to name just a few. These commodities have a current price, percentage change, high/low and an update period. Currencies are very much the same offering cross rates (between various currencies) and percentage changes, all providing a vast set of data. Much of this data is ideally suited for semantic mark-up exploiting ontologies representing the change and relationship between commodities and the market itself.

Environmental Controls / Building Management: in the proposed ubiquitous world, environmentally sensitive buildings will produce vast amounts of data on which time sensitive decisions will need to be made. Where the semantically enhanced pub/sub paradigm is best operational is within an environment in which rapid changes in the form of publications are matched to pre-existing sorted and stored subscriptions. The distributed filtering of vast amounts of semantically enriched data produced by environmental controls, and ability to perform long-running semantic queries over the data, is an ideal application scenario for a KBN.

Hospital equipment monitoring critically ill patients: there are many areas in hospital care and medicine in which the consumer and producer would benefit from the loose coupling offered through pub/sub systems. Patient care is something in which advances in computing have shown to have a positive effect with regard to

day-to-day care. If each patient, or the machines monitoring or treating the patient, were seen as publishers then this would allow a healthcare professional to subscribe to alerts when specific criteria are met. It is easy to see the obvious benefits of this scenario, however with such critical data it is crucial that this data would be delivered in an assured and timely manner, an area in which further research is required.

CONCLUSION AND FURTHER WORK

This chapter describes and discusses the implementation of two novel extensions to the Siena Content-based Networking system to extend it to become a new class of Knowledge-based Networking. One extension provides ontological concepts as an additional message attribute type, onto which subsumption relationships, equivalence, type queries and arbitrary ontological relationships can be applied. The second extension provides a bag type to be used allowing bag equivalence, sub-bag and super-bag relationships to be used directly in the subscription filters, composed with any of the Siena subscription operators or the ontological operators previously mentioned. This research has only just begun to explore applications for the expressiveness of the knowledge-based networking. As presented in the motivational case studies above, ongoing research by the authors is focussing on how our KBN implementation can be applied across a wide selection of application areas, including:

- Decentralised semantic service discovery (Roblek, 2006)
- Discovery and change notification of policies between federated communication service providers
- Sensor readings in a multi-domain heterogeneous ubiquitous computing application
- RSS extended with semantic mark-up in Web 2.0/Semantic Web (Keeney et. al., 2008b)

- Semantically rich notifications from heterogeneous network elements in Operational Support Systems (Lewis et. al., 2005; Keeney et. al., 2006b; Keeney et. al., 2007a)
- Distributed fault correlation using semantically rich notifications (Tai, 2007; Tai et. al., 2008)
- Semantically rich notifications about changes in financial markets
- Semantic routing of multimedia (MPEG) stream with semantic meta-data

One of the main questions that surround the use of ontologies deep in the network at the routing layer remains the evaluation of the resulting performance overhead. Previous small scale studies in this area (Lewis et. al., 2006; Keeney et. al., 2006b; Keeney et. al., 2007a; Keeney et. al., 2008a) show a definite performance penalty but this may be acceptable when offset against the increased flexibility and expressiveness of the KBN subscription mechanism. Further research is required to evaluate how the performance of "off-the-shelf" ontology tools will affect the scalability of KBNs within larger scales. These results point to the potential importance of semantic clustering for efficient network and performance scalability. Ongoing work is also focussing on extending the Knowledge-Based Network to incorporate semantic-based clustering. This work aims to provide a network environment in which routing nodes, publishers and subscribers are clustered based on their semantic footprint and interests (Keeney et. al., 2008b). The benefits of this are threefold: Firstly, this reduces the processing time involved in making routing decisions based on the messages content. Its take fewer hops to get from source to destination, as these are already closely linked based on the likelihood of there being a match between the two. Secondly, this allows for natural grouping of likeminded publishers and subscribers as seen in traditional web forums / newsgroups. Thirdly, it allows certain areas of the network to have specialised sub/super ontologies which do not need to contain the semantics of the whole network. This means that the knowledge base sizes can reduced and the knowledge base updates can be localised. This cluster-based approach to pub/sub networks turns the normal user-based search paradigm full circle as network data is passed from node to node towards those who are most likely to be interested in the data as opposed to those users searching out that same data. In our initial work clusters were statically designed and operated (Keeney et. al., 2008b). In this sense nodes were assigned to clusters without the possibility of changing clusters once they have joined, in a manner similar to the approach taken in (Baldoni et. al., 2007). This initial clustering method demonstrated how even inflexible and static clustering can have a substantial positive effect on overall performance. However, we expect that any practical system will need to adapt its clustering to reflect the constantly changing profile of semantics being sent and subscribed to via a KBN, thus creating a network environment in which messages are passed from node-to-node, cluster-to-cluster based not on the data's destination but based on the message's semantic data. Recently completed work focussed on allowing users and brokers to join and leave clusters dynamically and independently. Clusters can then be seen as organic structures in which users and brokers join and leave as their own personal interests drift, grow, reform and refine. Current work is also focusing on integrating policy-based cluster management for a KBN to support sophisticated clustering schemes. This will support overlapping clusters and hierarchies of clusters under separate administrative control (Lewis et. al., 2006). In addition, the effect of semantic interoperability in node matching functions and in inter-cluster communications is being assessed (GuoS et. al., 2007; GuoS et. al., 2008). This requires evaluation of different schemes for injecting newly discovered semantic interoperability mappings into the ontological corpus held by KBN routers, as well as how these mappings are shared between routers. Work is ongoing to build on these initial evaluations (GuoS et. al., 2007; GuoS et. al., 2008) to design and implement a flexible mapping strategy management framework. Work is also ongoing to investigate how mappings can be dynamically distributed around the network as the knowledge bases of clients joining and leaving the network affect the spread of knowledge across the network. It is foreseen that a KBN itself would be ideal for such a distribution mechanism.

REFERENCES

W3C: The Wine Ontology (2003). *The Wine Ontology*. Retrieved from http://www.w3.org/TR/owl-guide/wine.rdf

Baldoni, R., Beraldi, R., Quema, V., Querzoni, L., & Tucci-Piergiovanni, S. (2007). TERA: topic-based event routing for peer-to-peer architectures. In Proceedings of Distributed event-based systems, (DEBS2007) New York.

Borsje, J., Levering, L., & Frasincar, F. (2008, March 16-20). Hermes: a Semantic Web Based News Decision Support System. In *Proceedings of The 23rd Annual ACM Symposium on Applied Computing* Fortaleza, Ceará, Brazil.

Burcea, I., Petrovic, M., & Jacobsen, H.-A. (2003). I know what you mean: Semantic Issues in Internet-scale Publish/Subscribe Systems In *Proceedings of the International Workshop on Semantic Web and Databases* (SWDB03), Berlin, Germany.

Cai, M., & Frank, M. (2004, May). RDF Peers: A scalable distributed RDF repository based on a structured peer-to-peer network. In *Proceedings of WWW conference*, New York.

Carroll, J., Dickinson, I., & Dollin, C. (2004, May 17-22). Jena: Implementing the Semantic Web Recommendations. In *Proceedings of World Wide Web Conference 2004*, New York. Retrieved from http://jena.sourceforge.net/

Carzaniga, A. (2008). *Siena – Software*. Retrieved from http://www.inf.unisi.ch/carzaniga/siena/software/index.html

Carzaniga, A., Rosenblum, D. S., & Wolf, A. L. (2001). Design and Evaluation of a Wide-Area Event Notification Service. *ACM Transactions on Computer Systems, 19*(3). doi:10.1145/380749.380767

Chan, C. Y., & Ni, Y. (2006). Content-based Dissemination of Fragmented XML Data. In *Proceedings of International Conference on Distributed Computing Systems*, (ICDCS 2006).

Chirita, P.-A., Idreos, S., Koubarakis, M., & Nejdl, W. (2006, May 10-12). Publish/Subscribe for RDF-Based P2P Networks. In *Proceedings of the 1st European Semantic Web Symposium* (ESWS 2004), Heraklion, Greece.

Christian, H.-W., & James, H. (2007). Toward expressive syndication on the web. In *Proceedings of the 16th international conference on World Wide Web*. Alberta, Canada: ACM.

Cilia, M., Bornhövd, C., & Buchmann, A. P. (2003). An Infrastructure for Distributed, Heterogeneous Event-Based Applications. In *Proceedings of CoopIS 2003*. Catania, Italy: CREAM.

Clark, J., & DeRose, S. (1999). *XML path language (xpath)*. Retrieved from http://www.w3.org/TR/xpath

Diao, Y., Altinel, M., Franklin, M. J., Zhang, H., & Fischer, P. (2003). Path sharing and predicate evaluation for high-performance XML filtering. [TODS]. *ACM Transactions on Database Systems, 28*(4), 467–516. doi:10.1145/958942.958947

Eugster, P., Felber, P., Kenmarrec, A. M., & Guer-rout, R. (2003, June). The many faces of publish/subscribe. [CSUR]. *ACM Computing Surveys, 35*(2). doi:10.1145/857076.857078

Grossnickle, J., Board, T., Pickens, B., & Bell-mont, M. (2005, October). RSS - Crossing Into the Mainstream. In Proceedings of Yahoo! IPSOS Insight.

Guo, S., Keeney, J., O'Sullivan, D., & Lewis, D. (2007, November 27-29). Adaptive Semantic Interoperability Strategies for Knowledge Based Networking. In *Proceedings of the International Workshop on Scalable Semantic Web Knowledge Base Systems (SSWS '07) at OTM 2007.* Vilamoura, Portugal.

Guo, S., Keeney, J., O'Sullivan, D., & Lewis, D. (2008, April 7-11). Coping with Diverse Semantic Models when Routing Ubiquitous Computing Information. In *Proceedings of the Workshop on Managing Ubiquitous Communications and Services (MUCS2008) at NOMS 2008.* Bahia, Brazil.

Guo, Y., & Heflin, J. (2005). LUBM: A Bench-mark for OWL Knowledge Base Systems. *Journal of Web Semantics, 3*(2). doi:10.1016/j.websem.2005.06.005

Guo, Y., Heflin, J., & Pan, Z. (2004). An Evalua-tion of Knowledge Base Systems for Large OWL Datasets (Technical Report), Bethlehem, PA: CSE department, Lehigh University.

Gupta, A., & Suciu, D. (2003). Stream processing of xpath queries with predicates. In *Proceedings of 2003 ACM SIGMOD Intl conference on Man-agement of data* (pages 419–430).

Haarslev, V., & Moller, R. (2001). RACER Sys-tem Description. In *Proceedings of IJCAR 2001, volume 2083 of LNAI*, (701–706). Siena, Italy: Springer.

Jennings, B., van der Meer, S., Balasubramaniam, S., Botvich, D., O'Foghlu, M., Donnelly, W., & Strassner, J. (2007, October). Towards Autonomic Management of Communications Networks. *IEEE Communications Magazine, 45*(10). doi:10.1109/MCOM.2007.4342833

Keeney, J., Jones, D., Roblek, D., Lewis, D., & O'Sullivan, D. (2008). Knowledge-based Seman-tic Clustering. In *Proceedings of ACM Symposium on Applied Computing*, Fortaleza, Brazil.

Keeney, J., Lewis, D., & O'Sullivan, D. (2006, July 19-21). Benchmarking Knowledge-based Context Delivery Systems. In Proceedings of ICAS06, Silicon Valley, CA.

Keeney, J., Lewis, D., & O'Sullivan, D. (2007, March). Ontological Semantics for Distributing Contextual Knowledge in Highly Distributed Autonomic Systems. *Journal of Network and Systems Management*, 15.

Keeney, J., Lewis, D., O'Sullivan, D., Roelens, A., Wade, V., Boran, A., & Richardson, R. (2006, April). Runtime Semantic Interoperability for Gathering Ontology-based Network Context. In *Proceedings of Network Operations and Manage-ment Symposium* (NOMS 2006), Toronto, Canada.

Keeney, J., Roblek, D., & Jones, D. Lewis, & O'Sullivan, D., (2008). Extending Siena to sup-port more expressive and flexible subscriptions. In *Proceedings of the 2nd International Confer-ence on Distributed Event-Based Systems* (DEBS 2008), Rome, Italy.

Lewis, D., Keeney, J., O'Sullivan, D., & Guo, S. (2006, October 23-25). Towards a Managed Extensible Control Plane for Knowledge-Based Networking. In *Proceedings of Distributed Sys-tems: Operations and Management Large Scale Management, (DSOM 2006), at Manweek 2006*, Dublin, Ireland.

Lewis, D., O'Sullivan, D., Power, R., & Keeney, J. (2005, October). Semantic Interoperability for an Autonomic Knowledge Delivery Service. In *Proceedings of Workshop on Autonomic Communication* (WAC 2005), Athens, Greece.

Li, H., & Jiang, G. (2004). Semantic Message Oriented Middleware for Publish/Subscribe Networks. *Proceedings of the Society for Photo-Instrumentation Engineers, 5403,* 124–133.

Loser, A., Naumann, F., Siberski, W., Nejdl, W., & Thaden, U. (2003). Semantic overlay clusters within super-peer networks. In *Proceedings of Workshop on Databases, Information Systems and Peer-to-Peer Computing in Conjunction with the (VLDB '03).*

Lynch, D., Keeney, J., Lewis, D., & O'Sullivan, D. (2006, May). A Proactive Approach to Semantically Oriented Service Discovery. In *Proceedings of Innovations in Web Infrastructure (IWI 2006). at World-Wide Web Conf.*, Edinburgh, Scotland.

Martin, D., Burstein, M., Hobbs, J., Lassila, O., McDermott, D., McIlraith, S., Narayanan, S., Paolucci, M., Parsia, B., Payne, T., Sirin, E., Srinivasan, N., & Sycara, K. (2004, November 22). OWL-S: Semantic Markup for Web Services. *W3C Member Submission.*

Meier, R., & Cahill, V. (2005). Taxonomy of Distributed Event-Based Programming Systems. *The Computer Journal, 48*(5), 602–626. doi:10.1093/comjnl/bxh120

Motik, B., & Sattler, U. (2006). Practical DL Reasoning over Large ABoxes with KAON2. Retrieved from http://kaon2.semanticweb.org/

Mühl, G., Fiege, F., & Pietzuch, P. (2006). *Distributed Event-Based Systems*. Berlin, Germany: Springer-Verlag.

OWL. (2004). *W3C Recommendation: OWL Web Ontology Language Overview*. Retrieved June, 2008 from http://www.w3.org/TR/owl-features/

Pan, Z. (2005). Benchmarking DL Reasoners Using Realistic Ontologies. In Proceedings of Intl workshop on OWL: Experience and Directions (OWL-ED2005). Galway, Ireland.

Parsia, B., & Sirin, E. (2004). Pellet: An OWL-DL Reasoner. Poster at ISWC 2004, Hiroshima, Japan.

Pellet Performance. (2003). *Pellet Performance*. Retrieved from http://www.mindswap.org/2003/pellet/performance.shtml

Petrovic, M., Burcea, I., & Jacobsen, H. A. (2003). S-ToPSS: semantic Toronto publish/subscribe system. In *Proceedings of the 29th international conference on Very large data bases* (VLDB03), Berlin, Germany.

Petrovic, M., Liu, H., & Jacobsen, H.-A. (2005, September). CMS-ToPSS: Efficient Dissemination of RSS Documents. In *Proceedings of 31st International Conference on Very Large Data Bases* (VLDB).

Pietzuch, P., & Bacon, J. (2002). Hermes: A Distributed Event-Based Middleware Architecture. In *Proceedings of International Conference on Distributed Computing Systems.*

Pietzuch, P., & Bacon, J. (2003, June). Peer-to-Peer Overlay Broker Networks in an Event-Based Middleware. Distributed Event-Based Systems (DEBS'03). In *Proceedings of the ACM SIGMOD/PODS Conference*, San Diego, CA.

Raiciu, C., Rosenblum, D. S., & Handley, M. (2006, July 4-7). Revisiting Content-Based Publish/Subscribe. In *Proceedings of 26th IEEE international Conference on Distributed Computing Systems Workshop.*

Roblek, D. (2006). Decentralized Discovery and Execution for Composite Semantic Web Services. (M.Sc. Thesis, Computer Science), Dublin Ireland: Trinity College.

Roitman, H., Carmel, D., & Yom-Tov, E. (2008). Maintaining dynamic channel profiles on the web. In *Proceedings of the 34th Conference on Very Large Data Bases* (VLDB 2008), Auckland, New-Zealand.

Schlosser, M., Sintek, M., Decker, S., & Nejdl, W. (2002). HyperCuP—Hypercubes, Ontologies and Efficient Search on P2P Networks. In *Proceedings on the International Workshop on Agents and Peer-to-Peer-Systems*. Bologna, Italy: Springer.

Segall, B., Arnold, D., Boot, J., Henderson, M., & Phelps, T. (2000). Content-Based Routing in Elvin4. In Proceedings of AUUG2K, Canberra, Australia.

Skovronski, J., & Chiu, K. (2006, December 4-6). Ontology Based Publish Subscribe Framework. In *Proceedings of International Conference on Information Integration and Web-based Applications Services*. Yogyakarta, Indonesia.

Songlin, H., Muthusamy, V., Li, G., & Jacobsen, H.-A. (2008). Distributed Automatic Service Composition in Large-Scale Systems. In *Proceedings of the 2nd International Conference on Distributed Event-Based Systems* (DEBS 2008), Rome, Italy.

Tai, W. (2007, December). Fault Management System using Semantic Publish/Subscribe approach. (M.Sc. Thesis, Computer Science). Dublin Ireland: Trinity College.

Tai, W., O'Sullivan, D., & Keeney, J. (2008, April). Distributed Fault Correlation Scheme using a Semantic Publish/Subscribe system. In *Proceedings of Network Operations and Management Symposium* (NOMS 2008), Salvador, Brazil.

Tempich, C., Staab, S., & Wranik, A. (2004). REMINDIN': semantic query routing in peer-to-peer networks based on social metaphors. In *Proceedings of the International World Wide Web Conference* (WWW), New York.

Wang, J., Jin, B., & Li, J. (2004). An ontology-based publish/subscribe system. In *Proceedings of ACM/IFIP/USENIX International Conference on Middleware*.

Weisstein, E. W. (2002). *Multiset. MathWorld – Wolfram Resource*. Retrieved from http://mathworld.wolfram.com/Multiset.html

Zou, Y., Finin, T., & Chen, H. (2004, April). F-OWL: an Inference Engine for the Semantic Web. In *Proceedings of Workshop on Formal Approaches to Agent Based Systems*, (LNCS 3228).

Chapter 11
StreamMine

Christof Fetzer
Dresden University of Technology, Germany

Andrey Brito
Dresden University of Technology, Germany

Robert Fach
Dresden University of Technology, Germany

Zbigniew Jerzak
Dresden University of Technology, Germany

ABSTRACT

StreamMine is a novel distributed event processing system that we are currently developing. It is architected for running on large clusters and uses content-based publish/subscribe middleware for the communication between event processing components. Event processing components need to enforce application-specific ordering constraints, e.g., all events need to be processed in order of some given time stamp. To harness the power of modern multi-core computers, we support the speculative execution of events. Ordering conflicts are detected dynamically and rolled backed using a Software Transactional Memory.

INTRODUCTION

The goal of StreamMine project[1] (http://stream-mine.inf.tu-dresden.de/) is to develop a middleware that supports scalable, near real-time processing of streaming data. As the number of events and data sources increases exponentially, the processing power needed to cope with that information has to be scaled accordingly. For example, for real-time detection of call fraud in a telephone system, one would expect to process from 10,000s to 100,000s events per second. Each event carries a few hundreds of bytes of data and one needs to access several kilobytes of stored data to process an event. Hence, one needs to be able to spread the computation across multiple cores and multiple computers to keep up with this event rate.

The natural choice for scaling up such Event Stream Processing (ESP) applications (Babcock et al., 2002) is to distribute their workload (Cherniak et al., 2003). Such distribution can be performed locally – using multiple processors and process-

DOI: 10.4018/978-1-60566-697-6.ch011

ing cores available on a single machine (Welsh et al., 2001). It can also be performed in a distributed environment by using multiple machines connected by a network. Such machines can be either connected by a Local Area Network, forming a cluster (Sterling et al., 1995), or by a Wide Area Network, forming a cloud (Hwang et al., 2007). We architect StreamMine for distributed systems consisting of many-core CPUs. The major contribution of StreamMine is the ability to automatically distribute ESP applications across multiple cores and machines. A StreamMine user is presented with a simple interface that allows for an automatic distribution and parallelization of ESP applications. The parallelization and distribution are transparent to the user and are performed automatically by StreamMine.

An ESP application consists of a set of connected filter components. In order to harness the processing power of multiple cores per machine, StreamMine can automatically parallelize filters. If the operations to be performed only depend on the input events, i.e., they are stateless, one can trivially parallelize the processing by creating multiple instances of the filter component. However, if there are dependencies between the events, such a simple replication is not a valid approach: processing events out of order can result in incorrect state and incorrect outputs. In other words, StreamMine needs to – at least logically – process events in a given order. Parallelization is a non-trivial problem that can result in significant development costs and runtime overheads. In order to facilitate event processing in multi-core environments, we propose and investigate the speculative execution of events. Speculative execution allows us to process events in parallel that should normally be processed sequentially, even when they are received out-of-order. We use an underlying Software Transactional Memory (STM) (Herlihy & Moss, 1993; Shavit & Touitou, 1995) infrastructure to optimistically process the events in the context of transactions.

The distribution of components across multiple machines poses another set of challenges. The machines used by StreamMine are commodity PCs connected by an IP network. The combination of commodity components and unreliable communication links can result in messages being arbitrarily delayed or dropped. Moreover, there is a very high probability that one or more PCs will crash during an execution. For example, Google's MapReduce jobs, which run on a cutting edge computing infrastructure, suffered in March 2006 on average about 6 worker deaths per job. These jobs were using on average 268 machines and the average completion time was 874 seconds (Dean, 2006).

For ESP applications, we need to expect that the type of processing of the events and hence, the dataflow of the events, will change during the lifetime of a system. In other words, we need to be able to support incremental changes to the system. Besides supporting a simple way of composing event processing applications, we need to support the merging of multiple event processing applications to reduce the computing resources that are required.

Taking above factors into consideration when choosing the communication infrastructure for StreamMine, we have decided to use a Content-Based Publish/Subscribe (Eugster et al., 2003) system as our communication middleware. Content-based publish/subscribe (pub/sub) systems provide a decoupled and flexible communication infrastructure, which allows for easy composition of N-to-M communication patterns. Moreover, thanks to its decoupling properties, it naturally supports dynamic systems, and unlike point-to-point communication schemes, it allows for easy system reconfiguration and merging of computations. In order to cope with potentially large amount of events and components in the system, we propose a novel approach towards construction of the content-based publish/subscribe systems. We show how one can use Bloom filters (Bloom, 1970) to abstract away the content of messages in order to reduce the latency and increase the throughput of events in the system. We also demonstrate an approach to load balancing in publish/subscribe systems. Load balancing allows us to evenly distribute the load between the components without violating the decoupling properties of the publish/subscribe service.

RELATED WORK

Very large input data sets and the need to finish processing in a reasonable amount of time were the main driving factors behind batch processing systems such as MapReduce (Dean & Ghemawat, 2008). MapReduce is a programming model for processing large data sets. In the MapReduce model, a user specifies two functions: a *map* function and a *reduce* function. Data is first processed using a *map* function to generate intermediate key-value pairs. Subsequently, the *reduce* function merges all intermediate values associated with the same intermediate key. MapReduce implementation automatically parallelizes and executes user-supplied *map* and *reduce* functions on a cluster of machines.

However, in many application domains, the data source is continuous. A continuous data source implies that the size of the input data is unbounded. Such input data is constantly generated and is usually related to real-time, real-life events. An example of such input data might be the list of outgoing and incoming calls from all mobile phones operated by a certain telephone company. Another example might be the log of transactions performed with credit cards issued by a given bank. In order to be able to work with such data, e.g., to detect credit card fraud or calls made with stolen/hijacked phones, a continuous processing engine for very large amounts of real-time data is required. The approach towards distributed processing of high volume data proposed by the MapReduce system suffers from the fact that it cannot cope with such situations. The main reason

being that MapReduce is a staged architecture and, as a consequence, all input has to be processed by the first stage (creating intermediate key-value pairs) before it can be processed by the subsequent, reduce stage. The staged processing introduces unnecessary latencies and such latencies cannot be accepted when we need to guarantee very tight real-time bounds. This assumption is strengthened by the fact that in many applications, delays in the detection of certain conditions can result in monetary loss, e.g., a stolen credit card being used to make purchases or expensive calls made with a stolen cell phone.

The continuous data sources described above are a consequence of omnipresence of sensors, computers and networks and has motivated the establishment of Event Stream Processing (Babcock et al., 2002) as an active field of research. ESP applications target cases such as the telephone or credit card fraud detection described above and many others in fields such as sensor networks or system monitoring. There are many known ESP systems, for example, Borealis (Abadi et al., 2005), StreamFlex (Spring et al., 2007) and GSDM (Koparanova & Risch, 2004). The main difference between these and StreamMine is the way scalability is addressed. The scalability issue may be divided into two sub-problems, one refers to how individual components may be made scalable and the other refers to how to make the communication between components scalable.

Regarding component scalability, Borealis and StreamFlex assume components are normally stateless and, as a consequence, can be parallelized by the creation of replicas for the overloaded components. If components are stateful, Borealis uses load shedding to reduce their load. Nevertheless, if events cannot be dropped randomly, user assistance is required to decide which events to discard. In GSDM, costly stateful processors are parallelized following a partition-compute-combine approach. This approach, however, can be applied in a very limited set of cases in which two conditions are satisfied: (1) the semantics of the component allows the partition of the input stream and the recombination of the result stream; and, (2) the computational costs of the partition and recombination phases are small compared to the compute phase (otherwise the achievable parallelization would be very limited, according to Amdahl's Law). StreamMine uses an underlying Transactional Memory (TM) (Herlihy & Moss, 1993) as a mechanism for speculation (Brito et al., 2008). This speculation mechanism reduces end-to-end latency by processing events out of their normal order and increases throughput by allowing components to process events in parallel. In addition, these improvements are achieved without user assistance and without waiving sequential semantics. Sequential semantics guarantees that an execution is indistinguishable from a sequential execution, which is important in applications where repeatability and determinism are desired. In addition, it eases the development of the components by not needing the application developers to account for concurrency intricacies.

To achieve communication scalability, Stream-Mine uses a novel publish/subscribe system that uses Bloom filters to speed up message processing and asynchronous I/O to maximize throughput - see Section *Bloom Filter-Based Routing*. The idea behind the publish/subscribe systems has been first framed by the Information Bus system (Oki et al., 1993). Information Bus (and thus pub/sub paradigm) has been motivated by the need to provide continuous operation, dynamic system evolution and interoperability between the dynamic components of distributed systems. The Information Bus proposed a so called topic-based publish/subscribe system, similar in concept to the generative communication model of Linda (Carriero & Gelernter, 1989). The first content-based pub/sub system was proposed in (Rosenblum & Wolf 1997). Since then a number of content-based pub/sub systems has been developed and made available for the research community (Carzaniga et al., 2001; Fidler et al., 2005; Tarkoma, 2008).

One of the main bottlenecks of the publish/subscribe systems has been the speed of the content-based information routing. One of the first approaches towards providing a fast and scalable infrastructure for content-based routing has been presented in (Aguilera et al., 1999), where authors propose to search through the predicates of all subscriptions comparing them with the content of the incoming messages. The presented algorithm's complexity is however linear with the number of subscriptions stored in the brokers. Another approach has been proposed by (Carzaniga & Wolf, 2003). Authors use a version of a counting algorithm, which performs well in case of event forwarding, simultaneously exposing high over-head in case of subscription updates. Orthogonally, (Aekaterinidis & Triantafillou, 2007) proposed the use of Bloom filters to aggregate, filter, and match information in subscriptions, thus creating per broker subscription summaries compacting subscription information. However, the proposed approach requires a known, bounded publication and subscription language, thus limiting the ex-pressiveness of the content-based pub/sub system. The authors of (Cao & Singh, 2005) propose to combine content-based addressing with a multi-cast based approach. Presented solutions require however that every broker has to be aware of all other brokers in the network, thus breaking the required decoupling of pub/sub components. Authors in (Jerzak & Fetzer, 2007) proposed the use of prefix routing, in order to cope with the matching complexity. The proposed solution requires a consistent layout of the routing structures across all brokers - ensuring such a common layout might be too expensive in the distributed environment.

Another class of systems are the DHT-based pub/sub solutions (Pietzuch, 2004; Tryfonopoulos et al., 2007). Although they expose superior matching times, their design is based on topics. Moreover, a failure of a single node might seize the delivery of messages for all topics a given broker manages. A similar approach has been presented in (Cutting et al., 2008) where instead

Figure 1. Logical view of a typical event stream-processing network

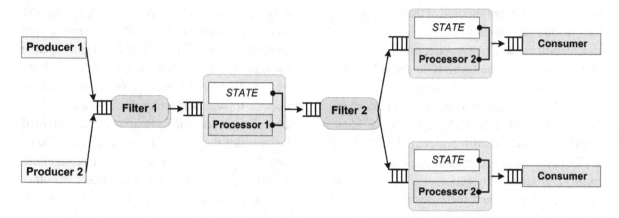

of explicit topics authors use tags, which designate the rendezvous nodes for the publications in the peer-to-peer system.

ARCHITECTURE

A typical event processing application is structured as a cascade as depicted in Figure 1. Producers (e.g., sensors) generate events that are processed by a sequence of components until a consumer is reached (e.g., a monitor dashboard, an actuator back in the observed system). During this process, the event streams may be joined or split. Events can also cause the generation of newer events, have some offline information added to them (e.g., from a database) or be filtered out of the stream.

In StreamMine, although the application is logically represented by a dataflow, the underlying physical connections resemble a cloud. Figure 2 depicts a possible physical materialization of the system from Figure 1. The elements inside the cloud are the brokers from the publish/subscribe (pub/sub) middleware, which are responsible for efficiently routing messages to and from each component.

The main advantage of using a pub/sub infrastructure is the loose coupling between components. In StreamMine, components can be inserted

and removed at runtime by other components. This feature is essential for building systems that can scale up and down by orders of magnitude. For example, a Filter component that executes a costly operation may need to be replicated such that its workload will be split between itself and the new replica. With pub/sub this load balancing is possible by simply initiating a new replica (either at the local site or in a remote one) and creating a subscription for the new replica.

Components can also be replicated for fault-tolerance purposes. In this case, the replicas will have the same subscription as the original so that they process the same subset of events. A resulting requirement is that events have ids that allow duplicated events to be discarded. By using these ids, when more than one replica of a fault-tolerant component is operational, the replicated results can be discarded when they reach the next component.

In event stream applications, a common type of stateful component is one that learns from the stream. Components learn by building (limited-size) sketches of the event stream or by fitting it to a model. This is an example of a case where the order that events are processed may affect the result. There are three main reasons why total ordering among events is necessary: (1) consistency, if such a stateful component is replicated for fault tolerance, all the components must process

Figure 2. Physical deployment of the logical network of Figure 1

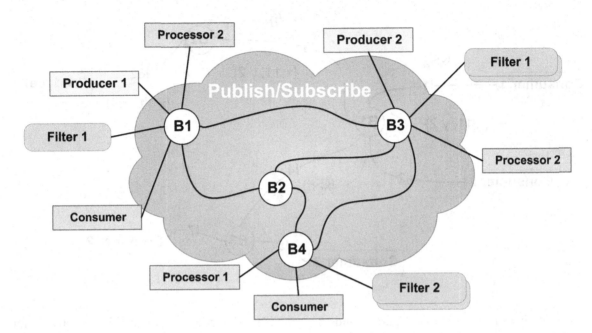

the events in the same order; (2) determinism, it is frequently desirable that the system can be replayed, i.e., the same sequence of events result in the same outputs; (3) correctness, for example, if the component computes a trend in the data, then, processing two events in the inverted order (e.g., because they were previously processed by different replicas of a preceding component and took different paths or different processing times, which resulted in an order inversion) leads to an incorrect result. StreamMine supports the use of unique timestamps for ordering events. These unique timestamps are typically given by the application but these can also be generated by StreamMine if needed.

BLOOM FILTER-BASED ROUTING

Overview

The ultimate goal of using a pub/sub service in the context of StreamMine is to provide a (1) fast,

(2) flexible, and (3) decoupled infrastructure for information exchange between components. The pub/sub communication model is well suited for the highly dynamic ESP processing systems. It is asynchronous: neither publishers (data producers) nor subscribers (data consumers) are blocked when producing or receiving data. Both subscribers and publishers are anonymous to each other and the system as a whole. There is also no requirement on the communication between any two components to take place simultaneously. In content-based publish/subscribe systems subscribers express their interest in a given data using subscriptions. Publishers use events to disseminate the information they have produced.

StreamMine follows a common convention (Carzaniga et al., 2001; Pietzuch, 2004; Zhao et al., 2004; Fidler at al., 2005; Tarkoma, 2008; Jerzak & Fetzer 2008), where subscription s is a conjunction of predicates. A predicate p is a function which evaluates to either true or false for a given argument. Predicates consist of attribute names and attribute constraints. An attribute

Figure 3. Routing of subscriptions and forwarding of publications

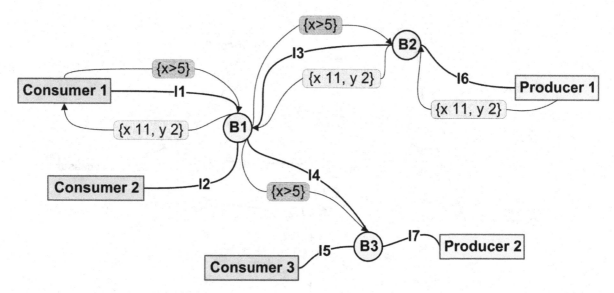

constraint, in turn, consists of an operator and an attribute value. An event *e* is a set of attribute name and attribute value pairs. Attribute name and value pairs forming an event are arguments for the predicate functions. A sample subscription can be represented as: $s \equiv \{ x > 5, y > 3 \}$. This sample subscription *s* consists of conjunction of two predicates *x>5* and *y>3*. The attribute name of the first predicate is *x*. The attribute constraint of the first predicate equals >5, where *5* is the attribute value and > is an operator. A sample event can be depicted as: $e \equiv \{ x 7, y 5, z 3 \}$. An important concept regarding the subscriptions is the binary coverage relation. We can say that subscription *s1* covers subscription *s2* if subscription *s1* matches the superset of events matched by subscription *s2*. We might also say that *s1* is more general than *s2*.

Subscriptions and events allow for dynamic information routing as the explicit source and destination addresses usually put on events are made obsolete by the content of the events. Hence, the potential failures of publishers or subscribers are naturally masked by the pub/sub communication middleware. A subscription *s* matches event

e if every predicate contained in the subscription evaluates to true over the content of the whole event.

An example of content-based information exchange in a publish/subscribe system has been presented in Figure 3. *Consumer 1* issues the subscription *{ x > 5 }* which is distributed into the network of brokers *B1*, *B2* and *B3*. Subsequently, Producer 1 issues the event *{ x 11, y 2 }* which is delivered by the broker network to the interested consumer. Please note that the subscription *{x > 5 }* has to be delivered to all brokers in the network so that if *Producer 1* changes its location from *B1* to *B2* or *B3* its events can still be transparently delivered to the *Consumer 1*.

In StreamMine, publishers and subscribers can be placed on different physical machines. Every machine runs a broker process which is responsible for the subscription routing and event forwarding. In the process of subscription routing every broker in the StreamMine network stores a set *S* of subscriptions which allows it to forward events to all interested subscribers from all potential publishers on the reverse paths of stored subscriptions.

Content Summarization using Bloom Filters

Current approaches (Carzaniga et al., 2001; Tarkoma & Kangasharju, 2006; Mühl et al., 2006) to content-based routing assume that events on their way from publisher to subscriber are matched based on their content at every broker they pass. If we refer to the example shown in Figure 3, we can observe that the event *{ x 11, y 2 }* needs to be matched against the subscription *{ x > 5 }* by both brokers *B1* and *B2*. Such an approach has two main issues: (1) the evaluation of predicate functions over events' content is slow, and (2) the slow evaluation has to be performed by every broker an event is forwarded to. Thus, the more brokers an event passes on its way, the higher the delay and the processing overhead.

We propose a new routing algorithms and accompanying data structures for use in StreamMine publish/subscribe communication middleware. The new approach allows us to route events evaluating the subscriptions' predicates over their content only once. This evaluation can take place at the first broker in the network or already at the publisher, so that no content-based matching needs to be performed at any of the brokers between the publisher and the subscriber.

We achieve this goal by using content summaries. Content summaries are compact content representation using integer values, thus greatly increasing the speed of event matching. Specifically, we do not use the rendezvous approach (Pietzuch, 2004; Cutting et al., 2008) and thus do not rely on any single broker in the network to be the matching point for a given class of events.

The routing processes requires the presence of subscriptions. Subscriptions arriving at the broker are stored for the purpose of matching against incoming events. An event arriving at the edge broker has to be matched based on its content with the stored subscriptions. The result of the content-based match is encoded in a Bloom filter subsequently attached to the event. Downstream brokers encountered by the event evaluate only the attached Bloom filter. Specifically, they do not consider the content of the event.

Let us once again consider the example shown in Figure 3. In case of event *{ x 11, y 2 }* it would be first forwarded to broker *B2*. Broker *B2* uses the content of the event in combination with the set of subscriptions it stores to compute a Bloom filter summarizing the content of the event. This Bloom filter is subsequently attached to the event by the broker *B2*. In the next step broker *B2* performs the matching process based only on the contents of the Bloom filter attached to the event. The result of the matching process is a set of brokers (in case of Figure 3 it is the broker *B1*) to forward the event *{ x 11, y 2 }* to. Broker *B1* upon reception of the event *{ x 11, y 2 }* with attached Bloom filter only needs to perform the matching process based on the contents of the attached Bloom filter. Therefore, broker *B1* does not need to consider the content of the event.

For the above scenario to work we have developed a novel data structures and algorithms for storing subscriptions at the publish/subscribe brokers and for matching of the incoming events. In the following sections we will describe in detail the data structures used to store the subscriptions in the publish/subscribe brokers, as well as the event matching algorithms.

Subscription Content Summarization

StreamMine brokers store subscriptions in the sbsposet data structure. The sbsposet stores subscriptions' predicates by abstracting away the conjunctive form of subscriptions, which allows for a more efficient storage structure - see Figure 4. The sbsposet stores predicates grouped by their attribute name. In the Figure 4 we can observe two attribute names: *x* and *y*. Every attribute name points to a partially ordered set (poset). Each poset begins with a virtual root node (*null*) and stores attribute constraints sharing the same attribute name. The structure of the attribute constraints re-

Figure 4. The sbsposet storing subscriptions fromTable 1

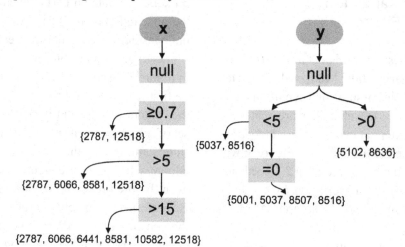

flects the partial order resulting from the covering relation between them. Every attribute constraint has a Bloom filter associated with it. Bloom filters are created when attribute constraints are inserted into the sbsposet. On Figure 4 Bloom filters of attribute constraints are represented by the indices of the bits set to one, enclosed in the curly braces.

The creation of Bloom filters for attribute constraints is performed by using the whole predicate containing the given attribute constraint as the input to the Bloom filter. Therefore, for the attribute constraint >5 the Bloom filter is constructed by hashing the $x>5$ string. The example subscriptions and corresponding Bloom filters are shown in Table 1. Throughout this chapter, unless otherwise stated, we use Bloom filters with k=2 hash functions with the width of m=2^{14} bits.

The Bloom filters of attribute constraints' in the sbsposet maintain the coverage relation between the attribute constraints belonging to a respective attribute name. Every attribute constraint covered by another attribute constraint reflects this fact in its own Bloom filter by performing an OR operation with the Bloom filter of the covering attribute constraint. A single Bloom filter reflects the whole chain of covering attribute constraints.

The sbsposet presented on Figure 4 contains subscriptions presented in Table 1. We can observe

Table 1. The set of subscriptions used in Figure 4, Figure 5 and Figure 6

Subscription	Source	Bloom filter
$\{x > 5\}$	f1@dot.com	6066, 8581
$\{x >= 0.7\}$	f2@dot.com	2787, 12518
$\{x > 15\}$	f3@dot.com	6441, 10582
$\{y < 5\}$	f4@dot.com	5037, 8516
$\{y = 0\}$	f5@dot.com	5001, 8507
$\{x > 15, y > 0\}$	f6@dot.com	5102, 6441, 8636, 10582

that every covered attribute constraint (e.g., >5) contains the bits from the covering attribute constraint (≥ 0.7) Bloom filter. An interesting property of the sbsposet is that by removing the conjunction between predicates of a single subscription it can reduce the number of stored predicates. A simple example are the subscriptions $\{x > 15, y > 0\}$ and $\{x > 15\}$ from the Table 1. The sbsposet needs only 2 predicates to store both subscriptions, while a naive approach would require the storage of all three of them.

The addition of a new subscription into the sbsposet is performed by decomposing it into single predicates and inserting the predicates on a one-by-one basis into the appropriate branches.

The removal of subscriptions from the sbsposet using a counting Bloom filter (Fan et al, 2000) is also straightforward, however for the sake of the presentation brevity we omit the details of the implementation.

The matching of events with the sbsposet must ensure that every event encodes in its Bloom filter all predicates matching its content. Therefore, for every predicate in the the sbsposet matching the event's content, the corresponding predicate's Bloom filter is added (using an OR operation) to the (initially empty) Bloom filter of the event. By storing the Bloom filters of all covering predicates in every predicate in the sbsposet we can optimize the process of addition by performing it only once per matching predicate and simultaneously encoding a possibly large number of covering predicates.

As an example let us consider an event $e \equiv \{$ $x\ 7,\ y\ 8$ } and the sbsposet presented in Figure 4. Initially the Bloom filter of the event is empty. The matching of the event starts with the $x\ 7$ attribute name and value. This attribute name and value selects the >5 attribute constraint from the x poset. The Bloom filter of the selected attribute constraint is ORed with the (so far empty) Bloom filter of the event, with result being stored in the Bloom filter of the event. Now the Bloom filter of the event reads *{2787, 6066, 8581, 12518}*. The matching of the $y\ 8$ attribute name and value pair is performed in a similar manner. The $y\ 8$ attribute name and value pair selects >0 attribute constraint from the y poset. Hence, the Bloom filter of the event e finally contains: *{2787, 5102, 6066, 8581, 8636, 12518}*.

Event Content Summarization

The sbsposet stores the content of the subscriptions without regarding the conjunctions between the predicates of a single subscription. The loss of this information could lead to a potentially large number of false positives, i.e., events delivered to subscribers which did not subscribe to them. Hence, there is a need for a data structure which

would represent the conjunction between the predicates of subscriptions. In this section we introduce such data structure: the sbstree. The main task of the sbstree is to represent the disjunction of conjunctions of predicate values. Therefore, in contrast to the sbsposet, the sbstree stores subscriptions in their conjunctive form. Specifically, the sbstree does not store any predicates, instead, it works exclusively with the Bloom filters representing subscriptions and events.

The disjunction of the predicates of a single subscription can be expressed with a single Bloom filter using an *OR* operation between the Bloom filters of single predicates. However, this approach cannot be used to represent a conjunction of subscription predicates. Therefore, we use the sbstree structure to cope with that issue - see Figure 5. The path from the root to the leaf of the sbstree forms a conjunction of the subscription predicates, or more precisely, the conjunction between the bits set in the subscriptions' Bloom filters.

A Bloom filter of a subscription is an OR between the Bloom filters of the subscriptions' predicates. For a given subscription the bits set in its Bloom filter form a path rooted at the virtual sbstree root node - *null*. The leaf of the path represents an interface on which the given subscription arrived. Hence, a path leading to the source interface forms a conjunction of all predicates of a given subscription.

The event matching process starts at the virtual root of the sbstree. For every integer representing a bit set in the Bloom filter of an event, a comparison with children of the root node is performed. In case of a match, the given path is followed until either: (1) the end of the path (indicated by the source interface) is reached or (2) a given node in the path does not have a corresponding value in the event's Bloom filter.

The above algorithm suffers however from the fact that for every bit set in the event's Bloom filter a path in the sbstree needs to be selected and possibly followed. From the sbstree traversal algorithm we can conclude that following a given

Figure 5. The sbstree storing subscriptions fromTable 1

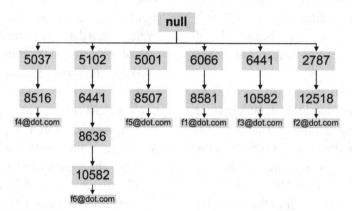

path does not guarantee the reaching of the source interface at the end of it. Additionally, the complexity of the sbstree traversal grows exponentially with the number of bits set in the event - as more and more paths need to be followed.

Therefore, we have developed an improved version of the sbstree, based on a counting algorithm (Pereira et al., 2000; Fabret et al., 2001). The new version of the algorithm stores all bits set in all subscriptions' Bloom filters in a linear fashion - top row in Figure 6. All bits from a Bloom filter forming a subscription are assigned a counter. The value of the counter is equal to the number of bits set in the subscription's Bloom filter - middle row. For the subscription which arrived at interface *f6@dot.com* this value equals 4, as it has two predicates, each composed of a Bloom filter having 2 bits set. The counter itself represents a set of bits which need to be present in the event in order to trigger the subscription it is connected to.

Upon arrival of an event its Bloom filter is parsed for the values of set bits. Every bit set in the event's Bloom filter triggers the corresponding bit in the counting sbstree. Triggering a bit corresponds to the decreasing of the counter which is connected to it. Whenever a counter reaches 0 the corresponding subscription matches the given event. Specifically, event can be forwarded to the

interface on which the matching subscription arrived - bottom row in Figure 6.

The above algorithm, in contrast to the original sbstree variant, is linear with the number of bits set in the event. In Section *Measurements* we present the evaluation of this approach and contrast it with the original sbstree.

STM-BASED PARALLELIZATION

Overview

As discussed earlier, stateful components cannot be parallelized by simple replication. As a consequence, stateful components are often the bottlenecks of event processing applications. A component is classified as stateful if the output for the input event e_t, where t is the timestamp of the event, may depend not only on the event itself, but also on the current state of the component, which is function of previous events. Having the application programmer design a component that is already parallel can increase too much the costs due to the decrease in productivity. For example, when developing a parallel component for a multiprocessor machine, a programmer can use coarse locks to control the concurrent access to the shared state, but that will result in small, if any, gains. On the other hand, if the programmer uses

Figure 6. The counting algorithm variant of the sbstree from Figure 5 storing three subscriptions from Table 1

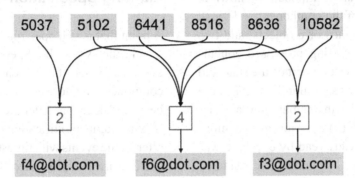

fine-grained locking or lock-free algorithms, the development gets too complex and too error-prone.

In some cases, stateful components that operate over single events can be transformed in stateless components that operate over a small set of events. In order for this to be possible, there must be no dependence between two sets of events, for example, computing the moving average of the value attribute of a stream of events using jumping windows. However, the same approach is not valid for sliding windows. In some other cases, known as partition-compute-combine, the stateful component can be replicated as long as the events processed by each replica are chosen by a partition algorithm and their results are merged by a combine algorithm (Koparanova & Risch, 2004). Unfortunately, there are no general algorithms for the partition and combine phases, what is required is the application programmer to provide these two components in addition to the original stateful component.

Thus, if the main goal is to provide an infrastructure that allows application programmers to build scalable applications in a productive way, we should not require that they understand all the intricacies of parallel and distributed computing, and then automatic parallelization is the only solution left. Parallelizing compilers have being investigated for a long time, but because the decisions are made in compilation-time, when much less information is available, they are pessimistic and, thus, explore less parallelism than indeed available.

Optimistic Parallelization of Stateful Components

The approach used in StreamMine is to use speculative execution to process events in parallel. The main idea is to use components that were not designed to be parallel and, thus, are easier to design. When two events are executed concurrently, they execute in sandboxes that isolate one from the other. During the processing, if they do not interfere with each other, they are said to be non-conflicting. After finishing processing, results of non-conflicting executions can be merged into memory. On the other hand, if they conflicted, the execution with less priority (e.g., the execution of the later event) is aborted and scheduled to be executed after the conflicting one. The main requirement is that all events are executed in a sandboxed environment and their modifications to the system state are made visible only in the order of their timestamps. As a consequence, the history of component states is the same as in a sequential execution.

This speculative execution is enabled by an underlying Software Transactional Memory (STM). The STM intercepts the memory accesses (reads, writes, memory allocations and releases) and check against interferences. Consider for example the stateful component *Processor1* in Figure 1, the internal structure of this component is shown in Figure 7. The state of the component is divided into 6 parts (e.g., 6 fields or 6 objects) and two events e_9 and e_{11} are being processed concurrently. If (a) e_9 accesses one part of the state and e_{11} accesses another, as shown in Figure 7, there is no interference and both computations are successful. Otherwise, if (b) e_9 writes to a memory position that was already read by e_{11}, there was interference because e_{11} should have considered the value written by e_9 and thus, e_{11} must abort and re-execute. However, if (c) e_9 reads from a position that was modified by e_{11} it can still retrieve the original values as modifications from e_{11} are still not incorporated to the component state yet (i.e., they are still confined in the sandbox). Further, (d) there is also no interference if both write to the same position (and none of them read), because the written values will only be incorporated to the component state at the timestamp order and, thus, the write from the later event will overwrite the write from the earlier one as expected. Finally, (e) there is no interference if both read from the component's state.

In addition, the illusion of a sequential execution is provided by having a counter that indicates the timestamp whose effects should be made visible next (shown as *NEXT=9* in Figure 7). Events that finished processing, but that cannot be made visible yet are stored in a waiting set, as is the case of e_{16} in Figure 7.

The benefits of the speculation affect both latency and throughput. First, consider the case that a stateful component (such as component *Processor1* in Figure 1) receives messages from replicated components (such as the replicated filters in the same figure). If the semantics and the workload are such that there is some parallelism available (i.e., there are cases in which two different events do not conflict), then processing these events out of their normal order will decrease their processing latency. Additionally, because more than one event can be processed at the same time, throughput is also improved.

Limiting Speculation

A point that is still open regards how speculative should the system be. Speculatively processing event e_{200} when the last changes written to the component's state were from event e_{10} may not be a good decision. After the processing from e_{200} is finished, its modifications will only be visible after all the events with timestamps in between 10 and 200 are also processed. Thus, if e_{200} conflicts with any of these missing events its processing will be discarded. To decide how much parallelism the combination component and workload has and how far should the system speculate, we use *conflict predictors* (Brito et al., 2008).

A conflict predictor looks at all events available, in other words, both the events not yet processed and the events processed but also aborted (e.g., because of conflicts) and decides based on the available resources which one should be processed next. The specification of the conflict predictor is a set of rules that groups events into classes and establishes a speculation horizon for each class. This set may combine user-provided rules, rules derived from static analysis and rules generated by runtime analysis. By default, no static analysis or user-provided rules are used, just a simple but useful predictor is available. This default predictor puts all events in the same class and tries to keep the number of aborts low by dynamically changing the speculation horizon. Through this approach, even if it the workload does not exhibit any parallelism, execution performance will still be as good as a sequential execution. However, if there is parallelism, at least part of it will be harnessed. This predictor works by keeping the numbers of aborts close, but not equal, to zero. If there are no aborts, it is an indication that perhaps more speculation can be done. Otherwise, if the number of aborts is too high, too much speculative work is being discarded and this impacts negatively on system performance. Although such predictor is likely to have a suboptimal performance, it

Figure 7. Structure of a speculative component

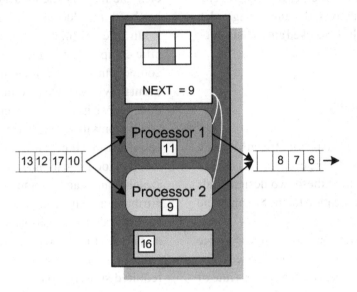

allows out-of-the-box sequential components to achieve immediate gain.

Distributed Speculation

Finally, the system gives the user the option to enable that speculative results are propagated to downstream components. If this is the case, speculative components will forward speculative results and will later send updates to these events as they get aborted and executed and when they reach a final version. It is up to the user to mark which components should accept speculative results. For example, if some component is a gateway to a physical system the application programmer may not want that actions are taken based on speculative results. By not marking a component to accept speculative events, the component will ignore all events that are not marked as final. Components that are marked as speculative, however, can handle speculative events very efficiently. Because the underlying STM monitors all the reads and writes from events it processes, the STM can efficiently detect if the update received for a previous speculative version really requires a re-execution. For example, if the final version

of an event differs from the previous version only by some attribute value and this attribute was not read by the current execution, then there is no need to abort and re-execute the current event as the results would be the same.

Because of the speculation support for stateful components, events can also be processed out of their normal order even in cases that the component semantics requires ordered processing. As mentioned in previously, if stateful components are replicated for fault tolerance, all the replicas must process the events in the same order and that requires costly atomic broadcast protocols. Speculation becomes then very helpful to hide such costs. On the one hand, if events have timestamps that describe a total order, these events can be simply submitted to component as they arrive and the speculation support will ensure that the component state and, consequently, the result of the computation will obey this total ordering and thus, be the same in all replicas. On the other hand, if the component does not require a total ordering the atomic broadcast protocol may use an optimistic delivery to submit events as speculative to the component, assign the optimistic order as the timestamps and later deliver the non-speculative versions with the correct order. Then, even if events were optimistically delivered in the wrong order, the computations will be checked and consistency guaranteed.

LOAD BALANCING

Two of the main design goals for StreamMine are scalability and near real-time event processing. In this section, we illustrate these two design goals using a real-world application for the StreamMine framework.

Our example application is a Quality of Service monitoring tool for prepaid telephone services. The goal of the tool is to supervise the success rate of reload transactions for prepaid mobile phones - see (Campanile et al., 2008) for more details. Typically, several distributed services participate in the reload transaction process. Such a transaction is considered to be successful if all participating services have executed their participating services with success in a timely fashion. Hence, the basic idea for the Quality of Service monitoring tool is to publish information about the reload process of all involved services and to analyze these events using a Timed Finite State Machine (TFSM). Those state machines have states and transitions for a successful or failed reload of one specific transaction and are automatically created for each new reload transaction. Obviously, there might be Quality of Service requirements that need near real-time processing of events in order to detect the success or failure of a specific ongoing transaction. Another problem arises with the huge amount of transactions that are executed in parallel. As mentioned before, each such transaction will have its own TFSM. This results in unacceptable high load if the state machines were to be executed on a single node. Hence, we need to support the scalability of the monitoring process to an arbitrarily high number of parallel transactions. To ensure real-time processing at very high event frequen-

cies, the load balancing scheme of StreamMine has been introduced.

In StreamMine, we balance the load of specific components on the nodes by replicating the component to another node and by partitioning the event flow toward these components. This could be done by partitioning the event domain over these components using an appropriate subscription for each state machine. However, this approach may result in too large number of subscriptions, is not scalable and can become complicated for some attribute domains, e.g., strings. Therefore, we have introduced a hash-based partitioning scheme as part of the publish/subscribe, which allows balancing the load while limiting the amount of required subscriptions.

Figure 8 presents the logical view on a deployment of a QoS monitoring network. *Source 1* and *Source 2* represent event sources of monitored services and these implicitly act as load balancing proxies. Boxes in the middle are monitoring components each one executed by different nodes hosting a couple of FSMs, in general cluster nodes. As mentioned above, each FSM is responsible for observing the QoS of one transaction. Monitoring results are published by the FSMs and are received by the consumers. Each event containing a new transaction identifier triggers the creation of a corresponding FSM by the monitoring component. The hash-based load balancing is used to partition the event flow between the two event sources and the two monitoring components at nodes N_i resulting in an approximately even load distribution. In the following, the StreamMine load balancing architecture will be presented.

As already mentioned, in StreamMine publish/subscribe is used as the underlying communication abstraction. Let $e = \{txid,...\}$ be an event issued by load balancing proxy and $s = \{hfn(txid) \bmod N\}$ be a subscription issued by cluster nodes, whereas *txid* is the key for the partitioning, here the transaction identifier and N is the total number of cluster nodes. Our initial approach only supports static

Figure 8. Logical view of a quality of service monitoring network

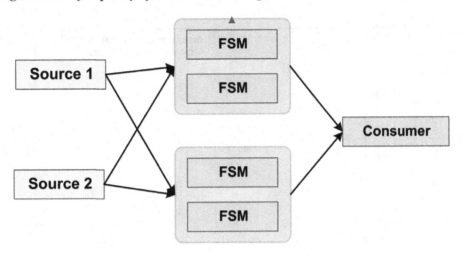

load balancing. In future, we will extend Stream-Mine to support dynamic load balancing too.

MEASUREMENTS

In order to evaluate our approach for building of the StreamMine processing system we have performed a series of measurements exposing the benefits and performance of the components of the StreamMine platform. Unless otherwise stated all measurements have been conducted using the StreamMine cluster of the Systems Engineering group at the Dresden University of Technology.

Microbenchmarks

In order to highlight the importance of the counting sbstree optimization we have performed a test in which we have varied the average number of bits set per event while maintaining a constant number of subscriptions in the original sbstree and counting sbstree - see Figure 9. The average number of bits set per event corresponds to the number of subscriptions matching a given event, divided by the average number of predicates per subscription times the number of bits per predicate.

We can observe that the matching time in case of the original sbstree (indicated as sbstree in the Figure 9) increases exponentially with the number of bits set in the event. The optimized variant (indicated as counting in the Figure 9), using a counting algorithm is linear with the number of bits set.

In order to compare the behavior of both variants of the sbstree with the traditional matching algorithm presented in (Tarkoma & Kangasharju, 2006) we have performed an experiment shown in Figure 10. In this tests we were varying the sizes of the respective matching structures - forest for (Tarkoma & Kangasharju, 2006), sbstree for the original variant of the sbstree and counting for the optimized version of the sbstree - by inserting increasing number of random subscriptions. For every new size of the routing structure we have matched 1000 random events and plotted the cumulative matching time of 1000 events.

The subscriptions and events were created using 20,000 unique attribute names. Attribute names for both events and subscriptions are selected from the Automatically Generated Inflection Database (AGID)[2], based on aspell word list, containing 112,505 English words and acronyms. Subscriptions consist of 2 to 5 predicates. The number of

Figure 9. sbstree and the counting sbstree data structures performance

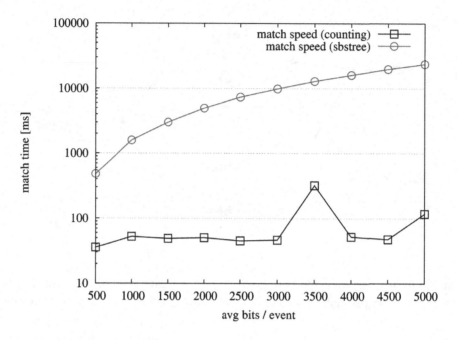

Figure 10. Comparison of matching times of different data structures

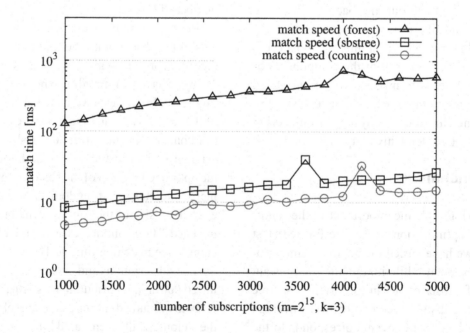

predicates is selected using a uniform distribution. The likelihood of subscription coverage and event matching has been set up using the Pareto distribution. The Pareto distribution used in all tests had the parameters α=1 and β=3[3].

Figure 11. The experiment setup

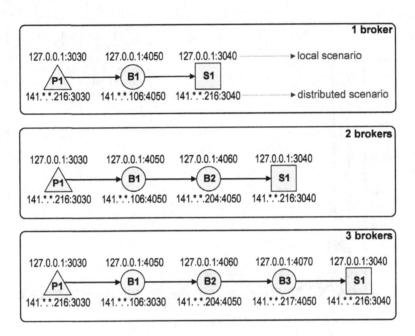

Figure 10 clearly illustrates the improvement in the matching times when using the content summarization techniques - especially in comparison to the traditional, full content-based approach - forest. We can observe that the counting version of the sbstree performs better in comparison to the original sbstree, due to the simpler data structure and the linear matching time algorithm.

Latency and Throughput

In the following tests we have focused on two aspects on the functioning of the publish/subscribe communication substrate - the (1) latency and (2) throughput as perceived by subscribers in a real, functioning deployment. For the following experiment we have created three networks: 1 broker, 2 brokers and 3 brokers - see Figure 11. In every network we have deployed one subscriber (*s1*) and one publisher (*p1*) connected by an increasing number of brokers between both of them.

Every network setup has been executed in two settings - local and distributed. In the local set-

ting we have created the networks one machine and connected their components via a loopback interface. For the distributed scenario every component was placed on a separate node of the StreamMine cluster.

Figure 12 compares the event forwarding latencies perceived by subscribers in different scenarios. We can observe that the increase in latency is proportional to the number of intermediate brokers and in case of the distributed scenario is dominated by the network latency. For the latency measurement we have assumed 200 different subscribers connected to each broker.

For the throughput measurements (see Figure 13) we have evaluated the counting sbstree routing structure performance without the influence of the networking components. We have connected a subscribers directly to the publish/subscribe broker - which implies direct event delivery using upcalls. The subscriber issued 200 different subscriptions. Subsequently, we have attached one publisher which issued events matching varying number of the subscriptions. Figure 13 shows the

Figure 12. Histogram of the event forwarding latencies

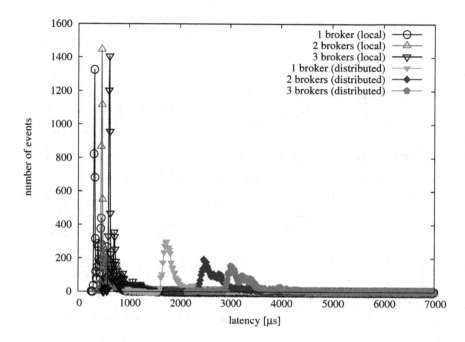

throughput of the whole setup in terms of events per second received by the subscriber. The three different distributions: uniform (unfrm distr), normal (norm distr) and Pareto (Pareto distr) determine the overlap between the events and subscriptions, i.e., they determine the likelihood of an event being matched by a subscription and subsequently delivered to the subscriber. In the upper part of the graph the total amount of events delivered to the subscriber has been plotted. We can observe that the throughput is proportional to the number of delivered events, as more and more upcalls need to be performed by the broker. The above test has been performed on a commodity PC. We can therefore conclude that the raw throughput of the content-based publish/subscribe middleware satisfies the real-time requirements of ESP applications (Stonebraker et al., 2005).

Parallelization and Speculative Execution

Regarding the STM-based parallelization, we have two scenarios that illustrate its benefits in common

cases. Both scenarios are based on the system of Figure 1. The first scenarios consider a case that there is no available parallelism in the workload (and/or in the semantics of the component), but there can still be some gain in latency. In this case we assume a system in which messages can be delivered out of the order of interest for the stateful component. Further, we assume the usage of a failure-awareness mechanism that can signal when a certain subset of the messages is guaranteed to be final (or a time-out that allows late messages to be discarded).

The end-to-end latency of individual events is shown in Figure 14. The behavior of the system depicted in the figure is periodic. In this example, events 1 to 9 (and then 11 to 19 and so on) are received but must wait until some information piggy-backed on event 10 (respectively, 20, 30 and so on) in order to be confirmed. As a consequence, the events modulo 10 are the ones that have higher latency as they need to wait a whole period before committing. Similarly, events with timestamp $ts\%10 = 9$, have the lowest end-to-end latency. Because the speculative system does

Figure 13. Event throughput

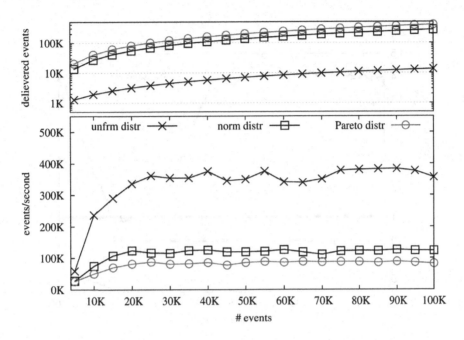

Figure 14. End-to-end latency comparison

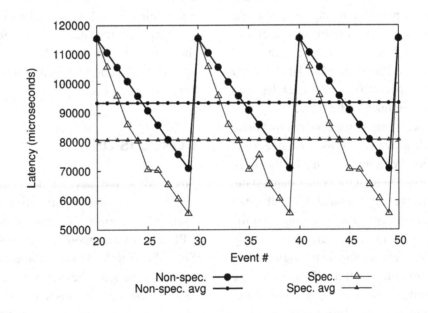

Figure 15. Throughput comparison for different levels of parallelism

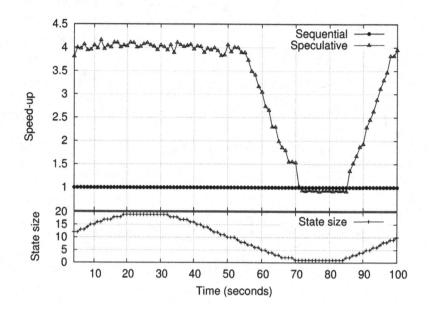

some processing in advance, events (except for the module 10 events) have a lower end-to-end latency. An example of a wrong ordering and the resulting extra cost of a re-execution can be seen for event 36. For this experiment, the non-speculative system has an average latency 16% greater than the speculative system, as depicted by the two average lines in the figure.

For the second scenario we consider that there is some parallelism available in the component and workload. More precisely, we consider that each event modifies one field of the components state. During runtime, we vary the number of unrelated fields in the component state, as depicted in the lower part of Figure 15: if the state contains only one field (the minimum value in the figure), any two events that execute concurrently will interfere because they modify the same field; on the other hand, if there are 20 fields (the maximum value), the probability that some events can execute in parallel without interfering is high. The speed

up of the speculative system in comparison to the non-speculative one can be seen in the upper part of Figure 15. As expected, when the state size is big enough the speculation is able to increase the throughput and when the state size is one, the speculative version performs similar to the non-speculative version.

CONCLUSION

StreamMine has several novel features. First, it uses speculation to parallelize the processing events. This permits us to spread the execution of CPU intensive components across multiple cores. Second, StreamMine uses a content-based pub/sub service to distribute the events among components. Our experience indicates that conventional event stream applications that are based on the pipe and filter paradigm are not easy to scale up. Using pub/sub, we can connect components via subscriptions instead of pipes. This increases the composability and scalability of StreamMine. Third, we extended the pub/sub mechanism to support load balancing across multiple machines.

StreamMine is under active development and we are currently implementing several applications on top of StreamMine to evaluate and evolve the APIs, the protocols, and our design decisions.

REFERENCES

Abadi, D. J., Ahmad, Y., Balazinska, M., Çetintemel, U., Cherniak, M., Hwang, J.-H., et al. (2005). The design of the borealis stream processing engine. In M. Stonebraker, G. Weikum. & D. DeWitt (Eds.), In *Proceedings of the 2005 Conference on Innovative Data Systems Research* (pp. 277-289). Asilomar, CA.

Aekaterinidis, I., & Triantafillou, P. (2007). Publish-Subscribe Information Delivery with Substring Predicates. *IEEE Internet Computing, 11*(4), 16–23. doi:10.1109/MIC.2007.90

Aguilera, M. K., Strom, R., E., Sturman, D., C., Astley, M., & Chandra, T., D. (1999). Matching Events in a Content-Based Subscription System. In *PODC '99: Proceedings of the eighteenth annual ACM symposium on Principles of distributed computing* (pp. 53-61). Atlanta, GA: ACM Press.

Babcock, B., Babu, S., Datar, M., Motwani, R., & Widom, J. (2002). Models and issues in data stream systems. In *PODS '02: Proceedings of the twenty-first ACM SIGMOD-SIGACT-SIGART symposium on Principles of database systems* (pp. 1-16). New York: ACM Press.

Bloom, B. (1970). Space/Time Trade-offs in Hash Coding with Allowable Errors. *Communications of the ACM, 13*(7), 422–426. doi:10.1145/362686.362692

Brito, A., Fetzer, C., Sturzrehm, H., & Felber, P. (2008). Speculative out-of-order event processing with software transaction memory. In R. Baldoni (Ed.), *Proceedings of the Second International Conference on Distributed Event-Based Systems* (pp. 265-275). New York: ACM Press.

Campanile, F., Coppolino, L., Giordano, S., & Romano, L. (2008). A business process monitor for a mobile phone recharging system. *The EUROMICRO Journal of Systems Architecture, 54*(9), 843–848. doi:10.1016/j.sysarc.2008.02.005

Cao, F., & Singh, J. P. (2005). MEDYM: Match-Early with Dynamic Multicast for Content-Based Publish-Subscribe Networks. In *Proceedings of Middleware 2005* (pp. 292–313). Berlin/Heidelberg, Germany: Springer Verlag. doi:10.1007/11587552_15

Carriero, N., & Gelernter, D. (1989). Linda in context. *Communications of the ACM, 32*(4), 444–458. doi:10.1145/63334.63337

Carzaniga, A., Rosenblum, D., & Wolf, A. (2001). Design and evaluation of a wide-area event notification service. *ACM Transactions on Computer Systems, 19*(3), 332–383. doi:10.1145/380749.380767

Carzaniga, A., & Wolf, A. (2003). Forwarding in a Content-Based Network. In *Proceedings of ACM SIGCOMM 2003* (pp. 163-174). New York: ACM Press.

Cherniack, M., Balakrishnan, H., Balazinska, M., Carney, D., Çetintemel, U., Xing, Y., & Zsonik, Y. (2003). Scalable distributed stream processing. In *Proceedings of the 2003 Conference on Innovative Data Systems Research*. Asilomar, CA.

Cutting, D., Quigley, A., & Landfeldt, B. (2008). SPICE: Scalable P2P implicit group messaging. *Computer Communications, 31*(3), 437–451. doi:10.1016/j.comcom.2007.08.026

Dean, J. (2006). Experiences with MapReduce, an Abstraction for Large-Scale Computation. In Proceedings of Keynote PACT06.

Dean, J., & Ghemawat, S. (2008). MapReduce: simplified data processing on large clusters. *Communications of the ACM, 51*(1), 107–113. doi:10.1145/1327452.1327492

Eugster, P., Felber, P., Guerraoui, R., & Kermarrec, A.-M. (2003). The many faces of publish/subscribe. *ACM Computing Surveys, 35*(2), 114–131. doi:10.1145/857076.857078

Fabret, F., Jacobsen, H. A., Llirbat, F., Pereira, J., Ross, K. A., & Shasha, D. (2001). Filtering algorithms and implementation for very fast publish/subscribe systems. In *SIGMOD '01: Proceedings of the 2001 ACM SIGMOD international conference on Management of data* (pp. 115-126). New York: ACM Press.

Fan, L., Cao, P., Almeida, J., Broder, A.Z (2000). Summary cache: a scalable wide-area web cache sharing protocol. *IEEE/ACM Transactions on Networking, 8*(3), 281-293.

Fidler, E., Jacobsen, H. A., Li, G., & Mankovski, S. (2005). The PADRES Distributed Publish/Subscribe System. In Proceedings of Feature Interactions in Telecommunications and Software Systems (pp. 12-30).

Herlihy, M., Eliot, J., & Moss, B. (1993). Transactional memory: Architectural support for lock-free data structures. In *Proceedings of the 20th Annual International Symposium on Computer Architecture, 1993* (pp. 289-300). New York: ACM Press.

Hwang, J.-H., Çetintemel, U., & Zdonik, S. Fast and Highly-Available Stream Processing over Wide Area Networks. In *Proceedings of the 24th International Conference on Data Engineering* (804-813) Washington, DC: IEEE Computer Society.

Jerzak, Z., & Fetzer, C. (2007). Prefix Forwarding for Publish/Subscribe. In *DEBS '07: Proceedings of the 2007 Inaugural International Conference on Distributed Event-Based Systems* (pp. 238-249). Toronto, Canada: ACM Press.

Jerzak, Z., & Fetzer, C. (2008). Bloom Filter Based Routing for Content-Based Publish/Subscribe. In *DEBS '08: Proceedings of the second international conference on Distributed event-based systems* (pp. 71-81). Rome, Italy: ACM Press.

Koparanova, M., & Risch, T. (2004). High-performance grid stream database manager for scientific data. In F. Fernández Rivera, Marian Bubak, A. Gómez Tato & Ramon Doallo (Eds.), *European Across Grids Conference* (pp. 86-92). Berlin/Heidelberg, Germany: Springer Verlag.

Mühl, G., Fiege, L., & Pietzuch, P. (2006). *Distributed Event-Based Systems*. New York: Springer-Verlag.

Oki, B., Pfluegl, M., Siegel, A., & Skeen, D. (1993). The information bus - an architecture for extensible distributed systems. In *Proceedings of the 14th Symposium on the Operating Systems Principles* (pp. 58-68). New York: ACM Press.

Pereira, J., Fabret, F., Llirbat, F., & Shasha, D. (2000). Efficient Matching for Web-Based Publish/Subscribe Systems. In *CoopIS '02: Proceedings of the 7th International Conference on Cooperative Information Systems* (pp. 162-173). London: Springer.

Pietzuch, P. (2004). *Hermes: A Scalable Event-Based Middleware*. (PhD Thesis) Cambridge, UK: Computer Laboratory, Queens' College, University of Cambridge.

Rosenblum, D., & Wolf, A. (1997). A design framework for Internet-scale event observation and notification. *SIGSOFT Software Engineering Notes, 22*(6), 344–360. doi:10.1145/267896.267920

Shavit, N., & Touitou, D. (1995). Software Transactional Memory. In *Proceedings of the Symposium on Principles of Distributed Computing*, (pp. 204-213).

Spring, J. H., Privat, J., Guerraoui, R., & Vitek, J. (2007). Streamflex: high-throughput stream programming in java. In *OOPSLA '07: Proceedings of the 22nd annual ACM SIGPLAN conference on Object oriented programming systems and applications* (pp. 211-228). New York: ACM Press.

Sterling, T., Becker, D. J., Savarese, D., Dorband, J. E., Ranawake, U. A., & Packer, C. V. (1995). BEOWULF: A parallel workstation for scientific computation. In D. P. Agrawal (Ed.), *Proceedings of the 24th International Conference on Parallel Processing*. London: CRC Press.

Stonebraker, M., Cetintemel, U., & Zdonik, S. (2005). The 8 requirements of real-time stream processing. *SIGMOD Record, 34*(4), 42–47. doi:10.1145/1107499.1107504

Tarkoma, S. (2008). Dynamic filter merging and mergeability detection for publish/subscribe. *Pervasive and Mobile Computing, 4*(5), 681–696. doi:10.1016/j.pmcj.2008.04.007

Tarkoma, S., & Kangasharju, J. (2006). On the cost and safety of handoffs in content-based routing systems. *Computer Networks, 51*(6), 1459–1482. doi:10.1016/j.comnet.2006.07.016

Tryfonopoulos, C., Zimmer, C., Weikum, G., & Koubarakis, M. (2007). Architectural Alternatives for Information Filtering in Structured Overlays. *IEEE Internet Computing, 11*(4), 24–34. doi:10.1109/MIC.2007.79

Welsh, M., Culler, D. E., & Brewer, E. A. (2001). Seda: An architecture for well-conditioned, scalable internet services. In *Proceedings of the 18th ACM Symposium on Operating Systems Principles (SOSP '01)*, (pp. 230-243). New York: ACM Press.

Zhao, Y., Sturman, D., & Bhola, S. (2004). Subscription Propagation in Highly-Available Publish/Subscribe Middleware. *Lecture Notes in Computer Science, 3231*, 274–293.

KEYTERMS AND DEFINITIONS

Bloom Filter: A probabilistic data structure able to test whether an element is contained within a set or not. A single Bloom filter is able to store a whole universe of elements.

Conflict Predictor: A module that controls the amount of speculation in a processing component that uses optimistic parallelization.

Publish/Subscribe: A communication paradigm where the communicating parties are decoupled in terms of time, space and synchronization. Obsoletes the need for source and destination addresses on data messages.

Stateful Component: A component that uses not only the current input, but also a state derived from previous computations, in order to execute its computations.

STM (Software Transaction Memory): A programming construct that allows blocks of code to be executed in a way that appears to be atomic.

ENDNOTES

[1] StreamMine is a part of larger EU project (FP7-216181) on Scalable Automatic Streaming Middleware for Real-Time Processing of Massive Data Flows (STREAM) - http://www.streamproject.eu/.

[2] For more details see: http://wordlist.sourceforge.net

[3] As defined in the umontreal.iro.lecuyer.randvar.ParetoGen package – see http://www.iro.umontreal.ca/~simardr/ssj/ for more details.

Chapter 12
Event Processing in Web Service Runtime Environments

Anton Michlmayr
Vienna University of Technology, Austria

Philipp Leitner
Vienna University of Technology, Austria

Florian Rosenberg
CSIRO ICT Centre, Canberra, Australia

Schahram Dustdar
Vienna University of Technology, Austria

ABSTRACT

Service-oriented Architectures (SOA) and Web services have received a lot of attention from both industry and academia. Services as the core entities of every SOA are changing regularly based on various reasons. This poses a clear problem in distributed environments since service providers and consumers are generally loosely coupled. Using the publish/subscribe style of communication service consumers can be notified when such changes occur. In this chapter, we present an approach that leverages event processing mechanisms for Web service runtime environments based on a rich event model and different event visibilities. Our approach covers the full service lifecycle, including runtime information concerning service discovery and service invocation, as well as Quality of Service attributes. Furthermore, besides subscribing to events of interest, users can also search in historical event data. We show how this event notification support was integrated into our service runtime environment VRESCo and give some usage examples in an application context.

INTRODUCTION

Following the Service-oriented Architecture (SOA) paradigm shown in Figure 1, service providers register services and corresponding descrip-

tions in registries. Service consumers can then find services in the registry, bind to the services that best fit their needs, and finally execute them. Web services (Weerawarana, Curbera, Leymann, Storey, & Ferguson, 2005) are one widely adopted realization of SOA that build upon the main stan-

DOI: 10.4018/978-1-60566-697-6.ch012

dards SOAP (communication protocol), WSDL (service description) and UDDI (service registry). Over the years, a complete Web service stack has emerged that provides rich support for multiple higher level functionalities (e.g., business process execution, transactions, metadata exchange etc.).

Practice, however, has revealed some problems of the SOA paradigm in general and Web services in particular. The idea of public registries did not succeed which is highlighted by the fact that Microsoft, SAP and IBM have shut down their public registries in the end of 2006. Moreover, there are still a number of open issues in SOA research and practice (Papazoglou, Traverso, Dustdar, & Leymann, 2007), such as dynamic binding and invocation, dynamic service composition, and service metadata.

One reason for these issues stems from the fact that service interfaces, service metadata and Quality of Service (QoS) attributes change regularly. Furthermore, new services are published, existing ones might be modified, and old services are finally deleted from the registry. This is problematic since service providers and consumers are usually loosely coupled in SOA. Thus, service consumers are not aware of such changes and, as a result, might not be able to access changed services any more. In this regard, the lack of appropriate event notification mechanisms limits flexibility because service consumers cannot automatically react to service and environment changes.

The current service registry standards UDDI (OASIS International Standards Consortium, 2005a) and ebXML (OASIS International Standards Consortium, 2005b) introduce basic support for event notifications. Both standards have in common that users are enabled to track newly created, updated and deleted entries in the registry. However, additional runtime information concerning service binding and invocation as well as QoS attributes are not taken into consideration by these approaches.

We argue that receiving notifications about such runtime information is equally important

Figure 1. Service-oriented architecture (Michlmayr, Rosenberg, Platzer, Treiber, & Dustdar, 2007)

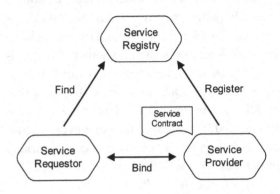

and should, therefore, be provided by SOA runtime environments. Furthermore, complex event processing mechanisms supporting event patterns, and search in historical event data are needed for keeping track of vast numbers of events. In this chapter, we focus on such runtime event notification support. Our contribution is threefold: firstly, we present the background of this work and describe the motivation based on a case study from the telecommunications domain. Secondly, we introduce the VRESCo runtime environment (Michlmayr, Rosenberg, Platzer, Treiber, & Dustdar, 2007) and describe its notification support in detail. This includes event types, participants, ranking, correlation, subscription, and notification mechanisms, as well as event persistence, event search, and event visibility. Finally, we show some usage examples and point to further application scenarios enabled by our work.

BACKGROUND

This section consists of two main parts. In the first part, we summarize several research approaches that are related to our work. In the second part, we briefly introduce the open source event processing engine Esper which we use as technical background for our prototype.

Related Work

Event-based systems in general, and the publish/subscribe pattern in particular have been the focus of research within the last years. This research has led to different event-based architecture definition languages, for instance Rapide (Luckham & Vera, 1995), and QoS-aware event dissemination middleware prototypes (Mahambre, Kumar, & Bellur, 2007). Moreover, data and event stream processing has also been addressed in various prototypes, such as STREAM (Arasu, et al., 2008) or Esper (EsperTech, 2008).

Approaches to integrate publish/subscribe and the SOA model resulted in the two specifications WS-Notification (Oasis International Standards Consortium, 2006) and WS-Eventing (World Wide Web Consortium, 2006). While WS-Eventing uses content-based publish/subscribe, WS-Notification provides topics (WS-Topics) as a means to classify events. In both specifications, publishers, subscribers, and the event infrastructure are implemented as Web services. However, event processing mechanisms besides topic- and content-based filtering of events are not addressed by these specifications. The combination of SOA and event-driven architectures is further addressed by Enterprise Service Bus (ESB) implementations (e.g., Apache Servicemix[1]). In contrast to our work, ESBs mainly focus on connecting various legacy applications by using a common bus that performs message routing, transformation and correlation.

Cugola and Di Nitto (Cugola & di Nitto, 2008) give a detailed overview of other research approaches combining SOA and publish/subscribe. Furthermore, they introduce a system that aims at adopting content-based routing (CBR) in SOA. Their approach is built on the CBR middleware REDS (Cugola & Picco, 2006), and provides notifications following WS-Notification. Service discovery is implemented according to the query-advertise style using UDDI inquiry messages. In this work CBR is mainly used to perform service discovery, while we focus on event processing

and notifications in service runtime environments. Additionally, we also provide support for dynamic binding and invocation, as well as QoS attributes and service metadata.

Service registries (e.g., UDDI, ebXML) represent one part of the SOA triangle that is responsible for maintaining a service repository including publishing and querying functionality. Both UDDI and ebXML provide subscription mechanisms to get notified if certain events occur within the service registry. However, these notifications are limited to the service data stored in the registry and do not include service runtime information. Notifications are sent per email or by invoking listener Web services. Other approaches such as Active Web Service Registries (Treiber & Dustdar, 2007) use news feeds such as Atom (Sayre, 2005) for dissemination of changes in the service repository content. News feeds enable to seamlessly federate multiple registries, yet, in contrast to our approach do not provide fine-grained control on the received notifications since they follow the topic-based subscription style. Furthermore, similar to UDDI and ebXML, these approaches do not include service runtime information.

There are several approaches that address search in historical events. Rozsnyai et al. (Rozsnyai, Vecera, Schiefer, & Schatten, 2007) introduce the Event Cloud system aiming at search capabilities for business events. Their approach uses indexing and correlation of events by using different ranking algorithms. In contrast to our approach, the focus of this work is on building an efficient index for searching in vast numbers of events whereas subscribing to events and getting notified about their occurrence is not addressed.

Li et al. (Li, et al., 2007) present a data access method which is integrated into the distributed content-based publish/subscribe system PADRES. The system enables to subscribe to events published in both the future and the past. In contrast to our work, the focus is on building a large-scale distributed publish/subscribe system that provides routing of subscriptions and queries.

Jobst and Preissler (Jobst & Preissler, 2006) present an approach for business process management and business activity monitoring using event processing. The authors distinguish between SOA events regarding violation of QoS parameters and service lifecycle, and business/process events building upon the Business Process Execution Language (BPEL). These events are fired by receive and invoke activities within BPEL processes. Unlike our approach, the focus is on search and visualization of business events whereas subscribing to events is not addressed. Furthermore, the different SOA events are not described in detail.

Esper

The open source engine Esper (EsperTech, 2008) provides event processing functionality and is available for both Java and C#. Esper supports several ways for representing events. Firstly, any Java/C# object may be used as an event as long as it provides getter methods to access the event properties. Event objects should be immutable since events represent state changes that occurred in the past and should therefore not be changed. Secondly, events can be represented by objects that implement the interface java.util.Map. The event properties are those values that can be obtained using the map getter. Finally, events may be instances of org.w3c.dom.Node that are XML events. In that case, XPath expressions are used as event properties.

Additionally, Esper provides different types of properties that can be obtained from events:

- Simple properties represent simple values (e.g., name, time).
- Indexed properties are ordered collections of values (e.g., user[4])
- Mapped properties represent keyed collections of values (e.g., user['firstname'])
- Nested properties live within another property of an event (e.g., Service.QoS)

In Esper, subscriptions are done by attaching listeners to the Esper engine, where each listener contains a query defining the actual subscriptions. These listeners implement a specific interface that is invoked when the subscription matches incoming events. The queries use the Esper Event Processing Language (EPL) which is similar to the Structured Query Language (SQL). The main difference is that EPL is formulated on event streams whereas SQL uses database tables: select clauses specify the event properties to retrieve, from clauses define the event streams to use, and where clauses specify constraints. Furthermore, similar to SQL there are aggregate functions (e.g., sum, avg, etc.), grouping functions (group by), and ordering structures (order by). Multiple event streams can be merged using the insert clause, or combined using joins. In addition to that, event streams can be joined with relational data using SQL statements on database connections. To give a simple example, the following EPL query triggers when a new service is published by 'TELCO1'.

```
select * from ServicePublishedEvent where
Service.Owner.Company = 'TELCO1'
```

EPL provides a powerful mechanism to integrate temporal relations of events using sliding event windows. These operators define queries for a given period of time. For instance, if QoS events regularly publish the QoS values of services, then subscriptions can be defined on the average response time during the last 6 hours as shown in the following simplified example.

```
select * from QoSEvent win:time(6 hours).
stat:uni('ResponseTime')
```

```
where average > 300
```

Finally, EPL supports subqueries, output frequency, and event patterns. The latter are used to define relations between subsequent events (e.g., representing 'followed by' relations). For

Figure 2. TELCO case study (Michlmayr, Rosenberg, Leitner, & Dustdar, 2008)

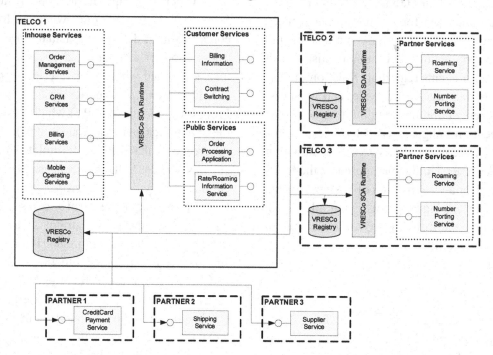

more information on Esper and EPL we refer to (EsperTech, 2008).

VIENNA RUNTIME ENVIRONMENT FOR SERVICE-ORIENTED COMPUTING

This section describes the VRESCo project[2] (Vienna Runtime Environment for Service-Oriented Computing) and its event notification support. Before we go into the details of this event notification support we give a motivating example for our work, followed by a brief introduction of the overall runtime architecture and the service metadata model of VRESCo.

Motivating Example

The case study shown in Figure 2 is adapted from (Michlmayr, Rosenberg, Platzer, Treiber, & Dustdar, 2007) and will be used for illustration purposes. In this case study, a telecommunication company (TELCO) consists of multiple departments that provide different services to different service consumers. *Inhouse services* are shared among the different departments (e.g., CRM services). *Customer services* are only used by the TELCO customers (e.g., view billing information) whereas *public services* can be accessed by everyone (e.g., get phone/roaming charges). Additionally, the TELCO consumes *partner services* (e.g., credit card service) as well as *competitor services* from other TELCOs (e.g., number porting service). Furthermore, service providers maintain multiple revisions of their services.

This case study shows several scenarios where notifications are useful. Consider for example that TELCO1 wants to get notified if new shipping services get available or if new revisions of TELCO2's number porting service are published. Furthermore, it is also important to know if services get unavailable or are removed from the registry (e.g., in order to automatically switch to another service). Besides these basic event notifications another concern for TELCO1 is to observe QoS

Figure 3. VRESCo overview

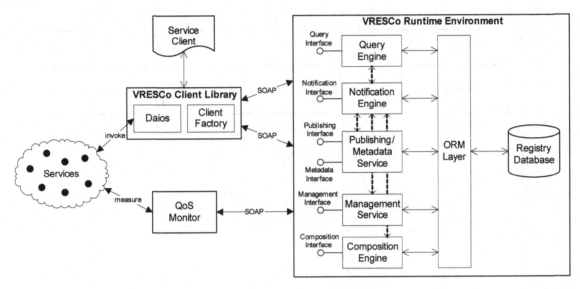

attributes. For instance, TELCO1 wants to react if the response time of a service falls beyond a given threshold. This implies that the environment considers runtime information of its services. To go one step further, TELCO1 also wants to get notified if the average response time of TELCO2's number porting service (measured within a time frame of 6 hours) falls beyond a given threshold since this might violate their Service Level Agreement (SLA).

In addition to subscribing to certain events of interest, TELCOs also want to search in the vast amount of historical events. In that way, stakeholders are enabled to observe the history of a given service or service provider within a given period of time, when deciding about the integration of external services into their own business processes.

In these scenarios notifications have clear advantages over traditional approaches using runtime exceptions, since service consumers can instantly react to failures or QoS changes. The power of events additionally opens up new perspectives and applications scenarios that can be built in a flexible manner. For instance, this includes SLAs and service pricing models as well as provenance-aware applications, which are discussed later.

VRESCo Overview

The event notification approach presented in this chapter was implemented as part of the VRESCo runtime introduced in (Michlmayr, Rosenberg, Platzer, Treiber, & Dustdar, 2007). Before going into the details of our eventing approach, we give a short overview of this project.

The VRESCo runtime environment aims at addressing some of the current challenges in Service-oriented Computing research (Papazoglou, Traverso, Dustdar, & Leymann, 2007) and practice. Among others, this includes topics related to service discovery and metadata, dynamic binding and invocation, service versioning and QoS-aware service composition. Besides this, another goal is to facilitate engineering of service-oriented applications by reconciling some of these topics and abstracting from protocol-related issues.

The architecture of VRESCo is shown in Figure 3. To be interoperable and platform-independent, the VRESCo services which are implemented in C#/.NET are provided as Web services. These services can be accessed either directly using the SOAP protocol, or via the client library that provides a simple API. Services and associated

Figure 4. Service revision graph (Leitner, Michlmayr, Rosenberg, & Dustdar, 2008)

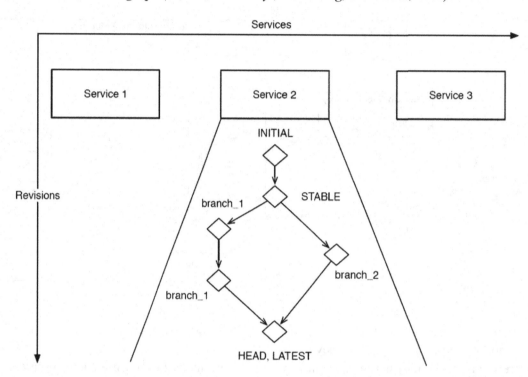

metadata are stored in the registry database that is accessed using the object-relational mapping (ORM) layer. The services are published and found in the registry using the publishing and querying engine, respectively. The VRESCo runtime uses a QoS monitor (Rosenberg, Platzer, & Dustdar, 2006) which continuously monitors the QoS, and keeps the QoS information in the registry up to date. Furthermore, the composition engine provides support for QoS-aware service composition (Rosenberg, Celikovic, Michlmayr, Leitner, & Dustdar, 2009). Finally, the event notification engine is responsible for notifying subscribers when events of interest occur.

Versioning, Dynamic Binding and Invocation

Web services evolve over time, which raises the need to maintain multiple service revisions concurrently. VRESCo supports service versioning by introducing the notion of service revision graphs (Figure 4), which define successor-predecessor relationships between different revisions of a service and support multiple parallel branches of the same service (Leitner, Michlmayr, Rosenberg, & Dustdar, 2008). Revision tags (e.g., INITIAL, STABLE, LATEST) are used to distinguish the different service revisions. Service consumers make use of versioning strategies to specify which revision of a service should be invoked (e.g., always invoke the newest revision, always invoke a specific revision, etc.).

To carry out the actual Web service invocations the Daios dynamic Web service invocation framework (Leitner, Rosenberg, & Dustdar, 2009) has been integrated into the VRESCo client library. Daios decouples clients from the services to be invoked by abstracting from service implementation issues such as encoding styles, operations or endpoints. Therefore, clients only need to know the address of the WSDL interface

Figure 5. Metadata model (Rosenberg, Leitner, Michlmayr, & Dustdar, 2008)

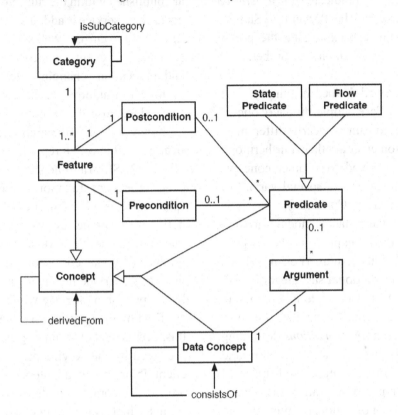

describing the target service, and the corresponding input message; all other details of the target service implementation are handled transparently. Besides dynamic invocation, VRESCo also supports dynamic binding of Web services. The aim is to dynamically bind to services offering the same functionality. The rebinding can either be QoS-based (using queries on QoS attributes) or content-based (using unique identifiers within different service categories). Rebinding strategies are used to define when the current binding of the service proxy should be evaluated (e.g., periodic, on demand, on invocation, etc.). We give an example for service invocations in VRESCo in Listing 1 below (see Section *Usage Examples*).

VRESCo Service Metadata Model

The VRESCo runtime provides a rich service metadata model capable of storing additional ser-

vice information in the registry. This is needed to capture the purpose of services to enable querying and mediating between similar services that perform the same task. The VRESCo metadata model presented in (Rosenberg, Leitner, Michlmayr, & Dustdar, 2008) is depicted in Figure 5. The main building blocks of this model are *concepts* that represent the definition of an entity in the domain model. We distinguish between three different types of concepts:

- *Features* represent concrete actions in the domain (e.g. PortNumber).
- *Data concepts* represent concrete entities in the domain (e.g., customers) which are defined using other data concepts and atomic elements such as strings or numbers.
- *Predicates* represent domain-specific statements that either return *true* or *false*. Each predicate can have a number of *arguments*.

For example, a predicate for a *feature* PortNumber could be Portability_Status_ Ok(PhoneNumber), expressing the portability status of a given phone number.

Concepts have a well-defined meaning specific to a certain domain. For example, the data Concept Customer in one domain is clearly different to the concept Customer in another. Furthermore, concepts may be derived from other concepts (e.g., PremiumCustomer is a special variant of the more general concept Customer).

Each feature in the metadata model is associated with one *category* expressing the purpose of a service (e.g., PhoneNumberPorting). Each category can have additional subcategories following the semantics of multiple inheritance to allow a more fine-grained differentiation. Features have *preconditions* and *postconditions* expressing logical statements that have to hold before and after the execution of a feature. Both types of conditions are composed of multiple predicates, each having a number of optional arguments that refer to a concept in the domain model. There are two different types of predicates: *Flow predicates* describe the data flow (i.e., the data required or produced by a feature) while *state predicates* express some global behavior that is valid either before or after invoking a feature.

Services in VRESCo can be mapped to this metadata model (e.g., services map to categories, service operations map to features, operation parameters map to data concepts, etc.). As a result, services that perform the same task but have different interfaces can be dynamically replaced at runtime. More information can be found in (Rosenberg, Leitner, Michlmayr, & Dustdar, 2008).

VRESCo Eventing Engine

This section presents the VRESCo notification support that was introduced in (Michlmayr, Rosenberg, Leitner, & Dustdar, 2008). The basic idea can be summarized as follows: notifications are published within the runtime if certain events occur (e.g., service is added, user is deleted, etc.). In contrast to current Web service registries, this also includes events concerning service binding and invocation, changing QoS attributes, and runtime information. Service consumers are then enabled to subscribe to these events.

Figure 6 depicts the architecture of the notification engine which represents one component of the VRESCo runtime shown in Figure 3. The event processing functionality is based on NEsper, which is a .NET port of Esper. Within the notification engine, events are published using the eventing service. Most events are directly produced by the corresponding VRESCo services (e.g., service management events are fired by the publishing service while querying events are fired by the querying service). In contrast to this, events related to binding and invocation are produced by the service proxies located in the client library. Event adapters are thereby used to transform incoming events into the internal event format which can be processed efficiently. The eventing service then forwards these events to the event persistence component that is responsible for storing events in the event database. Finally, the eventing service feeds incoming events into the Esper engine.

The subscription interface is used for subscribing to events of interest according to the methods proposed in the WS-Eventing specification. The subscription manager is responsible for managing subscriptions which are put into the subscription storage. In addition, subscriptions are translated for further processing. This is done by converting the WS-Eventing subscriptions into Esper listeners which are attached to the Esper engine.

The Esper engine performs the actual event processing and is, therefore, responsible for matching incoming events received from the eventing service to listeners attached by the subscription manager. On a successful match, the registered listener informs the notification manager that is responsible for notifying interested subscribers.

Figure 6. Eventing architecture (Michlmayr, Rosenberg, Leitner, & Dustdar, 2008)

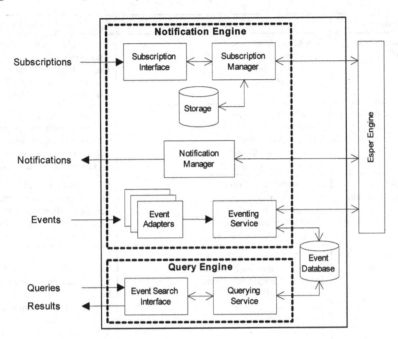

Depending on the listener type, the notification manager knows which notification type to use (e.g., email, listener Web service).

Finally, the search interface is used to search for historical events. The event database is implemented using a relational database and accessed via the ORM layer. The querying service returns a list of events that match the given query.

Event Types

The first step in developing such notification mechanism is to define all events supported by the engine. In the context of our work there are several events that can be captured at runtime. We have identified the events shown in Table 1 where events are grouped according to their event type. The event condition in the right column describes the situations when the event occurs. These event types form an event type hierarchy following the concept of class hierarchies (i.e., events inherit the properties of their parent event type) which is illustrated using colons.

The biggest group in this hierarchy is represented by the service management events that are triggered when services or service revisions and their associated metadata or QoS values change. Other event types include runtime information concerning binding and invocation, querying information and user information. All events inherit from the base type *VRESCoEvent* which provides a unique event sequence number and a timestamp measured during event publication.

Event Participants

Event-based systems usually consist of two types of participants that pose different requirements to the system, namely event producers and event consumers.

In general, events are produced by VRESCo components. However, different components are responsible for firing different kinds of events. These components, which mainly differ in their location, are described in this section. In this regard, we distinguish between *internal events* that are produced within the SOA runtime and

Table 1. VRESCo events

Event Type	Event Name	Event Condition
UserManagementEvent **: VRESCoEvent**	*UserAddedEvent* *UserModifiedEvent* *UserDeletedEvent* *UserLoginEvent* *UserLogoutEvent*	User is added to the runtime User is modified in the runtime User is deleted from the runtime User logs in using the GUI User logs out using the GUI
ServiceManagementEvent **: VRESCoEvent**	*ServicePublishedEvent* *ServiceModifiedEvent* *ServiceDeletedEvent* *ServiceActivatedEvent* *ServiceDeactivatedEvent*	New service is published into the runtime Service is updated (no new revision) Service is deleted from the runtime Service is activated in the runtime Service is deactivated in the runtime
VersioningEvent **: ServiceManagementEvent**	*RevisionPublishedEvent* *RevisionActivatedEvent* *RevisionDeactivatedEvent* *RevisionTagAddedEvent* *RevisionTagRemovedEvent*	New revision is published into the runtime Service revision is activated in the runtime Service revision is deactivated in the runtime Service revision tag is added by the owner Service revision tag is removed by the owner
MetadataEvent **: ServiceManagementEvent**	*ServiceCategoryAddedEvent* *ServiceCategoryModifiedEvent* *ServiceCategoryDeletedEvent* *FeatureAddedEvent* *FeatureModifiedEvent* *FeatureDeletedEvent* *MappingEvent*	Service category is added to the runtime Service category is modified in the runtime Service category is deleted from the runtime Feature is added to a service category Feature is modified in a service category Feature is deleted from a service category Service is mapped to a feature
QoSEvent **: ServiceManagementEvent**	*QoSRevisionEvent* *QoSOperationEvent* *RevisionGetsUnavailableEvent* *RevisionGetsAvailableEvent*	QoS value of service revision is published QoS value of service operation is published Service revision gets unavailable Service revision gets available again
BindingInvocationEvent **: VRESCoEvent**	*ServiceInvokedEvent* *ServiceInvocationFailedEvent* *ProxyRebindingEvent*	Specific service is invoked Service invocation failed Service proxy is (re-)bound to a specific service
QueryingEvent **: VRESCoEvent**	*RegistryQueriedEvent* *ServiceFoundEvent* *NoServiceFoundEvent*	Registry is queried using a specific query string Specific service is found by a query No services are found by a query

external events that are published by components outside the runtime. Most events are directly produced by the corresponding VRESCo services. For instance, service management events (e.g., *ServicePublishedEvent*) are fired by the publishing service. The same is true for versioning and metadata events. According to this, user management events are published by the user management service while querying events are produced by the querying service. All these event types have in common that they are produced as part of the VRESCo services and therefore represent internal events.

The application logic inherent to binding and invocation of services is located in the service proxies provided by the client library. As a result, the events concerning binding and invocation (e.g., *ServiceInvokedEvent*) are fired by this component. Therefore, VRESCo provides a notification interface in order to allow clients to publish binding and invocation events into the runtime. These client events represent external events that are then transformed into the internal event format by the runtime. Finally, the QoS monitor that regularly measures the QoS values of services is responsible for firing QoS events. Similar to the client library, the QoS monitor uses the notification interface to publish external events into the runtime.

Similar to event producers, we distinguish between *internal* and *external consumers*. Internal consumers reside within the runtime and register listeners at the Esper engine that are invoked when subscriptions match incoming events. External consumers outside the runtime are notified depending on the notification delivery mode defined in the subscription request.

In general, there are two main groups of external consumers: *humans* and *services*. Clearly, notification delivery mechanisms and the notification payload differ for these two groups. Humans are mainly interested in notifications sent per email, SMS or news feeds. In some scenarios, it might also be suitable to log the occurrence of events in log files that are regularly checked by the system administrator. In any case, notifications for humans might be less explicit since humans can interpret incomplete information. In contrast to this, service notifications can be sent using the Web service notifications standards WS-Eventing and WS-Notification. For our current prototype implementation, we have made use of the WS-Eventing specification since it represents a light-weight approach supporting content-based subscriptions.

Moreover, another distinction can be made between service providers and consumers that may be interested in different types of events. For instance, service consumers might not be interested in user management events or might not even be allowed to receive them. We introduce different event visibilities later.

Event Ranking

The importance and relevance of different events can be estimated by ranking them according to some fitness function. This is of particular interest when dealing with vast numbers of events. The following list describes several ways we have identified for ranking events:

- *Priority-based*: Event priority properties (e.g., 1 to 10 or 'high' to 'low') can be predefined according to the event model, or defined by the event producer when publishing the event. In the latter case, one problem might be that event producers do not know the importance of particular events related to others.

- *Hierarchically*: Events are ordered in a tree structure where the root represents the most important event while the leaves are less important.

- *Type-based*: All events are ranked based on the event type. That means each event has a specific type (possibly supporting type inheritance) that is used to define the ranking. However, the importance of some event might not always depend only on its type – sometimes the event properties will make the difference.

- *Content-based*: Events can be ranked based on keywords in the notification payload (e.g., the keyword 'exception' might be more important than the keyword 'warning' or 'info').

- *Probability-based*: In general, the event frequency depends on environmental factors. In this regard, one can assume that frequent events (e.g., *RegistryQueriedEvent*) might be less important than infrequent ones (e.g., *RevisionGetsUnavailableEvent*).

- *Event Patterns*: Finally, some events often occur as part of event patterns (e.g., proxy is bound to a specific service, followed by service is invoked using this proxy). The ranking mechanism could consider such event patterns.

VRESCo supports hierarchically, priority-, typed-, and content-based ranking. Probability-based ranking could be integrated by using the univariate statistic function provided by Esper. This mechanism calculates statistics over the occurrence of different events. In general, however, it should be noted that event ranking has one inherent problem: while some events can be critical for one subscriber, they might be only

Table 2. Event correlation sets

Event Correlation Set	Events	Correlation Identifier
User Management	Create, update & delete users	UserId
Service Lifecycle	Create, update, delete, bind, invoke & query services	ServiceId
Service Revision Lifecycle	Create, update, delete, bind, invoke, query & tag revisions	ServiceRevisionId
QoS	Correlate QoS measurements of one service revision	ServiceRevisionId
Service Category	Correlate events of services within one service category	ServiceCategoryId
Feature	Correlate events of services that provide one feature	FeatureId

minor for others. Yet, introducing event ranking mechanisms provides different ways to express the importance of events.

Event Correlation

Event-based systems usually deal with vast numbers of events that have to be managed accordingly. Event correlation techniques are used to avoid losing track of all events and their relationship. For instance, the work in (Rozsnyai, Vecera, Schiefer, & Schatten, 2007) describes the Event Cloud that provides different correlation mechanisms. Basically, the idea is to use event properties that have the same value as correlation identifier. For instance, two events (e.g., *ServicePublishedEvent* and *ServiceDeletedEvent*) having the same event attribute *ServiceId* are correlated since they both refer to the same service.

In the context of our work, we have identified a number of correlation sets summarized in Table 2, which shows the name of the correlation set, the events that are subsumed in this correlation, and the correlation identifier. The correlation sets cover three different aspects: user management using the *UserId* as correlation identifier, service (and service revision) lifecycle and *QoS* using *ServiceId* and *ServiceRevisionId*, and metadata information using *ServiceCategoryId* and *FeatureId*.

Besides correlating events using identifiers (e.g., the same *ServiceId*), we also consider temporal correlation of events. This is important since events that occur at the same time might be related. Furthermore, users are often interested in all events that occurred within a given timeframe. To accomplish temporal correlation of events, every event has a timestamp that is set during event publication. This timestamp can then be used to group events that happened within a given period of time (e.g., within the same hour, day, week, etc.).

The difference between event correlation sets and event types can be summarized as follows: while event types represent groups of events that occur in the same situations or indicate the same state change (e.g., some service is published), event correlation sets correlate all events that are related due to some event attribute (e.g., service revision X is published, deactivated, invoked, or the QoS value changes, etc.).

Subscription and Notification Mechanism

In general, event consumers can be enabled to subscribe to their events of interest in several ways (Eugster, Felber, Guerraoui, & Kermarrec, 2003). The most basic way is following the topic-based style that uses topics to classify events. Event consumers subscribe to receive notifications about that topic. Similar to topic-based subscriptions, the type-based style uses event types for classification. Even though these two styles are simple, they do not provide fine-grained control over the events of interest. Therefore, the content-based style can be used to express subscriptions based on the actual notification payload.

Since the VRESCo runtime is provided using Web service interfaces, the subscription interface should also be using Web services. WS-Eventing represents a light-weight specification that defines such an interface by providing five operations: *Subscribe* and *Unsubscribe* are used for subscribing and unsubscribing. The *GetStatus* operation returns the current status of a subscription, while *Renew* is used to renew existing subscriptions. Each subscription has a given duration specified by the *Expires* attribute. Finally, *Subscription End* is used if an event source terminates a subscription unexpectedly.

For implementing the event processing mechanism of the VRESCo runtime, we build upon an existing WS-Eventing implementation[3] that was extended for our purpose. WS-Eventing normally uses XPath message filters as subscription language that are used for matching incoming XML messages to stored subscriptions. The specification defines an extension point to use other filter dialects which we used to introduce the *EPLDialect* for using EPL queries as subscription language. The actual EPL query is then attached to the subscription message by introducing a new message attribute *subscriptionQuery*.

WS-Eventing distinguishes between subscriber (the entity that defines a subscription) and event sink (the entity that receives the notifications) that are both implemented using Web services. VRESCo additionally supports notifications sent per email and written to log files. Therefore, in addition to the default delivery mode *PushDeliveryMode* using Web services, we introduced *EmailDeliveryMode* and *LogDeliveryMode* which are attached to the subscription messages.

The subscription process is illustrated in Figure 7. When the subscription manager receives requests from subscribers, it first extracts the subscription and puts it into the subscription storage to be able to retrieve it at a later time. Then it extracts the EPL subscription query and the delivery mode

from the request and creates a corresponding Esper listener. This listener is finally attached to the Esper engine to be matched against incoming events. Furthermore, the subscription manager is responsible for keeping the subscriptions in the storage and the listeners attached to Esper synchronized. That means, when subscriptions are renewed or expire, the subscription manager re-attaches the corresponding listener or removes them, respectively.

Sending notifications can be done in several ways.

In the best-effort model, notifications are lost in case of communication errors. To prevent such loss, subscribers can send acknowledgements when receiving notifications. Besides pushing notifications towards subscribers, pull-style notifications enable subscribers to retrieve pending notifications from the event engine.

VRESCo notifications are sent push-style using emails or listener Web services. As shown in Figure 7, the notification manager knows which notification type to use depending on the listener attached to the Esper engine. On a successful match the notification manager first extracts this information from the listener. If the event sink prefers email notifications, the notification manager connects to an SMTP server. In case of Web service listeners, the notification manager invokes the corresponding listener Web service provided by the event sink. If the event sink cannot be notified, these pending notifications are stored in the event database and can be retrieved by the subscribers in pull-style.

Event Persistence and Event Search

Event notifications are often used when subscribers want to quickly react on state changes. Additionally, in many situations it is also important to search in historical event data. For instance, users might want to get notified if a new service revision is published into the registry while they

Figure 7. Subscription and event publication sequence

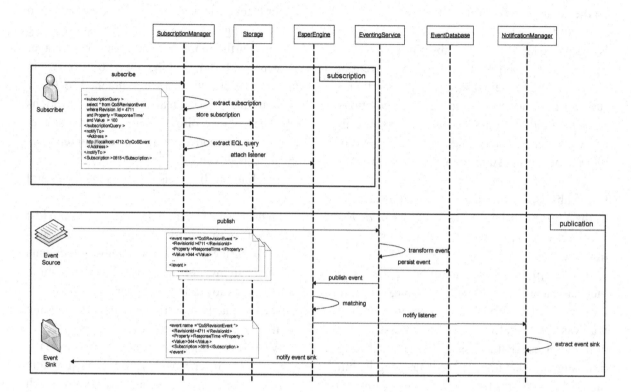

also want to search for the five previous service revisions.

To support such functionality, the VRESCo notification engine stores all events and provides an appropriate search interface for it. As illustrated in Figure 7, when events are published by an event source (e.g., QoS monitor), the eventing service first transform the events into the internal event format and then persists them into the event database. These events can be queried using the event search interface that is part of the querying interface which is used to search for services in the registry database. Data access in VRESCo is done via an ORM layer using NHibernate[4]. Therefore, the event search builds on the Hibernate Query Language (HQL).

Since event-based systems often deal with vast numbers of events, in some situations using relational databases might not be efficient enough. In such cases, building highly targeted and efficient index structures might be preferred. In this regard, we envision using the Vector space model in addition to a traditional relational event database. Following this model, documents (events) are represented by n-dimensional vectors where each dimension represents one keyword. The similarity of two vectors then indicates the similarity of the two corresponding documents (events) using these keywords. The advantage of the Vector space model compared to traditional database search is that the search returns a list of fuzzy matches together with a similarity rating. Furthermore, the search queries can be easily executed on multiple distributed vector spaces.

Event Visibility

In our first prototype, events were visible to all users within the runtime. However, this can be problematic in business scenarios. For instance, considering our TELCO case study shown in Figure 2, TELCO1 might agree that PARTNER1 can see events concerning service management and versioning, but might restrict that events related to binding and invocation are only visible for its own employees.

Mühl et. al. (Mühl, Fiege, & Pietzuch, 2006) discuss security issues in event-based systems by introducing different access control techniques such as access control lists (ACL), capabilities, and role-based access control (RBAC). ACLs, on the one hand, define the permissions of different users (principals) for specific security objects. Capabilities, on the other hand, define the permissions of a specific user for different security objects. The difference is that ACLs are stored for every security object while capabilities are stored for every user. Finally, RBAC extends capabilities by allowing users to have several roles that represent abstractions between users and permissions. Users can have one or more roles while permissions are directly granted to the different roles.

In the VRESCo notification engine, we have integrated an access control mechanism following RBAC (Michlmayr, Rosenberg, Leitner, & Dustdar, 2009). Therefore, VRESCo users are divided into different user groups. Event visibility can then be defined according to the event visibilities shown in Table 3.

It is interesting to note that in our work the event publisher is enabled to define the visibility of her events. While one publisher might not want that other users can see events ("PUBLISHER"), another might not define any restrictions ("ALL"). Additionally, user access to events can be granted only to specific users (e.g., "anton"). Finally, RBAC is introduced by either defining visibility for all users of a specific group (e.g., ":admins"), or all users within the same group as the publisher ("GROUP").

Besides defining event visibilities for different users and groups, more fine-grained access control is provided by allowing users to specify event visibilities for specific event types. Clearly,

Table 3. Event visibilities

Event Visibility	Description
ALL	Events are visible to all users
GROUP	Events are visible to all users within the same group of the publisher
PUBLISHER	Events are visible to the publisher only
<:GroupName>	Events are visible to all users within a specific group
<Username>	Events are visible to a specific user only

these definitions take the event type hierarchy into consideration: If no event visibility is defined for a specific event type, the engine takes the visibility of the parent type. If there is no visibility for any type the default visibility is chosen (i.e., ALL for type *VRESCoEvent*).

The access control mechanism is enforced by the eventing service and the notification manager shown in Figure 6. On the one side, the eventing service attaches both event visibility and name of the publisher to the event before feeding it into the Esper engine. While the name of the publisher can be directly extracted from the request message of the invoked VRESCo service (e.g., *QueryingService*), the event visibility of the publisher is queried from the registry database.

On the other side, when events match subscriptions the notification manager gets name and user group of the subscriber from the subscription storage and extracts publisher name and event visibility from the notification payload. Based on this information, the notification manager can verify if the current event is visible to the subscriber. If the event is visible the subscriber is notified, otherwise no notification is sent. Furthermore, the event search also follows the same principle: if events are not visible to the requester, they are removed from the search result.

In our approach, publishers are able to specify which subscribers can see which events by using event visibilities. Therefore, event access is mainly controlled by the publishers. Apart from that, however, subscribers are able to specify which

Figure 8. VRESCo runtime manager

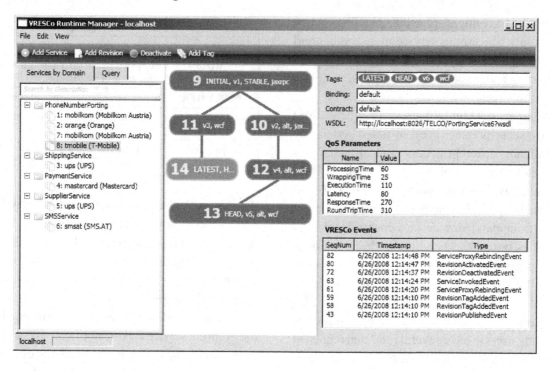

event producers they are interested in. This is done by specifying the event attribute *publisher* in the EPL subscription queries.

VRESCo Runtime Manager

Figure 8 shows a screenshot of the VRESCo Runtime Manager GUI (displaying the implementation of the TELCO case study). The service categories and their services are listed in the left part of the GUI that also provides a search interface for querying services within the registry database. The service revision graph of the selected service is illustrated in the middle, showing identifier and tags of the different service revisions. The initial revision is always placed on the top of the graph and the edges define the predecessor-successor relationship. The details of the selected service revision are shown in the right part including revision tags, URL of the WSDL document and current QoS parameters (e.g., response time, latency, etc.). The table in the bottom right corner depicts the service revision lifecycle represented

by all events related to this service revision (i.e., correlated using the same revision identifier) that are visible to the current user. The table shows sequence number, timestamp and type of the events.

Usage Examples

In this section, we use our motivating example to show how the VRESCo runtime is used to invoke services, as well as how the event notification support works in practice. The performance of the event engine and further examples illustrating the expressiveness of the subscription language can be found in (Michlmayr, 2010).

Listing 1 illustrates how service proxies are generated in VRESCo and how the Daios framework is used to invoke services. In lines 2-3, the VRESCo client factory is used to create a proxy for the querying service running on port number 8001 on localhost. Service proxies in VRESCo are defined using a search query that is constructed in lines 6-8. In this example, we want to access the latest service revision of category PhoneNumberPorting from TELCO2 having a response time of less than 500 milliseconds. In lines 12-13, the service proxy is created using this selection while the rebinding strategy PeriodicRebinding(10000) means that the binding should be evaluated every 10 seconds. For instance, if a new revision is published that better matches the selection, the service proxy should automatically rebind to this revision. In lines 16-22, the input message is built while the service is actually invoked using the request/response pattern in line 25. It should be noted that Daios also supports one-way and asynchronous invocation patterns. Finally, the number porting confirmation is extracted from the output message that is returned by the service (line 28).

Listing 1. Dynamic Binding and Invocation

```
01   // create proxy for the querying service
02   IVRESCoQuerier querier =
03       VRESCoClientFactory.
         CreateQuerier("localhost", 8001);
04
05   // select service and new provider
06   string selection = "Service.Category.Name
     like 'PhoneNumberPorting' and"
07       + " Service.Owner.Name like
         'TELCO2' and"
08       + "Tag.Name like 'LATEST' and QoS.
         ResponseTime < 500";
09   Provider prov = new Provider("TELCO1",
     "Main Street 1, A-1234 Vienna");
10
11   // create service proxy with periodic
     rebinding
12   DaiosProxy proxy = querier.
     CreateRebindingProxy(
13       selection, new PeriodicRebinding
         (10000));
14
15   // create input message
16   DaiosMessage inMsg = new DaiosMessage();
17   inMsg.SetString("NumberToPort", nr);
18   DaiosMessage provider = new
     DaiosMessage();
19   provider.SetString("address", prov.
     Address);
20   provider.SetString("name", prov.name);
21   provider.SetLong("id", prov.Id);
22   inMsg.SetComplex("NewProvider",
     provider);
23
24   // invoke service using request/response
     pattern
25   DaiosMessage outMsg = proxy.
     RequestResponse(inMsg);
26
27   // get response from output message
28   DaiosMessage conf = outMsg.
     getComplex("Confirmation");
```

The subscription procedure is shown in Listing 2. Again, it starts by creating a proxy for the corresponding VRESCo service; this time it is the subscription service running on localhost using port number 11111.

The code listing shows three subscription examples. The first one in lines 5-9 uses email notifications to root@localhost when the EPL query in line 6 is matched (i.e., every time a new revision for service 17 is published). The last argument represents the duration of the subscription in seconds (i.e., in this case, the subscription is valid for 30 minutes).

The second example in lines 12-18 declares interest if the availability of revision 23 is greater than 99 percent which is valid until 31.12.2009. In that case, a Web service notification should be sent to net.tcp://localhost:8006/OnVRESCo-Events. The second parameter in line 15 defines where *subscriptionEnd* messages should be sent.

The third subscription in lines 21-27 demonstrates statistical functions over event streams and the sliding window operator which are supported by Esper. In this example, the property Response-Time of *QoSOperationEvents* regarding service operation 33 of service revision 47 is inspected within a time frame of 12 hours. If the average response time is greater than 500 milliseconds any time before 20.09.2009 at 20:09, notifications should be sent per email.

In all three examples, the subscription identifier sid is returned by the subscription service. This identifier can be used to get the status, renew, or unsubscribe from this subscription. Furthermore, the notification payload also contains this identifier so that event consumers can correlate notifications to subscriptions.

Listing 2. Subscription Examples

```
01   IVRESCoSubscriber subscriber =
02       VRESCoClientFactory.
         CreateSubscriber("localhost", 11111);
03
04   // subscribe using email notifications
05   Identifier sid = subscriber.
     SubscribePerEmail(
06       "select * from RevisionPublishedEvent
         where Service.Id = 17",
07       "root@localhost",
08       60 * 30
09   );
10
11   // subscribe using Web service notifications
12   sid = subscriber.SubscribePerWS(
13       "select * from QoSRevisionEvent "+
14       "where Revision.Id = 23 and
         Property='Availability' and Value >
         0.99",
15       "net.tcp://localhost:8005/
         SubscriptionEndTo",
16       "net.tcp://localhost:8006/
         OnVRESCoEvents",
17       new DateTime(2009, 12, 31)
18   );
19
20   // use sliding window and statistics
21   sid = subscriber.SubscribePerEmail(
22       "select * from QoSOperationEvent"+
23       "(Revision.Id=47 and Operation.Id=33
         and Property='ResponseTime') "+
```

```
24       ".win:time(12 hours).stat:uni('Value')
         where average > 500",
25       "root@localhost",
26       new DateTime(2009, 9, 20, 20, 9, 0)
27   );
```

Finally, Listing 3 illustrates a concrete use case for the notification support demonstrating notification-based rebinding as opposed to the periodic rebinding exemplified in Listing 1. Using notifications the rebinding of service proxies can now be forced as soon as the given subscription matches (line 5). In VRESCo, event consumers have to implement the interface IEventNotification (line 1) that defines the event handler method Notify (lines 3-6). This handler method provides access to the subscription identifier and to the actual events.

Listing 3. Notification-based Rebinding

```
01   public class EventSink: IEventNotification
     {
02
03       public void Notify(VRESCoEvent[]
         newEvents, VRESCoEvent[]
         oldEvents,
04       string subscriptionId) {
05       proxy.ForceRebinding();
06       }
07
08   }
```

FUTURE TRENDS

In this chapter, we have presented the foundational work on event notification support in the VRESCo runtime environment. There are several application scenarios and research directions that are enabled by this work:

- **Provenance-aware Applications:** Provenance is an important issue that enables (especially in service-oriented systems) assertions on who did what in applications or business processes. Based on the availability of event data, provenance information can be gathered and used to proof compliance with certain regulations (e.g., laws, standardized processes, etc.) which is addressed by (Curbera, Doganata, Martens, Mukhi, & Slominski, 2008). Complementary to this work, we have introduced the notion of service provenance which defines provenance information of services (Michlmayr, Rosenberg, Leitner, & Dustdar, 2009).

- **SLAs and Service Pricing:** Service pricing models receive increasing attention as more and more services become available. In this regard, service usage can be automatically billed to the user account according to the agreed pricing model. The pricing is also influenced by the SLA defined between the interacting partners, possibly resulting in penalties if providers cannot meet the SLAs. Using event information stored in the event database, the billing information can be easily aggregated for given time periods by issuing queries over the event database. This allows flexible derivation of pricing models based on dynamically negotiated SLAs.

- **Event-based Composition:** The aim of SOA often is to achieve higher level business goals by composing multiple services possibly considering QoS attributes (Zeng, Benatallah, Ngu, Dumas, Kalagnanam, & Chang, 2004). Ideally, this composition should be dynamic in order to allow replacing services if there are alternative services performing the same task. The metadata model described earlier allows defining the differences between similar services that can then be used to mediate between services at runtime. In this regard, events may trigger the composition process. For instance, if the response time of some ser-

vice operation goes beyond a given threshold (which is highlighted by QoS events) the composition engine should restructure the composition using an alternative service providing the same operation. A complementary service composition approach using content-based publish/subscribe to automatically detect the compatibility of services is presented in (Hu, Muthusamy, Li, & Jacobsen, 2008).

CONCLUSION

In typical SOA environments, functional and non-functional properties of services change regularly. Since service providers and consumers are usually loosely coupled, the latter are not informed about such changes and may not be able to access changed services any more. Current registry standards provide basic support for event notifications when registry data changes. However, this does not include QoS attributes and runtime information concerning binding and invocation of services.

In this chapter, we have presented an event notification mechanism for service runtime environments that supports such information. Furthermore, temporal relation between events can be considered using sliding window operators and event patterns. Subscribers can be notified about events using emails or Web service notifications following WS-Eventing. Our approach was integrated into the VRESCo runtime which supports dynamic binding and invocation of services, service versioning, service metadata, and a registry database including publishing and querying services. Additionally, we have shown how the core VRESCo features and the notification support are used in practice. Finally, we have sketched different application scenarios and future research directions that are enabled by our approach.

REFERENCES

Arasu, A., Babcock, B., Babu, S., Cieslewicz, J., Datar, M., & Ito, K. (2008). STREAM: The Stanford Data Stream Management System. In Garofalakis, M., Gehrke, J., & Rastogi, R. (Eds.), *Data Stream Management: Processing High-Speed Data Streams*. Berlin, Germany: Springer.

Cugola, G., & di Nitto, E. (2008). On adopting Content-Based Routing in service-oriented architectures. *Information and Software Technology*, 50(1-2), 22–35. doi:10.1016/j.infsof.2007.10.004

Cugola, G., & Picco, G. P. (2006). REDS: a reconfigurable dispatching system. In *Proceedings of the 6th International Workshop on Software Engineering and Middleware (SEM'06)* (S. 9-16). Portland, OR: ACM Press.

Curbera, F., Doganata, Y. N., Martens, A., Mukhi, N., & Slominski, A. (2008). Business Provenance - A Technology to Increase Traceability of End-to-End Operations. In *Proceedings of the 16th International Conference on Cooperative Information Systems (CoopIS'08)* (S. 100-119). Monterrey, Mexico: Springer.

EsperTech. (2008). *Esper Reference Documentation*. Retrieved August 25, 2008, from http://esper.codehaus.org/

Eugster, P. T., Felber, P. A., Guerraoui, R., & Kermarrec, A.-M. (2003). The Many Faces of Publish/Subscribe. *ACM Computing Surveys*, 35(2), 114–131. doi:10.1145/857076.857078

Hu, S., Muthusamy, V., Li, G., & Jacobsen, H.-A. (2008). Distributed Automatic Service Composition in Large-Scale Systems. *Proceedings of the 2nd International Conference on Distributed Event-Based Systems (DEBS'08)* (S. 233-244). Rome: ACM Press.

Jobst, D., & Preissler, G. (2006). Mapping clouds of SOA- and business-related events for an enterprise cockpit in a Java-based environment. In *Proceedings of the 4th International Symposium on Principles and Practice of Programming in Java (PPPJ'06)* (S. 230-236). Mannheim, Germany: ACM Press.

Leitner, P., Michlmayr, A., Rosenberg, F., & Dustdar, S. (2008). End-to-End Versioning Support for Web Services. In *Proceedings of the International Conference on Services Computing (SCC'08)*. Honolulu, HI: IEEE Computer Society.

Leitner, P., Rosenberg, F., & Dustdar, S. (2009). Daios - Efficient Dynamic Web Service Invocation. *IEEE Internet Computing*, 13(3), 72–80. doi:10.1109/MIC.2009.57

Li, G., Cheung, A., Hou, S., Hu, S., Muthusamy, V., Sherafat, R., et al. (2007). Historic Data Access in Publish/Subscribe. In *Proceedings of the Inaugural International Conference on Distributed Event-Based Systems (DEBS'07)* (pp. 80-84). Toronto, Canada: ACM Press.

Luckham, D. C., & Vera, J. (1995). An Event-Based Architecture Definition Language. *IEEE Transactions on Software Engineering*, 21(9), 717–734. doi:10.1109/32.464548

Mahambre, S. P., Kumar, M. S., & Bellur, U. (2007). A Taxonomy of QoS-Aware, Adaptive Event-Dissemination Middleware. *IEEE Internet Computing*, 11(4), 35–44. doi:10.1109/MIC.2007.77

Michlmayr, A. (2010). Event Processing in QoS-Aware Service Runtime Environments, PhD Thesis, Vienna University of Technology.

Michlmayr, A., Rosenberg, F., Leitner, P., & Dustdar, S. (2008). Advanced Event Processing and Notifications in Service Runtime Environments. In *Proceedings of the 2nd International Conference on Distributed Event-Based Systems (DEBS'08)* (pp. 115-125). Rome: ACM Press.

Michlmayr, A., Rosenberg, F., Leitner, P., & Dustdar, S. (2009). Service Provenance in QoS-Aware Web Service Runtimes. In *Proceedings of the 7th International Conference on Web Services (ICWS'09)*, (S. 115-122). Los Angeles: IEEE Computer Society.

Michlmayr, A., Rosenberg, F., Platzer, C., Treiber, M., & Dustdar, S. (2007). Towards Recovering the Broken SOA Triangle - A Software Engineering Perspective. In *Proceedings of the Second International Workshop on Service Oriented Software Engineering (IW-SOSWE'07)* (pp. 22-28). Dubrovnik, Croatia: ACM Press.

Mühl, G., Fiege, L., & Pietzuch, P. (2006). *Distributed Event-Based Systems*. Berlin, Germany: Springer-Verlag.

OASIS International Standards Consortium. (2005a). *ebXML Registry Services and Protocols*. Retrieved August 22, 2008, from http://oasis-open.org/committees/regrep

OASIS International Standards Consortium. (2005b). *Universal Description, Discovery and Integration (UDDI)*. Retrieved August 22, 2008, from http://oasis-open.org/committees/uddi-spec/

OASIS International Standards Consortium. (2006). *Web Services Notification (WS-Notification)*. Retrieved August 22, 2008, from http://www.oasis-open.org/committees/wsn

Papazoglou, M. P., Traverso, P., Dustdar, S., & Leymann, F. (2007). Service-Oriented Computing: State of the Art and Research Challenges. *IEEE Computer*, *40*(11), 38–45.

Rosenberg, F., Celikovic, P., Michlmayr, A., Leitner, P., & Dustdar, S. (2009). An End-to-End Approach for QoS-Aware Service Composition. In *Proceedings of the 13th IEEE International Enterprise Computing Conference (EDOC'09)*. Auckland, New Zealand: IEEE Computer Society.

Rosenberg, F., Leitner, P., Michlmayr, A., & Dustdar, S. (2008). Integrated Metadata Support for Web Service Runtimes. In *Proceedings of the Middleware for Web Services Workshop (MWS'08)* (S. 24-31). Munich, Germany: IEEE Computer Society.

Rosenberg, F., Platzer, C., & Dustdar, S. (2006). Bootstrapping Performance and Dependability Attributes of Web Services. In *Proceedings of the IEEE International Conference on Web Services (ICWS'06)*, (pp. 205-212). Chicago: IEEE Computer Society.

Rozsnyai, S., Vecera, R., Schiefer, J., & Schatten, A. (2007). Event Cloud - Searching for Correlated Business Events. In *Proceedings of the 9th IEEE International Conference on E-Commerce Technology and The 4th IEEE International Conference on Enterprise Computing, E-Commerce and E-Services (CEC-EEE'07)* (S. 409-420). Tokyo, Japan: IEEE Computer Society.

Sayre, R. (2005). Atom: The Standard in Syndication. *IEEE Internet Computing*, *9*(4), 71–78. doi:10.1109/MIC.2005.74

Treiber, M., & Dustdar, S. (2007). Active Web Service Registries. *IEEE Internet Computing*, *11*(5), 66–71. doi:10.1109/MIC.2007.99

Weerawarana, S., Curbera, F., Leymann, F., Storey, T., & Ferguson, D. F. (2005). *Web Services Platform Architecture: SOAP, WSDL, WS-Policy, WS-Addressing WS-BPEL, WS-Reliable Messaging, and More*. Upper Saddle River, NJ: Prentice Hall.

World Wide Web Consortium. (2006). *Web Services Eventing (WS-Eventing)*. Retrieved August 22, 2008, from http://www.w3.org/Submission/WS-Eventing/

Zeng, L., Benatallah, B., Ngu, A. H., Dumas, M., Kalagnanam, J., & Chang, H. (2004). QoS-Aware Middleware for Web Services Composition. *IEEE Transactions on Software Engineering*, *30*(5), 311–327. doi:10.1109/TSE.2004.11

KEY TERMS AND DEFINITIONS

Service: Services are autonomous, platform-independent entities that can be described, published, discovered, and loosely coupled in novel ways. They perform functions that range from answering simple requests to executing sophisticated business processes requiring peer-to-peer relationships among multiple layers of service consumers and providers. Any piece of code and any application component deployed on a system can be reused and transformed into a network-available service.

Service Registry: Service registries provide repositories of services which contain service descriptions and additional service metadata. Services are published into registries by service providers, while service consumers query these repositories to find services of interest.

Service Provider: Services are provided and maintained by service providers which represent the owner of the service that define who is able to consume these services. Service providers may guarantee functional and non-functional Quality of Service (QoS) attributes which can be defined in Service Level Agreements (SLA).

Service Consumer: Services are invoked by service consumers in various ways. This can range from single invocations to invocations as part of a complex business processes. The technical service descriptions which are necessary to invoke services are found in service registries.

Event: Events represent situations, detectable conditions or state changes which trigger notifications (e.g., a service has changed in the registry).

Notification: Notifications are messages which are triggered by the occurrence of events.

These notifications are sent to all event consumers that have previously subscribed to the corresponding events.

Subscription: Subscriptions are used to declare interest in different events. This can range from simple topic-based subscriptions where events are grouped into different topics, over type-based subscriptions where events are part of event type hierarchies, to content-based subscriptions which enable fine-grained control over event attributes.

Event Producer: Event producers (also called event sources or publishers) are those entities that detect and finally publish events.

Event Consumer: Event consumers (also called event sinks) are those entities that receive notifications when certain events of interest occur. It should be noted that subscribers (i.e., the entity that creates subscriptions) and event consumers can represent different entities.

Event Engine: The event engine (also called event-based infrastructure) is responsible for managing subscriptions and matching of incoming events to stored subscriptions. If subscriptions match incoming events, the corresponding event consumers are notified.

ENDNOTES

[1]   http://servicemix.apache.org
[2]   http://vresco.sourceforge.net
[3]   http://www.codeproject.com/KB/WCF/WSEventing.aspx
[4]   http://www.nhibernate.org

Chapter 13
Event–Based Realization of Dynamic Adaptive Systems

André Appel
Clausthal University of Technology, Germany

Holger Klus
Clausthal University of Technology, Germany

Dirk Niebuhr
Clausthal University of Technology, Germany

Andreas Rausch
Clausthal University of Technology, Germany

ABSTRACT

Dynamic Adaptive Systems are the envisioned system generation of the future. These systems consist of interacting components from different vendors. As the components may join or leave the system at runtime, we have to provide the possibility to automatically compose the system at runtime. Using Request/Reply interaction among components enables us to compose the system based on the directed dependencies between caller and callee at runtime. However this leads to several problems, e.g. frequent polling. Event-based interaction can solve these problems but is missing explicit directed dependencies, which we need to compose the system at runtime. This paper describes our approach of realizing Dynamic Adaptive Systems using Event-based interaction among the components while maintaining automatic system composition. In addition, the paper presents an illustrative application example within a smart city.

INTRODUCTION

Nowadays the trend towards "everything, every time, everywhere" becomes more and more apparent. Electronic assistants, so called "information appliances", like network enabled PDAs, Internet capable mobile phones and electronic books or tourist guides are well known. The continuing progress of all IT sectors towards "smaller, cheaper, and more powerful" mainly enables this trend.

IT components are embedded in nearly every industrial or everyday life object. This trend is driven by new developments in the field of materials science like midget sensors, organic light

DOI: 10.4018/978-1-60566-697-6.ch013

emitting devices or electronic ink and the evolution in communications technology, especially in the wireless sector. As consequence of this trend, almost everyone has small, nearly invisible devices like mobile phones, PDAs, or music players in her or his adjacencies. Furthermore, network technologies like (W)LAN or Bluetooth moved mainstream. This facilitates the connection and combined usage of those devices.

However, in most domains, devices still provide only isolated applications with fixed hardware and software settings for their users. Unfortunately humans live in changing environments, carrying mobile devices, and have varying requirements over time. Hence, they need customizable applications that adapt dynamically to their specific needs in a constantly changing situation.

Ambient Intelligence, Ubiquitous Computing, and Pervasive Computing are research fields that utilize the trend towards "everything, every time, everywhere" and thereby aim for customizable applications. Numerous applications have already been developed in these field spanning diverse application domains like assisted living (Kleinberger et al., 2007; Hanak et al., 2007), assisted working (Zuehlke, 2008), or assisted training (Le et al., 2007) for example.

Dynamic Adaptive Systems consist of many devices, so called components, which together form the application itself. These components can join or leave the system during runtime, as they are not hard wired into the system, requiring a reconfiguration or so called adaptation of the system, due to the available components or even the user's needs or preferences.

Defining these adaptations explicitly at design time is impossible, since these applications are too large to be designed as a whole. Instead different vendors develop components for these systems based on interfaces defined in a common domain model. Performing these adaptations manually (as a user or as an administrator) during runtime is not possible as well, because of the high frequency of situation changes. Therefore, mechanisms are re-

quired which perform these adaptations automatically (Kephart et al., 2003). We call applications which are able to adapt to changing situations during runtime *dynamic adaptive applications*. The term "dynamic adaptive" will be abbreviated as *dynaptive* in the following. Components within a dynaptive application consequently are called *dynaptive components* according their ability to adapt themselves.

Ambient Intelligence, Ubiquitous Computing, and Pervasive Computing aim at performing these adaptations as unobtrusively as possible (Weiser, 1991). To do this, various approaches have been developed which enable the automatic adaptation of systems during runtime based on available components, users' preferences, and context information. In (Dey et al., 2000) for example, a framework has been presented which enables the development of context-aware applications focusing on gathering and processing context information. Furthermore, automatic service discovery mechanisms like described in (Vinoski, 2003) are needed. General component-based middleware solutions like OSGi (Lee et al., 2003; OSGi, 2003) for example already provide dynamic integration of components into a running system to a certain degree. However, they do not provide adaptation of components or systems, but are rather used to implement according infrastructures for dynaptive applications.

In order to be able to describe our autonomous adaptation mechanism for dynaptive applications in contrast to traditional applications, we take a look at the possible interaction patterns as described in (Buchmann et al., 2004) first (See Figure 1).

As depicted in the figure above, we have to regard two dimensions:

a. Who initializes the interaction and
b. The knowledge about the identity among the communication partners.

Figure 1. Possible interaction patterns (Buchmann et al., 2004)

		Initiator of Interaction	
		Consumer	Producer
Knowledge of	Yes	*Request/Reply*	*Messaging*
Counterpart	No	*Anonymous Request/Reply*	*Event-based*

In order to realize dynaptive applications, we have to consider Request/Reply and Event-based communication but extended by an autonomous mechanism to connect components automatically during runtime without user interaction. In the following, we first present an application example, which we carry throughout the chapter in order to show our approach to Request/Reply and Event-Based dynaptive applications. We then describe the Request/Reply interaction pattern with our autonomous adaptation approach to the example and its realization; followed by its drawbacks. Afterwards we present a straightforward approach to Event-Based communication for the example application and why this first approach is insufficient. We then compare both approaches and depict the advantages of having Event-Based interaction patterns for dynaptive applications followed by our approach with Event-Based interaction and autonomous adaptation and how this contributes to enhancing the application example. Finally A short conclusion and future work section round up the paper.

DYNAMIC ADAPTIVE APPLICATION EXAMPLE

In this section we present an example for a dynaptive application. This application example is used to illustrate assets and drawbacks of a Request/Reply and Event-based realization of those kinds of applications. Afterwards we present our approach which integrates the assets of both interaction styles.

Imagine the following situation. Paul and Susan spend their holidays in a small village in the mountains. One day they decide to make a trip to a city in the valley. They enter their car and start their navigation system because they are not familiar with the region. After entering the destination address, they start their trip. As the GPS signal is very good, the system guides them reliable through the streets by providing instructions on the display and via speech output.

They soon approach their destination city and reach a tunnel which is just in front of the city limits. As there is a confusing intersection behind the tunnel, it is important, that the navigation system continues guiding them. When driving through the tunnel unfortunately the GPS signal becomes unavailable and the system adapts to the new situation. Hence, the navigation system considers the car telemetry data like steering angle and speed, to determine the current position of the car; consequently the navigation system still provides useful instructions to Paul and Susan.

To reach their final destination, the system instructs them to drive to the other side of the town. After a while the GPS signal becomes available again. As the smart city also provides a so called *Traffic Message Channel (TMC)*, which offers traffic information like traffic jams or locations of accidents, the navigation system again reconfigures itself in order to consider this traffic information to compute the ideal route.

The navigation system identifies a traffic jam on their direct route and figures out an alternative route to their destination and also sends position and speed of Paul's and Susan's car to the TMC.

Thus, the TMC computes the traffic volume for specific streets and provides this information to navigation systems of other cars in town.

In addition accidents in the smart city can be detected by interpreting telemetry data of each car. This enables an automatic notification of the emergency service whenever an accident occurs. More ambitious scenarios using various kinds of available information like timetables of public transport, time of day, weather, and preferences of individual people are conceivable.

In our scenario we consider the following components; the GPS receiver, the navigation systems, the car telemetry, the TMC and the incident command. We show two approaches to realize such a dynaptive application; First using Request/Reply interaction and second Event-based interaction. Furthermore, we show how these components and consequently dynaptive application as a whole can adapt to different situations.

TWO OPPOSITE EXAMPLE REALIZATIONS

As mentioned before, we present two realization approaches using different interaction styles, especially the Request/Reply and Event-based interaction style. In particular we show how adaptation could be realized. Finally we compare both approaches regarding their potential of providing adaptive behavior.

Introduction to our Approach

Most dynaptive applications are built in a component-based approach (Vinosky, 1997; Szyperski, 1998). Thus, we identified five components for the example described in the previous section. These components are:

- *NavigationComponent*: This component implements the navigation functionality like computing the ideal route by taking

into account GPS data, car telemetry data, or traffic information.

- *GPSComponent*: This component provides the current position of the car.
- *TelemetryComponent*: This component provides information about car telemetry, like steering angles and current speed of the car.
- *TMCComponent*: This component gathers the position of all cars in town in order to provide traffic information to individual cars respectively their navigation systems.
- *IncidentCommandComponent*: The *IncidentCommandComponent* gathers information about accidents automatically, for example by directly accessing the *TelemetryComponent* of the cars in-town. It will then inform police and ambulance automatically.

In order to illustrate the structure and adaptation behavior of dynaptive components, we use the notation as depicted in Figure 2. A blue rectangle represents a single component. There are yellow bars associated to the component. A yellow bar is a so called *ComponentConfiguration*. The green circles represent provided *ComponentServices* and the red semi-circles indicate functionality, which is required by the component – so called *ComponentServiceReferences*. The *ComponentConfiguration* is simply a mapping between "required *ComponentServices*" (*ComponentServiceReferences*) and provided *ComponentServices*. If all "required *ComponentServices*" of a configuration are provided by other components, we call this configuration *active* or *current configuration*. We assume that only one configuration can be active at one time. Therefore, the component developer defines an order across the configurations of a component. In our notion, the most preferred configuration is depicted on top while the least preferred configuration can be found at the bottom. The reason for introducing *ComponentConfigurations* is that components can define

Figure 2. The structure of dynaptive components

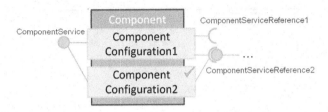

Figure 3. Initial system configuration with active GPSComponent

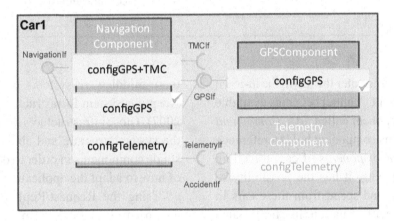

different behaviors for different configurations. Therefore, the behavior of a single component can be designed differently based on the availability of other components.

Request/Reply Realization of the Application Example

A running dynaptive application consists of various interacting components. In order to realize the application example with a Request/Reply interaction style, we are forced to provide each component with a reference to its required *ComponentsService* in order to call remote methods.

Figure 3 shows the initial system configuration when Paul and Susan start their trip (see previous section).

The *NavigationComponent* holds a reference to the *GPSComponent*. Thus, the *NavigationComponent* can get the current position by calling an according method at the *GPSComponent*. The

configuration called *configGPS* within the *NavigationComponent* is active and so the component uses GPS data in order to guide the user. The *TelemetryComponent* is available but inactive as the higher configuration *configGPS* is active, due to the available *GPSComponent*.

Figure 4 shows the situation when Paul and Susan drive through the tunnel. GPS becomes unavailable and therefore, the system as a whole as well as each component has to adapt to the new situation.

Instead of the *GPSComponent*, the *TelemetryComponent* is used to determine the current position. As a consequence, the *ComponentConfiguration* of the *NavigationComponent* changes to *configTelemetry*. Therefore, the internal behavior of the *NavigationComponent* changes accordingly because data like the steering angle and speed are analyzed in order to compute the current position of the car.

Figure 4. System configuration with active Telemetry Component

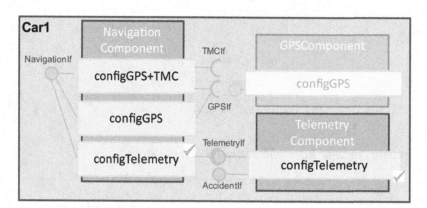

As Paul and Susan enter the town, the in-town Traffic Management Channel becomes available in form of a component called *TMCComponent* (see Figure 5). This component holds a reference to all *NavigationComponents* of cars travelling in-town. Consequently it has the possibility to analyze positions and speed from these cars in order to derive specific situations like traffic jams for example. Additionally each *NavigationComponent* holds a reference to the *TMCComponent* in order to retrieve traffic information. Using the Request/Reply interaction style, both components use remote method calls to retrieve the relevant information from their communication partners.

As mentioned before, an incident command within the town can recognize accidents. Figure 6 illustrates how the *IncidentCommand* interprets the telemetry information provided by the *TelemetryComponents* of the cars.

The *IncidentCommandComponent* has to poll this information regularly by calling an according method at the *TelemetryComponents*[1]. This demonstrates how the system adapts itself to different situations according to the availability of different components.

As components can appear or disappear at any time, a mechanism is needed which enables the adaptation at runtime without user action. Numerous middleware solutions are available which enable automatic adaptation at runtime. One of

these middleware systems is DAiSI (Dynamic Adaptive System Infrastructure) (Niebuhr et al., 2007). Those infrastructures use knowledge about the system structure and about the structure of single components in order to determine the way of how to adapt the application.

Using the Request/Reply interaction style, the structure of the system can be figured out by such a dynaptive system infrastructure. The result is a directed graph of components which can be established by a middleware for example by replacing references.

In the scenario described above, a middleware could automatically put a reference to the *TelemetryComponent* into the *NavigationComponent* if the *GPSComponent* becomes unavailable. Furthermore, the structure of single components can be extracted to determine the *ComponentConfiguration* for each component. All this can be done without requiring the user interaction.

While using the Request/Reply approach the following problems arise. Sensor components like the *TelemetryComponent* or the *GPSComponent*, described in our example, need to be polled by all components, which require their data. Therefore, the components *NavigationComponent* and *IncidentCommandComponent* have to start separate threads to poll for updated values periodically regardless whether values have changed. This leads to an unnecessary communication overhead.

Figure 5. Active system configuration for a smart city with multiple Cars and TMC

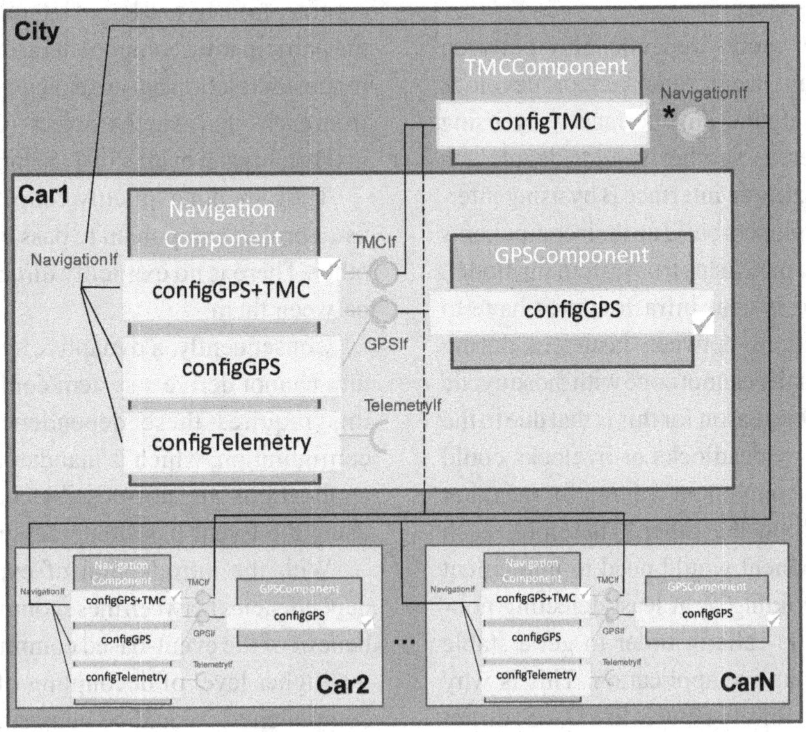

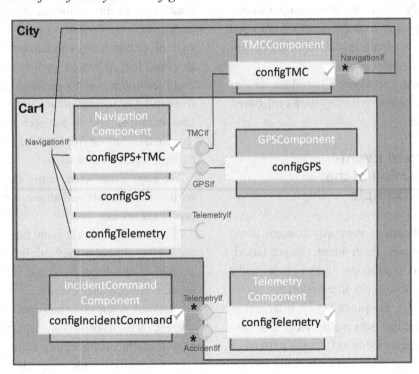

Figure 6. Overview of the final system configuration

Moreover managing concurrent executed threads is an error-prone issue.

Furthermore cyclic dependencies between components may occur; one vendor develops a component realizing an interface A by using interface B whereas another vendor develops a component realizing an interface B by using interface A, since developers build up their components based only on the interfaces from a domain model.

As dynaptive system infrastructures have to manage connections between instances during runtime, they usually cannot cope with those cyclic dependencies. One reason for this is that due to the cyclic dependency deadlocks or livelocks could occur, with each component waiting for the other respectively calling the other. Therefore, each dynaptive component would need to implement a mechanism avoiding or at least detecting non-terminating cyclic calls in order to get a stable and reactive dynaptive application. This is why dynaptive system infrastructures like DAiSI do not establish system configurations containing cyclic dependencies between component instances. By establishing only acyclic system configurations, the involved dynaptive components do not need these component spanning deadlock/livelock detection mechanisms.

In the following section we introduce how to develop the example above using the Event-based interaction style.

Straightforward Event-based Realization of the Application Example

Dynaptive applications normally contain components like sensors which interact Event-based. An Event-based interaction style is essentially different to Request/Reply interaction.

When using a Request/Reply interaction style, there is a clear service provider and service requestor relationship (cf. layers pattern in (Buschmann et al., 2001; Schmidt et al., 2000)) leading to a directed dependency between the

corresponding components. This relationship is contrary to an Event-based interaction style where the participating components are not in a provider requestor relation but are executed independently from each other as active objects (cf. broker pattern in (Buschmann et al., 2001; Schmidt et al., 2000).

They are not explicitly coupled. Instead, they use a broker component to pass messages to each other. There is no explicitly directed dependency between them.

Consequently, a dynaptive system infrastructure cannot derive a system configuration, since this requires these dependencies. Automatic configuration, which is mandatory for dynaptive applications are not possible in general, when using the Event-based interaction style.

With the introduction of explicitly directed dependencies between these components, the main benefit of the event-based communication model – a higher level of decoupling of the interacting components – would be negated.

A straightforward approach would be to connect all components producing or respectively consuming events within the application example to a common event bus (see Figure 7). In this case all components can exchange events using this bus. This approach leads to the disadvantage that the *Navigation-Component* gets *GPSEvents* from all *GPSComponents* within the application example and even irrelevant events like *AccidentEvents* or *NavigationEvents* from other *NavigationComponents*.

To sum it up when using the asynchronous realization as explained above, the following two problems arises:

First, using a single event bus all components are active although they might not be currently needed. Thus they consume power and send/receive unnecessary events.

Second, no clear system configurations can be derived by the dynaptive system infrastructure, so an automatic configuration is impossible.

Our approach addresses these problems by introducing several common typed event buses

Figure 7. Asynchronous realization of the application example

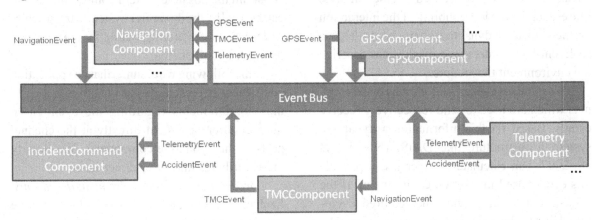

instead of one common bus delivering events of any type to solve the first problem. Additionally we add further specifications to the components describing their requirements to the typed event buses. These requirements can then be interpreted by a dynaptive system infrastructure providing the event channels and connecting components to them.

In the following section we compare both approaches regarding their applicability in realizing dynaptive applications. Afterwards we describe our solution in detail in the section 'Configurable, Event-based Interaction Style Realization'.

Comparing both Realizations of the Application Example

When realizing the application example using the Request/Reply interaction style, we can identify one big advantage: the explicit dependency between provider and requestor enables a dynaptive system infrastructure to derive and establish a system configuration at runtime. This automatic system configuration has been identified as an essential property of dynaptive applications.

Regarding to the above mentioned disadvantages the Request/Replay interaction style will not scale up for real dynaptive applications due to their size and complexity. Therefore, the usage of an Event-based interaction style in addition to

the Request/Reply interaction style is necessary for dynaptive applications.

Embedded systems – which are likely part of dynaptive applications – predominately use the Event-based, event-based interaction style (Hoffmeister et al., 2000; Bass et al., 2003). Furthermore, this enables n:m communication instead of single point to point connections.

However, dynaptive applications can only be configured automatically, if there are directed dependencies between the dynaptive components.

In this case a dynaptive system infrastructure can use these dependencies to derive and establish the system configuration, but these dependencies are not available when using the Event-base interaction style.

Our aim is to build a system, which uses the Event-based interaction style, where a dynaptive system infrastructure is still capable of deriving and establishing a system infrastructure automatically.

CONFIGURABLE, EVENT-BASED INTERACTION STYLE REALIZATION

The Event-based realization we propose follows a slightly different approach in order to be configurable. We do not introduce explicit dependencies between components, as this would contradict the advantages of the Event-based interaction style like especially the low coupling of the interaction partners. Instead, dynaptive components that use an Event-based interaction mechanism describe the environment they require.

This includes a specification of the event channels which are used to send respectively receive events. Based on this information a dynaptive system infrastructure like DAiSI (Niebuhr et al., 2007) can calculate whether it can provide this event-based interaction environment. If the infrastructure can provide the required event-based interaction environment it may connect the dynaptive component with this environment and activate the component.

To enable dynaptive components to specify their requirements for the event-based interaction environment we need to extend the component model introduced roughly in the section describing the Request/Reply realization of the application example. Thereby, each configuration of a dynaptive component is not only characterized by the provided and required services but also by a set of channel ports. These ports specify the required event channels, which the component uses to send respectively receive events. As shown in Figure 8, channel ports are characterized by the interface *InstanceComponentChannelPortIf*. A channel port will be connected to a specific event channel at runtime by the dynaptive system infrastructure, if the requirements of the channel port can be met. Therefore, each channel port specifies these requirements by several attributes.

These requirements are defined by the components themselves, since our view of dynaptive applications is, that they emerge from the direct interaction of several components instead of a separate application component, which explicitly orchestrates the producer and consumer components as it is defined for example in BPEL for workflows (OASIS, 2007). If such an application component exists, this component has requirements regarding producer and consumer components necessary for the application. These centralized requirements

constrain the possible system configurations and can be used by the dynaptive system infrastructure to derive a system configuration for the available dynaptive components.

In the following we assume, that no application component is present. Consequently we specify these requirements regarding the event-based interaction environment directly at the channel ports of the dynaptive components. We classify these attributes into three dimensions:

Port assurances and requirements definitions: Each component has ports, which describe if the port is either a consumer or producer.

Channel content definitions: Each port has channel content definitions.

Interaction partner definitions: Defines the consumer/ producer relationship.

We take a look at sample attributes from these three dimensions in the following. Figure 8 displays the new component model for event based dynaptive application including the defined sample attributes.

First, we take a closer look at attributes describing the *port assurances and requirements definitions*. The *portType* is an example of such an attribute. It can be a consumer or producer port. If a component specifies a channel port as consumer port, this means, that it requires an event channel connected to the port, from which it receives events. A producer port in contrary means that it requires an event channel connected to the port, where it may send events. This contains an assurance (the port processes respectively produces events) as well as an implicit asymmetric requirement (the port requires events from a producer respectively requires a consumer that processes his provided events). Next to the port type, several additional attributes may further specify a channel port. For example, it may be required that "an additional consumer is available as a safety mechanism" for a consumer port or that "an event is provided at least every 2 seconds" for a provider port.

Figure 8. Instance part of the DAiSI component model supporting event-based communication

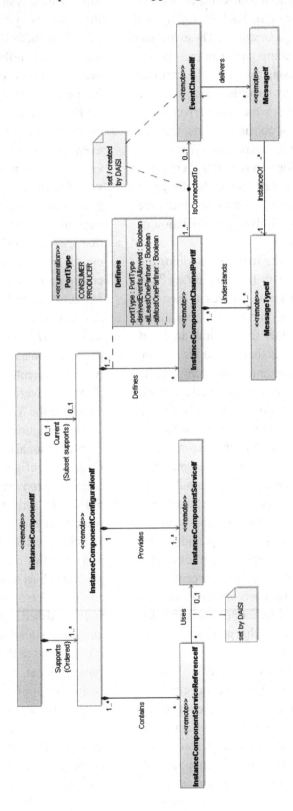

Secondly, we consider attributes describing the *channel content definitions*. Each port defines a set of event types it *understands*. These are the event types, that the component can provide respectively consume using this port. The attribute *derivedEventsAllowed* specifies, whether the port also accepts events, which are derived from another event type. For consumer ports, this defines whether they can be connected to a producer sending a specialization of a required event type. For producer ports this defines whether they can be connected to channels, where consumers are connected who process only a generalization of the produced event types.

At last, we investigate attributes describing the interaction partners. The attribute *atLeastOnePartner* specifies whether a port requires at least one asymmetric interaction partner. For consumer ports this defines whether they require an event producer in order to be activated and vice versa. This is important for example for components providing safety relevant information, like a crash detection component in a vehicle that needs to be sure, that another component is using its information. *AtMostOnePartner* defines that a component can only be connected to one or none asymmetric interaction partner. These attributes could also be parametrizable to specify at most 5 partners. *AcceptSubsets* defines whether a component accepts interaction partners that only understand or produce a subset of event types. In case of a producer port, this implies that it also accepts partners, which understand a subset of their generated events. In case of a consumer port, this implies that it also accepts partners that produce a subset of their required events.

All these specifications are necessary to enable components to specify valid connections to asymmetric interaction partners by connecting their channel ports to appropriate event channels. They can specify for example, that Paul's and Susan's *NavigationComponent* is supposed to receive events only from a *GPSComponent* within Paul's and Susan's car. This enables components to

influence the system configuration established by the dynaptive system infrastructure. The attributes introduced before are not complete but need to be investigated further to extract a minimal subset necessary to configure dynaptive asynchronously interacting systems.

For the realization of the application example in the following, we decided to use asynchronous, event-based interaction style only, although our enhanced component model introduced before would allow the combination with the synchronous interaction style as well. In the application example, we use these attributes e.g. to ensure that a *NavigationComponent* receives *GPSEvents* exactly from one *GPSComponent* in the *configGPS* configuration by specifying *atMostOnePartner* and *atLeastOnePartner* for the associated channel port. When only one *NavigationComponent* and one *GPSComponent* are present, this leads to a situation as depicted in Figure 9 where these two components are connected via a single event channel delivering *GPSEvents* from the *GPSComponent* to the *NavigationComponent*.

In the figure, the navigation component supports additional configurations. First of all, there is a lower configuration, which can be activated, if only *TelemetryEvent* is available. This configuration can be activated in the application example, when Paul and Susan enter a tunnel and the *GPSComponent* therefore, becomes unavailable. The port specifies that the *TelemetryEvents* must be generated by a single source. Therefore, a system configuration as depicted in Figure 10 is derived and established by the dynaptive system infrastructure in this situation. Now a *TelemetryComponent* is connected to the *NavigationComponent* by an event channel.

In the application example with Request/Reply interaction the TelemetryComponent is available but inactive, as telemetry data is not required. The only component that uses this data is the NavigationComponent, but it is not in the according configuration, as a higher configuration is available.

Figure 9. Derived system configuration with only one Navigation Component and one GPS Component present

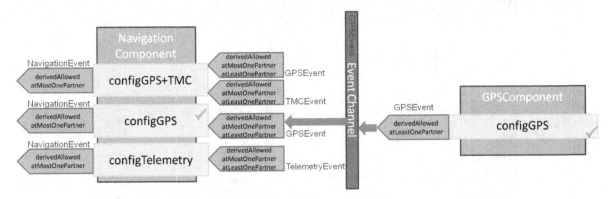

Figure 10. Alternate system configuration, when no GPSComponent is present

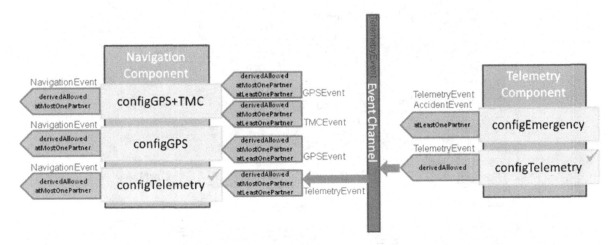

Figure 10 shows the *TelemetryComponent* with the activated *configTelemetry* configuration. The configEmergency configuration is inactive, as there is no component acting as a consumer for the set of *TelemetryEvents* and *AccidentEvents*.

Such a consumer is required for the *configEmergency* configuration, since its channel port does not specify *AcceptSubsets*. However, in the application example the *TelemetryComponent* does not specify this for the channel port of the configuration. We want to make sure that someone processes and understands the events it generates in case of an accident. This is important for the *TelemetryComponent* to be capable of reacting

in an alternate way (e.g. by directly setting up an emergency call to 911) if no such consumer component is available.

When Paul and Susan enter the city, their *NavigationComponent* receives *TMCEvents* from the smart city *TMCComponent*. The *TMCComponent* is connected to *EventChannels* of all in-car *NavigationComponents* (including the one of Paul and Susan) generating *NavigationEvents* containing the speed and direction of the cars. These events are used by the *TMCComponent* to generate *TMCEvents* containing information about traffic jams within the city. The *TMCEvents* are delivered to an own event channel (see

Figure 11. A system configuration within a city providing a TMC service

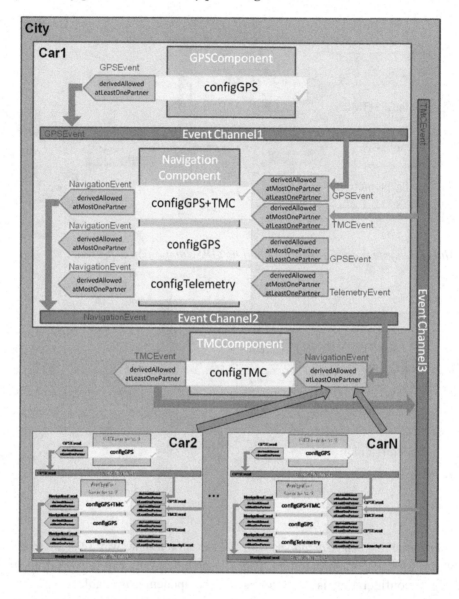

Figure 11, 'EventChannel3') and therefore, they are used to enhance the guidance quality of the *NavigationComponents* by activating their highest configuration (see Figure 11, 'configGPS+TMC').

The system configuration depicted in Figure 12 is established by the dynaptive system infrastructure. In case Paul and Susan are involved in an accident, their *TelemetryComponent* directly delivers *AccidentEvents* to an in-town *IncidentCommandComponent* of the police. The police

directly send an ambulance respectively a police patrol-car. As you can see in 12, the *TelemetryComponent* makes sure that a communication partner exists, who needs exactly *AccidentEvents* together with *TelemetryEvents*.

To sum up, as previously shown in Figure 8, a dynaptive component uses the *InstanceComponentChannelPortIf* with its attributes (e.g. portType, derivedEventsAllowed, atLeastOnePartner, …) to specify its requirements on the event-based

Figure 12. A system configuration from the application example established in case of an accident

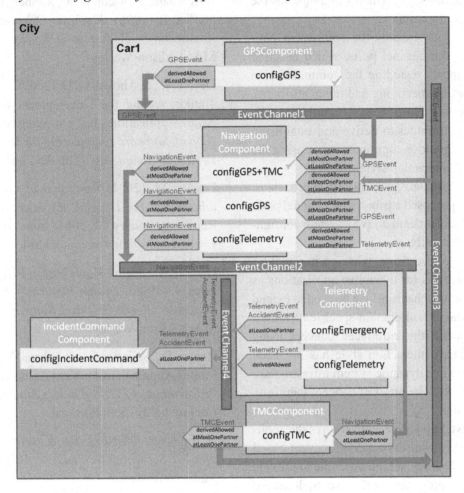

interaction environment. A dynaptive system infrastructure reflects all these requirements, selects the components to be executed, creates all required event-channels, and finally connects the dynaptive components via their ports to these event-channels.

CONCLUSION AND FUTURE WORK

In this chapter we show an example of a dynaptive application and describe the general properties of such dynaptive applications and demonstrate, that automatic reconfiguration is essential. We then introduce two realizations of the application example – one using a Request/Reply interaction style whereas the other one is using an Event-based interaction style. Both realizations have different advantages and drawbacks, which we explain in both sections.

As event-based interaction is common in dynaptive applications, we cannot assume that such an application may only use the Request/Reply interaction style. However, we identified, that a major drawback of the Event-based interaction style is the absence of directed dependencies between interacting components.

Directed dependencies are necessary for a dynaptive system infrastructure to derive a system configuration. Therefore, we identified an Event-based interaction style, which still maintains the automatic configurability.

Our approach extends the component configurations with channel ports, which describe the requirements regarding the communication environment by specifying additional attributes. This enables our dynaptive system infrastructure to use these attributes to derive and establish a system configuration for components, using the Event-based interaction style.

One open issue regarding our approach is, whether the proposed attributes for channel ports are sufficient in order to derive a system configuration automatically.

We need to investigate, what is a minimal set of these attributes for dynaptive applications as well as which mechanisms can be used to extend these minimal set with domain-specific attributes.

Furthermore, the introduced concept using an Event-based interaction style has not been implemented in a dynamic system infrastructure like DAiSI, yet. The proof of concept therefore, is still missing.

We showed a possible realization of the application example using our concept for event-based interaction. Thereby, we are now able to provide Event-based interaction style next to Request/Reply interaction style for dynaptive applications without losing the capability of automatic system configuration.

REFERENCES

Bass, L., Clements, P., & Kazman, R. (2003). *Software Architecture in Practice*. Reading, MA: Addison-Wesley.

Buchmann, A., Bornhövd, C., Cilia, M., Fiege, L., Gärtner, F., & Liebig, C. (2004). *DREAM: Distributed Reliable Event-based Application Management*. Berlin, Germany: Springer.

Buschmann, F., Meunier, R., Rohnert, H., & Sommerlad, P. (1996). *Pattern - Oriented Software Architecture - A System of Patterns*. Reading, MA: Addison-Wesley.

Dey, A. K., & Abowd, G. D. (2000). The Context Toolkit: Aiding the Development of Context-Aware Applications. In *Proceedings of the Workshop on Software Engineering for Wearable and Pervasive Computing*. Limerick, Ireland.

Fowler, M. (2004). *Inversion of Control Containers and the Dependency Injection Pattern*. Retrieved January 23, 2004, from http://martin-fowler.com/articles/injection.html

Gamma, E., Helm, R., Johnson, R. E., & Vlissides, J. (1995). *Design Patterns – Elements of Reusable Object-Oriented Software*. Reading, MA: Addison-Wesley.

Hanak, D., Szijarto, G., & Takacs, B. (2007). A Mobile Approach to Ambient Assisted Living. In *Proceedings of IADIS Wireless Applications and Computing*. Lisbon, Portugal.

Hoffmeister, C., Nord, R., & Soni, D. (2000). *Applied Software Architecture*. Reading, MA: Addison-Wesley.

Jordan, M., & Stewart, C. (2005). Adaptive Middleware for Dynamic Component-level Deployment. In *Proceedings of the 4th workshop on Reflective and Adaptive Middleware Solutions*. Grenoble, France: ACM.

Kephart, J. O., & Chess, D. M. (2003). The Vision of Autonomic Computing. *Computer, 36*(1), 41–50. doi:10.1109/MC.2003.1160055

Kleinberger, T., Becker, M., Ras, E., Holzinger, A., & Müller, P. (2007). Ambient Intelligence in Assisted Living: Enable Elderly People to Handle Future Interfaces. In Proceedings of Universal Access in Human-Computer Interaction. Ambient Interaction (LNCS Vol. 4555/2007, pp. 103-112,), Berlin, Germany, Springer.

Le, A., Jaitner, T., & Litz, L. (2007). Sensor-based Training Optimization of a Cyclist Group. In *Proceedings of the 7th International Conference on Hybrid Intelligent Systems* (pp. 265-270). Kaiserslautern, Germany.

Lee, C., Nordstedt, D., & Helal, S. (2003). Enabling Smart Spaces with OSGi. *Pervasive Computing, 2*(3), 89–94. doi:10.1109/MPRV.2003.1228530

Niebuhr, D., Klus, H., Anastasopoulos, M., Koch, J., Weiß, O., & Rausch, A. (2007). *DAiSI – Dynamic Adaptive System Infrastructure* (IESE-Report No. 051.07/E). Kaiserslautern, Germany: Fraunhofer Institut Experimentelles Software Engineering.

OASIS. (2007). *Web Services Business Process Execution Language*. OASIS Standard Version 2.0 from http://docs.oasis-open.org/wsbpel/2.0/wsbpel-v2.0.pdf

OSGi. (2003). *OSGi Service Platform: The OSGi Alliance*. Amsterdam: IOS Press.

Schmidt, D. C., Stal, M., & Rohnert, H. (2000). *Pattern-Oriented Software Architecture - Patterns for Concurrent and Networked Objects* (*Vol. 2*). Reading, MA: Addison-Wesley.

Szyperski, C. (1998). *Component Software – Beyond Object-Oriented Programming*. Reading, MA: Addison-Wesley.

Vinoski, S. (1997). CORBA: Integrating Diverse Applications within Distributed Heterogeneous Environments. *Communications Magazine*, 46-55.

Vinoski, S. (2003). Service Discovery 101. *Internet Computing*, 69-71.

Weiser, M. (1991). *The Computer for the 21st Century*. Scientific American.

Zuehlke, D. (2008). SmartFactory – from Vision to Reality in Factory Technologies. In *Proceedings of the 17th World Congress, The International Federation of Automatic Control*. Seoul, Korea.

ENDNOTE

[1] An alternate solution of letting the *TelemetryComponents* notify the *IncidentCommandComponent* instead is sketched in the comparison subsection by the keyword inversion of control.

Chapter 14
Event–Driven Mobile Computing with Objects

Tom Van Cutsem
Vrije Universiteit Brussel, Belgium

Wolfgang De Meuter
Vrije Universiteit Brussel, Belgium

ABSTRACT

We motivate why event-driven approaches are suitable to address the challenges of mobile and ubiquitous computing. In particular, we describe the beneficial properties of event-based communication in so-called mobile ad hoc networks. However, because contemporary programming languages feature no built-in support for event-driven programming, programmers are often forced to integrate event-driven concepts with a different programming paradigm. In particular, we study the difficulties in combining events with the object-oriented paradigm. We argue that these difficulties form the basis of what we call the object-event impedance mismatch. We highlight the various issues at the software engineering level and propose to resolve this mismatch by introducing a novel object-oriented programming language that supports event-driven abstractions from the ground up.

INTRODUCTION

This chapter focuses on programming abstractions for mobile computing (Mascolo, Capra, & Emmerich, 2002), a research domain that studies Weiser's vision of ubiquitous computing (Weiser, 1991) from a distributed systems' perspective. Mobile computing applications are deployed on mobile devices (e.g. cellular phones, PDAs, ...) equipped with wireless communication technol-

DOI: 10.4018/978-1-60566-697-6.ch014

ogy (e.g. WiFi, Bluetooth,...). Such devices form so-called *mobile ad hoc networks*, which are characterized by the fact that connectivity between devices is often intermittent (connections drop and are restored as devices physically move about) and the fact that there is little or no fixed support infrastructure, such that devices can often communicate only with physically proximate devices.

Event-based coordination is a natural fit for such networks. Events can be disseminated to multiple nearby interested parties (subscribers) without necessarily knowing the exact identity of these subscribers. This key property of event-based systems is crucial in a ubiquitous computing context, where the identity and number of nearby devices is not known at development time.

Contemporary software is not built using event-driven abstractions from the ground up. Rather, software development is predominantly object-oriented. In this paradigm, distributed applications are expressed in terms of distributed objects sending messages to one another. Thus, an application programmer using a mainstream object-oriented language will be forced to implement his or her own event infrastructure on top of the object-oriented infrastructure.

We will describe that combining object technology with event-based technology is not without problems. In a nutshell, the most important differences are the following. Objects communicate by means of messages, not by means of events. They do so via *remote object references*, which can only refer to a single object throughout their lifetime. Using an event-based system, however, a publisher can send messages to an arbitrary number of subscribers. On the other hand, objects introduce useful abstractions such as request/response interactions that cannot be directly expressed using pure event-based publish/subscribe communication. We have named the combination of these and a number of other issues the *object-event* impedance mismatch, by analogy with the *object-relational* impedance mismatch which describes the difficulties in combining objects with relational databases for the purpose of persistence (Carey & DeWitt, 1996).

We resolve the object-event impedance mismatch by means of a novel object-oriented programming language named *AmbientTalk* (Van Cutsem, Mostinckx, Gonzalez Boix, Dedecker, & De Meuter, 2007) that supports event-based programming from the ground up. For example, AmbientTalk provides asynchronous event notification between objects as a primitive operation. Rather than using a traditional multithreaded con-

currency model, the language features a reactive event loop concurrency model (Miller, Tribble, & Shapiro, 2005). Also, the language supports referencing abstractions that allow a single object to directly refer to an entire group of distributed, proximate objects. Messages sent via such references are automatically treated as events that are broadcast to all objects subscribed to the group.

After having briefly introduced AmbientTalk's main concepts, we show how the language enables one to program in an object-oriented yet event-driven way thus overcoming the object-event impedance mismatch.

The objectives of this chapter are threefold. Our goal is to:

1. give a detailed understanding of why event-driven communication is highly suitable in mobile ad hoc networks.
2. discuss the key differences between event-driven and object-oriented programming, and the consequences thereof for the developer.
3. propose a novel programming model that is both object-oriented and event-driven, allowing developers to use the strengths of both models without unnecessary complications raised by the object-event impedance mismatch.

BACKGROUND

Many systems designed specifically for mobile ad hoc networks (MANETs) adopt an event-driven communication paradigm. We will support this claim by discussing a number of concrete event-driven middleware systems designed for MANETs. Before doing so, we first describe why event-driven communication is so useful in MANETs by showing how it promotes loose coupling between communicating parties.

Coupling Properties of Event-Driven Communication

The advantages of event-driven communication lie in its loose coupling between communicating parties. In mobile ad hoc networks, such loose coupling is important because it allows communicating parties to abstract from the physical connectivity provided by the network, which is in constant flux because devices move in and out of communication range in unpredictable ways. By decoupling a communicating process from the underlying physical connectivity, communication is made more resilient in the face of temporary network disconnections, as will be explained later.

The following three properties are well-known in the literature, especially in the context of publish-subscribe architectures (P. Eugster, Felber, Guerraoui, & Kermarrec, 2003). They pertain to *decoupling* the communicating parties along three dimensions.

Decoupling in Time

Event-driven communication can be made decoupled in time, which implies that communicating processes do not necessarily need to be online simultaneously at the time an event is published. It is mostly achieved by buffering events in message queues, or by introducing a third-party "event broker" which stores events on behalf of the publisher. This allows the publisher's events to be delivered to subscribers even after the publisher has gone offline.

Decoupling in time makes it possible for communicating parties to interact across intermittent connections, because events may be stored (by the publisher or by an event broker) while the network connection is down and transmitted when the connection is restored.

Decoupling in Space

Event-driven communication can decouple processes in space. This means that a process does not necessarily need to know the identity or the total number of processes with which it is communicating.

Decoupling in space has a number of advantages over communication that is tightly coupled in space (such as an RPC call or a remote method invocation), especially in mobile ad hoc networks. First, it allows event-driven communication to be anonymous, in the sense that publishers do not need to know the exact identity of the subscribers interested in their events. When using an event broker, the publisher need only know the identity of the broker. If the publisher directly communicates events to interested subscribers, communication can still be decoupled in space if the publisher can simply broadcast events to all nearby processes, allowing receivers to filter the event themselves based on the relevancy of the event's type or content.

A second advantage of space-decoupling is that it enables publishers to abstract from the total number of subscribers, and that it enables subscribers to abstract from the total number of publishers. This enables applications to adapt more gracefully to changes in the ad hoc network. Publishers and subscribers may be registered or unregistered with an event broker (e.g. because devices move in or out of communication range) without requiring changes in already registered publishers or subscribers.

Synchronization Decoupling

Event-driven communication decouples publishers and subscribers in synchronization. This implies that the control flow of publishers is not blocked (suspended) upon publishing events (i.e. a publisher does not have to wait for subscribers to process the event), or from the point of view of the subscribers, that subscribers do not block the control flow of the publisher while processing an event. Publishers publish events asynchronously, allowing their thread of control to continue processing, while the event broker takes care of delivering the events to registered subscribers.

Synchronization decoupling is important in mobile ad hoc networks, where high network latencies and frequent network disconnections render synchronous RPC-style communication impractical. Asynchronous communication is favorable when network latencies are high, and as previously noted, asynchronously published events can easily be buffered when the network connection with the event broker or subscribers is temporarily down.

State of the Art in Event-Driven Middleware for MANETs

We previously argued that many systems designed for MANETs employ an event-driven communication paradigm. In this section, we demonstrate this by discussing a number of concrete systems. We divide related work into two broad categories: publish/subscribe systems and tuple space-based systems.

Publish/Subscribe Systems

In a publish/subscribe system, publishers and subscribers exchange data by means of event notifications. Subscribers may register to receive certain event notifications based on the type of the event (a.k.a. topic-based subscription) or on the event's contents directly (a.k.a. content-based subscription) (P. Eugster et al., 2003).

Many middleware systems for mobile ad hoc networks adapt the traditional Publish/Subscribe architecture with additional semantics that allow publishers and subscribers to define a physical range to scope the events they want to publish or receive. Two representative examples are

Location-based Publish/Subscribe (LPS) (P Eugster, Garbinato, & Holzer, 2005) and Scalable Timed Events and Mobility (STEAM) (Meier & Cahill, 2003).

Location-Based Publish/Subscribe

LPS is a content-based publish/subscribe architecture designed for nomadic networks (i.e. it is assumed that mobile devices can communicate via a shared infrastructure, e.g. a GSM or GPRS network). In order to scope interactions between devices, event dissemination and reception is bounded in physical space: a publisher defines a *publication range* and a subscriber defines a *subscription range*. Both are independent of the mobile devices' communication range. Only when the publication range of the publisher and the subscription range of the subscriber physically overlap is an event disseminated from the publisher to the subscriber. LPS decouples publishers and subscribers in time, space and synchronization. Decoupling in time is bounded by an event's *time-to-live*: after this timeout period has expired, an event is no longer published.

Scalable Timed Events and Mobility

STEAM is an event-based middleware designed for collaborative applications in mobile ad hoc networks. It shuns the use of centralized components such as lookup and naming services to avoid any dependencies of mobile devices on a common infrastructure. In STEAM, events can be filtered according to event type, event content and physical proximity. STEAM builds upon the observation that the physically closer an event consumer is located to an event producer, the more interested it may be in that producer's events. For example, in a Vehicular Ad hoc Network (VAN), cars can notify one another of accidents further down the road, traffic lights can automatically signal their status to cars near a road intersection or ambulances could signal their right of way to cars in front of them. To this end, STEAM allows events disseminated by producers to be filtered based on geographical location using *proximities* which are first-class representations of a physical range. Proximities may be absolute or relative (i.e. a relative proximity denotes an area surrounding a mobile node, changing as the node moves).

STEAM decouples publishers and subscribers in space and synchronization. It does not decouple them in time: published events are disseminated using multi-hop routing throughout their proximity, after which they disappear. Hence, if a subscriber is not in range at the time the event is disseminated, it will miss the event. Events must be made persistent by repeatedly publishing them.

One.world

One.world (Grimm et al., 2004) is a system architecture developed on top of Java, providing a common execution platform for pervasive computing applications. Again, an asynchronous publish/subscribe style of interaction is promoted because of its loose coupling, making communication decoupled in time, space and synchronization.

Epidemic Messaging Middleware for Ad Hoc Networks

Closely related to publish/subscribe systems are message queuing systems in which the event broker between publishers and subscribers is represented as an explicit queue. Popular message queuing systems, such as the Java Message Service (JMS) (Hapner, 2002) have been adapted for use in a mobile setting. One such adaptation is the Epidemic Messaging Middleware for Ad Hoc Networks (EMMA) (Musolesi, Mascolo, & Hailes, 2005) In JMS, Java components interact asynchronously by posting messages to and reading messages from message queues. This can be used both for point-to-point and publish/subscribe interaction. In JMS, queues are often managed by central servers. EMMA replaces such central servers by a discovery mechanism that allows queues to be discovered in the local ad hoc network.

EMMA, like JMS, distinguishes between durable and non-durable subscriptions to message queues. A durable subscription remains valid upon disconnection. Non-durable subscriptions are cancelled upon disconnection and the subscription must be made anew upon reconnection. In JMS, the server buffers events while a durable subscriber is disconnected. In EMMA, events for disconnected subscribers are not buffered but rather sent using an asynchronous *epidemic routing protocol*. Using this protocol, messages are broadcast to each host in range, which in turn sends them to all hosts in its range, and so on. Epidemic routing does not guarantee message delivery, but the delivery ratio increases as the number of nodes in the ad hoc network increases. If a message is flagged as *persistent*, the sender is notified of successful delivery via an acknowledgement.

Communication in EMMA is naturally synchronization-decoupled using message queues. It is space decoupled thanks to the use of topics to describe publish/subscribe queues. Thanks to its automatic discovery management of queues, it is suitable for use in pure ad hoc networks.

Tuple Space-Based Systems

Tuple spaces have originally been introduced in the coordination language Linda (Gelernter, 1985). In the tuple space model, processes communicate by inserting and removing tuples from a shared tuple space, which acts like a globally shared memory. Because tuples are anonymous, they are taken or copied from the tuple space by means of pattern matching on their content. Tuple space communication is decoupled in time because processes can insert and retract tuples independently. It is decoupled in space because the publisher of a tuple does not necessarily specify, or even know, which process will extract the tuple. This makes Linda ideal for coordinating loosely-coupled processes. The original Linda model has since been ported to contemporary languages such as Java resulting in artifacts such as IBM's TSpaces (Lehman et al., 2001) and Sun Microsystems' Javaspaces (Freeman, Arnold, & Hupfer, 1999).

Linda in a Mobile Environment

Tuple spaces have received renewed interest by researchers in the field of mobile computing. One shortcoming of the original tuple space model in light of mobile computing is the fact that synchronization decoupling is violated because there exist synchronous (blocking) operations to extract tuples from the tuple space. However, as the need for total synchronization decoupling became apparent for mobile networks, mobile computing middleware such as Linda in a Mobile Environment (LIME) (Murphy, Picco, & Roman, 2001) extends the basic model with *reactions* which are callbacks that trigger asynchronously when a matching tuple becomes available in the tuple space.

Naturally, a globally shared, centralized, tuple space does not fit the hardware characteristics of mobile ad hoc networks. Adaptations of tuple spaces for mobile computing, such as LIME, introduce agents which have their own, local *interface tuple space* (ITS). Whenever their host device encounters proximate devices, the ITS of the different agents is merged into a federated *transiently shared* tuple space, making tuples in a remote agent's tuple space accessible while the connection lasts.

Mobile Agent Reactive Spaces and Tuples on the Air

Other adaptations of the tuple space model for mobile ad hoc networks include Mobile Agent Reactive Spaces (MARS) (Cabri, Leonardi, & Zambonelli, 2000) and Tuples on the Air (TOTA) (Mamei & Zambonelli, 2004). These approaches circumvent the need for a globally shared tuple space by allowing agents (in the case of MARS) or tuples (in the case of TOTA) to migrate between connected hosts. Using migration, agents and tuples can be co-located to ensure a stable communication.

In MARS, each device hosts a tuple space and agents can only access that local tuple space. To access another tuple space, agents can migrate between hosts. MARS features a metalevel tuple space that allows programs to register reactions: callbacks that trigger whenever agents perform a read and/or write operation on the baselevel tuple space.

In TOTA, rather than merging local tuple spaces upon network connection, tuples are equipped with a *propagation rule* that determines how the tuple migrates from one tuple space to another. Hence, in TOTA, agents can access one another's tuples because it are the tuples themselves that propagate through the network as connections are established. Like LIME, it augments the tuple space model with a form of event notification to notify agents when certain tuples arrive in their tuple space.

Tuple spaces act as a middle man between different processes. As a result, there is no notion of a reference to any particular process. Tuple space-based communication is necessarily global to all processes sharing the tuple space, which may lead to unexpected interactions between concurrently communicating processes.

In the following section, we discuss the difficulties of combining event-driven communication abstractions with the object-oriented programming paradigm. Subsequently, we show how these difficulties can be overcome, by introducing a novel programming language named AmbientTalk, which successfully combines objects with events.

THE OBJECT-EVENT IMPEDANCE MISMATCH

In the previous section, we have discussed that event-driven communication is effective in ad hoc networks because it minimizes the dependencies between publishers and subscribers in time, space and synchronization. Alternative communication mechanisms (e.g. RPC or remote method invocation) do not engender such loose coupling (P. Eugster et al., 2003). Traditional remote method invocations couple participants in time (to perform the invocation, the receiver object must be online), space (the receiver must be known, and there is only one receiver) and synchronization (the invocation is performed synchronously). Saif and Greaves (2001) provide additional arguments against the use of RPC-based communication in ubiquitous computing systems.

Given the above arguments, we are forced to conclude that:

1. Object-oriented communication by means of traditional message passing schemes fails to provide a total decoupling of processes, which becomes a major obstacle when used in a distributed system connected by means of a mobile ad hoc network.
2. Publish/Subscribe and Tuple Space-based systems enable the best decoupling of processes, which makes them highly suitable for use in mobile ad hoc networks.

Combining these two facts, it appears that one is forced to abandon the object-oriented message passing abstraction if one wants to express scalable coordination between processes in a MANET. However, if the non-distributed part of an application is written in an object-oriented language, the overall application is then forced to combine two different paradigms: the object-oriented paradigm must interact with an alien communication paradigm (tuples or events).

A multi-paradigm approach is not necessarily a bad approach. However, its suitability depends on the difficulty of combining the paradigms involved. Below, we argue that combining objects with events is far from trivial. We claim that the lack of integration between the concepts from both paradigms leads to what we call the *object-event impedance mismatch*, analogous to the way object persistence suffers from the infamous *object-relational* impedance mismatch (Carey & DeWitt, 1996).

The object-relational impedance mismatch is caused by the fundamental differences between modeling data as objects and modeling data as tuples that are part of relations. For example, objects encapsulate their state, enabling operations to be polymorphic. Tuples expose state, enabling efficient and expressive filtering, querying and aggregation of data. Objects refer to one another via references, while tuples are associated with one another via foreign keys. Identity is fundamental to objects, while tuples lack any inherent form of identity, and so on. We discuss similar differences, but rather than contrasting objects with the relational model, we will contrast objects with event-driven communication models. *Figure 1* contrasts object-oriented with event-driven communication. The highlighted differences are discussed in each of the following sections.

Specific vs. Generic Communication

In object-oriented programming languages, objects communicate by means of message passing. The interface of an object usually corresponds to the set of visible methods it (or its class) defines. Messages encapsulate a "selector", which usually corresponds to the name of the method to be invoked on the object to which the message is sent. In object-oriented programming, therefore, communication between objects is expressed in terms of very *specific* operations. The generic acts of *sending* and *receiving* these messages are entirely hidden by the language.

In a distributed object-oriented language, sending and receiving messages to and from *remote* objects is most often done implicitly by means of the regular message passing semantics already provided by the language. In an object-oriented language, remote communication can be succinctly expressed as receiver.selector(arg). No explicit send operation is required. Likewise, it is often not necessary to introduce an explicit receive statement: received messages implicitly lead to the invocation of a method on the receiver object.

In event-driven systems and in tuple spaces, communication between processes is not in

Figure 1. Contrasting object-oriented communication (on the left) with event-driven communication (on the right)

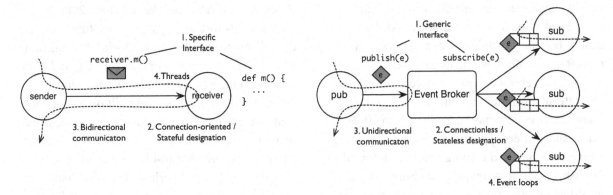

terms of specific operations, but rather in terms of very generic operations, whose arguments are the events or tuples to be communicated. This is very obvious in tuple spaces, where communication is performed in terms of explicit in(tuple) and out(tuple) operations. In an event-driven system, events are often dispatched to the event broker (the infrastructure between event producers and consumers) by means of a generic broker. publish(event) invocation. Likewise, event reception is often an explicit operation:

```
broker.subscribe(new Subscriber() {
  public void reactTo(Event e) {
    // handle incoming event
  }
})
```

One advantage of a generic communication interface is that it is much easier to express general patterns of communication. This is because such an interface already automatically abstracts from the details of the data being communicated. To express generic patterns via an object-oriented interface, one generally requires reflection to be able to intercept messages without reference to their specific selector and arguments (e.g. via Smalltalk's doesNotUnderstand: protocol (Foote

& Johnson, 1989) or Java's dynamic proxy classes (Sun Microsystems, 1999)).

In short, if objects are to communicate with one another via events, they must abandon their otherwise specific communication interface in favor of the generic interface promoted by the event system.

Connection-Oriented vs. Connectionless Designation

In a distributed object-oriented program, objects communicate via point-to-point channels known as (remote) object references. This abstraction provides limited or no support for space-decoupled (anonymous and one-to-many) communication. However, it remains a useful communication abstraction with two important properties. First, a remote object reference is a *connection-oriented* communication channel, which allows the sender to know the identity of the object to which it sends messages. Second, it is a *stateful* communication channel ensuring that multiple messages sent via the same reference are processed by the *same* receiver. Later messages are sent via the reference in the understanding that messages sent earlier are processed first.

Publish/subscribe systems and tuple spaces feature stateless one-to-many communication by default. In a mobile ad hoc network, this allows publishers to simply broadcast events to nearby interested subscribers without knowing their identity or their total number. At the same time it becomes difficult to express connection-oriented communication, as an event or tuple becomes accessible to *all* registered subscribers, or stateful communication, because multiple consecutive events may be received by a *different* set of subscribers. There exist ways of building connection-oriented communication on top of a publish/subscribe system (e.g. private event topics encoding a connection), but these abstractions are second-class, just like connectionless one-to-many communication is a second-class abstraction in an object-oriented programming language.

Note that the notion of a connection-oriented communication channel does not necessarily imply a *secure* communication channel. The above discussion deals with scoping and coupling issues on a software engineering level, not with issues such as one process intentionally "eavesdropping" on a communication channel between other processes.

In short, if objects are to communicate with one another via events, they must abandon the remote referencing abstraction in favor of connectionless communication via the event broker. This makes the receiver(s) of a message anonymous, which enables space-decoupled communication, but at the same time may lead to interference between communicating objects.

Bidirectional vs. Unidirectional Communication

Object-oriented programs communicate via message passing which fosters a request/reply style of communication. Even though this request/reply style is often implemented by means of synchronous (remote) method invocation, much research in concurrent object-oriented programming has been devoted to maintain the request/reply interaction pattern while relaxing the synchronization con-

straints (e.g. future-type message passing in ABCL (Yonezawa, Briot, & Shibayama, 1986), wait-by-necessity in ProActive (Baduel et al., 2006)).

Publish/subscribe systems and tuple spaces decouple processes by essentially introducing pure asynchronous one-way operations (e.g. publishing an event, writing a tuple). Request/response interaction can of course be built on top of such systems, e.g. by manually correlating request and response events/tuples by means of an identifier. Again, these are second-class abstractions.

Conversely - and perhaps less obviously - an asynchronous, unidirectional event notification also has to be represented by means of second-class abstractions in an object-oriented language. Signaling an asynchronous event is often performed by *synchronously* invoking a "notification" method that returns no result. The first-class asynchronous notification of an event system is thus represented confusingly as a synchronous method invocation in an object-oriented system.

In short, if objects are to communicate with one another via events, they must abandon the bidirectional request/response message passing abstraction in favor of unidirectional event notification.

Threads vs. Event Loops

In an event-driven system, events are usually delivered to an application by an *event loop* which is an infinite loop that accepts incoming events and dispatches them to the appropriate event handler. However, the integration of event delivery with multithreaded object-oriented languages often leaves much to be desired. The archetypical integration represents event handlers as "listener" or "observer" objects whose methods are invoked directly by the event loop. Because of the synchronous method invocation semantics predominant in object-orientation, it is the thread of control of the event loop itself that executes the method. The following code snippet depicts a canonical example of event notification in Java:

```
// code executed by application
thread
broker.subscribe(new Subscriber() {
  public void reactTo(Event e) {
    // code executed by event notifi-
cation thread
  }
});
```

This style of event notification has two important consequences:

- Event handler objects must be made multiple thread-safe, i.e. they require synchronization constructs to prevent data races when they concurrently access state manipulated by application-specific threads. Also, because thread management lies outside the control of the application, it is entirely implicit in the code whether multiple events are signaled to registered event handlers concurrently or sequentially. A change in the thread management may thus introduce race conditions into the application.
- If the event handler object uses the event loop's thread of control to perform application-level computation, it can make the event loop unresponsive. A testament to this is the documentation of the event-driven Java GUI construction framework Swing that advises developers to structure their applications such that listener methods relinquish control to the framework as soon as possible (Sun Microsystems, 2008).

Event loops provide the programmer with the ability to consider the handling of a single event as the unit of concurrent interleaving. The major strength of this model is that it significantly raises the level of abstraction for the developer: rather than having to consider the possible interleaving of each *basic instruction* in each method body, the programmer need only consider the interleaving of each distinct event. It has been argued even in the context of thread-based object-oriented concurrency models that complete mutual exclusion of each method of a concurrently accessible object ought to be the norm (Meyer, 1993).

Event loops also have their drawbacks. Event delivery is an asynchronous process, and most event-driven systems cannot succinctly express the overall control flow of an application. Rather, the control flow is dispersed across many different event handlers, a phenomenon known as *inversion of control* (Haller & Odersky, 2006) or *stack ripping* (Atul, Jon, Marvin, William, & John, 2002) in the literature. This often results in code and data that is fragmented across calls and callbacks.

Choosing between thread-based or event-based concurrency is the topic of long-standing debates in the literature, and we are certainly not the first to contrast these two systems. A well-known talk by Ousterhout provides a more general discussion comparing the assets and drawbacks of threads and events (Ousterhout, 1996). Miller (2006) studies the concurrency control properties of both models in detail.

In short, if objects are to communicate with one another via events, the event broker becomes an additional source of concurrency in the object-oriented program. In multithreaded languages, dealing with this additional source of concurrency is non-trivial, as it requires the programmer to carefully insert additional locks to ensure overall thread-safety. What is needed is an alternative model of concurrency control between objects that allows one to easily compose new sources of events with an existing application.

Reconciling Objects with Events

In this section, we have contrasted the communication properties of object-oriented and event-based publish/subscribe models. The simplest solution to resolve the object-event impedance mismatch is to discard either objects or events and to resort to a single-paradigm solution where only one of both is used. However, it should be clear from the above discussion that both paradigms have their merits. Neither does one solve the object-relational impedance mismatch by discarding objects or relational databases.

To resolve the object-event impedance mismatch, we will develop a novel programming model that tries to *unify* as much concepts as possible in the event-driven domain with concepts in the object-oriented domain. In the unified model, the programmer can seamlessly use event-driven communication in an object-oriented program. Our unified model does need to make tradeoffs, and will combine aforementioned properties of objects and events as follows:

- Objects should be able to communicate via events by means of the familiar, *specific* communication interface afforded by message passing. This makes communication more concise because there is no need for explicit publish and subscribe operations.
- Objects should be able to communicate via events by means of the familiar object referencing mechanism. While sometimes stateful connection-oriented communication is still required, object references must be augmented such that they can directly express stateless, *connectionless* communication.
- Objects should be able to communicate via events and still retain the ability to perform the *bidirectional* interactions afforded by method invocation.
- Objects should be equipped with an *event loop* concurrency model that can adequately cope with the additional sources of concurrency introduced by event brokers.

In the next section, we show how the Ambient-Talk programming language achieves the above requirements. Afterwards, we revisit the object-event impedance mismatch and how it is resolved by AmbientTalk by unifying object-oriented with event-driven communication.

AMBIENTTALK

AmbientTalk is a programming language embedded in Java. The language is designed as a distributed scripting language that can be used to compose Java components that are distributed across a mobile ad hoc network. The language is developed on top of the Java 2 Micro Edition (J2ME) platform and runs on handheld devices such as smart phones and PDAs. Even though AmbientTalk is embedded in Java, it is a separate programming language. The embedding ensures that AmbientTalk applications can access Java objects running in the same JVM. These Java objects can also call back on AmbientTalk objects as if these were plain Java objects.

The most important difference between AmbientTalk and Java is the way in which they deal with concurrency and network programming. Java is multithreaded, and provides either a low-level socket API or a high-level RPC API (i.e. Java RMI) to enable distributed computing. In contrast, AmbientTalk is a fully event-driven programming language. We discuss this is more detail below. Furthermore, network programming in AmbientTalk is only possible via an asynchronous API based on message passing. AmbientTalk is designed particularly for ad hoc networks:

- In an ad hoc network, objects must be able to discover one another without any infrastructure (such as a shared naming registry). Therefore, AmbientTalk has a service discovery engine that allows objects to discover one another in a peer-to-peer manner.
- In an ad hoc network, objects may frequently disconnect and reconnect because of network partitions. Therefore, AmbientTalk provides fault-tolerant asynchronous mes-

sage passing between objects: if a message is sent to a disconnected object, the message is buffered and resent later, when the object becomes reconnected. Other advantages of asynchronous message passing over standard RPC is that the asynchrony hides latency and that it keeps the application responsive (i.e. the event loop is not blocked during remote communication and is free to process other events).

Below, we first discuss AmbientTalk's event loop concurrency model in more detail. Subsequently, we explain AmbientTalk's support for publish/subscribe interaction by means of a new kind of object references named ambient references.

Event Loop Concurrency

In AmbientTalk, code is not executed by threads but rather by event loops. Each event loop perpetually processes events from an event queue and dispatches these events to appropriate event handlers. This works similar to how GUI frameworks (e.g. Java AWT or Swing) operate. All concurrent activities in the system are represented as events that are asynchronously handled by event loops. Unlike threads, event loops have no mutable shared state and communicate strictly by means of events. Because of this, event loops never have to lock state such that locks become unnecessary. Because there are no locks, event loops can never suspend on one. Therefore event loops cannot deadlock one another.

AmbientTalk maps the above concepts from event-driven programming onto concepts from object-oriented programming. The mapping is based upon the model of communicating event loops introduced in the E programming language (Miller et al., 2005). This model is itself based on the well-known actor model of computation (Agha, 1986). In AmbientTalk, event loops are known as *actors*. The event queue of an actor is

Table 1. Mapping event-driven concepts onto object-oriented concepts

	Event-driven Programs	**AmbientTalk**
Unit of concurrency	Event Loops	Actors
Communication via	Events	Messages
To send	Fire an event	Send a message asynchronously
To receive	Register a callback	Define a method
To handle	Invoke a callback	Invoke a method

called a *mailbox*. Events themselves are represented as *messages*. Firing an event is done by *asynchronously sending* a message to an object. The message is then enqueued in the mailbox of its actor. The actor handles the message by invoking the corresponding method on the receiver object. Hence, event handlers are represented as *methods* of the objects contained in the actor. Table 1 summarizes the mapping of event-driven onto object-oriented concepts in AmbientTalk.

In AmbientTalk, objects can communicate either by synchronous message sending (syntax: obj.m()) or by asynchronous message sending (syntax: obj<-m()). A single actor may contain multiple regular objects, and it is possible for objects contained in one actor to refer to objects contained in other actors. Such references that span different actors are named *far references* and only allow asynchronous access to the referenced object. Synchronous access to an object via a far reference raises a runtime exception. Any messages sent via a far reference to an object are enqueued in the message queue of the actor owning the object and processed by the owner itself. This is illustrated in Figure 2. The dotted lines represent an actor's event loop that perpetually takes messages from its mailbox and synchronously executes the corresponding methods on its objects. The control flow of an actor's event loop never "escapes" its actor boundary. When object

Figure 2. AmbientTalk actors as event loops

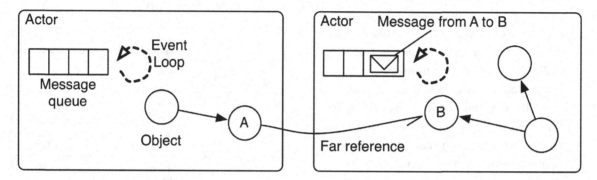

A sends a message to object B via a far reference, the message is enqueued in the message queue of B's actor, which eventually processes it.

The following code snippet illustrates how AmbientTalk can be used to query a WeatherService object representing a weather service in the ad hoc network to retrieve weather information for a given city:

```
when: weatherSvc<-
getWeather("Brussels, Belgium") be-
comes: { |info|
  // update weather information in
the user interface
}
// code hereafter is executed immedi-
ately after the message send
```

The above code consists of an asynchronous message send and an event handler to process the reply. We assume that the weather service object is accessible via the weatherSvc variable, which denotes a remote AmbientTalk object that wraps a Java component implementing the weather service.

When the getWeather message is received by the remote weatherSvc object, that object's getWeather method is invoked. The return value of this method is used as the reply to the message. This reply is signaled asynchronously to the caller. The when:becomes: control structure is used to install an event handler that can process this reply.

The return value is passed to this event handler (cf. the info variable in the example). Note that code following the when:becomes: control structure is executed immediately after sending the message, before the event handler code is fired.

To summarize, AmbientTalk is a scripting language embedded within Java that allows programmers to easily write distributed event-driven programs. As illustrated by means of the above example, event-driven concepts are represented in AmbientTalk by means of object-oriented concepts such as messages and method invocations. However, what is still lacking is a means to perform a publish/subscribe style of communication with remote objects. This is the subject of the following section.

Ambient References

In the weather service example, we assumed that the weatherSvc variable contained a reference to the remote weather service object. In this section, we discuss how such a reference can be easily constructed by means of *ambient references*. An ambient reference is a reference to one ore more service objects of a given type that are available in the ad hoc network. Usually this type corresponds to a Java interface that the service(s) implement. For example, a reference to a nearby weather service can be constructed as follows, assuming that WeatherService represents a Java interface:

```
def weatherSvc:= ambient: WeatherSer-
vice;
```

The variable weatherSvc contains an ambient reference, which is a proxy for any nearby object exported (made available) via the WeatherService interface. The AmbientTalk runtime automatically starts a service discovery request and keeps track of the available exported objects that can be referenced by the ambient reference.

Ambient references decouple sending and receiving objects in synchronization, space and time. Decoupling in synchronization is achieved because one may only send messages asynchronously via an ambient reference. The sender object does not wait until recipient service objects receive the message.

Decoupling in space is achieved because remote service objects are addressed by means of types (e.g. Java interfaces). A sender object does not need to know the identity of the service object to which the ambient reference refers, only its type. Ambient references can also be used to send a message to not just one, but to *all* available matching services in the ad hoc network. To broadcast a message, it must be annotated with an @All annotation, and results must be processed as follows:

```
def sensors:= ambient: Temperature-
Sensor;
whenAll: sensors<-getTemperature()@
All becomes: { |values|
   // values is an array of tempera-
ture sensor readouts
}
```

In this example, all nearby TemperatureSensor services are asked to sense the current temperature. Results from a broadcasted message may then be processed by means of the whenAll:becomes: control structure.

By default, when sending a message to an ambient reference for which no matching services

are available, the message is not received by any service and is discarded. To introduce time decoupling, messages may be annotated with an @Expires(timeout) annotation. Here, timeout is a timeout period (in milliseconds), describing how long the message should remain available to be received by remote services. The ambient reference will then buffer the message until it expires.

How is it that ambient references combine publish/subscribe communication with object-oriented communication? Objects communicate via object references. In publish/subscribe systems, components communicate via event brokers (the mediator between publishers and subscribers). Ambient references *represent the event broker as an object reference*. Therefore, an ambient reference can be regarded as a little publish/subscribe engine of its own. In a publish/subscribe system, publishers send events to an event broker, which is responsible for delivering those events to interested subscribers. With ambient references, the act of sending a message to an ambient reference represents the publication of an event for consumption by nearby interested objects, which implicitly subscribe to these messages by being exported under a universally agreed upon type.

Previous and Related Work

The AmbientTalk language discussed in this chapter is more accurately named AmbientTalk/2, because it is a revised version of an earlier language with the same name, which we shall refer to as AmbientTalk/1 (Dedecker, Van Cutsem, Mostinckx, D'Hondt, & De Meuter, 2006). AmbientTalk/1 is a programming language that distinguishes between active and passive objects. Active objects are the unit of concurrent and distributed computing and resemble AmbientTalk/2's actors. They contain regular (a.k.a. passive) objects. Unlike in AmbientTalk/2, in AmbientTalk/1 passive objects cannot be referred to from within another active object. Only active objects can be addressed remotely. Passive objects also cannot be passed by (far) reference between active objects, they are always passed by copy.

A novelty of AmbientTalk/1 in comparison to similar languages based on active objects is that the language introduces additional mailboxes to store outgoing messages and previously sent or received messages (each active object has an "inbox", an "outbox", a "sentbox" and a "receivedbox") and that these mailboxes are made accessible to the programmer. By manipulating these mailboxes, an application can monitor and change messages received in the past and to be processed in the future. This allows one to express among others synchronization, failure handling and replication strategies. AmbientTalk/2 does not feature these mailboxes explicitly, but if required they can be reconstructed by the programmer. AmbientTalk/1 features no embedding with Java, such that it cannot be used to compose Java objects across an ad hoc network.

A detailed explanation of ambient references is beyond the scope of this chapter. An extensive explanation of their design and implementation can be found elsewhere (Van Cutsem, 2008). The ambient reference abstraction is not the first one that tries to unify object-orientation with a publish/ subscribe interaction style. The ActorSpace model (Callsen & Agha, 1994) was an early attempt at augmenting actors with the space-decoupling properties of the tuple space model. Distributed Asynchronous Collections (P. T. Eugster, Guerraoui, & Sventek, 2000) combine objects with events by representing the event broker as an object-oriented *collection*, rather than as an object *reference*. Many to many invocations (M2MI) is a novel paradigm for communication in wireless ad hoc networks (Kaminsky & Bischof, 2002). M2MI introduces *handles*, special references akin to ambient references that enable one to unicast, multicast or broadcast invocations to proximate objects. As is the case with ambient references, Java interfaces are used to anonymously designate

the receivers of the invocation. M2MI handles have no equivalent for the @Expires annotation supported by ambient references, and thus do not feature any time decoupling.

Limitations

The main limitation of our language-centric approach using AmbientTalk is its lack of interoperability and its platform-dependence. AmbientTalk objects can only communicate with remote objects that are also written in AmbientTalk, using a protocol and message format specific to the language. Furthermore, the implementation is very dependent upon Java technology (although the language itself is technology-neutral) and difficult to configure (e.g. it is tied to a specific network protocol stack). In addition, the current implementation lacks extensive performance optimizations.

Ambient references currently lack a number of features common in commercial event queuing systems. Such features include ranking available services according to their properties, fuzzy matching of service descriptions, persistent events and transactions, and encrypted communication to ensure privacy. Also, the matching based on types assumes that all devices in the network know about and agree upon the names of these types. This assumption may not always hold in open, non-administered, ad hoc networks.

THE OBJECT-EVENT IMPEDANCE MISMATCH REVISITED

We now turn our attention once more to the issues in combining objects with events, which we previously named the object-event impedance mismatch. In each of the following sections, we revisit the problems described previously and show how they are dealt with in AmbientTalk.

Specific vs. Generic Communication

Previously, we contrasted the specific communication interface of objects with the generic communication interface that is often provided by publish/subscribe architectures. Ambient references maintain the specific interface of objects by representing events as asynchronously sent messages. This has both drawbacks and advantages. A drawback is that this makes the representation of events explicit in the code. In the weather service example, the event getWeather(city) is represented as a message send of which the message selector identifies the kind of event and the message arguments constitute the event data. Both sender and receiver need to be fully aware of the structure (selector and arguments) of the message.

Representing events as messages has two major advantages. First, event publication can be unified with object-oriented message sending. Rather than having to explicitly construct an event as an object of a certain type and then invoking a generic publish(Event) method, a message is asynchronously sent to an object representing the subscribers and the event's type becomes the selector of that message. Second, event handling can be unified with object-oriented method invocation. Rather than having to represent event notification by subscribing a generic reactTo(Event) callback method to an event type, it is represented by having an event loop invoke a method whose selector corresponds to the kind of event.

In short, AmbientTalk resolves the impedance mismatch by representing events as asynchronously sent messages, thus maintaining the specific communication interface of object-oriented message passing.

Connection-Oriented vs. Connectionless Designation

Recall from our previous discussion that while object-oriented referencing abstractions provide connectionless communication but no space decoupling, pure event systems provide space decoupling but do not cater to any connection-oriented communication.

AmbientTalk naturally supports connection-oriented designation by means of its far references. A far reference to an object provides a stateful communication channel to that object. However, this channel is not decoupled in space: the identity of the receiver object must be known to the far reference.

While far references cater to connection-oriented designation, ambient references enable connectionless designation: they designate any number of service objects anonymously by means of their type. Also, by means of the @All annotation, they provide direct support for one-to-many communication. As a result, one-to-many messages act as event notifications to nearby interested service objects (subscribers). Time decoupling can be introduced in the event system by means of the @Expires annotation. Communication across an ambient reference is stateless, so subsequent messages may be received by different service objects. This enables a mobile device to transparently access different service objects as the user roams.

One difference between ambient references and publish/subscribe systems is that in the latter systems, it is generally the subscriber that specifies what kind of *events* it wants to accept. Ambient references invert this relationship. Using ambient references, it is the sender of a message (the publisher) that specifies what kind of *receiver* (subscriber) can accept the message. The type of an ambient reference thus delimits what services are eligible to receive its messages. The downside is that, if a receiver object (subscriber) places additional constraints on the message contents, it will have to filter messages explicitly by means of conditional tests in its invoked method. In contrast, a content-based publish/subscribe system can perform some such conditional tests within the event broker. This introduces less overhead because the broker can filter out certain events before notifying the subscriber.

In short, AmbientTalk resolves the impedance mismatch by providing ambient references, which introduce a form of stateless and connectionless designation in the object-oriented paradigm. If stateful communication is required, far references are the abstraction of choice.

Bidirectional vs. Unidirectional Communication

Publish/subscribe systems are good at broadcasting information from publishers to subscribers. However, if subscribers need to pass information to event publishers, this can only be accomplished by turning the subscribers themselves into publishers and by turning event publishers into subscribers explicitly to gather the replies. AmbientTalk avoids this pattern by means of in-line event handlers via the when:becomes: and whenAll:becomes: control structures. The return value of an asynchronously invoked method can naturally serve as an implicit reply from receiver (subscriber) to sender (publisher).

In short, ambient references resolve the impedance mismatch by using event handlers to express bidirectional communication without giving up on the full synchronization decoupling afforded by event brokers.

Threads vs. Event Loops

Previously, we noted that event-driven frameworks are mostly incorporated into (multithreaded) object-oriented languages by means of listeners and their callback methods. However, because a callback method is invoked synchronously by a thread that is not managed by the application, the application developer must be aware of the resulting concurrency control issues. By unifying event notification with the asynchronous invocation of a receiver's method, such issues are avoided. In particular:

- because the AmbientTalk language ensures that incoming messages are processed serially by an actor, the receiver object does not need to guard against race conditions on its data. While the serial execution of incoming messages conservatively limits the overall concurrency of the system (i.e. some methods are safe to execute in parallel), the resulting system becomes safer and more compositional (additional sources of events and additional event types can be added without requiring any additional concurrency control in existing code).

- because methods are processed asynchronously by the actor owning the receiver object, the thread of control of the event broker remains responsive. Hence, a method that takes a long time to complete does not monopolize the resources of the entire event delivery subsystem.

The major drawback of event-based systems is that they suffer from an inversion of control. This drawback is mitigated to some extent in AmbientTalk because of its support for in-line event handlers. This allows the continuation of an asynchronous message send to be specified at the point where the send is performed, leading to less code fragmentation because the reply can be processed in the same computational context as the one in which the message was sent.

In short, because of the event loop architecture of AmbientTalk, service objects must not take any additional synchronization precautions when being designated by one or more ambient or far references. This is in contrast to multithreaded object-oriented programs where explicitly subscribing to an event broker introduces concurrency control issues.

Reconciling Objects with Events

Previously, we explicitly stated which properties of objects and events needed to be combined.

Here, we summarize how AmbientTalk achieves a unification of objects and events:

- AmbientTalk maintains the *specific* communication interface of object-oriented message passing by representing events as (asynchronously sent) messages.
- Ambient references, like event brokers, provide *connectionless* designation, catering to anonymous interactions with groups of proximate objects. Far references provide *connection-oriented* designation, but are not decoupled in space.
- AmbientTalk makes use of in-line event handlers to retain the *bidirectional* communication of message passing, without sacrificing the time and synchronization decoupling afforded by event-driven communication.
- AmbientTalk's *event loop* concurrency ensures that receivers of messages (subscribers) do not need to take any synchronization precautions when handling messages (events).

The combination of AmbientTalk's event loops, far references, in-line event handlers and ambient references together effectively bridge the gap between event-driven and object-oriented abstractions.

Alternative Approaches to Resolving the Impedance Mismatch

We already mentioned Distributed Asynchronous Collections (DACs) (P. T. Eugster et al., 2000), which combine objects with events by representing the event broker as a *collection*, rather than as an object *reference*. While this approach allows for connectionless communication in an object-oriented system, it still provides a publish/subscribe interface to the programmer. We go further in integrating publish/subscribe with object-orientation than DACs. For example, we unify events with messages and event delivery with method invocation. No such unification is provided by DACs. Also, DACs provide no support for dealing with replies to events. Adding an element to a collection is a unidirectional operation.

Eugster et al. (2001) propose another object-oriented language extension with support for events. Again, this language extension does not attempt to fully integrate the publish/subscribe style with object-oriented concepts. Rather, this extension provides *both* object-oriented and event-driven concepts, with little integration between the two. Events are represented as objects, but the extension still distinguishes event notification from message passing and objects maintain their thread-based execution model. Similarly, Matsuoka and Kawai (1988) have studied the incorporation of tuple space communication in an object-oriented language. Again, their system does not make any attempt at unifying concepts of tuple space-based communication with object-oriented language features. They represent tuples as objects, but they still distinguish interaction with a tuple space from sending messages to objects.

Summary

We have discussed the differences between event-driven and object-oriented communication. Because of these differences, event-driven communication is not always easy to integrate in an object-oriented application. We have named this phenomenon the object-event impedance mismatch.

Subsequently, we introduced AmbientTalk, a programming language intended for composing software components across a mobile ad hoc network. AmbientTalk is an object-oriented programming language but provides first-class abstractions to represent event-driven communication in terms of asynchronous messages. A publish/subscribe style of interaction is made possible by means of ambient references, which allow one to anycast or broadcast a message to available objects of a certain type in the ad hoc network.

Finally, we have shown how the abstractions provided by AmbientTalk resolve the object-event impedance mismatch. Objects still communicate by message passing, can still perform bidirectional communication and can engage in connectionless communication via ambient references. Multithreading is replaced with event loop concurrency, enabling additional event sources to be added to an application without change in terms of concurrency control.

FUTURE TRENDS

The emerging fields of mobile and ubiquitous computing form an ideal application domain for advanced distributed event-based systems. Applications in this domain have to deal with a high number of context changes, including changes in the physical environment (such as the availability or unavailability of particular services). The current state of the art to incorporate events into an application is via middleware, requiring the application developer to interact with the middleware via a traditional library API.

In the future, we hope to see this state of the art evolve into systems where the actual programming language in which applications are built is augmented with standard support for distributed event-based communication. Doing this requires language designers to think about alternative models of computing where event-based concepts can be mapped onto concepts that are already present in the language or the language's paradigm.

AmbientTalk is our first step towards such an event-based programming language. Our current prototype implementation serves as a proof-of-concept. Future work focuses in part on a more efficient implementation and on more elaborate applications. So far, AmbientTalk has been used to build among others a peer-to-peer instant messaging client, a matchmaking application that allows

proximate buyers and sellers to trade items with one another, a multiplayer version for ad hoc networks of the well-known video game Pong and a mobile social networking application that allows users to share their profiles and keep each other up to date of what they are or have been doing.

CONCLUSION

Like in many other application domains, the loose coupling afforded by event-driven communication has turned event-based techniques into a de facto standard for the design of applications that need to be deployed on highly volatile mobile ad hoc networks. Such ad hoc networks form the hardware substrate on top of which pervasive and ubiquitous computing applications are deployed.

While event-based programming is a crucial technique, it remains only a second-class abstraction in mainstream programming languages. This exposes a need for new models that combine event-driven communication with the first-class abstractions already present in today's programming languages.

We conclude this chapter with a summary of our objectives. We have:

1. discussed why event-driven communication is suitable in mobile ad hoc networks by describing its support for decoupling in time, space and synchronization.
2. highlighted the key differences between objects and events, the combination of which we have named the object-event impedance mismatch.
3. proposed a novel programming model to resolve the mismatch. This model is validated by means of a programming language named AmbientTalk. The language's event loop model allows event-driven concepts (events, event handlers, ...) to be mapped almost one-to-one onto object-oriented concepts (messages, methods, ...). The language

supports publish/subscribe communication in an object-oriented manner by means of ambient references, which represent a traditional event broker as a special kind of object reference.

REFERENCES

Agha, G. (1986). *Actors: a Model of Concurrent Computation in Distributed Systems*. Cambridge, MA: MIT Press.

Atul, A., Jon, H., Marvin, T., William, J. B., & John, R. D. (2002). Cooperative Task Management Without Manual Stack Management. In *Proceedings of the General Track of the annual conference on USENIX Annual Technical Conference*, (pp. 289-302).

Baduel, L., Baude, F., Caromel, D., Contes, A., Huet, F., & Morel, M. (2006). Programming, Deploying, Composing, for the Grid. In Cunha, J., & Rana, O. (Eds.), *Grid Computing: Software Environments and Tools*. Berlin, Germany: Springer-Verlag. doi:10.1007/1-84628-339-6_9

Cabri, G., Leonardi, L., & Zambonelli, F. (2000). MARS: A Programmable Coordination Architecture for Mobile Agents. *IEEE Internet Computing*, *4*(4), 26–35. doi:10.1109/4236.865084

Callsen, C. J., & Agha, G. (1994). Open Heterogeneous Computing in ActorSpace. *Journal of Parallel and Distributed Computing*, *21*(3), 289–300. doi:10.1006/jpdc.1994.1060

Carey, M. J., & DeWitt, D. J. (1996). Of Objects and Databases: A Decade of Turmoil. VLDB'96, In *Proceedings of 22th International Conference on Very Large Data Bases*, (pp. 3-14).

Dedecker, J., Van Cutsem, T., Mostinckx, S., D'Hondt, T., & De Meuter, W. (2006). Ambient-oriented Programming in AmbientTalk. In *Proceedings of the 20th European Conference on Object-oriented Programming (ECOOP), 4067*, (pp. 230-254).

Eugster, P., Felber, P., Guerraoui, R., & Kermarrec, A. (2003). The many faces of publish/subscribe. *ACM Computing Surveys*, *35*(2), 114–131. doi:10.1145/857076.857078

Eugster, P., Garbinato, B., & Holzer, A. (2005). Location-based Publish/Subscribe. In *Proceedings of the Fourth IEEE International Symposium on Network Computing and Applications*, (pp. 279-282).

Eugster, P. T., Guerraoui, R., & Damm, C. H. (2001). On objects and events. In *OOPSLA '01: Proceedings of the 16th ACM SIGPLAN conference on Object oriented programming, systems, languages, and applications*, (pp. 254-269).

Eugster, P. T., Guerraoui, R., & Sventek, J. (2000). Distributed Asynchronous Collections: Abstractions for Publish/Subscribe Interaction. In *ECOOP '00: Proceedings of the 14th European Conference on Object-Oriented Programming*, (pp. 252-276).

Foote, B., & Johnson, R. (1989). Reflective Facilities in Smalltalk-80. In *Proceedings of the 4th International Conference on Object-Oriented Programming Systems, Languages and Applications (OOPSLA 89)*, (pp. 327-335).

Freeman, E., Arnold, K., & Hupfer, S. (1999). *JavaSpaces Principles, Patterns, and Practice*. Essex, UK: Addison-Wesley Longman Ltd.

Gelernter, D. (1985). Generative communication in Linda. *ACM Transactions on Programming Languages and Systems*, *7*(1), 80–112. doi:10.1145/2363.2433

Grimm, R., Davis, J., Lemar, E., Macbeth, A., Swanson, S., & Anderson, T. (2004). System support for pervasive applications. *ACM Transactions on Computer Systems, 22*(4), 421–486. doi:10.1145/1035582.1035584

Haller, P., & Odersky, M. (2006). Event-Based Programming without Inversion of Control. In *Proc. Joint Modular Languages Conference, 4228*, (pp. 4-22).

Hapner, M. (2002). *Java Message Service Specification (Version 1.1)*. Redwood Shores, CA: Sun Microsystems, Inc.

Kaminsky, A., & Bischof, H.-P. (2002). Many-to-Many Invocation: a new object oriented paradigm for ad hoc collaborative systems. In OOPSLA '02: Companion of the 17th annual ACM SIGPLAN conference on Object-oriented programming, systems, languages, and applications, (pp. 72-73).

Lehman, T.J., Cozzi, A., Xiong, Y., Gottschalk, J., Vasudevan, V., & Landis, S. (2001). Hitting the distributed computing sweet spot with TSpaces. *Computer Networks, 35*(4), 457–472. doi:10.1016/S1389-1286(00)00178-X

Mamei, M., & Zambonelli, F. (2004). Programming Pervasive and Mobile Computing Applications with the TOTA Middleware. In *PERCOM '04: Proceedings of the Second IEEE International Conference on Pervasive Computing and Communications*, (pp. 263-276).

Mascolo, C., Capra, L., & Emmerich, W. (2002). Mobile Computing Middleware. In *Proceedings of Advanced lectures on networking* (pp. 20–58). New York: Springer-Verlag. doi:10.1007/3-540-36162-6_2

Matsuoka, S., & Kawai, S. (1988). Using tuple space communication in distributed object-oriented languages. *SIGPLAN Not. Special issue: 'OOPSLA 88 Conference Proceedings, 23*(11), 276-284.

Meier, R., & Cahill, V. (2003). Exploiting Proximity in Event-Based Middleware for Collaborative Mobile Applications. In *Proceedings of the 4th IFIP International Conference on Distributed Applications and Interoperable Systems (DAIS'03)*, (pp, 285-296).

Meyer, B. (1993). Systematic concurrent object-oriented programming. *Communications of the ACM, 36*(9), 56–80. doi:10.1145/162685.162705

Miller, M. (2006). *Robust Composition: Towards a Unified Approach to Access Control and Concurrency Control*. Baltimore, MD: John Hopkins University.

Miller, M., Tribble, E. D., & Shapiro, J. (2005). Concurrency among strangers: Programming in E as plan coordination. In. *Proceedings of the Symposium on Trustworthy Global Computing, 3705*, 195–229. doi:10.1007/11580850_12

Murphy, A., Picco, G., & Roman, G.-C. (2001). LIME: A Middleware for Physical and Logical Mobility. In *Proceedings of the The 21st International Conference on Distributed Computing Systems*, (pp. 524-536).

Musolesi, M., Mascolo, C., & Hailes, S. (2005). EMMA: Epidemic Messaging Middleware for Ad hoc networks. *Personal and Ubiquitous Computing, 10*(1), 28–36. doi:10.1007/s00779-005-0037-4

Ousterhout, J. (1996). Why Threads Are A Bad Idea (for most purposes), In *Proceedings of the Presentation given at the 1996 Usenix Annual Technical Conference*.

Saif, U., & Greaves, D. J. (2001). Communication primitives for ubiquitous systems or RPC considered harmful. In *Proceedings of the International Conference on Distributed Computing Systems*, (pp. 240-245).

Sun Microsystems. (1999). *Dynamic Proxy Classes*. Retrieved August 25, 2008, from http://java.sun.com/j2se/1.3/docs/guide/reflection/proxy.html

Sun Microsystems. (2008, February 14th 2008). *Concurrency in Swing*. Retrieved August 12th 2008, from http://java.sun.com/docs/books/tutorial/uiswing/concurrency

Van Cutsem, T. (2008). *Ambient References: Object Designation in Mobile Ad Hoc Networks*. Brussels, Belgium: Programming Technology Lab, Faculty of Sciences, Vrije Universiteit Brussel.

Van Cutsem, T., Mostinckx, S., Gonzalez Boix, E., Dedecker, J., & De Meuter, W. (2007). Ambient-Talk: object-oriented event-driven programming in Mobile Ad hoc Networks. In *Proceedings of the XXVI International Conference of the Chilean Computer Science Society (SCCC 2007)*, (pp. 3-12).

Weiser, M. (1991). The computer for the twenty-first century. *Scientific American*, 94–104. doi:10.1038/scientificamerican0991-94

Yonezawa, A., Briot, J.-P., & Shibayama, E. (1986). Object-oriented concurrent programming in ABCL/1. In *Conference Proceedings on Object-oriented programming systems, languages and applications*, (pp. 258-268).

KEY TERMS AND DEFINITIONS

Mobile Ad Hoc Network: A computer network without any fixed infrastructure, consisting of mobile nodes that communicate by means of wireless links.

Decoupling in Time: Two or more processes are decoupled in time if they do not need to be online at the same time while communicating.

Decoupling in Space: Two or more processes are decoupled in space if they can communicate without needing to know one another's identity.

Decoupling in Synchronization: Two or more processes are decoupled in synchronization if their control flow is not blocked upon sending or receiving information.

Event Broker: The middleman in a publish/subscribe architecture that decouples subscribers from publishers. Both publishers and subscribers register with the event broker, whose task it is to forward published events to the appropriate subscribers.

Connection-Oriented Communication Channel: A channel in which the sender process knows about the identity of the unique receiver process, and in which it is assumed that messages arrive in order. Example: communication via TCP.

Connectionless Communication Channel: A channel in which the sender process does not know about the identity and the number of receivers, and in which no assumptions are made that messages arrive in order. Example: communication via UDP.

Event Loop: A perpetual loop that accepts events from one or more event sources and dispatches these events (usually sequentially) to the appropriate event handlers.

Chapter 15
Event–Based System Architecture in Mobile Ad Hoc Networks (MANETs)

Guanhong Pei
Virginia Tech, USA

Binoy Ravindran
Virginia Tech, USA

ABSTRACT

The strong decoupling between information producers and consumers in event-based (usually publish/ subscribe) systems is attractive in the loosely coupled and dynamic network scenarios such as mobile ad hoc networks (MANETs). However, achieving end-to-end timeliness, reliability properties, with limited message overhead, is still an open problem in publish/subscribe (P/S) systems in MANETs. In this chapter, we cover the current state of the knowledge of interconnection topology, event routing schemes and innovative architectural support of P/S systems in MANETs with latest academic and industrial research practices and outcomes. We consider challenging issues from timeliness, reliability, message overhead, etc. with multi-publish-hop event delivery in typical use notional scenarios. Both theoretical analysis and performance evaluation of different solutions are afforded. We also examine and discuss a special issue on system re-configurability and "event causal dependencies."

INTRODUCTION

The *publish/subscribe* paradigm (Muhl, Fiege, & Pietzuch, 2006) communicates on the basis of either the message content or the message source being of interest to destinations – as opposed to the source specifying the recipient(s). P/S systems can be considered to be a form of event-based systems, in the sense that the information injected to and propagated through the system can be treated as events. A unit in the system can act either/both as information producers (*publishers*) or/and consumers (*subscribers*). Subscribers declare their interests via subscriptions to certain events, most commonly specified by the content or the topic of the events (with different expressive power), and publishers produce events of information to the system. The event routing mechanism implemented in the P/S system (usually *middleware*)

DOI: 10.4018/978-1-60566-697-6.ch015

then takes charge of the event delivery according to the subscription knowledge.

The strong decoupling between information producers' and consumers' identities in a P/S system is appealing in loosely coupled network background such as *mobile ad hoc networks* (MANETs) (Corson & Macker, 1999), because of the ease with which components can be added, removed or changed at runtime. What is more, P/S features the capability of supporting one-to-many connectivity along with redundant information producers and consumers. This is conducive to constructing fault-tolerant or high-availability applications with redundancy and fault detection/handling mechanisms, and also building reconfigurable applications in a dynamic or even intermittent environment (Castellote & Bolton, 2002).

A MANET is a collection of mobile devices with dynamically changing membership and multi-hop topologies composed of wireless links. By its nature, a MANET is a self-organizing adaptive network, and thus needs to be formed and maintained in a distributed manner without centralized support or fixed infrastructures (Sarkar, Basavaraju, & Puttamadappa, 2007).

However, the potential advantages of the P/S interaction model atop MANETs are not fully realized by the state of the art predominant products and research solutions, such as TIB/RV (Oki, Pfluegel, Siegel, & Skeen, 1993), SCRIBE (Castro, Druschel, Kermarrec, & Rowston, 2002), SIENA (Carzaniga, Rosenblum, & Wolf, 2001), REDS (Cugola, Nitto, & Fuggetta, 1998), Kyra (Cao & Singh, 2004), IBM's WS-Notification (Graham, Niblett, Chappell, Lewis, Nagaratnam, Parikh, Patil, Samdarshi, Sedukhin, Snelling, Tuecke, Vambenepe, & Weihl, 2004), RTI's DDS (Castellote & Bolton, 2002) based on OMG's DDS (Object Management Group, 2007).

To fully and yet carefully explore the applicability of publish/subscribe systems overlaid on MANETs, one must cope with network uncertainties and constraints, including (but not limited to):

- frequent link breakages and temporary network disconnections;
- temporary node unavailability and node joins or departures at unpredictable times; and
- mobility-induced resource constraints on the overall architecture, such as limits on bandwidth, latency, and energy consumption.

Some efforts have been made (Fiege, Gartner, Kasten, & Zeidler, 2003; Cao & Shen, 2007; Muthusamy, Petrovic, & Jacobsen, 2005) assuming either that only a subset of nodes in the network can roam and act as clients or that the clients are always at one-hop away from the fixed infrastructure, and therefore focus only on a restricted subset of the problem space. A more general and common scenario in MANET is that every node in the network has mobility and can access the publish/subscribe service (e.g., by running a publish/subscribe middleware), while also acting as brokers for message forwarding and event matching to make the service available.

With all these in mind, to realize an effective and efficient publish/subscribe architecture on top of MANETs for timely and reliable data delivery – with suitable reconfiguration mechanisms – is a nontrivial problem.

In the rest of this chapter, readers are advised with these questions:

- What interconnection architecture is appropriate for P/S service in MANETs concerning node mobility, information availability, and resource constraints?
- How to self-organize mobile nodes into an interconnection topology with support for reliable and timely delivery?
- How to increase system performance and mitigate system instability with reasonable overhead from system reconfiguration and maintenance?

A Notional Example Scenario of P/S in MANET

Above is a notional example scenario abstracted from the military application space. As illustrated by Figure 1, a commander either makes (i.e., invokes) or publishes a service request to "agents" to find out something about the situation in a particular combat region. Naturally, there is a mission-critical time constraint for him to obtain that information — the information's utility to the mission degrades after a certain amount of time following his service request. To obtain that information: two forms of imagery are collected by subscription from surveillance platforms, and fused by recipient agents; re-scoped down to the geographical region of interest (perhaps by publishing it to a service which does that); re-sized to fit on soldiers' PDA screen (perhaps by publishing the re-scoped information to a service which re-sizes it); published to, and annotated by, one or more soldiers in that region; then sent by those soldiers to a ground-to-space relay and from there to a satellite and then to the commander making the original service request (and probably other interested officers). Most, if not all, of these communications are effectively publish/subscribes, despite the historical tendency to over-emphasize a dichotomy between control and data.

Similar scenarios can be found in other application domains such as automotive (Marques, Goncalves, & Sousa, 2006), event-based sensor networks (Leguay, Lopez-Ramos, Jean-Marie, & Conan, 2008) and e-health (Lupu, Dulay, Sloman, Sventek, Heeps, Strowes, Strowes, Twidle, Twidle, Keoh, & Schaeffer-Filho, 2008) systems.

BACKGROUND

Typically, in MANETs, a *structured* interconnection topology is used; most of the recent representative work (Huang & Garcia-Molina, 2003; Picco, Cugola, & Murphy, 2003; Cao & Shen, 2007; Mottola, Cugola, & Picco, 2008; Pei, Ravindran, & Jensen, 2008) present algorithms for constructing and maintaining an event routing tree as the interconnection topology under a filter-based routing scheme. There are also *structure-less* approaches — e.g., the work by Baldoni, Beraldi, Cugola, Migliavacca, & Querzoni (2005) employs a form of informed flooding-based event routing using Euclidean distances; the adoption of rendezvous-based protocols first described in SCRIBE (Castro et al., 2002), has been studied in a MANET setting (Carvalho, Araujo, & Rodrigues, 2006), and also in an Internet setting by Bharambe, Rao, & Seshan (2002); Costa, Gavidia, Koldehofe, Miranda, Musolesi, & Riva (2008) propose a gossip-based protocol used in VANETs (Vehicular Ad Hoc Networks). It is also worth noting that various approaches used in P/S systems atop wired networks and may be potentially used in wireless ad hoc environments. For instance, a distributed hash table configuration is used to facilitate the interconnection topology (Ratnasamy, Francis, Handley, Karp, and Shenker, 2001; Stoica, Morris, Liben-Nowell, Karger, Kaashoek, Dabek, & Balakrishnan, 2003; Costa & Frey, 2005). Some approaches do not maintain any deterministic data structure on the topology at a peer. In this case, event routing is neither filtering-based nor rendezvous based, like in the work by Costa, Migliavacca, Picco, & Cugola (2004) and Datta, Quarteroni, & Aberer (2004); the authors employ probabilistic flooding and gossiping, respectively. Chakraborty, Joshi, Yesha, & Finin (2006) propose a service/control information dissemination architecture/protocol in a MANET pervasive computing environment, utilizing bounded advertisement, peer-to-peer dynamic caching and logical group-based selective forwarding of discovery requests.

Tree-Based vs. Non-Tree Approaches

We refer to a *tree-based interconnection* topology as an approach to interconnect mobile nodes

Figure 1. Example scenario

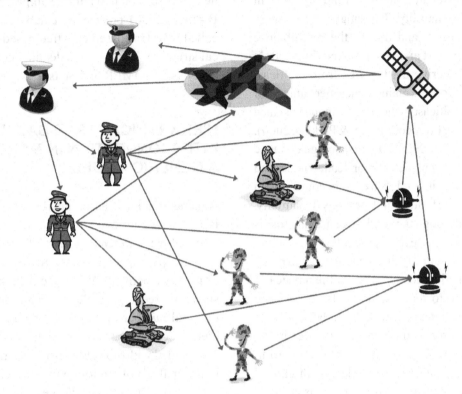

with an acyclic structure. We refer to *non-tree* approaches as the ones that use either rendezvous-based or flooding-/gossip-based approaches. The fundamental difference between the tree-based topology and the non-tree approaches is that the interconnection topology of non-tree approaches tends to weakly match the underlying physical topology, because of the loose geographical coupling among the nodes. The impact on the message delivery is that latency can be very high and sometimes unpredictable as information might pass across many nodes some of which might be slow as well as have long physical paths between them in the underlying network. Non-tree approaches may find their suitable applications in scenarios such as delay tolerant networks (Costa, Mascolo, Musolesi, & Picco, 2008), and thus deviate from part of our goal as to accommodate timing properties of the system. Also, large periods of message flooding in some peer-to-peer

networks can cause congestions and inefficient use of bandwidth.

By contrast, a tree-based topology can more efficiently and effectively meet the end-to-end timeliness requirement (Junginger & Lee, 2004; Carzaniga & Hall, 2006). However, in order to meet the demands of large scale MANETs, we need a more effective architecture to handle node failures, link breakage, node joins and departures, while still maintaining the same or improved timely message delivery.

Existing MANET Tree-Based P/S Architecture

To the best of our knowledge, for MANETs, most *P/S-tree-based interconnection* approaches employ a single P/S tree structure in the whole network or build one P/S tree for each publisher. The latter model can only accommodate a limited number of designated publishers and suffers from the lack of scalability. The single-tree strategy is typically used; and due to the resemblance between multicast and P/S in MANETs, Mottola et al. (2008) derive the P/S tree from multicast tree construction and maintenance mechanisms of MAODV (Multicast Ad hoc On-Demand Distance Vector Routing Protocol) (Royer & Perkins, 2000). In large scale MANETs, as the average distance increases between the root or leader node and the other nodes in the P/S tree, it becomes much harder to handle failures and topology dynamicity. Moreover, although the single-tree architecture is fully based on the geographical topology, there still exists a possibility that a publisher's message may take a path much longer than the physically shortest route to reach some of its subscribers, and sometimes the path stretch can be as high as $O(n)$ (in which n is the number of nodes in the network). Generally, low path stretch is hard to achieve in a spanning tree unless with careful design which may be running-time-consuming in decentralized settings. For this issue, interested readers are referred to Elkin, Emek, Spielman, & Teng S (2008) and Peleg (2002), etc., on construction of low stretch spanning tree.

In contrast, instead of maintaining a spanning tree over the network, a *hierarchical tree-based interconnection* in MANET self-organizes all the nodes into non-overlapping trees and has specific mechanisms for inter-tree communication. The roots of the trees function as the inter-tree brokers. Pei et al (2008) propose SOMER – a self-organizing and self-reconfiguring MANET event routing architecture which employs the hierarchical tree-based interconnection as the underlying network connection method. We will show later the comparison of the hierarchical tree-based interconnection and the single P/S tree approaches through both qualitative and quantitative analysis.

From now on, we mainly consider tree-based approaches for event delivery on top of MANETs. What we also need to keep in mind is that, the idea of using hierarchical architecture in ad hoc

networks is not new; for example, Pleisch & Birman (2006) present a scalable P/S system called SensTrac based on a tree-based hierarchical structure. We will give detailed comparative analysis of SOMER and SensTrac later.

HIERARCHICAL TREE-BASED INTERCONNECTION IN MANET: A GENERIC MODEL

A hierarchical tree-based interconnection model is advantageous over a single-tree strategy in terms of path stretch and network traffic, which may greatly impact the timeliness and reliability of message delivery. While the path stretch in a spanning tree is possible to reach to an order of the total number of nodes, the path stretch in a hierarchical tree-based interconnection model can be bounded with much lower-order parameters as in the analysis of the following network model.

Suppose that the system (i.e., a MANET) has already completed self-organization and remains stationary. To build an analytical model, we make the following assumptions:

- All nodes are homogeneous and evenly distributed within a 2-dimensional area; and
- Each node has the same number of immediate neighbors and thus each tree has the same number of immediate neighboring trees. Let the number of neighbors of any node be $N_{Neighbor}$.
- We do not consider the physical topology where nodes are placed to form a string or a circle.

Graph Theory Preliminaries

An r-tree is a tree with root r. Let $T(r)$ denote such a tree. The level of a vertex v in $T(r)$ is the length of the path rTv. Each edge of $T(r)$ joins vertices on consecutive levels, and it is convenient to think

Figure 2. Pentagon, octagon: Invalid model example

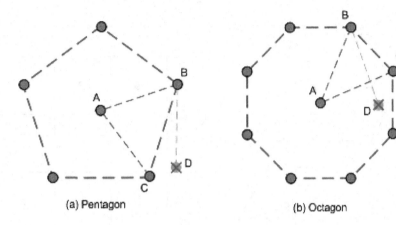

(a) Pentagon   (b) Octagon

of these edges as being oriented from the lower to the higher level, so as to form a branching. Each vertex other than *r* and *v* on the path *rTv*, is called an ancestor of *v*, and each vertex of which *v* is an ancestor is a descendant of *v*. Two vertices are related in *T* if one is an ancestor of the other. The immediate ancestor of *v* is its predecessor or parent, denoted *p(v)*, and the vertices whose predecessor is *v* are its successors or children. A leaf of a tree is the node which has no successors. We refer to the process of going from a non-root node to the root by way of its parent as "going upward," and going the reverse way as "going downward."

Model Analysis

Suppose $N_{Neighbor}$ is subject to a normal distribution with mean of 6 and standard deviation of 4. Only 4 and 6 are the possible number of neighbors that can produce an evenly distributed physical topology given that we rule out the case where $N_{Neighbor}=2$ as the third assumption states above; for example, if $N_{Neighbor}=5$, we would not be able to find a topology to maintain the equality of the distances between any two of neighboring nodes, thus violating the 2-dimensional even distribution assumption.

We provide a sketchy proof for finding valid $N_{Neighbor}$ values here. For $N_{Neighbor}=3$, it turns out this case is invalid because we will end up finding there are in fact 6 immediate neighbors. For $N_{Neighbor}=5$, as Figure 2(a) shows, suppose that node *A* has exactly 5 immediate neighbors placed on the corners of a pentagon. Then for node *B*, we should find at node *D*, where $|AB| = |BD|, \angle BAC = \angle ABD$, an immediate neighbor of node *B* due to the isotropy of our model. However, the fact that $|CD| < |AB| = |BD|$ makes the node *D* an *invalid* node in the graph because there should not be any edge that is shorter than the distance between two immediate neighbors. Similar statement applies to the cases where $N_{Neighbor}>6$ as shown in Figure 2(b). Therefore, we have

$$N_{Neighbor}=4,6. \qquad (1)$$

The network can be represented by a connectivity graph $N(V, E, P)$, where $V = \{v_1, ...,v_n\}$, is the set of nodes, *E* is the set of links, and $P = \{ p_1, ..., p_k \}$ is the set of publishers. Also, let *n* denote the number of nodes and *k* denote the number of publishers. Let D_T denote the density of root nodes among all the nodes in the area. Let the longest distance from the tree root to a

Figure 3. Network model structures

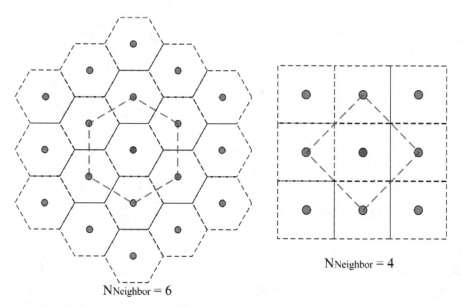

$N_{Neighbor} = 6$

$N_{Neighbor} = 4$

node in its territory be R (i.e., the radius). Let the distance between any pair of neighboring nodes be denoted as $dist_{Min}$. We assume that $0.5 dist_{CST} < dist_{Min} < dist_{CST}$, where $dist_{CST}$ is the one-hop wireless transmission range (carrier sensing range), such that a node should only be able to receive signals from an immediate neighbor.

From Equation 1, we can derive that the network connection links form either a grid-like ($N_{Neighbor}=4$) or a beehive-like ($N_{Neighbor}=6$) structure. Figure 3 illustrates this, where the solid dots denote nodes in the network, and the red hexagons and squares with nodes at the angles cover the set of immediate neighbors of the center node. Assume that a straight path can be established between any pair of neighboring trees' roots. It can be noticed that for any root node to reach another root node, if their trees are not neighboring, it needs a route by way of other root nodes, and this route takes at most one turn along the route. Let β denote the angle of the turn if there is one. Now,

$$\beta = \begin{cases} 120°, \text{ for } N_{Neighbor} = 6 \\ 90°, \text{ for } N_{Neighbor} = 4 \end{cases}. \quad (2)$$

Let $s(p_i, v_j)$ be one of p_i's subscribers which are associated to the tree rooted at $v_j \in V$. We estimate the average distance from a node to its tree root to be $\dfrac{R}{2}$. Let $length(u, v)$ denote the length of the path between two nodes u and v. Now, the length of the path between p_i and $s(p_i, v_j)$ is:

$$\begin{aligned} &length(p_i, s(p_i, v_j)) \\ &= length(p_i, r(p_i)) + length(r(p_i), v_j) + length(v_j, s(p_i, v_j)) \\ &= \frac{R}{2} + length(r(p_i), v_j) + \frac{R}{2} \\ &= R + length(r(p_i), v_j). \end{aligned} \quad (3)$$

Let $dist(u, v)$ denote the straight-line distance between the two nodes u and v. We can upper bound the length of the path between p_i and $s(p_i, v_j)$ as:

$$length(r(p_i), v_j) \leq \frac{dist(r(p_i), v_j)}{|\sin\beta|}$$

$$\Rightarrow length(p_i, s(p_i, v_j)) \leq \begin{cases} R + \dfrac{2}{\sqrt{3}} dist(r(p_i), v_j), \text{ for } N_{Neighbor} = 6 \\ R + \dfrac{2}{\sqrt{2}} dist(r(p_i), v_j), \text{ for } N_{Neighbor} = 4 \end{cases}. \quad (4)$$

For a large-scale MANET, $dist(p_i, s(p_{i,}v_j))$ is approximately $dist(r(p_i),v_j)$. Therefore,

$$length(p_i, s(p_i,v_j))$$

$$\leq \begin{cases} R + \dfrac{2}{\sqrt{3}} \, dist(p_i, s(p_i,v_j)), \text{ for } N_{Neighbor} = 6 \\[2ex] R + \dfrac{2}{\sqrt{2}} \, dist(p_i, s(p_i,v_j)), \text{ for } N_{Neighbor} = 4 \end{cases}$$

$$= R + \dfrac{2\sqrt{2}}{\sqrt{N_{Neighbor}}} \, dist(p_i, s(p_i,v_j)).$$

$$(5)$$

Path Overhead Ratio

We define the *path overhead ratio* (POR) based on the variation of the concept of *path stretch* (i.e., the ratio of the length of the route between two nodes to the length of the shortest path (both in Euclidean distance)) as the extra number of hops required by the path over that of the straight line distance. The number of hops is proportional to the path length, and hence:

$$POR_{SOMER} \leq \frac{length(p_i, s(p_i,v_j)) - dist(p_i, s(p_i,v_j))}{dist(p_i, s(p_i,v_j))}$$

$$= \frac{2\sqrt{2}}{\sqrt{N_{Neighbor}}} + \frac{R}{dist(p_i, s(p_i,v_j))} - 1.$$

$$(6)$$

By applying this POR in a large-scale network where $R << dist(p_i, s(p_{i,}v_j))$, we obtain:

$$POR_{SOMER} \leq \frac{2\sqrt{2}}{\sqrt{N_{Neighbor}}} - 1 = \begin{cases} \dfrac{2}{\sqrt{3}} - 1, \text{ for } N_{Neighbor} = 6 \\[2ex] \sqrt{2} - 1, \text{ for } N_{Neighbor} = 4 \end{cases}$$

$$(7)$$

Figure 4 gives an example showing the path from p_i to $s(p_{i,}v_j)$ with red lines. In the figure, the blue solid nodes are root nodes and the solid hexagons denote the abstract territories of the trees rooted at the center nodes.

Now we can observe that for the hierarchical tree-based architecture, POR can be bounded with small-order parameters (and in large networks it can be as small as $\sqrt{2} - 1$ or $\frac{2}{\sqrt{3}} - 1$); whereas for single-tree-based and structure-less approaches, the length of the path could sometimes be up to the order of the total number of nodes. Thus, the benefits of the hierarchical tree-based architecture include the following:

- event delivery can be completed in a more timely manner;
- shorter and bounded path length implies lower probability for a delivery to fail, despite frequent link breakages and node failures; and
- the multi-tree structure gives enough leeway to devise multi-path event delivery schemes through the use of multiple neighboring trees and multiple border routers (leaf nodes) among the trees, and may also balance the network traffic and alleviate congestion.

BASIC ARCHITECTURE DESIGN

SOMER (Pei et al, 2008) makes an example of how a typical P/S architecture with hierarchical tree-based interconnection is designed, while SensTrac (Pleisch & Birman, 2006) gives a specific application of the hierarchical tree-based interconnection in a specific wireless sensor networks. Next we explore the architecture design space with the example of SOMER.

As shown in Figure 5, SOMER employs a two-layer structure. At the lower layer (*inner-tree level*), nodes are self-organized into multiple P/S trees rooted at several root nodes across the network. At the top layer (*inter-tree level*), root

Figure 4. A path established between p_i and $s(p_i, v_j)$

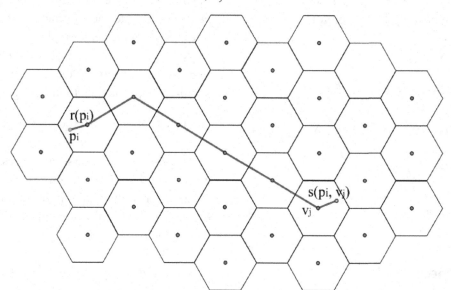

nodes act as super publisher/subscriber nodes and constitute an overlay network.

Inner-Tree Level Interconnection

The *publish/subscribe tree* (PST) is constructed in a distributed request/reply fashion. Any node that has not joined a PST broadcasts the *join request message* and waits for its immediate neighbors that are currently in a P/S tree to reply with a *join reply message*. After a timeout for the neighbors' reply expires, if the node gets any replies, it greedily selects the best candidate neighbor as its parent according to the *parent evaluation metric* (PEM); otherwise, it continues to broadcast new join request messages. Each node in the system can only join one PST.

Inner-Tree Subscription and Publication

Each node v_0 has its own subscription interests $s(v_0)$, termed *inherent subscription*. On joining the tree, node v_0 sends a subscription message containing $s(v_0)$ upward toward the *grafting node* (defined below) for each of its ancestors to

match, and subsequently publish information to v_0, which is now a leaf in that tree. We define a non-leaf node u_0's *effective subscription* $S(u_0)$ as the "combined" subscription formed by merging u_0's inherent subscriptions with all of its descendants' subscriptions. Thus, the *grafting node* of node v_0 is the closest node (in terms of number of hops) in the upward path, whose effective subscription overlaps with that of v_0. In this way, each non-leaf node maintains data structures for its successors' subscription interests, and in u_0's parent $p(u_0)$'s view, u_0 is a child that has the subscription interests of $S(u_0)$.

Parent Evaluation Metrics (PEM)

PEM is the key mechanism for constructing and maintaining the PST. There are various PEMs, including:

*   Shortest-path Metric (SM); and
*   Publication-overhead-aware Metric (PM).

SM is a most straightforward metric, in which a node chooses the neighbor at the lowest level as

Figure 5. Architecture overview

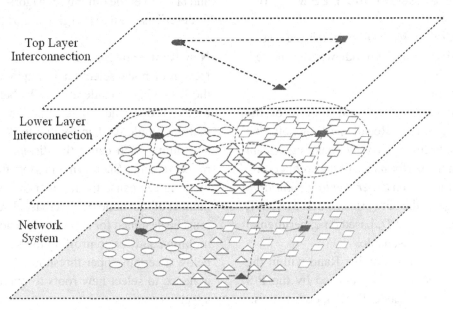

its parent. PM uses the subscription information of the other nodes in the tree to reduce the total number of messages needed to complete the delivery of publication events. Huang & Garcia-Molina (2003) propose a publication-overhead-aware metric that requires a' priori knowledge of the exact rate of invocation (publication) values for each category of subscription; otherwise the lack of that knowledge will cause high inaccuracy in a distributed system where the rates of invocation vary over time or even are unpredictable as is the case in the example scenarios. Thus, their metric not only suffers from the requirement to have that global information to function properly, but also implies that the PEM they use is limited to topic-based publish/subscribe systems. Note that in a content-based system, subscriptions are so finely grained that events are not classified according to any predefined categories but rather according to the properties of the events themselves. The PM proposed by Cao & Shen (2007) requires *periodic* messages to go upwards (towards the root) in the tree through a probabilistically chosen candidate parent until it reaches a new grafting node. However, in a large-scale network where

subscription interests vary greatly, especially for a content-based system, finding a grafting node may require traversing a significantly long distance and incurring much higher overhead than SM. Cao & Shen (2007) claim an advantage of less "overhead" over the MAODV-based PST approach. However, by "overhead," they only count the messages used for publication without considering the considerable overhead incurred for a better path detection initiated periodically by each node other than the root (level-0) and level-1 nodes.

Based on these observations, we choose Shortest-path Metric (SM), like the MAODV-based PST approaches and what Pleisch & Birman (2006) did. Generally, SM inherently favors the timing property of event delivery.

Inter-Tree Level Interconnection

A natural question that arises for this distributed architecture is that which nodes should be the roots, given no centralized administration. Another question is how the inter-tree communication mechanism is designed to facilitate event routing.

Root Node Selection and Tree Merging

A node's role (either root or non-root) in the system is designated when the P/S middleware is initiated.

Root Selection

We use a random-number-based strategy for distributed root selection. Recall that D_T denotes the density of roots. In fact, a fixed D_T has a significant influence on the performance of the architecture, as shown in the section *Experimental Evaluation*. Given a range of the network size of a MANET, we can select the best D_T based on experimental studies, which is reasonable.

A uniform random number Rand within the range of 0 to $Rand_{Max}$ is produced by the P/S middleware on each node. For a node v, that is:

$$\begin{cases} v \text{ is a root node, if } 0 < Rand(v) \le D_T \cdot Rand_{Max} \\ v \text{ is a non-root node, otherwise} \end{cases}$$

Tree Merging

Since we do not force geographical knowledge into root selection process, the roots selected may lie geographically too close (e.g., one or two hops away from each other). To deal with the performance degradation caused by that situation, a tree merging mechanism is employed.

A commonly used tree maintenance mechanism is a periodic *refresh message* broadcast with a sequence number, from the root throughout the tree. Every node rebroadcasting this message replaces the *level value* field with its own level value. Whenever a node v_0 receives a foreign tree node u_0's refresh message, it calculates the root-to-root distance between $r(v_0)$ and $r(u_0)$ from the level values of v_0 and u_0. If the root-to-root distance falls below a threshold *Merging_Thres*, either v_0 or u_0 sends a *merge request message* to the root, and the merging process is initiated. The tree merging process is similar to the partition merging in MAODV (Royer & Perkins, 2000), in which the root with a higher ID remains its role

and takes over the other tree. Algorithm 1(Figure 6) describes the tree merging algorithm.

New Root Node Selection

New root node selection is required based on the following considerations. To better address MANET dynamics, it is important to observe that the topology changes due to node mobility and failures can cause the distance among the roots to change. Such dynamics together with tree merging can cause the number of trees (i.e., the number of root nodes) to decrease. Thus, schemes must be designed for handling root node failures, mobility, and for controlling the tree sizes to be below a certain upper threshold. The basic idea is hence to select new roots to overcome those adverse conditions.

When a node v_0's level is above a certain predefined threshold *New_Root_Thres*, or it has not been a member of any tree for a certain time frame *Out_Period*, the node checks the possibility of advancing itself to a root. A new random number will be produced raising the probability of the node being accepted as a root. After the new root's "birth", it broadcasts invitations to "crop" tree members from other trees, until the level value of its leaf nodes is no smaller than that of at least one of the leaf nodes' neighboring foreign nodes. This algorithm is illustrated in Algorithm 2 (Figure 7).

Note that in a reliable wireless network with little topology change, the tree merging and new root selection strategies for system maintenance would hardly introduce any additional traffic overhead.

Inter-Tree Communication

Inter-Tree Route Establishment

Similar to the mechanism for tree merging, whenever a node v_0 receives a foreign tree node u_0's refresh message, it calculates the root-to-root distance between $r(v_0)$ and $r(u_0)$ from the level values of v_0 and u_0. If the new root-to-root distance

Figure 6. Algorithm 1: Tree merging algorithm

input: local node i's level number $level(i)$, level number $level(v)$ of foreign node v
            $r(v)$'s address

1   $r2rdistance \leftarrow level(i) + level(v) + 1$;
2   **if** $r2rdistance > Merging\_Thres$ **then**
3 | **if** $i \neq r(i)$ **then**
4 | | Send *merging message* with $r(v)$'s address to $r(i)$ to initiate tree merging
        | | process;
5 | **else**
6 | | Initiate $tree\_merging\_process(r(v)\text{'s address})$;

    /* Function $tree\_merging\_process()$ */
7   $tree\_merging\_process(r(v)\text{'s address})$:
8   **if** *My address is higher than $r(v)$'s address* **then**
9 | I will remain as a root;
10 | initiate the partition merging process;
11 **else**
12 | I will wait for the other root to initiate the partition merging and I will resign

Figure 7. Algorithm 2: New root node selection algorithm

input: local node i's level number $level(i)$

    /* For random number based strategy */
1   **if** $level(i) > New\_Root\_Thres$ or Out_Period-timer *expires* **then**
2 | $Rand(i) \leftarrow$ new random value;
3 | **if** $0 < Rand(i) < D_T \cdot Rand_{Max} \cdot \alpha$ **then** /* $\alpha > 1$ */
4 | | I become a root;
5 | **else if** Out_Period-timer *expires* **then**
6 | | start Out_Period-timer again;

is smaller than the current value, node v_0 records u_0 as the next hop destination for inter-tree routing, and sends a route report message upwards. On receiving the route report message, v_0's parent $p(v_0)$ checks its local root-to-root distance value. If the new value reported is better, $p(v_0)$ will also update its record and send a route report message upwards. In this way, the root can maintain the freshest feasible route for inter-tree communication. Further, a node only registers the next hop destination and the corresponding root-to-root distance for inter-tree routing, requiring only small memory usage. When u_0 receives an inter-tree

message (e.g., a publication or notification message) from a foreign node, u_0 will simply forward this message upward to its root. Now, we claim that the route established between two roots of the neighboring trees is the shortest. That can be proved by contradiction.

Inter-Tree Overlay Event Routing

Advertisement messages are widely used in P/S systems. On joining a P/S tree, a node sends both of its subscription and publication interests to the root node. From the top-layer view, root nodes can be treated as super publishers/subscribers, with the *effective subscription* and the *effective publication capabilities*. Event routing at the inter-tree level is based on periodic advertisement messages flooding across root nodes such that the root nodes' effective P/S interests are propagated through the top layer. Although the flooding among root nodes incurs message overhead, it is worth it when node mobility is high. (Experiments show that the amount of overhead is reasonable.) When a root node r_2 receives an advertisement message forwarded by a neighbor root r_1 and originally from r_0, r_2 updates r_0's P/S interests and the inter-tree route entry for r_0, and updates the corresponding next-hop root-address with r_1's address. In this way, the inter-tree routing uses the shortest reverse path to effectively propagate event notifications.

SOMER vs. SensTrac

In brief, SOMER is a more general architecture design comparing to SensTrac as rather an application specific design.

SOMER is different from SensTrac in several aspects including:

- Tree Construction. SensTrac uses static clustering via nodes' geographic location info from a GPS. In contrast SOMER uses dynamic clustering allowing nodes to be clustered on-the-fly;
- Inter-tree Communication. SensTrac uses AODV to find routes among root nodes (leaders), and use gossip to disseminate information among root nodes, whereas SOMER uses flooding among root nodes which is generally more suitable for a mobile environment;
- In SensTrac, the query (subscriber) node subscribes to the information of its area of interest (AOI) which is bounded by a given square, whereas SOMER's subscription model is more general;
- In SensTrac, the query node is the only subscriber, and does not act as a broker,

and the publishers are only the sensors in the AOI, whereas we do not have those limitations on subscribers or publishers;
- In SensTrac, nodes are stationary and only the query node can move, whereas we allow every unit to be mobile.
- SensTrac does not support system reconfiguration for causal dependencies which is address below (due to its static clustering mechanism).

SPECIAL ISSUE: EVENT CAUSAL DEPENDENCIES AND SELF-RECONFIGURING ARCHITECTURAL FACILITATION

Event Causal Dependencies

An important property of many emerging MANET-based applications that are suited for P/S-style communication is event causal dependencies. This refers to the existence of multi-publish-hops that are causally related (e.g., topic-wise). The causal relationship between two nodes is established by the causal dependency of their publication. We illustrate this above with a motivating scenario from the military domain. More use case scenarios with event causal dependencies can be found in other application domains such as event-based financial service (Tsai & Chen, 2008), event-based sensor networks (Leguay et al., 2008) and e-health systems (Lupu et al., 2008).

The conventional ideas about timely and reliable event delivery are always on a *single publish-hop* basis, by which we mean the travel of an event from its publisher to its subscribers, transparently through network devices such as routers. Thus that neglects a common phenomenon of *event causal dependencies*, which implies the existence of *multi-publish-hop* – i.e., a causal event's delivery triggers events that are causally dependent on that event, and later triggers other events with their deliveries, resulting in an event causal chain or an event causal graph comprised of causally-related nodes of the P/S system. To better understand these concepts, we illustrate them below through the event causal graph from the military domain corresponding to that in Figure 1.

Figure 8 depicts the underlying causal graph for the scenario in Figure 1. The information request made by Commander 1 maps to the logical links between Commander 1 and the agents, and will cause the agents' control and data event publication to aircrafts and soldiers on specific regions in the battlefield; and so on. The timeliness and reliability of the data delivery from the causal event initiator (Commander 1 in Figure 1) to the last step publishers in the whole chain (satellite in Figure 1), is critical and requires system reconfigurability due to the MANET dynamics; for example, any delay or failure that occurs for the event delivery between the commander and the agents will have an impact on every subsequent causal chain, whereas one soldier-relay link will only influence one causal chain.

Self-Construction of System Causal Graph

A causal graph is a graph comprised of causally-related nodes of the P/S system. The causal relationship between two nodes is established upon the causal dependency of their publication. The two necessary conditions for a node b to be a "causal child" of node a are:

- b subscribes to a's publication; and
- b's receipt of a's event publication message is the sufficient condition for b to generate a new publication to its subscribers.

In turn, a is b's "causal parent." To simplify the problem, we assume that there is at most one "causal parent" for each causally-related node, and none of a node's "causal descendants" can be its "causal parent". Hence, the causal graph boils down to a causal tree[1] or a causal forest with

disjoint causal trees. Now, advertising the nodes' subscription/publication interests among the P/S tree root nodes across the system enables the root nodes to build the causal graph for the P/S system.

Self-Reconfiguration with Awareness of Event Causal Dependencies

In SOMER, the system controls the physical P/S trees' self-organization: P/S trees stop their natural growth when each node has joined in a tree; the tree merging and new root selection mechanisms confine the sizes of the trees with *Merging_Thres* and *New_Root_Thres*. However, this may not be the "ultimate" solution for a system when event causal dependencies are exposed with time-criticality. Further performance improvements can be achieved based on the following observations:

- A physical tree of a comparatively larger size typically has more neighboring trees, implying increasing possibilities to establish direct inter-tree connection with more other trees. Consequently, the probability increases for a causally-involved node to get to its causal children through fewer intermediate forwarding nodes.

- One physical tree's expansion means other neighboring trees' contraction, as if trees are contending for nodes which can only belong to one tree at any given time. The policy for resolving trees' contention for nodes must be carefully designed. The goal is to shorten the event notification time along the P/S causal chains that originate at the "causal root" until they reach the "causal leaves" in the causal quasi-tree-like graph.

- For a given physical tree, if it contains more "causal nodes" at lower levels in the causal graph, or a larger number of causal nodes, it should be given greater preference in the resource contention resolution policy for physical expansion. That

Figure 8. Example causal graph

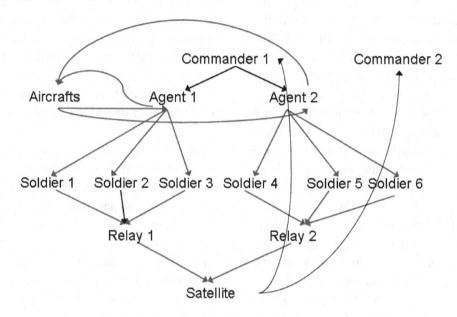

is because lower level causal nodes in the causal graph usually have greater influences on the subsequent event notification deliveries that directly or indirectly depend on those nodes in the corresponding causal chains. For instance, in Figure 1, any delay that occurs for the event notification delivery between the commander and the agents will have an impact on every subsequent causal chain, whereas the soldier-relay link will only have an influence on one causal chain.

• Furthermore, for the causal nodes covered by a physical P/S tree, if a large number of their causal descendants are already covered by the same tree, there is a higher possibility that the tree's expansion will no longer contribute to the timely and reliable event delivery across the causal graph.

Here, we define a tree's *expansion potential* (EP) as the metric for evaluating the potential benefit of a tree's expansion for event delivery

across the causal tree. For a physical P/S tree T, let $N_{Causal}(T)$ be the number of causal nodes that T covers, $Avg\_lev_{Causal}(T)$ be the average causal trees level number of those causal nodes contained in T (e.g., the causal root's level number is 1), and γ be the total percentage of causal descendants coverage in T for all the causal nodes in T. Now, $EP(T)$ is given by:

$$EP(T) = \frac{(N_{Causal}(T(v)) + \eta)}{(Avg\_lev_{Causal}(T) + \eta) \cdot (\gamma + \eta)}.$$

(8)

In Equation 8, η is a parameter with a constant positive value to offset the variables that could be zero. It is the leaf nodes that actually perform the expansion (i.e., fighting) through periodically requesting neighbor foreign nodes for a comparison of each other's *local expansion potential* (LEP). Because we need to also constrain the tree's branches from abnormal growth, a node v's LEP is:

$$LEP(v) = \frac{EP(T(v))}{(level(v) + \eta)}$$
$$= \frac{(N_{Causal}(T(v)) + \eta)}{(Avg\_lev_{Causal}(T(v)) + \eta) \cdot (\gamma + \eta) \cdot (level(v) + \eta)}.$$

$$(9)$$

For two nodes u and v involved in such a fighting, the one with the higher LEP will win, and the loser will join the winner tree as a child of the winner. It is possible that a node may constantly change its affiliation. To avoid that situation, we stop two nodes from fighting when:

$$\frac{1}{1+\delta} \le \frac{LEP(v)}{LEP(u)} \le 1 + \delta. \qquad (10)$$

In Equation 10, δ is a tunable parameter to control the fighting severity.

The result of the fighting may change the size of a tree. The fighting is allowed to take place only when the merging threshold or new root selection threshold is not violated.

The causality awareness module can be integrated as an add-on to the existing event routing hierarchical architecture, as done in SOMER. Yet we should note that only reconfigurable architecture can support such dynamic self-architectural-facilitation to causal dependencies.

EXPERIMENTAL EVALUATION

We conducted sets of simulation experiments using NS-2, with the Random Trip Mobility Model package (Boudec & Vojnovic, 2005) to generate sets of random MANET topologies. In this section, we first take a look at the hierarchical architecture's performance from different respects by considering the effects of parameter settings, timeliness and reliability of event delivery under different mobility, node failure and network topology settings, improvement on causal event delivery, and

show the results of the comparative experiments on various typical event routing models.

The simulation environment is built with each node having its own subscription interest. A randomly selected set of nodes act as publishers. We used the model of Number Intervals (Huang & Garcia-Molina, 2003) for P/S pattern generation. We used a large number interval as the interest pool, and a node's subscription interest is represented by a random subset of the interest pool. We call it a match when the number associated with a published event falls into the range of a node's interest. Table 1 details the simulation settings.

Timely and Reliable Event Delivery

We compared SOMER's timeliness and reliability against significant past MANET P/S protocols including:

- SP-COMBO (Huang & Garcia-Molina 2003), a PST protocol with a combination of shortest path and publication-overhead-aware metrics;
- DSAPST (Cao & Shen, 2007), a distributed subscription-aware PST protocol; and
- PS-MAODV (Mottola et al., 2008), a MAODV-based PST protocol.

The network size was varied with a constant D_T in the experiments. We omitted SensTrac from the comparison due to its highly specific application space.

Timeliness

We measured the timeliness performance by reciprocating the measured time cost for event delivery. Then we normalized each protocol's performance to the worst performance. In this way, we can also observe the trend of relative performance as the network size changes.

Figure 9 illustrates SOMER's advantage in timely delivery over others. DSAPST performs

Table 1. Simulation settings

Parameter	Range/Value
Simulation Area	1000m × 1000m to 3300m × 3300m
Network Size	50 to 250 nodes
Simulation Period	1000s
Node Wandering Velocity	0 to 10 m/s
Pause Time	1s to 3s
MAC Protocol	IEEE 802.11 w/ 2Mbps Bandwidth
Root Node Density (D_r)	1% to 18%
Wireless Transmission Range	200m
Proportion of Publishers	30% to 40%
Proportion of Causal Nodes	5% to 15%
Proportion of Causal Nodes	10%
Tree Refresh Period	10s
Advertisement Period	10s
Transmission Jitter	5% to 10%
Advertisement Period	10s
Merging_Thres	1 to 3
New_Root_Thres	5 to 10
Out_Period	15s
Tree Contention Tuner (δ)	0.2 to 0.4

Figure 9. Comparison for timely delivery

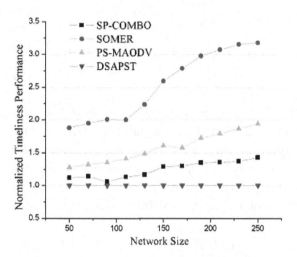

the worst because, the way it constructs the PST is mainly based on a publication-overhead-aware metric.

By normalizing with respect to DSAPST's performance, we clearly observe SOMER's performance gain over others. SOMER outperforms PS-MAODV, which is the second best, with at least a 30% improvement. As the network size scales up, SOMER's improvement over PS-MAODV increases up to around 100%, illustrating SOMER's superior scalability. This is due to increasing number of trees that can be used as top-layer brokers to make the route closer to the straight line between two nodes.

Reliability

We measured reliability through delivery ratio, which is calculated as the number of copies of event messages successfully received by their subscribers over the number of copies of event messages that should arrive at all their subscribers, given no failures or no topology changes. We used node failure rate as a variable to assess system reliability, with node roaming speed randomly ranging from 0 to 10m/s. Figure 10 shows that SOMER can survive a 12% node failure rate with over 75% event delivery ratio. The highest performance improvement of SOMER is over 10% higher delivery ratio than that of PS-MAODV.

Note that SOMER also provides 30% higher delivery ratio over the other two rivals. SOMER's improvements are due to its inherent distributed multi-tree structure, and also due to its tree merging and new root selection strategies that effectively counter network unreliability.

More interesting results on event delivery with awareness of causal dependencies, network traffic load, and system configuration parameters can be found in (Pei et al, 2008).

Tree Merging and New Root Selection

Recall that through the tree merging and new root selection strategies, the system interconnection is self-reorganized, thereby optimizes the tree

Figure 10. Delivery ratio vs. node failure rate

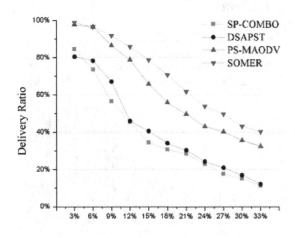

Figure 11. Effectiveness of tree merging and new root selection

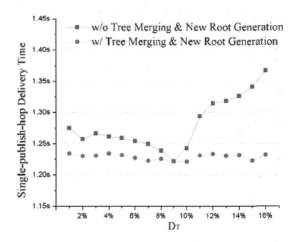

distribution toward timely event delivery. We tested the effectiveness of the tree merging and new root selection mechanisms.

From Figure 11, we observe that after these strategies are used, the average time cost for single-publish-hop event delivery is effectively lowered and the system gives a stable performance as the initial density of root nodes (D_T) varies all the way from 1% to 16%. (We used 200-node networks for these experiments.) This implies that those mechanisms are stable. The decrease in the event delivery time is due to the new root selection strategy when the initial D_T is small. Furthermore, it is the tree merging strategy that reduces the delivery time while the initial D_T is comparatively large.

Event Delivery with Awareness of Causal Dependencies

To evaluate the effectiveness of SOMER's causality awareness design, we performed experiments with network topologies of varying size. Figure 12 illustrates that, for causal event deliveries across causal graphs, SOMER achieves around 10% performance gain than that without the reconfiguration for causal dependencies, and even higher performance gain when the network size scales up. This trend corresponds to Equation

6, which states that the upper bound of the path overhead ratio decreases for larger size networks. Therefore, the experiment shows that SOMER's design is scalable and effective toward reducing the total event delivery time across a causal graph.

Network Traffic Load

For a structured architecture, the traffic load of a P/S system is from two sources: Structure Maintenance Messages and P/S Messages. The number of messages is counted whenever a message is produced or forwarded.

We observed that, among PS-MAODV, DSAPST, and SP-COMBO, only PS-MAODV have relatively acceptable timeliness and reliability. Thus, we compared SOMER and PS-MAODV with respect to the network traffic load incurred under the same simulation scenarios and publication patterns. Figure 13 shows that SOMER only introduces slightly more maintenance messages due to the neighbor-tree route setup and tree merging and new root selection processes while highly enhances timely event delivery.

Figure 12. Timeliness gain for causal event deliveries with varying network size

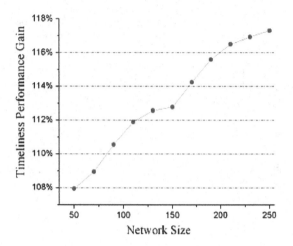

Figure 13. Effectiveness of tree merging and new root selection

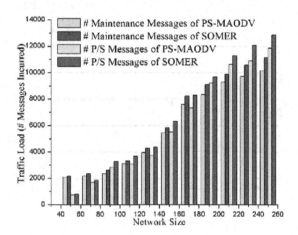

CONCLUDING REMARKS AND FUTURE TRENDS

In this Chapter, we introduce various event routing architecture models, and give design details of a typical latest model called SOMER, a self-organizing MANET event routing architecture with dynamic system self-reconfiguring ability for event causal dependencies. The design facilitates the event delivery with timeliness and reliability properties. Effective strategies and mechanisms for P/S system to function in a MANET are provided. The experimental study demonstrates the evaluation results of multiple P/S models from different design aspects.

So far, most of the current MANET P/S systems designs are focused on basic event routing architecture. More work could be done in developing more sophisticated mechanisms to either resolve specific design constraints or to provide support to specific needs of various systems. Advanced mechanisms are still needed to (1) recognize causal links; (2) build causal graphs; (3) determine the critical causal links or chains; (4) tune system performance based on the knowledge of causal dependencies. For example, the system causal graph model is now limited to one causal tree or independent causal trees; new schemes should be invented to extend the architectural support to allowing processing arbitrary causal graphs (i.e., allowing multiple causal parents) to facilitate causally-related event deliveries. Besides, now that we have only dealt with deterministic time-sensitive causal dependencies, there are also probabilistic causal dependency links, delay-tolerant causal dependency links, etc. By probabilistic, we mean that an event may probabilistically cause a subsequent event to successfully occur. By delay-tolerant, we mean that a causal descendent event may take some time to occur, or may not have a stringent timeliness requirement. It is likely that application-specific approaches may be required for certain cases.

When it comes to generic design concerns like real time support or energy-aware P/S systems, there is still plenty of design space to explore. For instance, we may have to address system extensibility for event routing protocols to integrate features like real-time scheduling strategies and multi-path schemes. For a P/S system in MANET, that will either result in enhancement of existing protocols or brand new system architecture designs. What is more, scalable design opens another important research area for MANET P/S systems to survive large-scale networks.

While gaining the timeliness and reliability from tree-based approaches, we should never forget that the inherent congestion issue and the thus-incurred problems on energy efficiency and fairness. In this sense, it is also necessary and promising to further study structure-less topology architectures as alternatives to existing tree-based models with experiments and understanding of their tradeoffs.

Last, but not least, we notice that current MANET P/S technologies typically explicitly or implicitly embrace the use of connectionless protocols. Although *"the use of connectionless protocols like UDP and multicast enables systems to scale up higher than with connection-oriented (point-to-point) protocols like TCP"* (Castellote & Bolton, 2002), tradeoff among reliability, scalability, timeliness and other properties deserve deeper study, which may be application-specific.

REFERENCES

Baldoni, R., Beraldi, R., Cugola, G., Migliavacca, M., & Querzoni, L. (2005). Structure-less content-based routing in mobile ad hoc networks. In *Proceedings of the IEEE International Conference on Pervasive Services*, (pp. 37-46).

Bharambe, A. R., Rao, S., & Seshan, S. (2002). Mercury: a scalable publish-subscribe system for internet games. In *Proceedings of the 1st Workshop on Network and System Support for Games,* (pp. 3-9).

Boudec, J.-Y. L., & Vojnovic, M. (2005). Perfect simulation and stationarity of a class of mobility models. *Proceedings - IEEE INFOCOM, 2005,* 2743–2754.

Cao, F., & Singh, J. P. (2004). Efficient event routing in content-based publish-subscribe service networks. *Proceedings - IEEE INFOCOM, 2004,* 929–940.

Cao, X., & Shen, C.-C. (2007). Subscription-aware publish/subscribe tree construction in mobile ad hoc networks. In *Proceedings of the 13th IEEE International Conference on Parallel and Distributed Systems*, (pp. 1-9).

Carvalho, N., Araujo, F., & Rodrigues, L. (2006). Reducing Latency in Rendezvous-Based Publish-Subscribe Systems for Wireless Ad Hoc Networks. In *Proceedings of the 26th IEEE International Conference Workshops on Distributed Computing Systems*, (p. 28).

Carzaniga, A. & Hall., C. P. (2006). Content-based communication: a research agenda. In *Proceedings of the International Workshop on Software Engineering and Middleware*, (pp. 2-8).

Carzaniga, A., Rosenblum, D. S., & Wolf, A. L. (2001). Design and evaluation of a wide-area event notification service. *ACM Transactions on Computer Systems, 19*(3), 332–383. doi:10.1145/380749.380767

Castellote, G.-P., & Bolton, P. (2002, February). Distributed real-time applications now have a data distribution protocol. *RTC Magazine*. Retrieved August 23, 2008, from http://www.rti.com/docs/RTC_Feb02.pdf

Castro, M., Druschel, P., Kermarrec, A., & Rowston, A. (2002). Scribe: A large-scale and decentralized application-level multicast infrastructure. *IEEE Journal on Selected Areas in Communications, 20*(8), 1489–1499. doi:10.1109/JSAC.2002.803069

Chakraborty, D., Joshi, A., Yesha, Y., & Finin, T. W. (2006). Toward Distributed Service Discovery in Pervasive Computing Environments. *IEEE Transactions on Mobile Computing, 5*(2), 97–112. doi:10.1109/TMC.2006.26

Corson, S., & Macker, J. (1999). Routing protocol performance issues and evaluation considerations (RFC 2501). *Network Working Group*.

Costa, P., Gavidia, D., Koldehofe, B., Miranda, H., Musolesi, M., & Riva, O. (2008). When cars start gossiping. In *Proceedings of the 6th Workshop on Middleware for Network Eccentric and Mobile Applications*, (pp. 1-4).

Costa, P., Mascolo, C., Musolesi, M., & Picco, G. P. (2008). Socially-aware Routing for Publish-Subscribe in Delay-tolerant Mobile Ad Hoc Networks. *IEEE Journal on Selected Areas in Communications*, *26*(5), 748–760. doi:10.1109/JSAC.2008.080602

Costa, P., Migliavacca, M., Picco, G. P., & Cugola, G. (2004). Epidemic algorithms for reliable content-based publish-subscribe: an evaluation. In *Proceedings of the 24th International Conference on Distributed Computing Systems*, (pp. 552-561).

Cugola, G., Nitto, E., & Fuggetta, A. (1998). Exploiting an event-based infrastructure to develop complex distributed systems. In *Proceedings of the 20th International Conference on Software Engineering*, (pp. 261-270).

Datta, A., Quarteroni, S., & Aberer, K. (2004). Autonomous gossiping: A self-organizing epidemic algorithm for selective information dissemination in wireless mobile ad-hoc networks. In *Proceedings of 1st International IFIP Conference on Semantics of a Networked world*, (pp. 126-143).

Elkin, M., Emek, Y., Spielman, D. A., & Teng, S. (2008). Lower-Stretch Spanning Trees. *SIAM Journal on Computing*, *38*(2), 608–628. doi:10.1137/050641661

Fiege, L., Gartner, F., Kasten, O., & Zeidler, A. (2003). Supporting mobility in content-based publish/subscribe middlewares. In *Proceedings of ACM/IFIP/USENIX International Middleware Conference 2003*, 103-122.

Graham, S., Niblett, P., Chappell, D., Lewis, A., Nagaratnam, N., Parikh, J., et al. (2004). *Publish-Subscribe Notification for Web services*. Retrieved August 23, 2008, from http://www.ibm.com/developerworks/webservices/library/specification/ws-pubsub/

Huang, Y., & Garcia-Molina, H. (2003). Publish/subscribe tree construction in wireless ad-hoc networks. In *Proceedings of the 4th International Conference on Mobile Data Management*, (pp. 122-140).

Junginger, M., & Lee, Y. (2004). A self-organizing publish/subscribe middleware for dynamic peer-to-peer networks. *IEEE Network*, *18*(1), 38–43. doi:10.1109/MNET.2004.1265832

Lupu, E., Dulay, N., Sloman, M., Sventek, J., Heeps, S., & Strowes, S. (2008). Amuse: autonomic management of ubiquitous e-health systems. *Concurrency and Computation*, *20*(3), 277–295. doi:10.1002/cpe.1194

Marques, E. R. B., Goncalves, G. M., & Sousa, J. B. (2006). The use of real-time publish-subscribe middleware in networked vehicle systems. In *Proceedings of the 1st IFAC Workshop on Multivehicle Systems*.

Mottola, L., Cugola, G., & Picco, G. P. (2008). A self repairing tree topology enabling content-based routing in mobile ad hoc networks. *IEEE Transactions on Mobile Computing*, *7*(8), 946–960. doi:10.1109/TMC.2007.70789

Muhl, G., Fiege, L., & Pietzuch, P. R. (2006). *Distributed event-based systems*. Berlin-Heidelberg, Germany: Springer-Verlag.

Muthusamy, V., Petrovic, M., & Jacobsen, H. (2005). Effects of routing computations in content-based routing networks with mobile data sources. In *Proceedings of the 11th Annual International Conference on Mobile Computing and Networking 2005*, (pp. 103-116).

Object Management Group. (2007). *Data Distribution Service for Real-time Systems specification, V1.2*. Retrieved August 23, 2008, from http://www.omg.org/technology/documents/formal/data_distribution.htm

Oki, B., Pfluegel, M., Siegel, A., & Skeen, D. (1993). The information bus – an architecture for extensive distributed systems. In *Proceedings of the 1993 ACM Symposium on Operating Systems Principles (SOSP)*, (pp. 58-68).

Pei, G., Ravindran, B., & Jensen, E. D. (2008). On A Self-organizing MANET Event Routing Architecture with Causal Dependency Awareness. In *Proceedings of the 2nd IEEE International Conference on Self-Adaptive and Self-Organizing Systems*, (pp. 339-348).

Peleg, D. (2002). Low stretch spanning trees. In *Proceedings of 27th International Symposium of Mathematical Foundations of Computer Science*, (LNCS, pp. 68-80).

Picco, G. P., Cugola, G., & Murphy, A. L. (2003). Efficient content-based event dispatching in the presence of topological reconfiguration. In *Proceedings of the 23th International Conference on Distributed Computing Systems*, (pp. 234–243).

Pleisch, S., & Birman, K. (2006). Senstrac: Scalable querying of sensor networks from mobile platforms using tracking-style queries. In *Proceedings of the 3rd IEEE International Conference on Mobile Ad-hoc and Sensor Systems*, (pp. 306-315).

Ratnasamy, S., Francis, P., Handley, M., Karp, R., & Shenker, S. (2001). A scalable content-addressable network. In *Proceedings of the ACM SIGCOMM*, (pp. 161-172).

Real-Time Innovations, Inc. (2007). *RTI Data Distribution Service*. Retrieved August 23, 2008, from http://www.rti.com/products/data_distribution/RTIDDS.html

Royer, E. M., & Perkins, C. E. (2000). *Multicast Ad hoc On-Demand Distance Vector (MAODV) Routing (INTERNET DRAFT)*. Mobile Ad Hoc Network Working Group.

Sarkar, S. K., Basavaraju, T., & Puttamadappa, C. (2007). *Ad hoc mobile wireless networks: Principles, protocols and applications*. Boca Raton, FL: Auerbach Publications. doi:10.1201/9781420062229

Stoica, I., Morris, R., Liben-Nowell, D., Karger, D.R., Kaashoek, M.F., Dabek, F., & Balakrishnan, H. (2003). Chord: a scalable peer-to-peer lookup protocol for internet applications. *IEEE/ACM Transactions on Networking, 11*(1), 17-32.

Tsai, W.-C., & Chen, A.-P. (2008). Service oriented architecture for financial customer relationship management. In *Proceedings of the 2nd international conference on Distributed event-based systems*, (pp. 301-304).

KEY TERMS AND DEFINITIONS

Wireless Networks: A kind of network formed by nodes which communicates with each other via wireless channels over the air, in contrast to those networks partly or entirely connected by electric wires or optical fibers.

Wireless Ad Hoc Networks: A kind of decentralized wireless network, in which each node may potentially perform packet forwarding/routing, in contrast to wired networks or one-hop wireless networks.

Mobile Ad Hoc Network (MANET): A kind of self-organizing wireless ad hoc network, with a network of mobile nodes dynamically connected through wireless links.

Publish/subscribe Tree: An acyclic tree-like network inter-connection topology which structures the nodes into a P/S system with a root node and other node directly or indirectly connected to the root.

Event Causal Dependency: The causality between two events – an event is triggered by another event. In a P/S system, the arrival or processing of Event *a* will actuate Event *b*'s publication that is causally dependent on Event *a*.

Single Publish-Hop: Causal event delivery from a node to the other, which triggers an event that is causally dependent on the event delivered on the recipient node.

Multi-Publish-Hop: An event causal chain or even an event causal graph constructed based on the event causal dependencies.

Self-Organization: A process of forming into a coherent unity or functioning whole, by developing an organic structure of the networked system, usually increasing complexity without being guided or managed by any outside entity.

Self-Reconfiguration: A process of system restructuring or rearrangement, in order to adapt to dynamic environment or changing specification without being guided or managed by any outside entity.

ENDNOTE

[1]   To avoid confusion against "physical P/S trees", we always say "causal tree" when we refer to one.

Chapter 16
mTrigger:
An Event–Based Framework for Location–Based Mobile Triggers

Ling Liu
Georgia Institute of Technology, USA

Bhuvan Bamba
Georgia Institute of Technology, USA

Myungcheol Doo
Georgia Institute of Technology, USA

Peter Pesti
Georgia Institute of Technology, USA

Matt Weber
Georgia Institute of Technology, USA

ABSTRACT

Location-based triggers are the fundamental capability for supporting location-based advertisements, location-based entertainment applications, personal reminders, as well as presence-based information sharing applications. In this chapter, we describe the design and the implementation of mTrigger, an event-based framework for scalable processing of location-based mobile triggers (location triggers for short). A location trigger is a standing spatial trigger specified with the spatial region over which the trigger is set, the actions to be taken when the trigger conditions are met, and the list of recipients to whom the notification will be sent upon the firing of the location trigger. The mTrigger framework consists of three alternative architectures for supporting location triggers: (1) the client-server architecture, which allows mobile clients to register and install location triggers of interest on the mTrigger server system; the server being responsible for processing location triggers, performing associated actions and sending out notifications upon firing of triggers; (2) the client-centric architecture, which enables mobile users to manage and process location triggers on their own mobile clients; and (3) the decentralized peer-to-peer architecture, which allows mobile users to collaborate with one another in terms of location trigger processing. The server-centric architecture is particularly suitable for supporting public and shared location triggers, enabling effective sharing of location trigger processing among

DOI: 10.4018/978-1-60566-697-6.ch016

multiple users. The client-centric architecture is more suitable for users possessing mobile clients with high computational capacity and more sensitive to the location privacy of their location triggers. The decentralized peer-to-peer architecture provides on-demand and opportunistic collaboration in terms of location trigger evaluation. Clearly, the performance optimizations for server-centric architecture should focus on efficient and scalable processing of location triggers by reducing the bandwidth consumption and the amount of redundant computation at the server; whereas, the performance optimizations for client-centric architecture and decentralized architecture should also take into account energy efficiency of mobile clients in addition to computational efficiency. In addition, processing of location triggers with moving target of interest requires the knowledge of position information of the moving target and may not be suitable for the client-centric architecture. This chapter will describe the design principles and the performance optimization techniques of the mTrigger framework, including a suite of energy-efficient spatial trigger grouping techniques for optimizing both wake-up times and check times of location trigger evaluations.

INTRODUCTION

Location-based services such as location-based advertisement, location-based entertainment and location-based personal assistant are emerging business applications that demand location-based mobile triggers. Location triggers are standing spatial triggers. Similar to time-based triggers that are used to remind us of the arrival of a future reference time point, location triggers are set on a spatial location of interest, which subscribers of the trigger will travel to at some future time instant. Companies or merchants may use location triggers to support location-based advertisements; for example, Bloomingdale's may send a 20% sale coupon to its medallion member shoppers who are located within five miles of its stores. Individuals use location-based triggers to set up personal reminders indicating the arrival to a spatial location of interest. For instance, a user could set a spatial alarm (Spatial Alarms Project, n.d.) on her mobile client, which alerts her whenever she is near the dry cleaner or her favorite grocery store in her neighborhood, reminding her to pick up or drop her dry cleaning items, or automatically retrieving her stored grocery list.

Dey and Abowd (Dey, A., & Abowd. G., 2000) describe a context-aware system for supporting reminders in order to provide appropriate signals at appropriate times. For example, a reminder for bringing a paper for a meeting is most effective when the user is leaving her office to head for the meeting room. (Dey, A., & Abowd. G., 2000) primarily deals with building a context-aware toolkit for supporting reminder delivery at appropriate times. The ability to locate users using GPS, cell phone positioning and other navigational systems makes context-aware reminder systems feasible. *PlaceMail* (Ludford, P., Frankowski, D., Reily, K., Wilms, K., & Terveen, L., 2006). studies issues related to location-based information systems in order to support useful location-based reminder systems and functional place-based lists. The study determines that effective reminder delivery depends on people's movement patterns through an area and the geographic layout of the space. However, none of the previous work emphasizes the ability to process reminders (or more generally, location-based triggers) *efficiently* from a systems-based perspective. This chapter focuses on optimization techniques which should be deployed for efficient and scalable processing of location-based triggers in different systems-based architectural settings, including a server-centric, client-centric and decentralized architecture. We also consider bandwidth and energy constraints on the client side, which are resource-constrained devices despite significant enhancements in the past few years, for efficient trigger evaluation. More concretely, we develop safe period and safe region-based optimizations in order to facilitate efficient and scalable processing of location-based triggers.

A mobile location-based trigger (location trigger for short) is defined as a standing spatial trigger with four mandatory components: (1) the spatial region over which the trigger is set, (2) the list of subscribers or recipients to which the notification will be sent when the trigger condition is met, (3) the actions to be taken upon firing of the trigger, and (4) the stop condition which specifies the termination constraint for the trigger. Mobile users can define and register their location triggers with the mTrigger (mTrigger, n.d.) system. Once a location trigger is installed, the system will start monitoring the spatial region of the trigger and whenever any mobile subscribers of the trigger enter the spatial region, the location trigger is fired and the specified actions are executed. The result of the action will be sent to the subscriber (recipient) of the location trigger. For example, "*inform me whenever I am within two miles of the dry cleaner at the crossing of Druid Hill Road and Briarcliff Road in the next four weeks*" is a location trigger installed on the spatial region two miles around the dry cleaning store at the crossing of North Druid Hill and Briarcliff with a stop condition of four weeks. The action is simply a notification sent to the owner of this location trigger.

Location triggers are classified based on the scope of their recipients and the mobility characteristics of their monitoring targets and their subscribers. We define three classes of location triggers based on the scope of the recipients: Private, Public and Shared. A location trigger is private if only the owner of the trigger is its recipient. In contrast, public location triggers are those triggers that are installed by their respective owners and are shared by all users of the system. Notifications of public triggers are sent to all users

who have subscribed to the triggers and are online at the time of notification. Traffic notifications, or notifications related to hazardous road conditions are typical examples of public location triggers. While defining a location trigger, the owner may specify a list of people with whom she wants to share the trigger. We call this class of location triggers shared triggers. In this case, only people who are specified on the recipient list and confirm as subscribers of the trigger will receive a notification whenever the location trigger is fired.

Location triggers are also categorized based on the mobility characteristics of their subscribers and monitoring targets. Three types of location triggers are considered in the design of mTrigger: (Mobile *Subscriber*, Static *Monitoring Target*), (Mobile *Subscriber*, Mobile *Monitoring Target*), and (Static *Subscriber*, Mobile *Monitoring Target*). A location trigger is of the type (Mobile *Subscriber*, Static *Monitoring Target*) if it has a static monitoring target, such as the dry cleaning store in the previous example, and a moving subscriber. A location trigger is of the type (Mobile *Subscriber*, Mobile *Monitoring Target*) if its monitoring target, as well as its subscriber are both on the move. For example, "*notify me when all my friends are within a five mile vicinity on highway I-85 North*" is a typical example of this type of trigger. A location trigger is said to be of the type (Static *Subscriber*, Mobile *Monitoring Target*) if its monitoring target is moving but its subscriber is still. In this case, the spatial region of the location trigger moves as the monitoring target moves, but the position of the subscriber remains unchanged. "*Notify me whenever bus No. 5 is five minutes away from the bus-stop near my office*" is an example of a Static *Subscriber* Mobile *Target Trigger*.

In this chapter, we develop an event-based framework, called mTrigger, for scalable processing of location-based mobile triggers. The mTrigger framework consists of three alternative architectures for supporting location triggers:

- The *client-server architecture*, which allows mobile clients to register and install location triggers of interest on the mTrigger system server. The server is responsible for processing location triggers, performing associated actions and sending out notifications upon the firing of triggers. The server-centric architecture is particularly suitable for supporting public and shared location triggers, enabling effective sharing of location trigger processing among multiple users. Clearly, the performance optimizations for the server-centric architecture should focus on efficient and scalable processing of location triggers by reducing the bandwidth consumption and the amount of redundant computation at the server (Bamba, B., Liu, L., Yu, P. S., Zhang, G., & Doo, M., 2008), (Bamba, B., Liu, L., Yu, P. S., & Iyengar, A., 2009).
- The *client-centric architecture*, which enables mobile users to manage and process location triggers on their own mobile clients. The client-centric architecture is particularly suitable for supporting private location triggers and for mobile users who are more sensitive to the location privacy (Bamba, B., Liu, L., Pesti, P., & Wang, T., 2008), (Gedik, B., & Liu, L., 2008) of their location triggers. The performance optimization for client-centric architecture should take into account energy efficiency of mobile clients in addition to computational efficiency (Murugappan, A., & Liu, L., 2008).
- The *decentralized peer-to-peer architecture* is built on top of a client-centric architecture, and is suitable for processing location triggers shared among a group of mobile users. The decentralized architecture relies on collaboration among mobile users in close vicinity and with common interests, to share location trigger processing costs.

In addition, processing of location triggers with moving target of interest requires the knowledge of positioning information of the moving target and thus is not suitable for the client-centric architecture.

In this chapter, we focus on the event-based framework of mTrigger in terms of both client-server architecture (Section 2) and client-centric architecture (Section 4) and the set of server-side optimization techniques for scalable processing of location triggers (Section 3). We evaluate the mTrigger optimization techniques in terms of accuracy and scalability (Section 6), and conclude with the summary and a discussion of future work.

MTRIGGER CLIENT-SERVER ARCHITECTURE

The server-centric architecture consists of five main components.

- The first component is the mobile client module, which handles the interaction of mTrigger Engine with its mobile clients as well as the communication with the mTrigger application server.
- The second component is the mTrigger application server, which supports a variety of application plug-ins that enable the mTrigger engine to support different location services such as spatial alarms, mGraffiti, location-based games, location-based advertisements, and so forth. Each application plug-in typically implements the transformation of application specific service requests to location trigger expressions, which in turn registers with the mTrigger engine through the Trigger Manager.
- The third component is the location trigger manager which handles the trigger registration, mobile user registration, user account management and mTrigger administrative

Figure 1. The mTrigger Client-Server Architecture

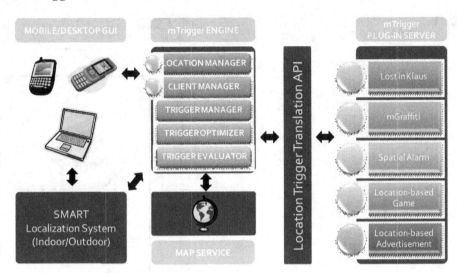

assistance functions. All location triggers and user information are maintained in the mTrigger database.

- The fourth component is the optimization module, which offers a suite of server-side optimization techniques, such as spatial trigger indexing techniques, safe period and safe region techniques for optimizing both wake-up times and check times of location triggers shared by multiple subscribers.

- The fifth component is the location trigger processing module, which evaluates location triggers closer to the mobile subscribers of interest based on the safe region or safe period prediction techniques. Concretely, a mobile client will wake up whenever she moves outside the safe region of one of her location triggers or whenever the safe period for one of her location triggers has expired. Furthermore, a location trigger will be checked through its trigger condition evaluation whenever a mobile user enters the corresponding location trigger monitoring region.

Figure 1 shows an overview of the client-server architecture of the mTrigger system. In this architecture, only a thin client of mTrigger is required to run on the mobile clients, which communicates with the mTrigger engine through the mTrigger application plug-in services.

Positioning: We assume that each mobile device either has a GPS unit or is equipped with Wi-Fi capability. The GPS unit provides longitude, latitude, and altitude of current location. Given the poor performance of GPS indoors, we use Wireless Positioning Systems, such as SMART, to address indoor positioning problems. SMART uses area localization techniques to provide an approximate location of a mobile user with a Wi-Fi enabled device, within an area (for example, the Georgia Tech campus). One of the positioning techniques used in SMART is the Wi-Fi Access Points-based location fingerprinting technique. There are typically two approaches for location acquisition. One approach is to require all mobile users interested in receiving location-based services to update their current position periodically or use one of the inference-based location update techniques, such as dead reckoning. The second location acquisition technique is to apply loca-

Figure 2. Bounding box for a trigger region and installed triggers

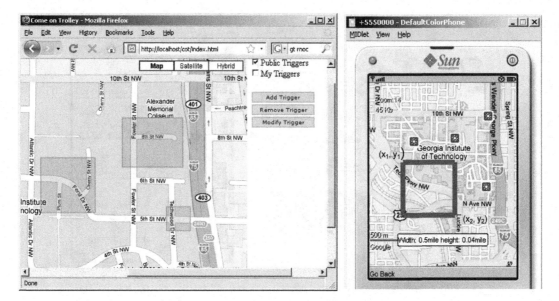

tion determination methods such as triangulation, trilateration, and so forth.

Mobile Device: A thin mTrigger client will be required for all users in the mTrigger client-server architecture. In the mTrigger thin client, we allow users to install a location trigger through both text interface and map-enabled GUI. By enabling a map-based interface, the mobile user can simply mark the spatial area of interest on the map to define the monitoring region of a location trigger. An example usage scenario of such a thin client is shown in Figure 2.

Location Triggers: A mobile user can install many location triggers that are shared among a group of mobile users. On the other hand, public location triggers can be installed by authorized users only. The mTrigger (mTrigger, n.d.) location trigger evaluation engine consists of the client manager, trigger manager, wakeup and trigger evaluation optimization module and location trigger processing module. Each mobile user can register her location triggers directly with the mTrigger engine or through the mTrigger application server plug-in services. As mentioned earlier, each plug-in handles the transformation of the application specific location service request, such as a spatial alarm, or a location-based advertisement request to a semantically equivalent location trigger expression, which in turn is submitted to the mTrigger system for registration, processing and notification.

One of the design goals of the mTrigger event-based framework is to provide middleware support for many location-based services and applications that use location triggers as a fundamental capability. This design framework enables applications to provide the proper mTrigger plug-in services. Typically, different applications may use different compositions of location trigger elements and specification. Some use all the elements of location triggers and others use only some of the elements. Typically the service request format converter may add or remove some elements, or overcome the mismatch between the mTrigger engine and plug-in server. A database is used to provide persistent storage for mobile users along with their positioning information and corresponding safe regions as well as the location triggers registered with the mTrigger system.

Formally, a location trigger *t* is specified in terms of eight elements:

t = (tID, Owner, Region, Event, Subscriber, ActionType, Action, Start, Stop).

- *tID* is a unique identifier that is assigned to each location trigger upon its registration.
- *Owner* is specified by the user id of the owner/creator of this location trigger.
- *Region* is specified by a landmark and a spatial monitoring area nearby or around the landmark, on which the location trigger is set. The monitoring area of the region could be represented by a rectangular bounding box. For simplicity, in the rest of this chapter, we represent each of the trigger regions by a rectangular bounding box, denoted by (x_1, y_1, x_2, y_2), where (x_1, y_1) and (x_2, y_2) represent the bottom-left and top-right vertices of the bounding box, as shown in Figure 2.
- *Event* specifies non-spatial event-based trigger conditions associated with the installed trigger, represented by *<monitored attribute, condition predicate>*. For example the price of gasoline at a gas station is one such non-spatial condition.
- *Subscriber* specifies a list of mobile subscribers to which the trigger notifications should be delivered.
- *ActionType* defines the type of action to be performed when the location trigger is evaluated to be true. *ActionType* is specified by a three element vector *<format, file name, application name>*: (1) *format* specifies the multimedia data format, such as text, URL, image, audio or video; (2) *file name* is the multimedia data file to be sent along with the notification; and (3) *application name* specifies the application plug-in name, such as mGraffiti, Spatial Alarm, or Location-based advertisement.

- For each type of action, the *Action* field contains the delivery and display method, e.g., text or other multimedia message, or an executable method.
- *Start* defines the condition when the location trigger monitoring should start. A simple format for *Start* is a time point, such as 10AM on October 20, 2008.
- *Stop* defines a termination condition for the location trigger. It can be specified as either a time duration, such as 30 minutes (half an hour) from the installation time, or a time point, such as 8AM on October 20, 2008 (two weeks from the installation).

Recall the example location trigger, "*Notify me whenever I am within 2 miles of the dry cleaner at the corner of Druid Hill Road and Briarcliff Road in the next four weeks*". The trigger region is defined by the landmark object - the dry cleaner at the corner of Druid Hill and Briarcliff, and the spatial monitoring region (two miles) around the given landmark object. The subscriber and the owner are the same for this location trigger. The type of the action in this case is simply a notification message sent to the subscriber with Spatial Alarm as the plug-in name. To make the notification more user-friendly, the mTrigger system allows the users to supply photo images or symbols in addition to the text message they wish to receive when the trigger condition is met. In the context of this example, the owner of the location trigger may provide the dry cleaning shop name and a recent photo of the store in JPEG format at the time of trigger installation. The action, by default, will be a method call for displaying the store name and the photo image on the screen of the mobile client. When there is an absence of user-supplied notification information, the text description of the location trigger will be sent to the subscriber.

Location Trigger Processing Engine: Figure 3 provides an overview of the mTrigger engine design. The mTrigger engine manages the mobile

Figure 3. mTrigger engine architecture

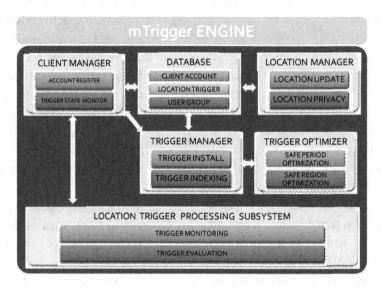

users' accounts and their service requests for installation, removal and modification of location triggers through the Client Manager and the Location Trigger Manager. The *Client Manager* handles all communication between mobile devices and the mTrigger Engine. We provide two ways to interface with mobile clients: one is to allow mobile clients to directly request registration, location update and trigger installation. Another is to allow mobile clients to communicate with the mTrigger system indirectly through the application specific plug-in service. In addition, the Client Manager manages the user account for all registered users, through account activation/ deactivation functions. User accounts may include the notification destination address of the user to which the notification of her location triggers will be delivered. Users

can revise the notification recipients and the notification address for each location trigger installed, as some triggers are shared by a number of users. The client manger will also handle the group creation and maintenance. We use the mutual agreement protocol to establish the sharing of location triggers among multiple users in order to alleviate any location privacy concerns.

For example, assume Alice wants to send a greeting video clip to Bob when Bob gets back home from a business trip. In order to be informed about Bob's arrival by the mTrigger system, Alice needs to obtain Bob's agreement that allows Alice to install a location trigger around the neighborhood of Bob's home. We plan to incorporate stronger location privacy protection (Bamba, B., Liu, L., Pesti, P., & Wang, T., 2008), (Gedik, B., & Liu, L., 2008) in our second release of the mTrigger system. The Location Trigger Manager consists of the Trigger Installation module, Trigger Indexing module and Trigger State Management. The Location Trigger State Management module tracks the state of location triggers and serves requests on active triggers, terminated triggers and number of evaluations performed for each location trigger. The third component of the mTrigger engine is the *Position Manager*, which consists of two key components: Position Update Module and Location Privacy Module. The Position Manager handles the location tracking and update functions for mobile clients according to their location privacy constraints. Two alternative types of location acquisition methods are supported in the mTrigger system: location determination through the

Figure 4. Location triggers example 1: CampusTrolleyEnRoute, campus shuttle tracking service offered on Georgia Tech Campus

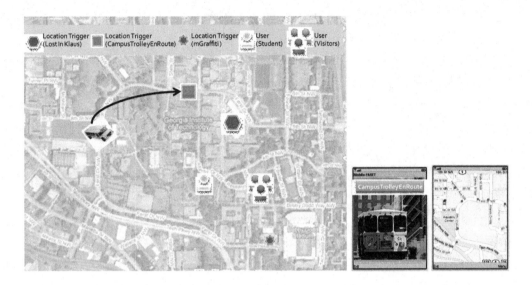

SMART localization system, or a mobile client reporting model, where mobile users report their current location periodically or using a system supplied location derivation method to reduce the energy consumption and the server load due to frequent location updates. *Trigger Optimizer* is the fourth component of mTrigger and consists of the main optimization modules such as Safe Period Optimization and Safe Region Optimization (to be discussed in Section 3). Finally, the fifth component of the mTrigger engine is the *Location Trigger Processing Module*, which consists of Trigger Monitoring and Trigger Evaluation. The mTrigger engine uses a database to maintain user account information, location trigger registration and state management, group creation and maintenance, and for position updates of mobile users.

In principle, upon receiving a location update, the system should evaluate the relevant location triggers. If a trigger condition is met upon a location update of a mobile client, then a trigger notification will be sent to the registered subscriber(s). Clearly, when a mobile user is far away from all her installed location triggers, her location updates would not cause any of her active location

triggers to be true. In order to reduce the energy consumption and the amount of unnecessary location trigger evaluation, the location trigger optimizer provides various data structures such as R-Tree, BR-Tree and Grid index structure. A suite of safe period and safe region-based optimization techniques offer significant savings by reducing the number of unnecessary wakeups and checks to be performed at the mTrigger server. We will describe some of the optimization techniques in Section 3.

It is important to note that the location trigger expression is designed as an internal modelling construct for location triggers in the mTrigger framework. For different applications, one can use application specific plug-ins to make the mTrigger engine transparent to its end users. We now illustrate this abstraction through a number of example applications.

CampusTrolleyEnRoute: The first example is a Georgia Tech Campus bus tracking application. Figure 4 depicts a map of Georgia Institute of Technology and the first prototype of the mTrigger system. Red rectangles represent previously installed location triggers. There are shuttles on

Figure 5. Location trigger example 2: mGraffiti – location-based virtual graffiti

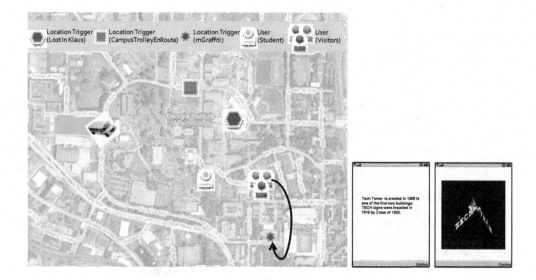

the campus to transfer students and faculty from one building to another within the campus. The CampusTrolleyEnRoute application provides a valuable service in terms of providing the current bus location and an estimate of the time of arrival of the bus to the nearby bus stop. By building CampusTrolleyEnRoute using mTrigger, one can provide a much richer collection of services without additional development effort. For example, one can support spatial alarm type of services such as *"notify me when the yellow bus is five minutes away from the Klaus Advanced Computing Building (KACB)"* or *"when the yellow trolley is about half a mile away."* One can also support services like *"when the trolley approaches the KACB bus stop, ring CERCS office director to notify that IBM visitors are arriving"*. Figure 4 shows the current location of the Trolley buses being monitored.

mGraffiti: The second example application is called mGraffiti; a sample screenshot is shown in Figure 5. The main idea of mGraffiti (mGraffiti, n.d.). is to provide location-based virtual graffiti capability for PDA, Smartphone and Tablet PC users. With virtual graffiti, anyone can become a graffiti writer, including hikers, holiday-goers, and frequent travellers. In the context of the Georgia Tech campus, visitors can benefit from location-dependent virtual Graffiti to walk through the campus and visit the major buildings of the campus, such as the Tech Tower, stadium, recreation center, etc. Instead of writing down messages on walls or carving them on trees, visitors can express their impressions of the campus and leave graffiti or photos of the campus locations of interest using mGraffiti. Alternatively, we can also use Spatial Alarms to implement virtual graffiti as a public trigger or a shared location trigger with a simple action of displaying the text message and the multimedia attached with the text message. When visitors arrive at certain buildings or places of interest, the graffiti message and attached photos will pop up on the mobile client. Figure 5 shows an example of a textual graffiti message at Tech tower.

Spatial Alert: The third example application is Spatial Alert. Spatial Alerts are also called spatial reminders, used widely for reminding mobile users when they approach some location of interest, defined either by themselves in the form of spatial reminders, or defined by their friends through spatial alerts (Bamba, B., Liu, L., Yu, P. S., & Iyengar, A., 2009). In fact, spatial alerts are a specific location trigger, in which notification is the only type of action being taken when the location trigger conditions are met. Examples of spatial alerts include *"Remind me when I am within two miles of the dry cleaning store at the crossing of Druid Hill Road and Briarcliff Road"* or *"Remind me when any of my friends are present in the Starbucks coffee store across the Ferst Drive from KACB"*. The main distinction between mGraffiti and the Spatial Alert service lies in the usage model and the context in which a location trigger is utilized.

Lost in Klaus: This is an indoor navigation system. Most new buildings on the Georgia Tech campus have complex building layouts and it is often difficult to navigate and find specific office locations (Lost in Klaus, n.d.). This is especially true for visitors, including students or staff members that do not have an office in the building. This project provides indoor shortest path-based navigation on mobile devices. The application utilizes Wi-Fi based localization to locate a mobile user in the building, and shortest path-based navigation to show how to get to a targeted office or meeting room through the shortest travel path. There are several ways one can employ location trigger capabilities to provide value added services. For instance, the KACB building manager can set a public location trigger on the entry points of the building. Upon entering the building, the Lost in Klaus client application will be launched on the client devices of mobile users. It determines the current position of a mobile user and provides the shortest path navigation to the destination office or classroom or meeting room. One can also extend this navigation system by allowing visitors or students to leave a note at a particular office in case the professor or the meeting organizer is absent at the time of the meeting. This can be easily supported by directly employing the mGraffiti plug-in service. Another desired capability is to monitor a specific room and notify the subscriber whenever the professor enters the room, or the door of the office being monitored is open. This can be provided through the Spatial Alarm plug-in to the mTrigger system.

SCALABLE PROCESSING OF LOCATION TRIGGERS: SERVER-SIDE OPTIMIZATIONS

Processing of location triggers requires meeting two demanding objectives: *high accuracy*, which ensures no location triggers are missed, and *high scalability*, which guarantees that location trigger processing is highly efficient and can scale to a large number of triggers and the growing base of mobile users.

A naïve approach to location trigger processing is periodic evaluation at a high frequency. Concretely, each location trigger is evaluated periodically by testing whether the subscriber has entered the spatial region of the location trigger and whether the non-spatial trigger conditions are also met. High frequency in periodic evaluation is essential to ensure that none of the location triggers are missed. However, a proper setting of such a period is a hard problem. If the period is set too large, mobile clients may pass trigger monitoring regions without firing the location triggers in time, incurring a higher miss rate. On the other hand, if the update period is kept too small, it would make the mobile client wake up too frequently, which is energy inefficient for mobile clients and unnecessarily increases the processing load at the server. As a result, the periodic evaluation approach to processing location triggers suffers from a number of drawbacks.

- First, the miss rate of location trigger evaluation is unpredictable as there is no appropriate technique for the system to determine the ideal trigger evaluation period. In the case of a high alarm miss rate, the system fails to meet the desired high accuracy requirement of location trigger processing.

- Second, the periodic trigger evaluation approach is expensive as it performs a large number of unnecessary evaluations at high frequency; hence, it is not scalable in the presence of a large number of location triggers and a large number of mobile users. In fact, the number of unnecessary evaluations increases as mobile users move farther away from the monitoring regions of their location triggers.

Consider an example scenario where a mobile user is moving in an area (say Atlanta) that is far away from her registered location triggers (say Miami). Obviously, it is unnecessary to evaluate her location triggers upon her location updates until she has traveled to a location that is near the city border of Miami. Motivated by this observation and the drawbacks of the periodic trigger evaluation, we argue that in a client-server platform, we need to optimize the location trigger evaluation by reducing the number of wakeups that the mobile clients need to perform for the purpose of location trigger evaluation. We also need to minimize the amount of unnecessary trigger processing the server needs to perform upon each location update of a mobile client. We now outline two optimization techniques used in the mTrigger system to enhance scalability of the location trigger processing engine while maintaining desirable accuracy: *safe period optimizations* and *safe region optimizations*.

Safe Period Optimizations

Safe period is defined as the duration of time for which it is safe not to check a given location trigger for a given mobile subscriber, as the probability that the location trigger condition is met for the subscriber is zero. Consider a subscriber S_i and a location trigger T_j. The safe period of trigger T_j

with respect to subscriber S_i, denoted by $sp(S_i, T_j)$ can be computed based on the distance between the current position of S_i and the trigger monitoring region R_j, taking into account the motion characteristics of S_i and the monitoring target of location trigger T_j. Concretely, the two factors that influence the computation of safe period $sp(S_i, T_j)$ are (i) the velocity-based motion characteristic of the subscriber S_i, and (ii) the distance from the current position of subscriber S_i to the spatial region R_j of location trigger T_j. Thus the safe period $sp(S_i, T_j)$ can be computed as follows:

$$sp(S_i, A_j) = \frac{d(S_i, R_j)}{f(V(S_i))}$$

Clearly, the distance measure between the current location of the mobile subscriber and the trigger region R_j is the first important parameter for safe period computation. The second important parameter is the velocity measure of the mobile subscribers or the mobile targets of the location trigger. *Euclidean distance* and *road network distance* are among the most commonly used distance measures. In order to ensure 100% location trigger accuracy, we use *maximum velocity* of the mobile clients for the velocity measurement even though it provides a pessimistic estimate for the safe period. Other measures like *expected speed* may be used with precaution. We refer interested readers to (Bamba, B., Liu, L., Yu, P. S., Zhang, G., & Doo, M, 2008) for a more detailed discussion on these topics.

Concretely, the motion-aware safe period approach optimizes the location trigger processing through two basic mechanisms. First, we introduce the concept of safe period to minimize the number of unnecessary trigger evaluations, increasing the throughput and scalability of the system. We show that our safe period-based trigger evaluation techniques can significantly reduce the server load for location trigger processing, compared to the periodic evaluation approach, while preserving the accuracy and timeliness of location triggers. Second, we develop a suite of location trigger grouping techniques based on spatial locality of the triggers and motion behaviour of the mobile subscribers, which significantly reduces the safe period computation cost for location trigger evaluation at the server side. Only when the mobile client is entering a location trigger group region, say Group 3 in Figure 6, will all her location triggers within the group 3 be evaluated. By grouping similar location triggers based on their spatial locality and motion behaviour of the mobile users, we can, to some extent, reduce the number of trigger evaluations required.

Basic Safe Period Optimization: The safe period-based approach processes a location trigger in three stages. First, upon the installation of a location trigger, the safe period of the trigger with respect to each authorized subscriber is calculated. Second, for each trigger-subscriber pair, trigger evaluation is invoked upon the expiration of the associated safe period, and a new safe period is computed. In the third stage, a decision is made regarding whether the location trigger should be checked or one should wait for the new safe period to expire. If the new safe period is smaller than a threshold t_δ, it means that the mobile client is entering the trigger monitoring region and the trigger condition is checked. Compared to periodic trigger evaluation, the safe period approach for location trigger processing reduces the amount of unnecessary evaluation steps, especially when the subscriber is far away from all her location triggers. On the other hand, the main cost of the basic safe period approach described in this section is due to the excessive amount of safe period computations.

Subscriber Specific Spatial Locality Grouping-based Safe Period Optimization: We present one such technique here which groups location triggers according to the subscriber-specificity at the first level, followed by the spatial locality of the alarms at the next level. Figure 7(a) displays

Figure 6. Location trigger grouping

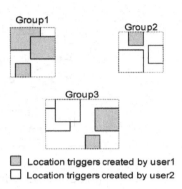

the monitoring regions for a set of installed location triggers. The triggers for user 1 are marked by shaded trigger monitoring regions. Basic safe period evaluation computes the distance from each of the six triggers $\{T_i \mid 1 \leq i \leq 6\}$. Subscriber-specific spatial locality-based grouping performs a two level grouping: the first level grouping is on all subscribers and the second level grouping is on spatial alarms relevant to each subscriber. We use a B-Tree based implementation to speed up search on subscribers and an R-Tree implementation to capture spatial locality of location triggers for each subscriber in order to speed up the search of triggers. The underlying data structure is a hybrid structure which uses a B-tree for subscriber specific search at the first level and an R-tree for subscriber specific spatial alarm search at the second level. Figure 7(b) shows an example of this grouping. Triggers installed by user 1 are grouped together in TG_1 and TG_4 and may be fired only when the user is entering the MBRs of TG_1 or TG_4. Subscriber specific spatial locality-based grouping has two advantages over the basic safe period computation approach. First, the number of safe period computations is significantly reduced. Second, each alarm group contains alarms relevant to a single user, thus no irrelevant processing is performed. Our experimental results show that this approach is efficient in the presence of a large number of subscribers, and for a large number of private and shared alarms.

Figure 7. Safe period-based alarm evaluation techniques

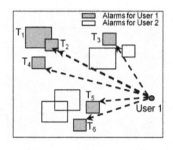

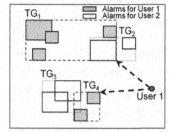

(a) Per trigger safe period
computation

(b) Subscriber specific trigger
grouping-based safe period
computation

We evaluate the scalability and accuracy of our safe period-based optimization using a road network simulator. Our experimental results show that our motion-aware safe period-based approach offers significant performance enhancements, while maintaining high accuracy of location triggers, especially compared to the conventional periodic trigger evaluation approach.

Safe Region Optimizations

In contrast to safe period-based optimization, the safe region-based optimization computes a safe region for each subscriber, instead of each subscriber and trigger (or trigger group) pair. As long as the subscriber remains inside its computed safe region Ψ_S, the probability of any relevant triggers of this subscriber being triggered is zero. The server computes a safe region for each subscriber and communicates this safe region to the subscriber. The subscriber is responsible for monitoring its position within the safe region. Once the subscriber moves out of its safe region, the subscriber will provide a location update to the server which recomputes the safe region. A simple grid structure is overlaid on top of the *Universe of Discourse (or the geographical map of interest)* while computing the safe region. By limiting the

safe region computation to the current grid cell of the subscriber position, we can drastically reduce the safe region computation cost.

We have developed a number of techniques for safe region computation in (Bamba, B., Liu, L., Yu, P. S., & Iyengar, A., 2009). In this section, we briefly review *Bitmap Encoded Safe Region (BSR)* techniques for safe region computation, which can express larger safe regions using a simple bitmap. We first describe a simple *Grid Bitmap Encoded Safe Region (GBSR)* computation technique and show that it does not provide an optimal way to compute safe regions efficiently and accurately. Then we develop an extension to the *GBSR* approach using a *pyramid* (Samet, H., 1990) data structure, referred to as the *Pyramid Bitmap Encoded Safe Region (PBSR)* approach.

Bitmap Encoded Safe Region Computation: A bitmap encoded safe region represents a safe region Ψ_S for subscriber S using a bitmap B of length n. A bit value of 1 indicates that a predefined region (cell) belongs to the safe region; whereas a 0 bit indicates the negation. BSR techniques exhibit the following advantages: (i) for low trigger density regions, it allows for the reduction of location updates from the clients to the server when compared to other known safe region approaches (Bamba, B., Liu, L., Yu, P. S.,

& Iyengar, A., 2009), (ii) it supports different granularity of safe region computations for different subscribers thus supporting heterogeneity among client capabilities, and (iii) clients can determine their position with respect to the safe region using a predefined (worst case) number of computations.

Consider the example in Figure 8(a). The mobile user S located at position P has four location triggers, each is specified by the spatial region of the trigger, denoted as $R(S, T_i)$ (i=1, 2, 3, 4). Thus the monitoring region for the mobile user S at point P is the shaded rectangle with four relevant trigger regions intersecting the grid cells. Considering a 3×3 grid overlaid on top of the monitoring region, there are only three shaded cells, which can be used as the safe region for S as shown in Figure 8(b). The server communicates this safe region to the client S. By providing each mobile client with its safe region, the server is virtually distributing the monitoring of when to wake up and check the relevant location triggers across all its mobile clients.

Grid Bitmap Encoded Safe Region Computation Technique: The safe region for a subscriber S can be represented by the set of grid cells as shown in Figure 8(b). We use a grid bitmap scheme to represent the safe region of subscriber S as shown in Figure 8(b). The cell $C_{k,l}$ is represented by a single bit $B(C_{k,l})$. If the following condition,

$$C_{k,l} \cap \sum_{m=1}^{|A_S|} R(s, A_m) = \varphi$$

is true, we set $B(C_{k,l}) = 1$, denoting that the entire cell $C_{k,l}$ belongs to the safe region Ψ_S. Otherwise we set $B(C_{k,l}) = 0$ and split $C_{k,l}$ into $U \times V$ smaller equi-sized cells. The same encoding procedure is used for each smaller cell. This bitmap encoding technique provides a compact representation for safe region Ψ_S. Figure 8(b) shows the safe region representation for the safe region of Figure 8(a) using a bitmap encoding scheme. There are no

trigger regions intersecting with the three shaded cells. Thus these cells are represented by 1's in the bitmap. For those cells intersecting with some trigger regions, the corresponding bitmap cells are represented by 0's. The safe region is represented using a simple bitmap $B = 0000011010$ which represents the cell bit values in a raster scan fashion. The first zero bit corresponds to the entire cell, indicating that the cell does not belong to the safe region and has location trigger monitoring regions intersecting with it. However, this approach has two obvious drawbacks. First, the size of the grid cell constrains the grid cell-based safe region computation to be limited to the cell size and fails to take into account other finer granularity spatial areas which do not overlap with any trigger regions in the safe region computation. As visible from Figure 8(b), this safe region representation is able to represent only a small portion of the monitoring region thus providing a poor estimate of the actual safe region. Second, this approach can be expensive in terms of communication costs incurred due to more frequent broadcasting of the newly calculated safe region to each mobile client whenever a subscriber moves out of her current safe region.

Figure 8(c) presents a 9×9 split of the cell at a finer resolution which allows for more accurate representation of the safe region. However, this approach is inefficient in representing safe regions for two reasons: (i) it unnecessarily uses a much larger bitmap than required to represent the safe region, and (ii) different regions will have different trigger densities thus making it difficult to select a uniform grid cell size.

One approach to overcome the grid-based bitmap safe region computation is to use a *pyramid* (Samet, H., 1990) data structure, and we refer to it as the *Pyramid Bitmap Encoded Safe Region (PBSR)* approach. The pyramid approach allows for more accurate representation of the safe region and more efficient safe region computation while keeping the bitmap size small. Furthermore, the *PBSR* approach provides flexibility by allowing

Figure 8. Bitmap encoded safe region computation

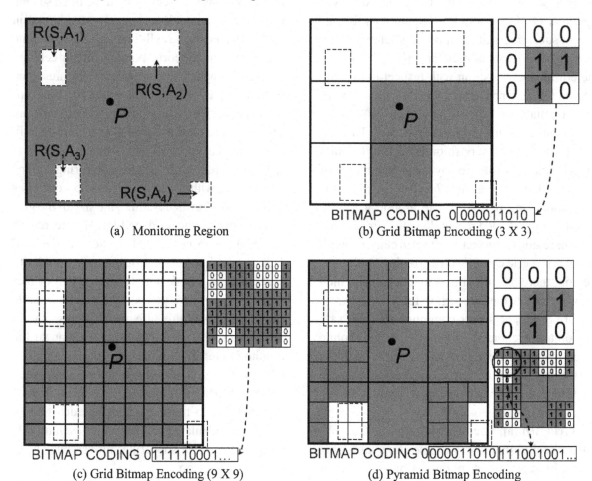

(a) Monitoring Region

(b) Grid Bitmap Encoding (3 X 3)

(c) Grid Bitmap Encoding (9 X 9)

(d) Pyramid Bitmap Encoding

clients to adjust the granularity of their safe region based on their computing capability. By allowing the safe region computation for each client to be personalized according to its capability, the *BSR* computation offers greater flexibility in safe region computation by providing larger, complex safe regions for clients with higher computational capacity.

Pyramid Bitmap Encoded Safe Region Computation Technique. The pyramid representation only splits those cells in the *base* grid (level $L=0$) into $U \times V$ smaller cells when $B(C_{i,j}^0$

$) = 0$, where U, V are system defined parameters. The process may be further repeated for several

iterations to form smaller cells at each level, thus forming a pyramid data structure of height h. Figure 8(d) shows the use of a pyramid structure with $h=2$. By further splitting cells with $B(C_{i,j}^0)$

$= 0$ into a 3×3 grid we obtain a much more accurate representation for the safe region of S. Recall that the grid-based approach, not only fails to represent the safe region accurately (3×3 grid in Figure 8(b)) when the grid cell size is large, but also fails to use a much larger bitmap (9×9 grid in Figure 8(c)) when the grid cell is small, due to the limited capacity at the mobile client and the high communication cost of broadcasting the larger bitmap to the client. In contrast, the

Figure 9. Possible locations of mobile user

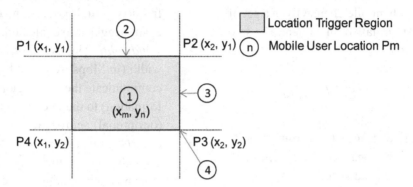

PBSR approach provides flexibility in computation of the safe region while being economical in terms of bitmap size. For example, the *GBSR* approach requires 82 bits, 1 bit for the entire cell and 81 bits for the 9×9 grid, to represent the safe region in Figure 8(c). In comparison, the *PBSR* approach requires only 64 bits, 1 bit for the entire cell, 9 bits for the cells at level 1 and 54 bits for the cells at level 2, to represent the same safe region, as shown in Figure 8(d). We refer interested readers to (Bamba, B., Liu, L., Yu, P. S., & Iyengar, A., 2009) for a more detailed description of the *PBSR* approach and the algorithmic details.

MTRIGGER CLIENT-CENTRIC ARCHITECTURE

The client-centric architecture includes the GUI interface on a mobile device, the mTrigger engine running on the mobile client, and a suite of energy-efficient location trigger evaluation techniques for optimizing both wake-up times and check times of location triggers.

In the client-centric architecture, all the necessary location trigger processing modules will reside on a mobile device such as cell phones, smart phones, and PDAs. Due to space constraints, we omit the client-centric architecture discussion and only describe briefly the optimization strategies used in our client-centric version of mTrigger. It

is important to note that similar to the server side optimization, if a mobile client is far away from any of her location trigger monitoring regions, then it should be possible to compute a safe wakeup period during which the client can sleep, while guaranteeing that none of the location triggers would be missed. There are two factors that are critical in determining a wakeup period: (a) the speed of the mobile client; and (b) the size of the location trigger region.

To compute the speed of the mobile client and her safe period with respect to her installed location triggers, we need to measure the distance between a mobile user and the trigger region for each of her location triggers. Based on the Euclidean distance measure, there are four possible user locations P_m as described in Figure 9. We use $d_{m,R}$ to denote the distance between one of the four possible user locations (P_m) and the location trigger region R, and we can compute $d_{m,R}$ as follows:

$$d_{m,R} = \begin{cases} 0 & (x_1 \leq x_m \leq x_2 \quad and \quad y_1 \leq y_m \leq y_2) \\ \min(|x_m - x_1|, |x_m - x_2|) & (x_1 \leq x_m \leq x_2) \\ \min(|y_n - y_1|, |y_n - y_2|) & (y_1 \leq y_n \leq y_2) \\ \min(D_{m,1}, D_{m,2}, D_{m,3}, D_{m,4}) & Otherwise \end{cases}$$

where $D_{m,1}, D_{m,2}, D_{m,3},$ and $D_{m,4}$ represent Euclidean distance between P_m and four rectangle vertices $P_1, P_2, P_3,$ and P_4 in R.

Based on the Euclidean distance and the expected speed, computed using maximum speed and the most recent speed, we can compute the safe period for each mobile client. Given a set of *n* location triggers installed on a client, her safe period *t* is defined by

$$t = \min(d_{m,R_1}, d_{m,R_2}, ..., d_{m,Rn}) / v_{expected},$$

where $d_{m,Rn}$ is the Euclidean distance from a mobile user's location to the location trigger region R_n and $v_{expected}$ is the expected speed.

EXPERIMENTAL EVALUATION

In this section, we provide a brief evaluation of the safe period (*SP*) and safe region optimization techniques compared to periodic alarm evaluation (*PRD*) and a theoretical optimal (*OPT*) approach. The theoretical optimal approach assumes no restrictions on resource availability, namely all relevant location triggers within the monitoring region can be broadcast to the client. This implies the client is fully aware of all relevant alarms in its vicinity. We measure the performance of all approaches based on four different evaluation metrics.

- *CPU Load/Capacity:* This factor measures the scalability of the system. It is measured as the ratio of the amount of CPU time used by the system to perform trigger processing and safe region or safe period computations to the amount of time available to the system to perform this processing. CPU load/capacity of greater than 100% indicates the failure of the system to scale to the desired configuration.
- *Wireless Communication Cost:* This is measured by the number of updates sent to the system by the mobile clients. We measure this parameter as a ratio of the communication costs required by a particular approach to the communication costs in-

curred by periodic trigger processing at a frequency high enough to ensure all relevant triggers are evaluated.
- *Bandwidth:* This is the downstream bandwidth (in Mbps) required by the system to communicate the safe region (or trigger information) to the clients for the safe region (or optimal) approaches.
- *Client Computation Cost:* This metric indicates the cost incurred by clients to check their position relative to the safe region in terms of average number of computations performed per client per second.

We do not measure location trigger evaluation accuracy as the parameters adopted for each processing approach ensure 100% of the location triggers are evaluated in all scenarios. The sequence of triggers to be evaluated is determined by a very high frequency trace of the motion pattern of the vehicles. We below briefly describe the experimental setup used to evaluate our system.

Experimental Setup: Our simulator generates a trace of vehicles moving on a real-world road network using maps available from the National Mapping Division of the U.S. Geological Survey (USGS, n.d.) in Spatial Data Transfer Format (SDTS (Spatial Data, n.d.)). Vehicles are randomly placed on the road network according to traffic densities determined from the traffic volume data in (Bamba, B., Liu, L., Yu, P. S., Zhang, G., & Doo, M, 2008). The simulator simulates the motion of vehicles on roads with appropriate velocity information; at intersections, vehicles may move in any direction with attached probability values. We use a map of Atlanta and surrounding region which covers an area around 1000 km² in expanse, to generate the trace. Our experiments use traces generated by simulating vehicle movement for a period of one hour, results are averaged over a number of such traces. Default traffic volume values allow us to simulate the movement of a set of 10,000 vehicles. Each vehicle generates a set of position parameters during the simulation which are evaluated against the generated spatial alarm information. Default values require each vehicle to generate updates with a period of less than a second for periodic processing. The default spatial alarm information consists of a set of 10,000 spatial alarms installed uniformly over the entire map region. Uniform distribution assumption is commonly considered in spatial database research. We vary the fraction of public alarms installed in the system to vary the number of alarms relevant to each client. This simulator setup allows us to the test the robustness of our framework under realistic mobility patterns.

Experimental Results: We conduct two sets of experiments detailing the performance of the *BSR* technique and present a comparison of the different approaches to location trigger evaluation.

1. Performance Evaluation of BSR Approach: This set of experiments is designed to evaluate the performance of the BSR approach. We vary the height of the pyramid from $h = 1$ (for *GBSR*) to $h = 7$ and observe the performance as shown in Figure 10. Figure 10(a) displays the wireless communication costs incurred as we increase the pyramid height from $h = 1$ to $h = 7$. It can be observed that the *GBSR* approach is highly inefficient as it limits safe region computation to a very high granularity. The safe region computed using this approach provides a very coarse representation of the actual safe region forcing the clients to frequently update their location as a result of which *GBSR* approach incurs high communication costs. As we increase the pyramid height, more accurate safe region representations can be computed and consequently wireless communication costs experience a sharp drop. Another observation is that *BSR* approaches display high sensitivity to alarm density levels; the performance deteriorates sharply for higher fraction of public alarms which implies higher density of relevant alarms. On the other hand, the bandwidth required by the server to broadcast the safe regions to the clients increases with pyramid height as shown in Figure 10(b). For higher level pyramids, larger bitmaps

are required to represent the safe region and hence higher bandwidth is required. For pyramid height $h = 7$, with high alarm density the downstream bandwidth requirement goes up to 3.2 Mbps, but for $h = 5$ this value remains below 250 Kbps even when the fraction of public alarms is increased to 0.2. Figure 10(c) displays the average number of computations performed per client per second to determine its position within the safe region. Clients need to perform the safe region containment detection check to determine if they need to update their position. For the *GBSR* approach the clients need to perform an average of 2-3 computations per second. This cost does not experience a significant increase with pyramid height for low fraction of public alarms. For higher fraction of public alarms the costs rise to 6-7 computations per second for a pyramid of height $h = 7$. As seen from Figure 10(d), for low pyramid height, safe region computation costs are low as relatively simpler computations are involved. On the other hand, alarm processing costs are high as a large number of updates are received from clients. On increasing pyramid height, alarm processing costs drop due to fewer client position updates. The safe region computation costs increase due to high complexity of safe region computation. Even despite the fewer number of safe region computations being performed at a higher pyramid height, the increase in cost of a single safe region computation is such that a net increase in safe region computation load is experienced. However, this cost can be significantly offset by using precomputed bitmaps for public alarms as described earlier. For $h = 4$ or $h = 5$, the overall CPU load is at its lowest point. With increasing the fraction of public alarms the system experiences an increase in overall CPU load.

2. Performance Comparison of Alarm Evaluation Techniques: Now we compare the performance of the safe region approach (*PBSR*) with periodic processing, safe period-based processing and the optimal approach. As seen in Figure 11(a), the safe region approach (*PBSR with h=5*)

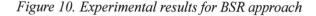

Figure 10. Experimental results for BSR approach

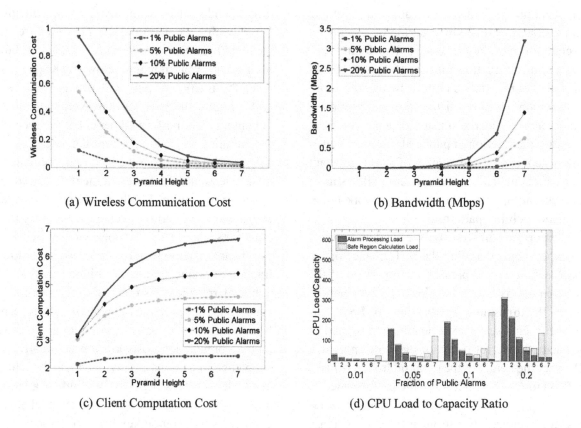

(a) Wireless Communication Cost

(b) Bandwidth (Mbps)

(c) Client Computation Cost

(d) CPU Load to Capacity Ratio

incurs very low wireless communication costs. Periodic processing requires clients to transmit each location update to the server incurring a wireless cost of 1 and is not shown in the figure. The safe period approach experiences significantly higher communication costs, approximately 2-3 times the cost incurred by the safe region approach. This is largely due to the pessimistic assumptions required to ensure that the safe period approach evaluates all triggers with a 100% success rate. For lower trigger density levels the gap between the optimal and safe region approach is much lower. The optimal approach would require clients to transmit updates only when the spatial constraints for one or more relevant location triggers are met. The CPU load experienced by each approach is as shown in Figure 11(b). The periodic approach (PR) has much higher trigger processing costs as

each update needs to be processed by the client and the CPU load does not scale. The processing load does not increase much at higher trigger densities, as each update is processed by this approach for all fractions of public triggers. The safe region approach, denoted by *PB* in the figure, experiences lower CPU load due to a much lower trigger processing load. With an increasing fraction of public alarms, the safe region computation, as well as the trigger processing load, rises. However, the total load incurred by the system is much lower than the periodic approach for all configurations. The safe period (SP) approach experiences a higher CPU load compared to the safe region approaches. This is a direct result of the larger number of location updates that need to be processed by the safe period approach. Results for the optimal approach are plotted to show that

Figure 11. Experimental comparison of different alarm evaluation techniques

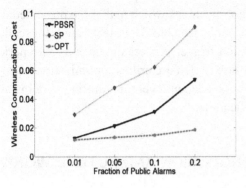

(a) Wireless Communication Cost

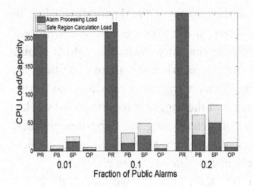

(b) CPU Load to Capacity Ratio

the safe region approaches do not incur much higher CPU load except for the situation when the fraction of public alarms is very high.

RELATED WORK

There are a number of research areas related to the mTrigger system, including Human Computer Interaction (HCI), Spatial-temporal databases, Geographical Information Systems (GIS), games programming, to name a few.

In the HCI area, location reminder systems have been studied by several projects (Dey, A., & Abowd. G., 2000), (Kim, S., Kim, M., Park, S., Jin, Y., & Choi, W., 2004), (Ludford, P., Frankowski, D., Reily, K., Wilms, K., & Terveen, L., 2006)., (Marmasse, N., & Schmandt, C., 2000), (Sohn, T., Li, K., Lee, G., Smith, I., Scott, J., & Griswold, W., 2005), focusing on usability of such applications. However, all existing work in the HCI area has been conducted from the usability perspective of the location reminder systems and applications. Geominder (Geominder, n.d.). and Naggie (Naggie 2.0, n.d.) are location reminder systems providing useful location reminder services using cell tower ID and GPS technology, respectively. With the development of mobile

platforms like Android (Android, n.d.) and iPhone SDK (Apple Developer Connection, n.d.), we see similar applications made available on a multitude of smart phones. However, none of these existing approaches deal with the system level performance optimization issues, which we believe are critical for wide deployment of any location trigger-based systems and applications.

In the realm of information monitoring, event-based systems have been developed to deliver relevant information to users on demand. User defined triggers can be initiated when new relevant information which is of personal interest to the user is detected by the system (Bazinette, V., Cohen, N., Ebling, M., Hunt, G., Lei, H., Purakayastha, A., Stewart, G., Wong, L., & Yeh, D., 2001), 23. In addition to monitoring continuously changing user information needs, the mTrigger system also needs to deal with the complexity of monitoring user location data in order to trigger relevant alerts in a non-intrusive manner.

Periodic reevaluation is commonly used for the continuous monitoring of moving objects (Jensen, C. Lin, D., & Ooi. B., 2004), (Mokbel, M., Xiong, X., & Aref, W., 2004), (Yu, X., Pu, K., & Koudas, N., 2005). Some work exists on monitoring continuous queries, which applies the concept of safe region directly or indirectly

(Cai, Y., & Hua, K., 2002), (Hu, H., Xu, J., & Lee, D., 2005). Spatial alarms differ from this work, as they do not demand periodic evaluation or reevaluation like continuous queries; instead they require one shot evaluation which should result in a trigger when the alarm conditions are satisfied. Once a trigger is activated, no further evaluation of the trigger is required. Our work is focused on determining the opportune moment for evaluating spatial alarms relevant to a client by seeking cooperation at the client end.

Furthermore, numerous works have dealt with the problem of energy conservation in mobile devices (Flautner K., & Mudge. T., 2002), 18, (Mudge, T., 2001). To the best of our knowledge, none have systematically addressed the processing of location triggers using an event-based framework. mTrigger is the first system that utilizes the concept of location triggers as an abstraction to model a variety of location-based monitoring and notification services, such as spatial alarms, (Bamba, B., Liu, L., Yu, P. S., & Iyengar, A., 2009), (Murugappan, A., & Liu, L., 2008), (mGraffiti, n.d.), and location-based advertisements (GPS Daily, 2008).

Other areas which deal with spatial regions are Geographic Information Systems (GIS) and game programming. In principle, Geographical Information Systems (GIS) do not deal with mobile objects and their movement characteristics. Instead, GIS focuses on map visualization and spatial data analysis.

In the area of game programming, detailed animated objects interact not only with their complex environments but also with each other. Real-time collision detection Ericson, C. (2005) is required in order to maintain a believable simulation status. It is essential to consider the approximate shapes of complex objects for collision detection in such environments. Additionally, rendering of objects is essential to maintain reasonable frame rates. Dead reckoning optimizations are commonly applicable to handle network delay impact (Baughmann, N. E., & Levinne, B. N., 2001) in such environments.

Our event-based system does not demand such continuous knowledge of position information of mobile clients which makes the presented optimizations feasible. Thus, techniques for scaling location trigger processing, such as safe period and safe region are critical optimizations for location dependent event monitoring systems and applications.

CONCLUSION AND FUTURE WORK

We have presented mTrigger – an event-based framework for providing a scalable infrastructure for supporting location-based event systems and applications in the future pervasive computing environment. The mTrigger development has two unique features. First, it provides an event-based location trigger framework for supporting large scale location trigger based applications through mTrigger plug-in modules. We demonstrate the usage and the benefits of mTrigger middleware framework through a number of mobile applications developed for the Georgia Tech campus. Second, we develop the safe period and safe region based optimization techniques as one of the building blocks of mTrigger's design for providing scalable and highly efficient location trigger evaluation. Our experimental evaluation shows that the mTrigger framework and its safe period and safe region optimizations can significantly reduce server processing load and application development cycle.

Our ongoing and future work continues along three directions. First, we are actively studying the optimization techniques for location triggers of mobile users traveling on road networks, especially how road network and travel time constraints can be utilized to optimize the processing of location triggers and enhance accuracy and performance of event based applications built on top of location triggers, such as location based advertisements, location based entertainment, and location-based tracking. Second, we are currently developing concrete mobility-based user group formation methods to be used as the building blocks for co-operative location trigger evaluation in decentralized geographical overlay networks (Zhang, J., Zhang, G., & Liu, L., 2007). Third but not the least, we are interested in location privacy and location security issues (Bamba, B., Liu, L., Pesti, P., & Wang, T., 2008), (Gedik, B., & Liu, L., 2008), (Wang, T., & Liu, L., 2009) for safeguarding user privacy in the location trigger supported event systems and applications.

ACKNOWLEDGMENT

This work is partially sponsored by grants from NSF CISE NetSE program, CyberTrust program, an IBM faculty award, an IBM SUR grant, and an Intel research council grant. The authors are also grateful to many discussions with members of the Distributed Data Intensive Systems Laboratory (DiSL) at Georgia Institute of Technology.

REFERENCES

Android. (n.d.). *An Open Handset Alliance Project*. Retrieved from http://code.google.com/android/

Apple Developer Connection. (n.d.). *iPhone SDK*. Retrieved from http://developer.apple.com/iphone/program/download.html

Bamba, B., Liu, L., Pesti, P., & Wang, T. (2008, April). Supporting Anonymous Location Queries in Mobile Environments with PrivacyGrid. In *Proceedings of 17th International World Wide Web Conference* (WWW). Beijing, China

Bamba, B., Liu, L., Yu, P. S., & Iyengar, A. (2009). Distributed Processing of Spatial Alarms: A Safe Region-based Approach. In *Proceedings of IEEE Int. Conf. on Distributed Computing*, June 22-26, Montreal, Quebec, Canada.

Bamba, B., Liu, L., Yu, P. S., Zhang, G., & Doo, M. (2008, December). Scalable Processing of Spatial Alarms. In *Proceedings of 15th International Conference on High Performance Computing*. Bangalore, India.

Baughmann, N. E., & Levinne, B. N. (2001). Cheat-proof playout for centralized and distributed online games, In *Proceedings of IEEE INFOCOMM*.

Bazinette, V., Cohen, N., Ebling, M., Hunt, G., Lei, H., Purakayastha, A., Stewart, G., Wong, L., & Yeh, D. (2001). *An Intelligent Notification System*. IBM Research Report RC 22089 (99042).

Cai, Y., & Hua, K. (2002). An Adaptive Query Management Technique for Efficient Real-Time Monitoring of Spatial Regions in Mobile Database Systems. In *Proceedings of the IEEE IPCCC*, (pages 259–266).

Daily, G. P. S. (2008, February 6). *NXP Fuels Rise of Mobile Location-Based Services*. Retrieved from http://www.gpsdaily.com/reports/NXP_Fuels_Rise_Of_Mobile_Location_Based_Services_999.html

Dey, A., & Abowd, G. (2000). CybreMinder: A Context-Aware System for Supporting Reminders. In *Proceedings of the Second International Symposium on Handheld and Ubiquitous Computing*, (pages 172–186).

Ericson, C. (2005). *Real-Time Collision Detection*. San Francisco: Morgan Kaufmann.

Flautner, K., & Mudge, T. (2002, December). Vertigo: Automatic Performance-Setting for Linux. *Operating Systems Review*, *36*(5S), 105–116. doi:10.1145/844128.844139

Flinn, J., & Satyanarayanan, M. (1999). *SOSP* (pp. 48–63). Energy-Aware Adaptation for Mobile Applications. In Proceedings of.

Gedik, B., & Liu, L. (2008, January). Protecting Location Privacy with Personalized k-Anonymity: Architecture and Algorithms. In *Proceedings of IEEE Transactions on Mobile Computing*, 7(1), 1–18. doi:10.1109/TMC.2007.1062

Geominder (n.d.). *Geominder*. Retrieved from http://ludimate.com/products/geominder/

Hu, H., Xu, J., & Lee, D. (2005). A Generic Framework for Monitoring Continuous Spatial Queries over Moving Objects. In *Proceedings of ACM SIGMOD*, 2005.

Jensen, C. Lin, D., & Ooi. B. (2004). Query and Update Efficient B+-Tree based Indexing of Moving Objects. In *Proceedings of VLDB*, (pages 768–779).

Kim, S., Kim, M., Park, S., Jin, Y., & Choi, W. (2004). Gate Reminder: A Design Case of a Smart Reminder. In *Proceedings of the Conference on Designing Interactive Systems*, (pages 81–90).

Liu, L. Pu, C., & Tang, W. (2000). WebCQ - Detecting and Delivering Information Changes on the Web. In *Proceedings of CIKM*, (pages 512–519).

Lost in Klaus. (n.d.). *Lost in Kluas*. Retrieved from http://www.cc.gatech.edu/~ulee3/LostInKlaus/

Ludford, P., Frankowski, D., Reily, K., Wilms, K., & Terveen, L. (2006). Because I Carry My Cell Phone Anyway: Functional Location-Based Reminder Applications. In *SIGCHI Conference on Human Factors in Computing Systems*, (pages 889–898)

Marmasse, N., & Schmandt, C. (2000). Location-Aware Information Delivery with Com-Motion. In *Proceedings of HUC*, (pages 157–171).

mGraffiti (n.d.). *mGraffiti*. Retrieved from http://www.cc.gatech.edu/projects/disl/mGraffiti/

Mokbel, M., Xiong, X., & Aref, W. (2004). Scalable Incremental Processing of Continuous Queries in Spatio-Temporal Databases. In *ACM SIGMOD* (pp. 623–634). SINA.

mTrigger (n.d.). *mTrigger Project*. Retrieved from http://www.cc.gatech.edu/projects.disl/mTriggers/

Mudge, T. (2001). Power: A First-Class Architectural Design Constraint. *Computer*, 34(4), 52–58. doi:10.1109/2.917539

Murugappan, A., & Liu, L. (2008, June 30-July 2). An Energy Efficient Approach to Processing Spatial Alarms on Mobile Clients. In *Proceedings of the ISCA 17th International Conference on Software Engineering and Data Engineering (SEDE-2008)*.

Naggie 2.0 (n.d.). *Naggie 2.0: Revolutionize Reminders with Location!* Retrieved from http://www.naggie.com/

Samet, H. (1990). *The Design and Analysis of Spatial Data Structures*. Reading, MA: Addison-Wesley.

Sohn, T., Li, K., Lee, G., Smith, I., Scott, J., & Griswold, W. (2005). A Study of Location-Based Reminders on Mobile Phones. In *Proceedings of UbiComp*. Place-Its.

Spatial Alarms Project. (n.d.). *Spatial Alarms Project*. Retrieved from http://www.cc.gatech.edu/projects/disl/SpatialAlarms/

Spatial Data. (n.d.). *Spatial Data Transfer Format*. Retrieved from http://www.mcmcweb.er.usgs.gov/sdts/

USGS. (n.d.). *U.S. Geological Survey*. Retrieved from http://www.usgs.gov

Wang, T., & Liu, L. (2009, September). Privacy-Aware Mobile Services over Road Networks. In *Proceedings of the 35th International Conference on Very Large Data Bases*. PVLDB 2(1): 1042-1053 (2009).

Yu, X., Pu, K., & Koudas, N. (2005). Monitoring k-Nearest Neighbor Queries over Moving Objects. In *Proceedings of ICDE*, (pages 631–642)

Zhang, J., Zhang, G., & Liu, L. (2007, September). GeoGrid: A Scalable Location Service Network. In *Proceedings of the 27th IEEE International Conference on Distributed Computing Systems*.

Chapter 17
Mobile Push for Converged Mobile Services:
The Airline Scenario

Sasu Tarkoma
Helsinki University of Technology, Finland & Nokia NRC, Finland

Jani Heikkinen
Helsinki University of Technology, Finland

Jilles van Gurp
Nokia NRC, Finland

ABSTRACT

In this chapter, we outline an event-based infrastructure that combines the Session Initiation Protocol (SIP) and Web technologies in order to realize secure converged mobile services. Our main service scenario has been supporting proactive information delivery to airline customers. This application built on top of the infrastructure is designed to help customers in several airport activities, such as check-in and boarding, and to enable highly crowded airports to offer consumer services. We focus on the two key technologies, namely SIP and Web technologies, and outline a distributed system architecture that was developed for the airline scenario. The scenario involves supporting the identification and tracking of users across different networks, enabling proactive information delivery (push) to the devices, and supporting rich multimedia interactions with users. The push system architecture is based on a lightweight access network consists of access points and edge proxies. The edge proxies connect the mobile users to the information backbone in a secure way using TLS or DTLS connections.

INTRODUCTION

Travelling is an activity that revolves around timely information, such as pertaining to locations, timetables, costs, and resources. Many of the traditional processes at airports can be automated by allowing a user to monitor and interact with information pertaining to flights and other topics of interest for the passenger. This motivates the design and implementation of a framework that supports proactive notifications and flexible user interaction.

DOI: 10.4018/978-1-60566-697-6.ch017

As a starting point for the system, message oriented middleware has long been common in the enterprise world where it is used to decouple components. Activities such as sending an order to a third party system, etc., are handled asynchronously. This allows applications to stay responsive while the requested work is being done in the background. An important advantage of asynchronous communication is that the communicating parties do not have to wait for the completion of tasks and can thus engage in other activities. In order to meet the demand for flexible user interaction, our system is based on standards based Web services technology. Both asynchronous communication and Web services technology present possibilities and challenges in a mobile and wireless environment. Asynchronous communication over the network is a user-friendly way to deal with some of the limitations of the environment, such as low bandwidth and high latency links and frequent disconnections. The problem at hand is the proactive delivery of Web content to mobile users in an efficient, secure, and standards based way.

In this chapter, we outline an event-based infrastructure for supporting mobile users that combines two frequently used systems, namely the *Session Initiation Protocol (SIP)* framework (Schulzrinne et al. 2000) and the Web services architecture. The aim of the chapter is to discuss the design and implementation issues of mobile push systems, especially from the viewpoint of the chosen scenario, and the role of SIP as a control plane protocol for various services, including Web services. SIP is a standardized part of beyond 3G networks, and it is used to setup multimedia communications in these networks. This makes SIP a promising candidate for sending Web-based updates to mobile devices. Our system uses SIP to realize mobile push using a lightweight access network consisting of access points and edge proxies. The edge proxies connect the mobile users to the information backbone, namely SIP

and Web servers, in a secure way using transport layer security (TLS or DTLS).

The system has been developed with emphasis on a particular scenario, namely supporting mobile airline customers. This application has been developed using an implementation of the infrastructure outlined in this chapter, and the aim is to help customers in several airport activities, such as check-in and boarding, and ultimately to enable highly crowded airports to offer better service (Tarkoma et al. 2007). The airline scenario involves supporting the identification and tracking of users across different networks, enabling proactive information delivery (push) to the devices, and supporting rich multimedia interactions with users. Many of these requirements can be found also in other mobile scenarios.

This chapter is structured as follows. First, we give an overview of the relevant push technologies, and discuss the requirements and current technological landscape for implementing the required converged communications functionality. Then the design and implementation of the converged mobile service infrastructure is presented. We then consider the airline scenario as an application for the system. Towards the end of the chapter, we discuss the benefits and limitations of the approach before the conclusions.

BACKGROUND

In this section, we give an overview of protocols and technologies that have been developed for asynchronous communications. We consider a number of push, pull, and hybrid communications schemes. Towards the end of this section, we discuss the SIP protocol and Web-based solutions. SIP and Web-based solutions are part of the recent trend towards convergence in the telecommunications world and therefore they are candidate technologies for proactive mobile services.

Delivery Strategies: Push

There are many strategies for supporting information delivery from publishers to subscribers. The two basic mechanisms involve *pull* and *push.* In the former, a subscriber polls a server or publisher and retrieves one or more events. In push, the publisher or a server proactively delivers events to subscribers. In *hybrid* push-based models small amounts of data are pushed after the actual, larger amount of data are pulled from the data source. Alternatively, popular data items can be pushed and unpopular data items pulled. Another variant of the push-based interactions is the *broadcast push* in which data are periodically broadcast to clients. Furthermore, a two-step hybrid broadcasting model has been proposed (Hauswirth, 1999). In the first step, broadcaster sends out a description of the content. In the second step, the receiver decides whether to pull the actual content according to the description sent by the broadcaster.

Beaver et al. (2006) have also considered the problem of scalable data delivery in the wireless Internet and proposed a hybrid model containing three channels. One channel is used by mobile clients to request documents over unidirectional, limited-bandwidth channel. The requests are not responded to immediately, but instead, the server uses one of the two remaining channels to broadcast the responses. The third channel periodically broadcasts the most popular documents. Beaver et al. also concludes that data broadcasting and selective data delivery methods offer solutions to the energy limitations faced by the mobile clients. Boukerche et al. (2004) have analyzed the quality-of-service factors of the *hybrid push-pull* algorithms in wireless networks.

Tosi (2003) described an integrated push architecture consisting of components standardized by IETF and WAP Forum. The aim of the architecture is to interwork with the IMS architecture. Thus, they have analyzed 3GPP requirements for push service development. The requirements include the ability to authorize push service usage and prevent unauthorized push content distribution. The dynamic IP of the terminal is associated to SIP identity in the standard SIP registration procedure. Moreover, the proposed architecture leans towards a complex *Push Proxy*, which queries the capabilities of the user on a request from a *Push Initiator* and uses *SIP redirect server* to forward request to the dynamic IP address of the terminal. Despite the set security requirements, Tosi discussed little the security towards the target of the push.

There has been a considerable amount of research searching for ways to deliver data to mobile users in wireless hotspots. Wireless hotspots typically cover a limited geographical area, yet they provide much higher bandwidth than cellular packet-switched networks. Mobile users stay in the range of wireless hotspots only a limited time period, namely the *dwell time*. Thus, the data delivery system should deliver the data in this period of time or provide data delivery continuation at the next hotspot. Herwono et al. (2006) have proposed a hierarchical architecture for a media point system. The media point system is centrally controlled in order to provide personalized push services for mobile users within WLAN hotspots. However, in reality, their system applies a pull approach. Nevertheless, they have provided scenarios in which the speed of the terminal is varied, including stationary, walking, moving car, and public transport scenarios. Furthermore, they concluded that most recent research efforts have aimed at providing the mobile terminal a fixed IP address, which would allow TCP connections to be kept alive while moving. Instead, Herwono et al. (2006) took an application-level approach to mobility management discarding the need for TCP keep-alive connections. Furthermore, they have analyzed the security requirements for a push-based media point system. The requirements include an analysis of choosing end-to-end or hop-to-hop security mechanisms, avoidance of public-key cryptography, and the possibility to reuse established security contexts.

Podnar et al. (2002) have also analyzed features of a mobile push service in three scenarios including *stationary*, *nomadic*, and *mobile* users. They concluded that services required for the mobile push service include *location management*, *content adaptation*, and *content presentation*, in addition to subscription management, content management, user profiles, and queuing strategy. They proposed a four-layer architecture for mobile push systems, which employs publish/subscribe middleware. In addition, Caporuscio et al. (2003), Huang et al. (2004), and Mühl et al. (2004) have researched architectural approaches to deploy the publish/subscribe paradigm in mobile, wireless environments.

Highly interactive services and applications for mobile devices require means to communicate using interoperable protocols in a way that minimizes the use of resources such as CPU, network and energy. A popular protocol in this context is the SIP protocol defined in RFC 3261. The SIP architecture and specifications is being developed by the IETF. There are many SIP extensions including an *event framework extension*. The event extension enables SIP user agents to subscribe to resources and then receive asynchronous notifications associated with the resources when they become available for the user. Furthermore, the SIP event framework enables application developers to apply the event-based design paradigm through this interoperable communication protocol. Another approach is to utilize existing *instant messaging (IM)* protocols for various kinds of asynchronous notifications, for example the XMPP protocol is one candidate protocol for general XML-based communications.

Delivery Strategies: Pull-Based Push

Non-push-based techniques based on *RSS (Really Simple Syndication)* feeds use a simple pull-based mechanism where the browser or some other program simply downloads the RSS feed periodically or when the user manually reloads the feed. However, not all of the push use cases can be supported this way. For example, instant messaging requires a push mechanism to deliver messages to the user, which was until recently not something that could be done easily in a browser (except with the help of plugins). In recent years, asynchronous communication has become a crucial requirement for Web applications and the common way to implement this functionality is to use *AJAX (Asynchronous JavaScript and XML)*. AJAX is based on issuing HTTP requests from JavaScript code executed by the Web browser, and it thus realizes a poll-based communication model.

Web applications use AJAX to provide a much more interactive and less page-oriented experience. AJAX is asynchronous in the sense that it decouples the request and response from the programmer's point of view. It is often used to support features such as form validation, fetching suggestions for auto-completion from the server, saving data on a server, etc. With AJAX applications, JavaScript code executed in the browser sends requests to the server. The response message that comes back (e.g. XML or *JSON*, a popular JavaScript-based data structure) is passed to a callback method that may dynamically update the browser window. Consequently, the script does not have to wait for the server response but can handle it later when the call back method is called. Meanwhile, the browser is available to handle user input and to execute other scripts in the page.

AJAX is not truly asynchronous since it still uses a request-response paradigm: the server cannot send an HTTP response without a prior request. To address this, a variation of AJAX, called *Comet* (Mahemoff 2006), has emerged that circumvents this limitation. Comet uses a similar mechanism of callback functions to handle the response from the server. However, instead of being a request-response mechanism, Comet instead uses HTTP requests to keep the connection to the server open. Figure 1 illustrates the client-server interactions in Comet. A long-lived connection is established that is then used to send and receive event data.

Figure 1. Example Comet interactions

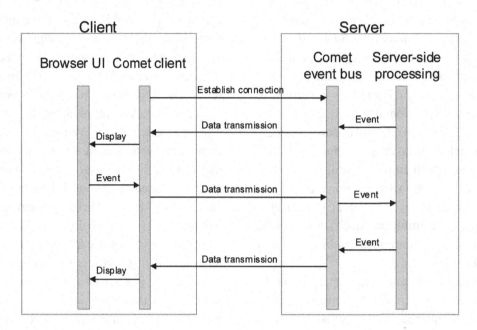

Before the connection times out, the client-side Comet system closes the connection and sends a new request. The main purpose of this technique is to allow the server to send a response at any time to the client. Furthermore, it is quite effective for implementing Web-based instant messaging and other client software that requires asynchronous interactions such as those found in Meebo, GMail, and Facebook.

Mobile Push Challenges

Recently, the mobile Web has started to mature and Web browsers in mobile phones are increasingly more feature rich and more often use the same rendering engines as desktop browsers. Thus, in principle, they can be used to run AJAX and Comet based Web applications or widgets as well. However, several problems exist that make using such solutions impractical:

• Mobile networks might be similar to normal networks from a pure functional perspective in the sense that they support TCP

and HTTP traffic. However, from a non-functional perspective they are quite different: setting up connections can take several seconds on some networks, latency is generally much higher and network reliability is an issue (for example when roaming). Additionally, there are several reachability issues that make communications in wireless and mobile networks challenging. The mobile devices typically have private IP addresses and as a consequence cannot be reached from outside the private network. A number of solutions can be used in such environments to traverse *NAT (Network Address Translation)* devices that are used to create the private addressing domains. AJAX and Comet solve reachability issues by relying on client-initiated interactions. Indeed, this is a simple way to circumvent the problem of how to reach the mobile client from the server-side; however, it results in frequent polling operations from the client or the use of a long-lived connection that is blocked at the server (as in Comet).

Figure 2. Airline scenario events and information push

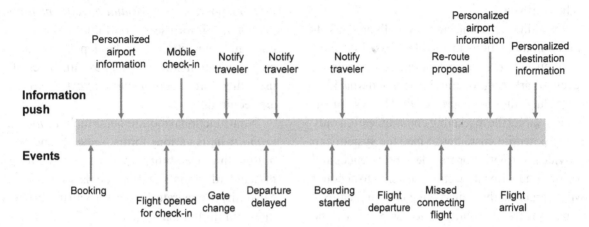

- While many high-end mobile browsers support the protocols and solutions needed by rich Web applications, they are running on hardware that is several magnitudes slower than a normal PC. Additionally, these devices have limited available memory, and AJAX applications tend to use much more memory than Web pages without JavaScript.
- Mobile devices need to conserve energy in order to make the battery last. Unnecessary network communication or client-side processing waste energy. The request-reply interactions typical of AJAX applications can be seen as an expensive way of implementing interactions for mobile Web applications.

Of course, when using an AJAX application for only a few minutes these problems are not that important. Indeed, AJAX techniques are now commonly used for mobile Web applications and widgets. However, an obvious use for asynchronous communication in mobile devices is to run applications in the background that are able to respond to messages all the time. This is already problematic for just one application. With multiple applications each keeping an HTTP connection open to the server, the problem only gets

worse. Therefore, more sophisticated solutions are needed for mobile services.

SCENARIO REQUIREMENTS

In this section, we present an overview of the mobile airline service environment and outline the central requirements for the service framework. Figure 2 illustrates the various events that can happen during travel starting from the booking of flights, check-in, gate open, departure, and ultimate arrival to the destination. These different events can then trigger a proactive information push to the passenger's mobile device. The key goal of information push is to provide the passenger with the latest information pertaining to the airports, flights, and destinations. Some of the information is personal and only intended for the particular passenger, and some of it is pertinent for a group of passengers sharing the same preferences or going to the same destination. Therefore, information targeting in a secure way is a fundamental requirement.

The information push needs to be secured on the level of the individual or group, and any subsequent interactions need to be secured as well. One example of a follow-up interaction with a passenger happens when a new route and seating is proposed to the passenger and it needs to be confirmed.

In addition, an airline may deliver marketing communications to travelers based on their preferences. Travelers can define if they want to receive marketing communications and what kind of products they are interested in. The system enables to target the marketing communication only to the potential customers, since the system can provide information on travelers' preferences and their journey. Location can be taken into account when targeting these marketing communications to passengers. Proximity to access points can be used to position the passenger within an airport; the location of the gate implicitly determines the routes the passenger will take within the airport complex.

Example interactions that can be provided by such a system include:

- Proactive mobile check-in.
- Notification of gate changes.
- Notification of last minute changes to their flight.
- Notification of disruptions or potential problems identified regarding the journey.
- Notifications pertaining to schedule changes, lost baggage, and other events of interest.
- Online Web service for information pertaining to rerouting and seating.
- Sending informative messages regarding the check-in gates, timetables, connecting flights, and local services.

The key functional requirements that need to be met in order to be able to realize the above functionality can be summarized as follows. First, *user profile management* is required for being able to track users, and match and tailor information for them. This involves a logically centralized database for storing the user profiles. Second, proactive *information push* is needed for the system to be able to inform the user and trigger

user interactions. Third, the system needs to be able to support *interactive multimedia content* and *personalize* it for users. Fourth, the information push and interactions may be dependent on the current location and other context attributes of a user, therefore *location and context awareness* are required.

Non-functional requirements include *interoperability* in its various forms, *security*, and *scalability*. Interoperability involves both the usage of standard formats and protocols, and also the capability to integrate the system with the airline's internal information systems.

DESIGN AND IMPLEMENTATION

In this section, we present the infrastructure for supporting converged communications for mobile users, and outline how the secure push communications is realized. As discussed in the previous sections, Web applications typically apply HTTP and polling in order to realize interactive usage experience. We observed that although push functionality can be implemented in this way, it is not very efficient and may result in delays in processing and presenting event-based information. Given that the SIP protocol framework has been designed with the requirements of mobile communications in mind, it is a natural enabler for mobile event-based systems. In the design and implementation of the mobile airline service framework, we have adopted the converged communications environment that combines SIP signaling with Web-based resource access. The former can be seen as the control plane and the latter as the data plane. The aim of the presented system is to investigate techniques for converged communications and how to realize proactive services with current and emerging standard-based solutions. In the airline scenario, SIP provides flexibility since communications can be established through a commercial mobile network, dedicated edge proxies operated by an airline or

Figure 3. Architectural view of the applications

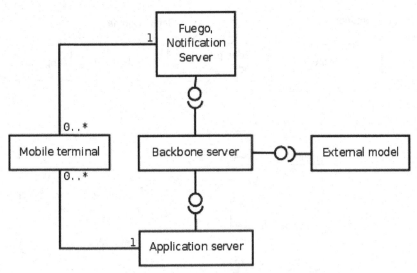

service provider, or a combination of these.

The system has been designed and developed for the airline service scenario; however, it can be applied to other domains as well. The development of new mobile services in this environment is motivated by the emergence of mobile devices with multi-mode communications, such as UMTS and WLAN, and interactive multimedia capabilities.

Figure 3 illustrates the architectural view of the applications, while some of the details have been omitted for brevity. The external model presents an existing real-world system, such as airline ticketing or scheduling system, that provides relevant information to the clients through the backbone server. The backbone server adheres to the *model-view-controller* pattern in which the view module closely interacts with the mobile terminal browser which provides a dynamic user interface. In most cases, the information updates are delivered to the user proactively through a push capability provided by the notification server. The application server is separated from the rest of the system in order to allow modular architecture and support for several types of applications.

The Fuego notification server provides the push notification capability. The notification service translates messages between the two technology domains, namely SIP and Web technologies, and provides content-based routing and forwarding service. In order to support scalability, the server functions can be distributed across multiple machines, for example a server farm. The push notification capability is realized using specific edge proxy servers, which are connected with the service backbone and terminate client-initiated, secure tunnels. The SIP terminal and user mobility support through the home SIP registrar is then used to keep track of the current edge proxy for a given user. In Figure 4, the edge proxy and registrar are shown as one entity. Alternatively, the Identity Provider (IdP) could include the registrar functionalities. Any service or application can query the registrar for the current location of a mobile user, and subsequently send SIP messages to the user's device.

Implementing Secure Mobile Push

In the following, we elaborate the mechanisms we have implemented in order to realize secure mobile push. Figure 4 gives an overview of the proposed lightweight access network with push capability. The system model of the secure push architecture consists of base stations, terminal

Figure 4. Overview of client-initiated secure communications channel

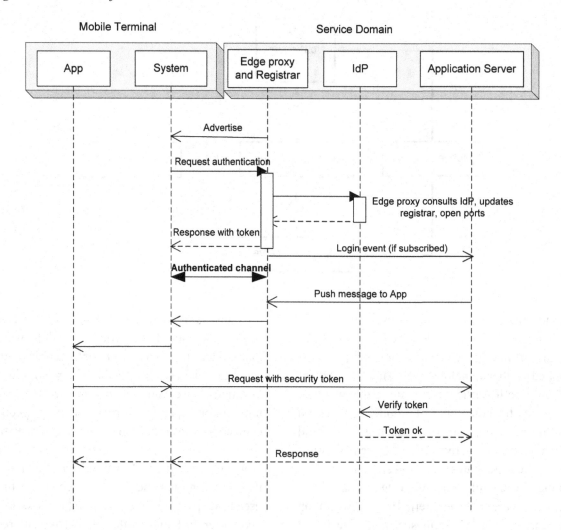

equipment, *edge proxies or gateways, identity provider (IdP)*, and services. The figure illustrates the two core parts of the system, namely mobile terminal domain and the service domain. Base stations can be deployed in a flexible manner at the airport and on different vehicles, such as airplanes and buses, given that they have connectivity to the access network.

Edge proxies advertise service access and services using the *Service Location Protocol (SLP)* defined in RFC 2608. Terminals access services through edge proxies or gateways which allow content to be pushed to terminals. Since a secure tunnel is established between the terminal

and an edge proxy, the access network can be a hotspot or cellular network. The benefit of this approach is that it copes well with different types of access networks; on the other hand, it requires per-connection state to be kept by the edge proxy. The edge proxy can issue a security token (or tokens) to the mobile terminal that can be used later to access services. Since the connection between the terminal and the edge proxy is authenticated, any messages from the client can be assumed to be proper. Thus computationally expensive signature checks are not needed for every message. The security domain may be federated with other domains through the IdP, including SIP-based *IP Multimedia Subsystem (IMS)* domains.

Connectivity and security problems introduced by the wireless network environment with private addressing space are being solved by requiring a secure client-initiated connection between terminals and edge proxies. Client-initiated connections solve problems related to terminal reachability due to NATs and firewalls over heterogeneous networks. However, security connection mechanisms themselves may introduce new security vulnerabilities such as the possibility for denial-of-service attacks. Therefore, the state maintained at the edge proxy should be minimized especially for the first phases of connection setup.

Current work at IETF is investigating client-initiated connections (Jennings et al., 2008). The key idea in this specification is to re-use the connection that was used to send the REGISTER request. This connection can be a bidirectional stream of UDP datagrams, a TCP connection, or some other type of transport protocol. It is the responsibility of the User Agent (UA) to maintain connectivity. The UA may also employ multiple flows to the proxy or registrar. In addition, a keep alive mechanism is included so that failed flows can be detected. Our prototype implementation is based on the Internet Draft and implements the client-initiated connection using TLS.

Our architecture is security protocol agnostic in sense that several protocols may be employed in this context, including TLS (RFC 4346), DTLS (RFC 4347), and HIP (RFC 4423). TLS does not have a puzzle computation for the connection initiator, but DTLS and HIP do have a puzzle mechanism to prevent DoS attacks. The HIP protocol has also mobility and multi-homing support. It is also possible to use SCTP with TLS (RFC 3436).

Basic security is therefore provided by TLS (or DTLS) and server-side certificates. Client authentication is performed using a lightweight browser-based login that involves an airline-issued identity number and a shared password. Assuming prior registration with the server-side system, push messages can be secured by utilizing shared keys. Given that SIP is used it is also possible to integrate the system with mobile operator issued identities.

Experimental Results

Initial experimental results with the system indicate that client-initiated persistent TLS connections help to reduce overhead with message-specific encryption and signature operations. If the messages are disseminated only within a trusted domain, a TLS connection with the edge proxy is sufficient to meet the basic security requirements.

We experimented with the implemented edge proxy in a heterogeneous network environment during the registration phase of the secure push communication. This phase is interesting, because it involves setting up the forwarding state in the edge proxy and the registrar. The edge proxy and registrar were hosted by Linux and run on Pentium 4 CPU 3.00GHz processor with 1.5 GB of memory. A NAT was used between them that used OpenBSD with similar processor and with 1GB of memory. Moreover, the registrar, edge proxy, and NAT communicated through Gigabit Ethernet. The link between NAT and wireless bridge was 100Mb full-duplex Ethernet. IEEE 802.11g was used in the wireless network. Our workload settings conform to the SPEC SIP design and workload specification (SPEC SIP Committee 2007) referring to a recently published IETF Internet-Draft for SIP end-to-end performance metrics.

Figure 5 presents experimental results with the edge proxy implementation. The results suggest that scalability of our SIP outbound enabled edge proxy is memory-bound when secure transports are used. The average CPU loads for both TLS and DTLS at the edge proxy are well below 20%. A single edge proxy was able to serve up to 4500

Figure 5. Edge proxy memory and CPU usage

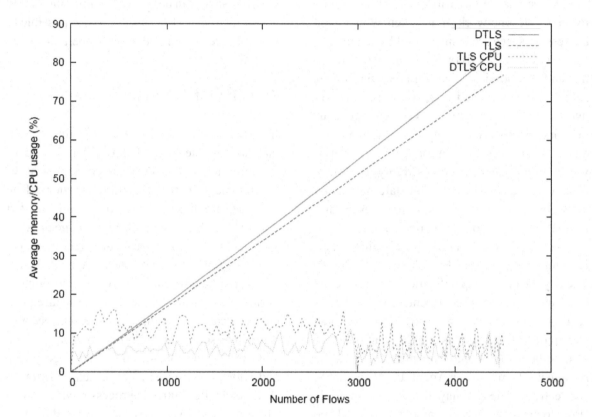

wireless user agent clients after which the memory was exhausted. The initial results suggest that adding more memory increases the number of concurrent sessions, and that the edge proxies are an instrument of scalability. Given the low CPU utilization, we believe that an edge proxy can be configured to handle tens of thousands of concurrent sessions (assuming low to moderate signalling needs).

IMPLEMENTING THE MOBILE AIRLINE SCENARIO

We implemented the mobile airline scenario using the presented infrastructure. The scenario was developed in co-operation with an airline company

aiming at more proactive role in customer-oriented operations. The airline company has an operational data store which receives data from other data sources, such as reservation, inventory, ticketing, and loyalty data. The data are consolidated inside the data store, hence providing a unified source of aggregated operational data to the airline. This data store stands for the external model of the system.

Given the scenario, the system can be divided into two important parts, namely the signalling part, and the content part. In our system, SIP is used for control and content push messages, and HTTP for content requests. Since SIP has been designed for session creation and control, it provides a suitable basis for asynchronous, message-based communications. The secure control plane, realized using SIP and TLS/DTLS, is used to send *electronic tickets* to users, and keep them up-to-date. In this section, we outline the interactions of mobile users with the backend, and consider the implementation of the electronic tickets using W3C XML technologies.

System-wise, each passenger with at least one open ticketed flight segment has an associated *session* with the mobile service. The travel session begins when the passenger books a flight. The system asks the passenger to check-in before a flight as illustrated in Figure 6. The request can be sent based on the passenger's location (e.g., when entering the airport) or on time (fixed period before departure). The system can also provide this information for all passengers even though they have not checked in yet. This can be used to notify passengers, for example, of a delayed flight.

After the check-in, an electronic boarding pass is generated for the user. Figure 7 presents a screenshot of the boarding pass from our implementation. The boarding pass is a resource that is proactively updated for the user, and thus it provides timely information regarding the flight. Other system messages during the session pertain to updates to flight information. The travel session ends when the trip has been completed and there are no remaining system messages for the traveler. A possible post-trip message could be for instance a baggage lost message, in which the system informs the traveler about lost baggage and later about arrival time of the baggage. Post-trip messages also include personalized information about the destination and related marketing information.

Pushing the Electronic Ticket

The electronic ticket is implemented using *SMIL 2.0*, a language specified by W3C, and the *X-SMILES* open source XML-browser. SMIL allows the design of interactive multimedia rich pages that can be updated at runtime. Information push to the mobile device is realized by sending a SIP message to the terminal with XML content. The SIP push message is processed by the electronic

Figure 6. Example of a check-in message

Figure 7. Example of an electronic boarding pass

ticket SMIL code by using callbacks and an asynchronous programming model. XML content can therefore be pushed to the electronic ticket and the ticket is updated in real-time when connection with the airline service has been established. The connection can be established either using a local access point or through a 3G/B3G access network.

The XML content that is pushed to the mobile browsers can represent a whole XML document or a part of an XML document. The latter case requires a way to position the incoming content fragment with some existing document. The system uses the REX format specified by W3C to do this in an interoperable way. A REX Generator translates given data into the REX format, which is an XML-based format for representing *Document Object Model (DOM)* events, which in turn refer to a data structure inside the browser and that can be interpreted by it. The browser interprets the REX message and then updates its internal DOM presentation of the user interface content accordingly. This approach allows the service to send only relevant DOM events to the browser thus resulting in compact messages and less intensive XML processing and presentation.

DISCUSSION AND FUTURE TRENDS

Because of the issues outlined previously, the mobile services may need a different solution than HTTP-based polling to support asynchronous communication efficiently. Devices need to be able to function for several days without needing to be re-charged. And, ideally, they should also be able to handle incoming messages all the time. As discussed in previous sections, the SIP protocol framework provides asynchronous signaling capability, especially intended for session setup, modification, and teardown. Therefore, the integration of these two technology domains, namely SIP and Web technologies, becomes interesting. In recent years, this so called converged communications has been under active research and development. In this article, we outlined a system that combines SIP and Web-based services.

As mentioned earlier, AJAX and Comet are essentially solutions around the fact that HTTP is not an asynchronous protocol. These techniques are needed because HTTP is the only protocol that Web browsers understand, and HTTP was designed as a synchronous protocol. An alternative way of supporting asynchronous communication is to use a non-browser-based approach. Native applications that are not restricted by the functions available on a browser can be designed and implemented to support asynchronous communications in a better and more efficient way. The challenges with native applications pertain to deployment, interoperability, and security to name some key issues. Given that Web applications are interoperable and easily accessible, it becomes increasingly attractive to introduce support for asynchronous communications and publish/subscribe in the browsers or middleware.

Being communication devices, many mobile phones already support some forms of asynchro-nous communication. For example, SMS can be considered non-Internet-based asynchronous communication form. Additionally, architectures such as IMS that enable asynchronous communication (e.g. MMS) in operator networks are actually IP based, but generally kept isolated from the real Internet by operators that deploy it for security and economical reasons (e.g. monetization through billing). Similarly, SIP is used for *voice over IP communication (VoIP)* and related signaling.

Lessons Learned from the Design

The lessons learned during our design and implementation pertain to the combination of SIP and Web protocols, realizing proactive notification capability for mobile services, utilizing XML-based data formats, and supporting security using a client-initiated secure tunnel to a proxy. Table 1 presents the previously discussed requirements and our lessons learned in meeting the requirement. We continue by discussing the lessons learned and alternative solutions in more detail.

In our infrastructure and the airline scenario SIP was used for control and content push messages. These solutions do have the advantage of working reasonable well across a wide range of phones and especially SMS has been successful as a (primitive) asynchronous messaging system (e.g. voting, notification and payment systems have been implemented using SMS). In addition, some mobile device vendors bundle instant messaging (IM) clients with their devices. Since IM clients compete with revenue from SMS and IMS based services, market adoption of such mobile IM clients is currently very low and many vendors that do support them disable them when the device is connected to a mobile network.

The mobile service architecture presented in this paper is an example of a system that allows the introduction of push-based communications in an XML browser. The system utilizes client-initiated connections to edge proxies. We implemented the push using SIP and relied on HTTP for the Web

Table 1. Requirements and solutions

Requirement	Solutions
User profile management	User profile management is implemented in the core servers and external servers. The model-view-control and multi-tier patterns are used to provide separation of concerns.
Information push	Application of the event notification pattern. Reachability is achieved using the proxy pattern and client-initiated connections or external push service (SMS, 3G,..).
Interactive multimedia content	Our solution is to use XML-based formats, namely SMIL for describing the user interactions. It is expected that more and more XML fragments are being sent over the wireless link. This requires mechanisms for doing delta updates for XML content. We used the REX format for representing DOM events, which is a standards compliant solution.
Location and context awareness	We considered both logical and physical location. Coarse level of physical locality was achieved by pinpointing the access point (base station). In addition, we considered in-door positioning technologies; however, these were not implemented in the prototype. It is expected that a third party in-door positioning service can be integrated with the system. Logical location is supported by allowing clients to issue content-based subscriptions. This feature can also be used to realize context-based messaging.
Interoperability	This was addressed by utilizing standards based solutions, such as SIP, SMIL and XML, REX, and HTTP.
Security	Security challenges were addressed by using TLS or DTLS tunnels from mobile devices to the edge proxies. This achieves security for the wireless hop given that the proxies have trusted certificates. The user authentication was done using browser-specific login.
Scalability	The model-view-control and multi-tier patterns provide scalability in the backbone system. The edge proxies are a building block for the access networks and their number and performance affects the scalability of the system. Experiments suggest that the number of simultaneous TLS/DTLS connections needs to be monitored and load balancing is needed when system resources are low.

requests and responses. The system combines SIP and Web services, and leverages SIP mobility support features to keep track of the current location of the mobile device. Our approach relies heavily on XML for data representation, and the combination of XML and REX allow the delivery of specific parts of an XML document. This is especially useful in the mobile and wireless environment, in which XML processing is to be avoided due to its computationally heavy nature and memory requirements. The applications need to be aware of the push system in the sense that they support the callbacks and can properly handle incoming events.

The favorable characteristics of the presented system include the ability to support communications over arbitrary access networks, interoperability with SIP and IMS, and the ability to push XML fragments over the air. The limitations of the system include the requirement for edge proxies and modifications needed to the SIP and Web client

stacks. Moreover, simple asynchronous communications, such as triggering an application, can be realized today using an SMS message. On the other hand, SMS is an old technology and offers a very limited communications capability. The outlined system is expected to be useful when asynchronous update of resources, namely XML documents, is a frequent activity.

In addition to the standard Web-based solutions and SIP, another strategy is to utilize an IM protocol for communications. IM clients are ideally suited for supporting a wide variety of asynchronous communication forms, other than just instant messaging. Generally, IM clients maintain a connection to the messaging server all the time. Instead of just delivering messages to users, the connection might also be used to deliver messages to programs, effectively turning the IM network into a message oriented middleware system.

Especially *XMPP (Extensible Messaging and Presence Protocol)* (RFC 3920) is suitable for this since it can support extensions for different types of messages communication styles. XMPP was developed originally as a flexible instant messaging protocol. It uses XML as its message format. A key benefit of XMPP is that it is possible to extend the protocol with custom XML messages. The XMPP Standards foundation hosts a few dozen of extension specifications that support a wide range of asynchronous communication patterns including: publish/subscribe mechanisms, presence and status updates, alerts, etc. Additionally, extensions have been defined for feature negotiation, service discovery, and many other features that make it useful as an asynchronous middleware solution.

Technology aside, XMPP is emerging as an increasingly widely supported standard protocol (standardized by IETF which also standardized HTTP). Currently, several tens of millions of users are using XMPP clients for messaging. More importantly, XMPP appears to be increasingly popular with Internet companies such as Google, Facebook, and Twitter, which use XMPP as an API that may be used by third party clients and Web sites.

Therefore XMPP appears to be a good candidate for mobile communications. One challenge with XMPP is that it was not designed for mobile networks. Particularly problematic are reconnects to the XMPP server which are more frequent in a mobile network. The main issue with that is that XMPP needs to re-establish a new session on every connection. This involves exchanging presence data in XML and parsing it. Both the amount of data exchanged and the effort of parsing it on reconnection cause mobile devices to waste many resources. Google has cited this as a reason for implementing a binary, non-XML-based protocol for the Android protocol that avoids the need for creating a new session on each connection. They have suggested various improvements that involve simplifying the protocol, using a more compact binary encoded data format, and providing a way of allowing the XMPP session to survive reconnections. It is expected that W3C's forthcoming

Efficient XML Interchange (EXI) standard for binary XML is a good candidate format for XMPP.

CONCLUSION

In this chapter, we presented and discussed an infrastructure for realizing converged mobile push-based services that builds on the Session Initiation Protocol (SIP) framework and W3C's Web services standards. Proactive information notification is an integral part of the framework and crucial for today's mobile services in general. We observed that one of the central challenges it to ensure reachability of the mobile device. In practice, mobile devices are not reachable due to the ubiquitous nature of firewalls and NATs in access networks. This means that special solutions are needed in order to ensure reachability. These solutions include client-side polling and establishing client-initiated connections to server-side components.

The favorable characteristics of the presented system include the ability to support secure communications over arbitrary access networks, interoperability with SIP and IMS, and the ability to push XML fragments over the air. Especially the capability of sending only the necessary information, for example only the changed parts of a document, is useful from the viewpoints of bandwidth usage and processing efficiency.

The limitations of the system include the requirement for edge proxies and modifications needed to the SIP and Web client stacks. The edge proxies are building blocks for access networks and a mechanism is needed to forward incoming connections to a new proxy when system resources are low. Moreover, simple asynchronous communications, such as triggering an application, can be realized today using an SMS message. The outlined system is expected to be useful when asynchronous update of resources, namely XML documents, is a frequent activity. The limitations of the system include the requirement for edge proxies and modifications needed to the SIP and Web client stacks.

More technically, the implemented system uses SIP for control and content push messages, HTTP for content requests, and utilizes client-initiated connections to be able to push messages to mobile devices. Content messages are described using XML, namely as DOM events using the REX format.

In this chapter, we have provided a bird's eye overview of combining various asynchronous communication mechanisms that are commonly used in today's Internet in a mobile Internet context. None of the mechanisms we discussed fully address the issues with particularly excessive battery usage. Currently, SIP and XMPP are promising technologies for future mobile event-based services; however, whereas SIP has been designed for the environment, XMPP needs to be modified to better support the mobile environment.

REFERENCES

Beaver, J., Chrysanthis, P. K., Pruhs, K., & Liberatore, V. (2006). To Broadcast Push or Not and What? In *Proceedings of the 7th International Conference on Mobile Data Management* (MDM 2006).

Boukerche, A., Dash, T., & Pinotti, M. C. (2004). Performance analysis of a hybrid push-pull algorithm with QoS adaptations in wireless networks. In *Proceedings of the IEEE Symposium on Computers and Communications* (ISCC).

Caporuscio, C., Carzaniga, A., & Wolf, A. L. (2003). Design and Evaluation of a Support Service for Mobile, Wireless Publish/Subscribe Applications. *IEEE Transactions on Software Engineering, 29*(12). doi:10.1109/TSE.2003.1265521

Extensions, X. M. P. P. (n.d.). *XMPP Extensions.* Retrieved from http://www.xmpp.org/extensions/

Hao-Ping Hung & Ming-Syan Chen. (2005). A general model of hybrid data dissemination. In *Proceedings of the 6th international conference on Mobile data management,* (pp. 220–228), Ayia Napa, Cyprus.

Hauswirth, M. (1999). *Internet-Scale Push Systems for Information Distribution - Architecture, Components, and Communication.* (Dissertation). Wien, Austria: Technischen Universitet Wien.

Herwono, I., Sachs, J., & Keller, R. (2005). Provisioning and performance of mobility-aware personalized push services in wireless broadband hotspots. *Computer Networks, 49*(3), 364–384. doi:10.1016/j.comnet.2005.05.010

Huang, Y., & Garcia-Molina, H. (2004). Publish/Subscribe in a Mobile Environment. *Wireless Networks, 10*(6), 643–652. doi:10.1023/B:WINE.0000044025.64654.65

IETF Network Working Group. (2004). RFC 3920 Extensible Messaging and Presence Protocol (XMPP): Core. *IETF.* Retrieved from http://www.ietf.org/rfc/rfc3920.txt

Jennings, C., & Mahy, R. (2008). *Managing Client Initiated Connections in the Session Initiation Protocol (SIP). IETF.* Internet Draft.

Mahemoff, M. (2006, March). Comet: A New Approach to Ajax Applications. *Ajaxian.* Retrieved from http://ajaxian.com/archives/comet-a-new-approach-to-ajax-applications

Mühl, G., Ulbrich, A., Herrmann, K., & Weis, T. (2004, May). Disseminating information to mobile clients using publish/subscribe. *IEEE Internet Computing, 8*(3), 46–53. doi:10.1109/MIC.2004.1297273

Podnar, I., Hauswirth, M., & Jazayeri, M. (2002). Mobile Push: Delivering Content to Mobile Users. In *Proceedings of the International Workshop on Distributed Event-Based Systems* (ICDCS/DEBS'02).

Schulzrinne, H., & Wedlund, E. (2000). Application-layer mobility using SIP. *SIGMOBILE Mob. Comput. Commun. Rev.*, *4*(3), 47–57. doi:10.1145/372346.372369

Tarkoma, S., Heikkinen, J., & Pohja, M. (2007). Secure push for mobile airline services. *Telecommunication Systems*, *35*(3-4), 177–187. doi:10.1007/s11235-007-9048-y

Tosi, D. (2003). An Advanced Architecture for Push Services. In *Proceedings of the Fourth International Conference on Web Information Systems Engineering Workshops*, (pp 193–200).

Chapter 18
Event–Based Interaction for Rescue and Emergency Applications in Mobile and Disruptive Environments

Katrine Stemland Skjelsvik
University of Oslo, Norway

Vera Goebel
University of Oslo, Norway

Thomas Plagemann
University of Oslo, Norway

ABSTRACT

Event-based interaction is suitable for rescue and emergency applications because the filtering capabilities can help to prevent information overload, and such interaction may be offered by an Event Notification Service (ENS). We focus on ENS in sparse Mobile Ad-hoc Networks (MANETs), since an incident may occur e.g., in a deserted place lacking infrastructure, the density of nodes may be low, and there may be physical obstacles limiting the transmission range. The asynchronous communication provided by the ENS is suited for an environment where there may be long-lasting network partitions. In this chapter, we describe characteristics of rescue operations and use this as a basis for discussing ENS design choices such as subscription language, architecture and routing. Afterwards, we present our own ENS solution, the Distributed Event Notification Service (DENS), which is tailored for such an application domain.

INTRODUCTION

Rescue operations may occur in environments where there is no communication infrastructure or where the infrastructure has been damaged by the incident itself. In addition to, or as a replacement to radio and telephones, handheld PDA-like devices can be used at the rescue site for communication. Such devices run applications and can spontaneously form a mobile ad-hoc network (MANET) using their wireless network interfaces. The purpose of setting up such a network is to exchange

DOI: 10.4018/978-1-60566-697-6.ch018

information. However, it is important to prevent information overload; the rescue personnel should be able to focus on the task of saving lives and not browse through a lot of information or search for information. They should be interrupted or notified only when something occurs that is of interest to them, and this type of interaction is suited to *event-based interaction*. Event-based communication may be offered by an event notification service (ENS) using the publish/subscribe communication paradigm. At the application level, instead of using classical client-server interaction, the application can describe events of interest in a *subscription* and is notified when such an event takes place by the ENS via a *notification*, i.e., the communication is asynchronous. This asynchronous communication service can be used by application developers to enable applications to adopt or react to certain events, or to directly notify the end-user about the event.

The main entities in an ENS are subscribers and publishers. Subscribers subscribe for information while publishers publish information concerning an event. They are decoupled by the ENS and can therefore communicate asynchronously. It is the task of the ENS to match subscriptions and notifications describing published events and deliver them to interested subscribers. The service itself may be centralized or distributed. ENSs differ in how events are observed, how subscriptions are expressed, i.e. the subscription language, how notifications are disseminated, and their overall architecture. The design is governed by the application domain and its requirements and the kind of network they are tailored for.

Important characteristics of rescue operations in this context are the heterogeneity inherent in the diversity of devices used, organizations, communication lines (information flows), applications, and the rescue operations themselves. These characteristics should be considered when deciding the requirements for the ENS for rescue and emergency in sparse MANETs. Some questions that arise in this context are:

- *Reliability and Availability:* To what extent must the service be available and what delivery semantics is needed?
- *Persistency:* For how long should notifications about events be stored?
- *Resource awareness:* What are the implications of the limited resources on handheld devices and in MANETs?
- *Expressiveness:* How to describe events and where to filter events?
- *Adaptability:* In what way may rescue operations differ and what impact may these differences have on the usage of the service and the network characteristics?
- *Security:* Does the presence of sensitive data such as medical records result in extra need for security compared to other application domains?

The rest of this chapter is organized as follows: We first describe the application scenario and network environment. We proceed by discussing design choices and related work. Next, we go a step beyond the state-of-the-art and present our approach, DENS, for use in rescue operations. We describe how its design and features take into account the design considerations. Finally, we make some concluding remarks and present future work.

BACKGROUND: APPLICATION SCENARIO AND NETWORK ENVIRONMENT

In the introduction, we motivated the usage of event-based interaction in mobile environments for rescue and emergency applications. In the following, we look more closely at the environment describing both the network itself and the characteristics of rescue operations. Then we revisit the questions mentioned in the introduction.

Rescue Operations

In this section, we discuss some general characteristics of rescue operations, and how rescue operations in Norway are organized (For more details we refer to (Sanderson et al. 2007) and (Munthe-Kaas et al. 2006)). Rescue and emergency operations are characterized by very hectic and dynamic environments, where time is a critical factor. There is a lot of movement and activity on the site as personnel may arrive and leave at different times. Typically, several organizations are involved in the operation, e.g., paramedics, fire fighters and police, in addition to a number of other organizations:

- **Fire brigade:** The main tasks are to control fire, as well as to monitor areas in danger of fire or explosion. Other typical tasks are to cut loose trapped people, help where needed, and place sensors to aid the monitoring of dangerous areas.
- **Medical personnel:** The main task is medical care, including registration of patients and evaluation of medical state. If sensors are used in aiding patient monitoring, medical personnel will place these on patients.
- **Police:** tasks include gathering evidence and securing the area.
- All personnel are involved in evacuating people and in general cooperating and supporting the rescue operation as needed.

Each organization has its own set of rescue operation procedures and guidelines which it must follow. The cooperation incentive in this situation is very strong since all participants share the same overall goal: rescue people and limit the impact of the disaster. Additionally, cross-organizational interaction procedures involving governmental organizations and other authorities have been defined. These procedures can be expected to include rules to establish a command and coordination structure for the scene of the incident. The coordi-

nation structure to some extent shapes the flow of information at the scene. Operational command is organized on three levels: the on-scene coordinator/command (OSC) – operational direction and coordination on the scene; the Rescue Sub-Centre (RSC) – local resource coordination; and the top-level coordination of the entire operation by a (Joint) Rescue Coordination Centre (RCC). In most rescue operations, only the first two levels will be needed. The main role and responsibility of the OSC is to coordinate resources and support at the site, and to accommodate the efforts of personnel from the participating organizations. It is of vital importance that the OSC has full overview of all available resources contributed from all participating teams and organizations at all times. For larger operations there is usually an on-scene coordinating team consisting of a police officer in charge of public order, a fire officer in charge of fire control, and a medical officer in charge of medical treatment, all reporting directly to the OSC acting as the leader of on-scene coordination team (OSC-team). There may also be a level of team leaders in each organization (e.g., medical) reporting to the officers that are members of the OSC-team. The main role of the RSC is that of assisting and relieving the OSC-team, as well as coordination of resources, e.g., when there is more than one rescue operation in the area. The role of the RCC is mainly to monitor the operation and give advice. In Figure 1, we give an example model illustrating the organizational structure of rescue operations, including role hierarchy and lines of reporting.

Even though every rescue operation is different, there are some commonalities that can be extracted into a set of general rescue scenario phases tailored for using MANETs in rescue operations. In the Ad-Hoc InfoWare project, we have identified the following six rescue scenario phases (Munthe-Kaas et al. 2006).

Phase 1 – A priori: This phase is before any incident takes place, when organizations exchange information on data formats and shared vocabu-

Figure 1. Organization and structure in rescue operations (taken from (Sanderson et al. 2007))

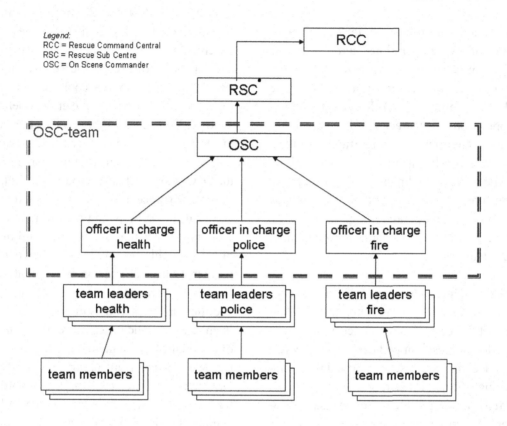

laries, and make agreements on procedures and working methods. Required certificates would be installed in this phase, and applications can be installed and configured.

Phase 2 – Briefing: This phase starts once the incident has been reported. The briefing involves gathering of information about the incident, e.g., weather, location, number of people involved, and facilities in the area. Some preliminary decisions about rescue procedures and working methods are also made at this stage. Based on information gathered during this phase, further configurations of the applications can be done and security levels be chosen.

Phase 3 – Bootstrapping the network: This phase takes place at the rescue site, and involves devices joining and registering as nodes in the network on arrival. In addition, rescue leaders are appointed in this phase.

Phase 4 – Running of the network: This is the main phase during the rescue operation. Nodes may join and leave the network and network partitions and merges may occur. Information is collected, exchanged and distributed. There may be changes in the roles different personnel have in the rescue operation, e.g., change of a rescue site leader. New organizations and personnel may arrive and leave the rescue site, new groups of an ad-hoc, task-oriented kind may form, possibly involving people from different organizations. Applications can communicate about available resources and capabilities of the nodes in the network. Replicas and proxies can be placed at strategic nodes in the network, and nodes can receive event notifications based on relevance and priority.

Phase 5 – Closing of the network: At the end of the rescue operation all services must be terminated. Applications can adapt to the closing of the network by acting on received information about degradation of the capabilities and resources of the network.

Phase 6 – Post processing: After the rescue operation, operation specific data, e.g., resource use, user movements, and how and what type of information was shared, may be analyzed to gain knowledge for future situations. Depending on the nature of the application, it may have gathered statistical or other information for post scenario analysis or for future use.

Preparations in the opening phases, as there are e.g., no restrictions on the network availability, can to some degree compensate for a lack of resources during the rescue operation. It would be optimal if all necessary information could be uploaded during these phases on the devices. However, this is generally impossible, because some information cannot be accessed by all organizations due to, for example, privacy concerns or there is just not enough time in the briefing phase to identify and upload the information. Examples for this type of information include security codes of doors at an emergency site, detailed building plans, specification of freight on a vehicle or in a storage, or medical records of persons that are known to be involved in the emergency. Furthermore, there is information generated at the emergency site during the operation, like sensor readings of room temperatures, information about how many injured persons have been detected at which location, readings of health monitors attached to injured persons, or information indicating the causes of an emergency situation, etc. Possible sources for information include both mobile devices carried by personnel, stationary devices, PCs in ambulances and rescue helicopters, and sensors. The information from these sources can be shared, but there are cases when sharing of sensitive information is not desired. As sensitive information, we can classify medical records of injured persons, environmental data, layout of buildings and installations, information about dangerous goods, collected evidence, available resources, and status reports.

Communication Infrastructure and Routing

There are a variety of state-of-the-art wireless technologies for portable devices including Bluetooth, Zigbee, the IEEE 802.11 specifications (IEEE-802.11), and cellular networks like UMTS and GSM. They differ in many aspects like transmission range, mobility support, bandwidth, and architecture (with or without infrastructure). Bluetooth and Zigbee have relatively short transmission range and therefore may not be suitable in a rescue operation where there are tens or hundreds of nodes spread over a possible large area. The usage of infrastructure-based networks through UMTS, GSM or Wireless Fidelity (WiFi) access points may be beneficial, but it is important to have a solution that also works for the worst-case since an incident may take place anywhere and the incident itself may have damaged the infrastructure. We therefore focus on 802.11 in ad-hoc mode. The available bandwidth for 802.11b is 11Mbps and for 802.11a and g is 54 Mbps. The communication range depends on the WiFi radio, the antennas, and the environment; the range is typically much longer in an open environment (up to 300 meters or more) than in a building (about 25-45 meters).

An important characteristic of wireless transmission is the usage of a shared medium; all nodes within one hop can hear the traffic which makes broadcasting efficient, but the disadvantage is problems such as hidden terminal and relatively low bandwidth compared to wired networks. In a MANET, all nodes may act as routers and the topology may change frequently. Routing protocols are therefore needed to allow multi-hop-routing. In a fully connected MANET, any two nodes can either communicate using some other nodes as routers, or directly, if they are one-hop neighbors. Information can also be flooded into the network. The task of finding paths between two nodes is performed by a routing protocol. There are two main groups of routing protocols: reactive and proactive. The reactive protocols find a path between a source and a receiver when the source wants to send a message; the proactive protocols keep routing tables always up-to-date. The receiver could be *one* node or a group of nodes, or all nodes. Additionally, the nodes may route messages based on a known receiver address or based on the content.

In sparse MANETs or intermittently connected MANETs, one cannot always assume an end-to-end path between a sender and a receiver and other routing schemes are therefore needed. *Store-carry-forward* protocols such as epidemic routing and gossiping take advantage of the mobility of the nodes and use them as relays to deliver messages. This approach is therefore *delay-tolerant*. Epidemic routing used in a partially connected ad hoc network was proposed by Vahdat and Becker (2000). The goal of epidemic routing it to deliver a message with high probability to a particular host(s), and also to minimize the set of other hosts that carry or transmit a particular message and by this limit the resource usage. When two hosts meet, they start to exchange information about stored messages, and based on this, send copies of messages they are missing. Zhengsheng Zhang provides an overview in (Zhang 2006) of different routing approaches for delay tolerant networks and sparse MANETs. The routing approach is based on whether the future topology is known; if a future topology may be estimated; or if nothing can be assumed concerning future topology. If no assumption concerning future topology can be made, all nodes can be considered candidates as carriers of a message to the destination. A less resource-intensive solution is to replicate to only a sub-set of the nodes, i.e., a probabilistic approach. If possible, historic data concerning mobility patterns and partition membership can be used to find potentially good carriers of data.

Requirements

In the introduction, we listed some questions related to ENS requirements. In this application domain, information could be life-saving, and the service should hence be highly *available*. Network partitions may occur because the density of nodes is too low to cover the area. Unstable conditions are the norm and not the exception; nodes may get turned off or get out of reach. It may be impossible to set up a route "in space" to reach a destination because of possible lasting network partitions. However, depending on the information value of the event, it should be possible to reach subscribers in other partitions and provide *at-least-once semantics* (Coulouris et al. 2001). Some events may have information value even though they are not delivered immediately while others such as health monitoring data may not. The design of the service must take into account the *resource constraints* of wireless networks and small devices, however, the duration of rescue operations are most often relatively short meaning that reducing battery consumption may not be too vital. The shared nature of wireless medium and the low bandwidth offered compared to wired networks makes it important to design the service while trying to lower the bandwidth consumption. This may have an impact on the filtering strategy of the ENS. The service should not set limitations by e.g., tailoring the design to a specific and simple subscription language in order to support different applications. Rescue operations may be quite diverse with respect to number of nodes, size of the area, devices used, network environment, and the service should be *adaptable* with regard to these parameters. An incident may happen in a deserted place where there is a lack of infrastructure, and the personnel need to put up their own MANET. For other incidents, the infrastructure is perhaps only half-functioning, and there may be different means for people to communicate. Weather conditions may also have an impact; devices may not work properly in extreme weather. Security is an important issue as there may be sensitive data such as health information etc., so access control to differentiate between which subscribers can subscribe for what kind of data may be needed.

DESIGN CHOICES

The applications make use of an ENS to filter, gather and distribute information in a disruptive environment. In this section, we discuss various design choices. The network environment and application scenario provide us with a set of *assumptions* and *requirements* that is used as a basis for this discussion. The offered subscription language is related to the application domain, while the network environment, such as degree of connectivity, frequency of topology changes, available bandwidth etc., has an impact on the architecture of the service and how routing of messages can be done.

Events and Subscription Languages

We start by discussing what kinds of events are of interest in our scenario. Applications or service users may be interested in real-world events such as a *fire*, a patient having a *heart attack*, or a *traffic jam*. The application developer must know what data and formats are relevant to detect such events in order to be able to write subscriptions. To what level of detail an event of interest may be described and the syntax of the subscription depends on the subscription language. A service that keeps track of the data sources may be needed, depending on the filtering approach of the ENS (at the publisher or subscriber).

From Real-World Events to Event Notifications

The question of what events may be of interest is closely related to the application domain. One possible application could be for coordinating

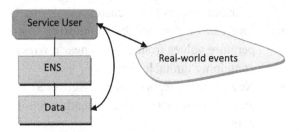

Figure 2. Real-world events and event notifications

and organizing the operation; dispatching of personnel and equipment, on-site identification of passengers, and registering casualties for medical treatment. Such an application would need to access information from devices belonging to the participants from different organizations. Not only updates on information, but also new available information such as a new map, a new picture taken or downloaded by a participant on his or her device could be regarded as a new event of interest to others. Another example is patient monitoring applications. If there are many casualties the rescue personnel can prioritize based on the severity of the injury after a first examination. However, the condition of a patient may not be stable, and it would therefore be beneficial to use sensor monitoring of patients that could alarm the rescue personnel if a critical health situation occurs, such as a heart attack.

For applications to be able to subscribe to events, knowledge about the data formats is needed, in other words, there need to be a mapping from the real-world events to the data that is stored in the network, as shown in Figure 2. Application events such as "fire" could be linked to a concept term "temperature"; an event "heart attack" could be composite and contain several concept terms such as pulse, blood pressure etc. The application developers therefore need to have this knowledge when designing the application and information available at start-up-time to be used for configuration of the service. Examples of real-world events and related concept terms are:

- Heart attack can be captured by pulse, heartbeat values from sensors.
- Fire can be detected by monitoring temperature values from sensors near a wreck containing ignitable liquid.
- New weather report can be discovered by an update of meta-data for weather report data.
- Dynamic neighborhood can be detected by frequent updates in the routing table.
- Progress in working plan can be captured by an update in a table due to a rescue personnel worker that "ticks" off a task.

Describing, Locating and Filtering Events

How detailed a subscriber can specify its interest depends on the expressiveness of the subscription language. The main types are subject/topic-based and content-based. In subject-based systems, subscribers subscribe to specific subjects and thus receive *all* notifications concerning this subject. Content-based languages are more expressive since subscribers can specify the content of the events they are interested in by using a filter. For instance, instead of receiving all notifications concerning the subject *pulse_readings,* subscribers may specify that they are only interested when the pulse readings are below a certain threshold (e.g., "pulse < 30"). In addition, geographical location or other context information of the subscriber or the publisher may be used as constraints. An additional filtering mechanism may be access control.

There are two "levels" when it comes to describing the events mentioned in the previous section. First, the name, or the kind of event, connected e.g., to a name of a table in a database or a file ("body_temp" from sensors monitoring patients, "battery-resources" etc.). Second, the content itself which the subscriber may provide filters for in content-based subscription languages, describing which "body_temp"-measurements or *events* are actually of interest. Using only subject-

based languages where a subscriber receives all notifications concerning all events about a specific subject would be too limiting for the scenario we have described. Also, the content interest may not be known before the operation and should therefore not be fixed. Take for instance an application keeping track of injured persons and their health status; an injured person is registered and given priority "code yellow", a medical person may e.g., put on a "wrist-band" having a bar-code and register the patient using a scanner and insert information such as "elderly woman", "diabetes" etc., and what kind of sensor has been attached. Based on these values the application may generate a content-based subscription, or the medical doctor can provide these values herself.

However, content-based subscription languages provide various degrees of expressiveness, i.e., how detailed an application can be when describing an event of interest. The filter may consist of several predicates depending on the number of attributes of the event, and operators connecting them. Usually the filters are Boolean functions, however some content-based languages also support *uncertainties*, i.e., fuzzy operators like "~" stating interest in events where the attribute value(s) are more or less equal to the value expressed in the subscription (Lekova et al. 2007). Some subscription languages support expressing interest in stateful events such as in SQL-like queries, like in an extension of the Gryphon system (Jin et al. 2003) or continuous queries for Data Stream Management Systems (Golab et al. 2003). This is interesting for instance in relation to events related to health data trends. All in all, being able to precisely describe exactly the events that are of interest may be beneficial from a bandwidth perspective as less event notifications are sent through the network, but the disadvantage is that the filtering and matching process may be more resource-intensive.

There are different models for content-based subscriptions and notifications. Mühl et al. (2006) define that the structure of a notification is dependent on the *data model* or *event model* and the subscription is defined by the *filter model* or *subscription model*. The subscription model is usually dependent on the event model, this implies, as described in (Baldoni et al. 2006), that the entities are aware of the syntax of the notifications and the semantics of the attributes. The subscribers and publishers are homogenous in the sense that the syntax of the subscription language matches the syntax of events. For instance, for structured records notifications have named set of attributes, and the attribute names are unique, and the subscriptions consist of one or several predicates (i.e., *attribute, operator, value*) combined by Boolean operators. Concerning semantics, the subscribers and publishers usually share a common vocabulary, meaning they have the same interpretation of the events. However, this may not always be the case. In (Petrovic et al. 2003), the authors suggest to re-write subscriptions to a default vocabulary using a synonyms list, and use concept hierarchy knowledge to find matching subscriptions that contain more generalized concepts of the notification's concept and using mapping functions. Cilia et al. (2004), describe a design where heterogeneous publishers and subscribers can interact. A concept-based software layer is used for mapping of subscriptions and notifications and providing correct context to ensure correct data and event interpretation. In the survey of (Baldoni et al. 2006), this is described as a *concept-based* language.

The ENS may have to support a number of applications, possibly running on devices belonging to different organizations. Therefore it would be limiting to only support a simple subscription language. If there are different applications using the same service and they use different subscription languages, either a meta-language is needed, i.e., every subscription is transformed by the service into this meta-language, the meta-language has to be the most expressive; or the service must support all the subscription languages. In (Jung et al. 2004), a meta-subscription language is suggested that enables applications to use different subscription languages. It is difficult or unrealistic to assume one specific event model as the data format for the events of interest may range from being a column in a database, variable in a kernel table, continuous data from sensor readings etc. We assume that the application developers know the data formats, and also know the "subjects" or "concept terms". However, concept terms belonging to different applications and possibly different organizations may be semantically similar or equal. A translation service or a knowledge manager bridging the semantic gap of the organizations could therefore be useful.

The applications do not need to know where the events may take place; it is the task of the ENS to either observe events and hence locate publishers (pull), or receive notifications on events from the publishers (push). In the first case, this means that the ENS must have knowledge about where events of certain concept terms may take place; this information must be gathered throughout the operation. Since all kinds of events may be of interest, it may not be desirable that all are published and reported to the ENS for filtering. The nodes could therefore exchange information on where and what kind of events may take place in order to make it possible to filter events according to relevant subscriptions close to the source. The nodes could actively report or advertise about local kinds of events to a specific service or the ENS (pro-active); or the ENS may send a query like a route request or file request for where such an event may take place (reactive); or browse a directory.

In short, with respect to the issue of subscriptions and events, we argue that for the emergency and rescue application domain expressive subscription languages should be supported and the event space should be flexible.

Event Notification Service Architecture

We have established that a variety of events could be of interest and that information on kinds of events and where events may take place must be available. The ENS uses this information or gathers this information, filters events based on received subscriptions and delivers information about events, event notifications, to interested subscribers. In this section, we discuss possible architectures of such a service, and subscription and notification routing approaches.

The event broker (also called server, mediator, router, dispatcher) is the ENS entity that matches and delivers subscriptions and notifications. For a centralized implementation, subscribers and publishers connect to the same entity. A distributed broker or ENS implementation is used to achieve scalability, fault tolerance and/or higher availability. Some possible distributed architectures are *broker overlay* or *P2P*. The broker overlay is usually hierarchical or flat, and the P2P architecture structured or unstructured (Baldoni et al. 2006).

Distributed content-based ENSs are usually implemented by a set of brokers forming an overlay. One approach for handling topology changes at the network layer or topology changes due to administrative changes (e.g., adding and removal of brokers) is the *strawman approach*, where a link removal and a link insertion are handled as un-subscriptions from the link being removed and sending subscriptions from the broker having a new link. Picco et al. (2003) suggest some optimizations for this approach, but assume that there is an algorithm that builds and re-builds a connected, loop-free tree at the routing level. Parzyjegla et al. in (2006), have a solution that ensures that message ordering is preserved and prevents notification loss after known administrated reconfigurations. A solution for how to repair a tree-shaped overlay in a MANET used for content-based routing due to topology changes, is presented in (Mottola et al. 2005).

The ENS performs routing of subscriptions, advertisements and notifications. The main difference with respect to other types of interaction is that the messages are routed based on their content and not sent to a specific destination address. The routing is dependent on the subscription language supported and the architecture used. For subject-based languages a network multicast routing protocol could be used by mapping subjects to groups, or if the nodes are members of a structured P2P system, the routing provided by the P2P system can be used, so-called rendez-vous-based routing (e.g., Hermes (Pietzuch 2004) and Scribe (Rowstron et al. 2001)). Another possibility is flooding: Notifications can be flooded and then filtered at the subscriber side or subscriptions can be flooded and filtered on the publisher side. For an ENS having a broker overlay architecture supporting a content-based language, there are variants of content-based routing. Brokers do notification forwarding by storing subscriptions in routing tables and then match notifications and subscriptions, which results in a forwarding set containing the addresses of brokers/subscribers which the notification should be forwarded to. Routing optimizations such as the usage of covering, i.e., a subscription A covered by a subscription B already in the routing table of a broker is not further forwarded in the overlay, which minimizes the routing table size. The subscriptions may be stored in e.g., a simple table structure such as JEDI (Cugola et al. 2001) or a partially ordered set (posets) such as in Siena (Carzaniga 1998) for faster insertions and deletions of subscriptions.

In a static, wired, non-mobile network, a fixed overlay topology that optimizes routing of subscriptions and notifications can be used, and as mentioned, there are solutions for handling changes due to e.g., administrative actions. Even though the topology is an application-level overlay, changes in the network topology due to mobility may require overlay topology changes. In a dynamic and disruptive environment such as sparse MANETs, there may be frequent topology changes at the network layer either resulting in a non-optimized or non-working broker topology because of network partitions and nodes being suddenly out of reach e.g., due to drained batteries or obstacles blocking the wireless signals. Maintaining multicast trees on the network layer could also be too expensive for a dynamic network. For these reasons, administering a broker overlay or structured P2P architecture may be too expensive and not feasible. So, for highly dynamic networks, an unstructured architecture may be the best option.

For dissemination of subscriptions and notifications in a MANET, there are several options. The simplest alternative is to use the broadcasting nature of the wireless medium and broadcast the messages to one-hop neighbors. Information can be flooded by re-broadcasting; several approaches for minimizing the number of re-transmissions are described in the literature (such as (Ni et al. 1999)). One protocol level higher, subscriptions and notifications can be routed by a network-layer routing protocol, and a multicast routing protocol can be used by establishing multicast groups for event subjects. Routing in the overlay may rely directly on the MAC-layer or cross-layering approaches can use information from the routing protocol for routing in the overlay. The choice of approach depends on the degree of dynamicity, if there are running network routing protocols, and the kind of delivery-QoS needed.

From these considerations, we can summarize the following: A distributed service that is robust and can handle network partitions is needed; in the presence of only mobile nodes and possibly frequent topology changes, it is costly to re-configure an overlay to reflect the underlying changes; the routing of messages should take into account the network dynamicity; and delay-tolerant delivery should be supported.

ANALYSIS OF RELATED WORK

In this section, we discuss related work based on the previous sections on requirements and design choices. Areas that are of interest are ENS and routing in disruptive environments.

ENS for Mobile Environments

Most ENSs are made for fixed networks; however, there are also systems for mobile environments. STEAM (Scalable Timed Events and Mobility) (Meier et al. 2002) is an event-based middleware for Wireless Ad-hoc Networks. All nodes run the service and there are no nodes having a broker role. The publishers announce their type of events and in which area this type is relevant. To receive events, the subscribers must be in the geographical area corresponding to the event, and join the corresponding group. Subject and proximity filters are located on the publisher side and content filters on the subscriber side. The solution does not fit if there are lasting network partitions and subscribers and publishers are located in different partitions.

Huang and Garcia-Molina present in (Huang et al. 2003), an algorithm for how to maintain and construct publish/subscribe trees where the publisher node is the root. A node chooses its parents in the publish/subscribe trees of interest. Each node sends beacons regularly, so a node can change parent if it gets information about e.g., a parent node closer to the root of the tree, or if it stops receiving beacons from its former parent. This may not work in a setting where all kinds of events can be of interest which in turn could result in a high number of multicast trees possibly having only a few members. Yoneki et al. describe a content-based publish/subscribe system for MANETs in (Yoneki et al. 2005). Event source brokers establish a multicast group based on the least constrained subscription concerning a named event. Again, this could result in a high number of multicast trees, however, as the system is content-based only notifications that are of interest to at least one of the multicast tree members are sent. They use an extension of the multicast routing protocol On-Demand Multicast Routing Protocol. Brokers that have matching subscriptions join this group. Bloom filters are used to compress information. A MANET version of JECho (Zhou et al. 2001), targeted for applications where publishers continually send events such as data collection applications, is described in (Chen et al. 2005). The broker overlay topology and thus the routing of events to subscribers is changed when the physical network changes and/or a broker is overloaded. An event generated at the publisher is sent to its home broker and afterwards sent to the subscriber(s) home brokers that filter the event. The home broker may be replaced after network topology changes. Frequent changing of brokers requires resending subscriptions and a handover process between new and old brokers. The brokers have up-to-date information about the broker network by requesting information from its routing protocol and by using a topology update protocol. If a neighbor broker moves away this overlay link is removed, and if a new broker enters the vicinity, a new overlay link may be added. Q (Avvenuti et al. 2005) uses a cross-layering approach by managing the path from publisher to subscriber using information from the routing table. Q is a type-based publish/subscribe infrastructure, which floods advertisements that contain address and event type information into the network to build a routing table. Depending on the size of the event space this could result in a large number of flooding operations. A node subscribes by sending a subscription following the route set up by the advertisements. Subscriptions are covered, so if an event dispatcher has already sent a subscription that covers the new one, it does not send it further. Every publisher sends PING messages that follow the event flow to subscribers. Each dispatcher adds its ID, so the dispatcher on the way knows the path. This information is checked with the routing table information, and if the path is not compliant with the real topology, the route is changed. Disconnections are handled indirectly by the PING messages; if a node stops receiving PING messages then it re-subscribes to initiate a new path.

During the rebuilding of trees or changing the topology overlay, event notifications may be lost. One way of dealing with this is damage repair: An approach for recovering from event losses in dynamic environments due to unreliable transport protocols and reconfiguring of the overlay is given in (Costa et al. 2003). The idea is to use epidemic routing to receive missed events. Periodically a broker picks a subscription filter and makes a gossip message containing this ID together with an ID of the events that matches this subscription. It sends it to some of the neighbors and they may then exchange missed events. In (Costa et al. 2005), a solution for a highly mobile scenario based on probability and forwarding subscriptions to parts of the network is presented. Subscriptions are propagated only to the brokers close to the subscriber. An event is routed along the link a matching subscription was received from. Since subscriptions are not replicated among all brokers, events are also forwarded along a randomly chosen subset of the links. If a link between two neighboring brokers disappears this is handled the same way as a un-subscription. The routing is hence a combination of content-based routing and gossiping.

A topic-based publish/subscribe system is described in (Baehni et al. 2005). The topics are arranged in a hierarchy. However, in most rescue operations there is more than one organization involved, including a number of applications, and it might therefore be a challenging task to structure the events into one hierarchy. The system is completely decentralized, and assumes no underlying routing protocol. Subscriptions are sent to neighbors using beacons, and events received are sent to neighbors based on this subscription knowledge. Each event that is published has a validity period, after this period the notification about the event is not propagated. Nodes send heartbeat messages to its neighbors that include a list of its subscriptions. Each node has a neighborhood table containing the IDs of the nodes and the subscriptions to those nodes whose subscriptions intersect with its own. The nodes also keep an event table of still valid events (garbage collection removes stale events). When a node discovers a neighbor with a similar subscription that lacks an event (each event is identified by a unique ID), it sends this event. The goal of this system is to achieve a high level of reliability of disseminating events to interested subscribers, while not using too much memory and battery power. In (Costa et al. 2008), an interest-based routing protocol for delay tolerant communication is presented. The support for delay-tolerant communication is of interest for a network where nodes may be sparsely distributed at the rescue scene. A node subscribes to one or more topics or *interest* and each message is tagged with an interest and routed based on this interest. The idea is that subscribers that have social ties usually have similar interest and eventually will meet again. Carriers for a message are chosen based on their *utility*; a node is a good carrier for a specific interest message if the node has been often collocated with other interested subscribers previously. They use forecasting techniques for predicting future movement. The routing has three phases that are repeated periodically. In the first phase, the nodes broadcast a message containing its interest plus utility values. In the second phase, the utility of the local node is computed for all interests, these values are compared with values of the neighbors. In the third phase, messages are forwarded to interested nodes and/or best carriers. Epidemic Messaging Middleware for Ad Hoc Networks (EMMA) (Musolesi et al. 2005) is based on JMS adapted for MANETs. To subscribe for a topic a subscriber registers at the client holding the topic. The subscriptions may be durable or non-durable. If a subscriber is not reachable, and the subscription is durable, epidemic routing is

used. The service provides *at-most-once* delivery semantics by having a list of identifiers to avoid delivery of message duplicates. A JMS message can be persistent or non-persistent. If buffer space is limited, persistent messages are prioritized. Each host holds a list of messages successfully delivered, so when these lists are exchanged during the anti-entropic epidemic protocol the list is updated: if a message has a single recipient and it has been delivered, it is deleted from the buffer; if a message has multiple recipients the identifiers of the delivered hosts are deleted from the associated list of recipients.

Information Dissemination

The problem of too many (unnecessary) retransmissions is most severe in *dense* MANETs and less resource intensive and therefore more scalable schemes than the simple epidemic routing approach and flooding have been suggested. For broadcasting of messages in MANETs, Ni et al. (1999) have suggested different schemes for reducing redundant re-broadcasting: One way is to limit the number of retransmissions by only forwarding a message with a probability P. Another way is to use a counter-based scheme where a node waits for a random time before broadcasting, and during this waiting period it counts the number of times it hears the same message. If this number is higher than a defined threshold C, then the node cancels the re-broadcasting. In the distance-based scheme, a node re-broadcasts a message only if the distance between itself and the closest node that has sent the same message is larger than a distance threshold D.

Instead of gossiping to all nodes to deliver a message, messages can be copied or sent to some of the nodes, either picked at random or using historic data related to e.g., which nodes have met which other nodes, and previous movement. *Random* approaches include e.g., the following: In *spray and wait* (Spyropoulos et al. 2005), a message is first "sprayed", i.e., copied, to L nodes that carry the message and try to deliver it. If the destination is not reached during the spray phase, the message is delivered directly by any of these nodes when possible. If available, *history data* or *prediction* can be used. In (Lindgren et al. 2003) historic data is used, the nodes remember which nodes they have met, how many times they have met, and when the meetings took place. This data is used to pick nodes that have higher probability of meeting the destination node as relays. The Context-Aware Routing (CAR) protocol (Musolesi et al. 2005b) delivers a message using the underlying routing protocol if the receiver is in the same partition, if not, it aims to determine which hosts are most likely to carry the message to the partition in which the destination resides. It is thus more scalable then the purely epidemic protocol if it is successful in its predictions. Different metrics such as mobility rate, mobility patterns, amount of battery energy and other type of context information is used.

In the ZebraNet project (Juang et al. 2002), the goal is to track zebras. The animals have attached sensors, which form a low density network. Radio transmission is low, so to collect data mobile base stations that come close to the zebras are used. Even though the base station does not get in contact with every sensor, it can receive data from non-reachable sensors because of replicated data. They suggest protocols for choosing which nodes to relay data to, e.g., by using history such as frequency on previous encounters with base-stations. A similar approach is used in the Data Mule project (Shah et al. 2003). In this architecture, however, there are three different devices: sensors that are spread around, *mules* (Mobile Ubiquitous LAN Extensions) and access points. The mules could be attached to e.g., animals, people or vehicles and they collect data from sensors and deliver data to access points.

Another way to overcome disconnections is to use dedicated nodes that move from one partition to another to "bridge" the partitions. In the Message Ferrying approach (Zhao et al. 2004),

dedicated nodes have the role of a postman and deliver messages to the destination. The authors suggest different approaches; one or multiple ferries, having ferries interact or not, and if so, different ways of cooperation. The ferry either follows a known route and nodes initiate to go to the ferry, or the ferry goes to meet the node if the node sends a service request using long range radio. In the node-initiated scheme, the node decides at some point to move towards a ferry and send and receive messages. The ferry can then deliver messages to the destination.

Hypergossiping (Khelil et al. 2007), aims to cover both dense and sparse MANETs; one scheme is used to gossip within a network partition, another is used to distribute messages across partitions. The generalized strategy is adapted to the mobility scenario, and the nodes choose a configuration based on its number of neighbors, i.e., its local view of the density. No global knowledge is therefore assumed. The challenge is to detect when there is network merging, so re-broadcasting can be performed and messages are sent to the new nodes. A network merging is detected by nodes sharing a list of recently received broadcast messages, and if a node receives a list that differs more than a threshold value, it concludes that they have not been part of the same partition lately, and re-broadcasting is started.

A NOVEL ENS APPROACH FOR EMERGENCY AND RESCUE APPLICATIONS

DENS (Skjelsvik 2008) is an ENS that is part of a middleware for rescue and emergency applications in sparse MANETs (Plagemann et al. 2006). It combines and utilizes many of the mentioned features of the described related work. The novelty lies in its flexibility with respect to subscription language, and its adaptive message delivery protocols handling the tradeoff of high availability verses low resource consumption by

Figure 3. DENS roles

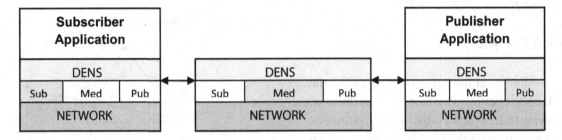

dynamic scaling of the service to the network environment. This is partly achieved by using other services in the middleware.

There are three main components in DENS: A subscriber, publisher and a mediator component as shown in Figure 3. The subscriber component receives subscriptions from local subscriber applications; the publisher component initiates monitoring of events based on subscriptions interested in events taking place on the local node; and the mediator components runs on some of the nodes in the network and form an overlay containing replicated subscriptions and notifications. Subscriptions are stored on the local subscriber node, at the relevant publisher node or nodes, and the mediators. DENS runs on all nodes, and it is a distributed service. The DENS overlay is unstructured.

Subscription Language Independence in DENS

An important DENS property is subscription language independence. We argue that for such a service to be useful in a rescue operation it must be suitable for different kinds of applications, especially due to the heterogeneous environment. Separating the choice of subscription language from how notifications are disseminated allows DENS to always use the most suitable language(s). DENS, uses language-specific plug-ins developed by the subscription language developer because this limits the overhead with respect to resource

usage. If an application does not need a complex subscription language a simple filtering and matching function for these subscriptions can be used. Another important aspect is the event space itself. The kinds of events may be tied to an ontology for instance, but which nodes are sources for the event could very well change during the course of an operation. To support a particular subscription language, DENS needs to be able to parse the subscription to find relevant concept terms. The concept terms are used to find the potential publisher nodes, unless stated explicitly in the subscriptions. A Knowledge Manager (KM) (Sanderson et al. 2005) keeps track of which nodes have data related to the concept terms. DENS on the publisher nodes need to be able to filter the events and therefore understand the subscription language. In addition, DENS must be able to match notifications and subscriptions, and the matching procedure must therefore have knowledge about the syntax of the language. The three language specific plug-ins that are needed per subscription language are parsing(), filtering() and matching(). The developer of a new subscription language and the KM enable DENS to resolve the conflict of supporting multiple subscription languages and decoupling subscribers and publishers at the same time. Subscription language IDs (SL-IDs) are used by DENS to identify which particular plug-in has to be invoked when a subscription or notification arrives. The subscription language independence is further described in (Skjelsvik et al. 2006) and the usage of two different subscription languages are given in (Skjelsvik et al. 2007) and (Lekova et al. 2007).

DENS Overlay

The mediators in the network form an overlay, the DENS overlay. The mediators perform the following tasks: receive subscriptions; receive notifications; match subscriptions and notifications; deliver notifications; synchronize information; and receive network monitoring information from the Resource Manager (Drugan et al. 2005) to do scaling of the overlay. Subscribers and publishers send subscriptions and notifications respectively, to a close-by mediator if the local node itself is not a mediator. The DENS overlay is used to:

- deliver subscriptions to publishers, and notifications to subscribers, if the underlying routing protocol cannot be used due to partitioning,
- as a repository of replicated subscription and notifications (See Figure 4).

In case of network partitioning, DENS supports information sharing in different network partitions, and it should work as good as possible if arbitrary nodes are switched off, including mediators. A subscriber may be in another partition than the publisher at the time an event of interest happens. Therefore, DENS has to provide a sufficient degree of redundancy through replication to assure a reliable service. The mediators may perform their own overlay routing using *store-carry-forward*. If a mediator fails to deliver a notification to a subscriber because it is not reachable, either because the subscriber node is in another partition or turned off, the notification is stored and possibly replicated, to be delivered when the subscriber node is in reach.

Combining routing-in-time and routing-in-space has been suggested in previous works, such as the mentioned EMMA (Musolesi et al. 2005),

Figure 4. Subscription repository

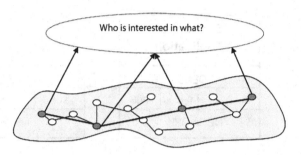

but DENS focuses more on adapting the degree of replication and ways of synchronizing information to the dynamicity of the network during the operation. DENS aims to handle the tradeoff between high availability and low resource consumption. This is achieved by scaling the size of the overlay, i.e., the number of mediators, and using the most suitable delivery protocol.

The number of mediators is dependent on the size of the network and its dynamicity. Information about the network can be extracted directly from the routing table. A pro-active routing protocol aims to have up-to-date routes to all nodes at all times, so if there are for instance relatively few route changes, DENS can assume a stable network. In addition, the Resource Manager uses routing information both to predict future network partitions and to discover clusters. Such information is used by DENS as e.g., cluster heads are potential mediators since they are close to many nodes. Other types of nodes that are potential mediators include nodes that move fast and cover a larger part of the incident area.

The number of mediators and which nodes are part of the overlay, may change e.g., if a network partition without any mediator is created, or there are too few or too many nodes running as mediators. Also, how the synchronization of information is done and the degree of replication depends on the mobility scenario, i.e., the density and the mobility of the nodes. Information dissemination is thus based on gossiping and the underlying network protocol.

Using DENS for Remote Patient Monitoring

One example of an emergency and rescue operation application based on DENS is remote patient monitoring or trauma care. In many cases, there may not be enough medical personnel to monitor all the injured persons and they could be distributed over a large area. Biomedical sensors can therefore be used for patient monitoring. A medical person can register, e.g., by using a wrist-bar-tag, examine and put on sensors, and then specify subscriptions specific for this patient on events of interest based on sensor values. Afterwards, he may continue registering other injured persons or help out in more acute situations. The sensors continuously monitor the status of the patient, and DENS can be used to inform the medical personnel when a critical event occurs. A small computing device could be attached to the patient and this device can collect data from the sensors. Examples of signs of interest are pulse rate, respiration rate, temperature, blood pressure and oxygen saturation. In (Osigma 2007), the development and evaluation of a patient monitoring application running on a small device, a Nokia N800 PDA, is presented. The application displays vital signs received from biomedical sensors on a screen such as blood pressure values shown in Figure 5; stores vital sign data on demand; patient; and event data; and sends an alert when pre-defined conditions occur. The alert mechanism uses a publish/subscribe service to send the alert messages to the medical personnel.

The events may be described by values of one sensor stream where a temperature or pulse value is above or below a certain threshold or they may be more complex. Some examples of complex events are combination of several sensor streams, average values or trends, or events expressed in

Figure 5. Patient monitoring application displaying blood-pressure (Osigma 2007)

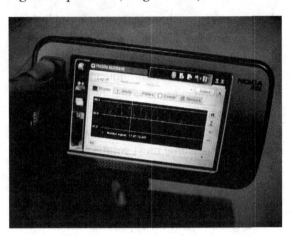

fuzzy instead of crisp values. Analysis of blood pressure or detecting a heart attack requires more than just observing one value at a time. This is thus an example of where the support for more than one subscription languages in DENS is useful: Simple events may be described using tuples or attribute-operator-value triples; Monitoring sensor streams over time to find trends or average values etc., could be achieved by having a Data Stream Management System processing the stream and express the subscription in a continuous query; and the medical person may in certain cases prefer to just indicate fuzzy values such as "high" or "low" instead of indicating exact value ranges and combinations of sensor values.

In some cases, the network may not be well-connected and there may be long-lasting network partitions. Reasons for this are low density of nodes, e.g., because the injured persons are spread over a large area, there may be obstacles blocking the wireless signals and there may be frequent topology changes due to movement. If the subscribers are in another network partition when an event of interest happens, replicating the notification in the DENS overlay and using store-carry-forward increases the chance of delivering the notification.

CONCLUSION

The usage of event-based interaction and its benefits have been described in a number of papers. In this chapter, we have motivated the usage of an ENS for rescue operation applications in sparse MANETs. However, as noted by (Raiciu et al. 2006), the usage of wide-area content-based publish/subscribe has not been widely deployed. One reason for this is that most solutions are tailored for a specific application and have a combined matching and event-delivery/routing design. We have in this chapter taken a practical approach by trying to describe the application scenario and requirements to discuss various design choices such as subscription language, ENS architecture and routing: We argue that the service should not be fixed to one specific subscription language, and that the architecture chosen must handle a dynamic environment where there may be long-lasting network partitions.

A lot of the reasoning we have given here was used as guidelines and design recommendations for DENS. Some important characteristics of DENS are: (1) DENS filters events at the source instead of sending every update. This is done by a DENS sub-component that monitors and filters types of events that a subscriber has expressed interest in. (2) DENS is subscription language independent and can support different languages at run-time using language specific plug-ins. (3) An underlying network routing protocol is assumed to route subscriptions and notifications, and DENS also uses information from the routing table. (4) All nodes run the service. In addition, some of the nodes act as mediators and form an overlay where subscriptions and possibly notifications are replicated to improve chances of reaching the destination.

Future work includes a more formal description on how the ENS should adapt to its environment, the tuning of the service and how to use information from e.g., routing protocols and other services to achieve this.

REFERENCES

Avvenuti, M., Vecchio, A., & Turi, G. (2005). A cross-layer approach for publish/subscribe in mobile ad hoc networks. In MATA'05, Proceedings of Mobility Aware Technologies and Applications: Second International Workshop (pp. 203-214). Monteral, Canada.

Baehni, S., Chabra, C. S., & Guerraoui, R. (2005). Frugal Event Dissemination in a Mobile Environment. In *Proceedings of the 6th ACM International Middleware Conference.* Grenoble, France.

Baldoni, R., Querzoni, L., & Virgillito, A. (2006). Distributed Event Routing in Publish/Subscribe Communication Systems: a Survey (Technical report – MIDLAB 1/2006). Rome: Dipartimento di Informatica e Sistemistica "A.Ruberti", Università di Roma la Sapienza.

Bellavista, P., & Corradi, A. (Eds.). (2006). *The Handbook of Mobile Middleware.* Boca Raton, FL: Auerback Publications. doi:10.1201/9781420013153

Carzaniga, A. (1998). *Architectures for an Event Notification Service Scalable to Wide area Networks.* (PhD thesis), Politecnico di Milano.

Chen, Y., & Schwan, K. (2005). Opportunistic Brokers: Efficient Content Delivery in Mobile Ad-Hoc Networks. In *Proceedings of the 6th ACM/ IFIP/USENIX International Middleware Conference, Volume 3790* (pp. 354-374, Lecture Notes in Computer Science), Berlin, Germany: Springer.

Cilia, M., Antollini, M., Bornhovd, C., & Buchmann, A. (2004). Dealing with Heterogeneous Data in Pub/Sub Systems: The Concept-Based Approach. In *Proceedings of DEBS'04, International Workshop on Distributed Event-Based Systems.*

Costa, P., Mascolo, C., Musolesi, M., & Picco, G. P. (2008). Socially-aware Routing for Publish-Subscribe in Delay-tolerant Mobile Ad Hoc Networks. In IEEE Journal on Selected Areas in Communications (JSAC). 26(5), 748 – 760.

Costa, P., Migliavacca, M., Picco, G. P., & Cugola, G. (2003). Introducing Reliability in Content-Based Publish-Subscribe through Epidemic Algorithms. In *Proceedings of DEBS'03, International Workshop on Distributed EventBased Systems.*

Costa, P., & Picco, G. P. (2005). Semi-probabilistic Content-Based Publish-Subscribe. In [*th IEEE International Conference on Distributed Computing Systems.*]. *Proceedings of ICSCS, 05*, 25.

Coulouris, G., Dollimore, J., & Kindberg, T. (2001). *Distributed Systems, Concepts and Design* (3rd ed.). Reading, MA: Addision Wesley.

Cugola, G., Nitto, E. D., & Fugetta, A. (2001). The JEDI Event-Based Infrastrucutre and its Application to the Development of the OPSS WFMS. [TSE]. *IEEE Transactions on Software Engineering, 27*(9), 827–850. doi:10.1109/32.950318

Drugan, O. V., Plagemann, T., & Munthe-Kaas, E. (2005). Building resource aware middleware services over MANET for rescue and emergency applications. In *Proceedings of the 16th Annual IEEE International Symposium on Personal Indoor and Mobile Radio Communications, International Congress Center (ICC).*

Eugster, P. (2007). Type-Based Publish/Subscribe: Concepts and Experiences. *ACM Transactions on Programming Languages and Systems, 29*(1).

Golab, L., & Tamer Özsu, M. (2003). Issues in Data Stream Management. *SIGMOD Record, 32*(2), 5–14. doi:10.1145/776985.776986

Huang, Y., & Garcia-Molina, H. (2003). Publish/Subscribe tree construction in wireless ad hoc networks. In *MDM'03, Proceedings ACM International Conference on Mobile Data Management* (pp. 122-140). Melbourne, Australia.

IEEE. (n.d.). *IEEE 802.11*. Retrieved from http://grouper.ieee.org/groups/802/11/index.html

Jin, Y., & Strom, R. (2003). Relational Subscription Middleware for Internet-Scale Publish-Subscribe. In *DEBS '03, Proceedings of the 2nd international workshop on Distributed event-based systems.*

Juang, H. Oki, Wang, Y., Martonosi, M., Peh, L., & Rubens, D. (2002). Energy-Efficient Computing for Wildlife Tracking: Design Tradeoffs and Early Experiences with ZebraNet. In ASPLOS'02, Proceedings of Architectural Support for Programming Languages and Operating Systems.

Jung, D., & Hinze, A. (2004). A Meta-Service for Event Notification. *Lecture Notes in Computer Science, 3290*, 283–300.

Khelil, A., Marrón, P. J., Becker, C., & Rothermel, K. (2007). Hypergossiping: A Generalized broadcast strategy for mobile ad hoc networks. *Ad Hoc Networks, 5*(5), 531–546. doi:10.1016/j.adhoc.2006.03.001

Lekova, A., Skjelsvik, K. S., Plagemann, T., & Goebel, V. (n.d.). Fuzzy Logic-Based Event Notification in Sparse MANETs. In *AINA'07, Proceedings of the 21st International Conference on Advanced information Networking and Applications Workshop* (Vol. 2, pp. 296-301), Niagara Falls, Canada.

Lindgren, A., Doria, A., & Schelén, O. (2003). Probabilistic Routing in Intermittently Connected Networks. In *MobiHoc'03, Proceedings of the Fourth ACM International Symposium on Mobile Ad Hoc Networking and Computing.*

Meier, R., & Cahill, V. (2002). STEAM: Event-Based Middleware for Wireless Ad Hoc Networks. In *DEBS'02, Proceedings of the 1st International Workshop on Distributed Event-Based Systems*, Vienna, Austria.

Mottola, L., Cugola, G., & Picco, G. P. (2005). *A Self-Repairing Tree Overlay Enabling Content-Based Routing on Mobile Ad Hoc Networks (Tech. Rep.)*. Politecnico di Milano.

Mühl, G. (2002). *Large-Scale Content-based Publish/Subscribe Systems*. (PhD thesis), Darmstadt, Germany: Darmstadt Univ. of Technology.

Mühl, G., Fiege, L., & Pietzuch, P. R. (2006). *Distributed Event-Based Systems*. Berlin-Heidelberg, Germany: Springer-Verlag.

Munthe-Kaas, E., Drugan, O., Goebel, V., Plagemann, T., Pužar, M., Sanderson, N., & Skjelsvik, K. S. (2006). Mobile Middleware for Rescue and Emergency Scenarios. In Bellavista, P., & Corradi, A. (Eds.), *Mobile Middleware*. Boca Raton, FL: CRC Press. doi:10.1201/9781420013153.ch48

Musolesi, M., Hailes, S., & Mascolo, C. (2005). Adaptive Routing for Intermittently Connected Mobile Ad Hoc Network. In *WoWMom'05, Proceedings of the IEEE 6th International Symposium on a World of Wireless, Mobile, and Multimedia Networks*, Taormina, Italy.

Musolesi, M., Mascolo, C., & Hailes, S. (2005). EMMA: Epidemic Messaging Middleware for Ad hoc Networks. *Personal and Ubiquitous Computing, 10*(1), 28–36. doi:10.1007/s00779-005-0037-4

Ni, S. Y., Tseng, Y. C., Chen, Y. S., & Sheu, J. P. (1999). The broadcast storm problem in a mobile ad hoc network. *In MobiCom'99, Proceedings of the 5th Annual ACM/IEEE International Conference on Mobile Computing and Networking* (pp. 151-162), Seattle, WA.

Osigma, T. (2007). *Evaluation of MIDAS Middleware in a Mobile Healthcare Scenario*. (Master thesis), Enschede, Netherlands: University of Twente.

Petrovic, M. Burcea, I. & Jacobsen, H. A. (2003). S-ToPSS: Semantic Toronto Publish/subscribe Systems. In *Proceedings of CLDB* (pp. 1101-1104).

Picco, G. P., Cugola, G., & Murphy, A. L. (2003). Efficient content-based event dispatching in the presence of topological reconfiguration. In ICDCS 03. In *Proceedings of the 23rd International Conference on Distributed Computing Systems* (pp. 234-243).

Pietzuch, P. R. (2004). *Hermes: A Scalable event-based middleware* (Tech. Rep. No. 590). Cambridge, UK: University of Cambridge.

Plagemann, T., Andersson, J., Drugan, O., Goebel, V., Griwodz, C., Halvorsen, P., et al. (2006). Middleware Services for Information Sharing in Mobile Ad-hoc Networks – Challenges and Approach. In *Proceedings of the Workshop on Challenges of Mobility, IFIP TC6 World Computer Congress*, Toulouse, France.

Raiciu, C., Rosenblum, D. S., & Handley, M. (2006). Revisiting Content-Based Publish/subscribe. In *ICDCSW'06, Proceedings of the 26th IEEE International Conference on Distributed Computing Systems Workshops*.

Rowstron, A., Kermarrec, A., Castro, M., & Druschel, P. (2001). SCRIBE: The design of a large-scale event notification infrastructure. In *Proceedings of the Networked Group Communication: Third International COST264 Workshop* (pp. 30-43, NGC 2001), London.

Sanderson, N., Goebel, V., & Munthe-Kaas, E. (2005). Metadata Management for Ad-Hoc Info-Ware – A Rescue and Emergency Use Case for Mobile Ad-Hoc Scenarios. In R. Meersman and Z. Tari (Eds.), *ODBASE'05: International Conference on Ontologies, Databases and Applications of Semantics*. (LNCS 3761, pp. 1365-1380).

Sanderson, N. C., Skjelsvik, K. S., Drugan, O., Pužar, M., Goebel, V., Munthe-Kaas, E., & Plagemann, T. (2007). *Developing Mobile Middleware – An Analysis of Rescue and Emergency Operations* (Tech. Rep. No. 358. ISBN 82-7368-317-8, ISSN 0806-3036). Oslo, Norway: Department of Informatics, Unversity of Oslo.

Shah, R. Roy, S., Jain, S., & Brunette, W. (2003). Data MULEs: Modeling a Three-tier Architcture for Sparse Sensor Networks. In *IEEE SNPA'03, Proceedings of the 1ˢᵗ IEEE International Workshop on Sensor Network Protocols and Applications* (pp. 30-41).

Skjelsvik, K. S. (2008). *A Distributed Event Notification Service for Sparse Mobile Ad-hoc Networks.* (Ph.d thesis), Oslo, Norway: University of Oslo.

Skjelsvik, K. S., Lekova, A., Goebel, V., Munthe-Kaas, E., Plagemann, T., & Sanderson, N. (2006). Supporting Multiple Subscriptions Languages by a single Event Notification Overlay in Sparse MANETs. In *Proceedings of ACM MobiDE*.

Skjelsvik, K. S., Søberg, J., Goebel, V., & Plagemann, T. (2007). Using Continuous Queries for Event Filtering and Routing in Sparse MANETs. In *FTDCS'07, Proceedings of the 11th IEEE International Workshop on Future Trends of Distributed Computing Systems* (pp. 138-148).

Spyropoulos, T., Psounis, K., & Raghavendra, C. S. (2005). Spray and Wait: An Efficient Routing Scheme for Intermittently Connected Mobile Networks. In *SIGCOMM'05, Proceedings of ACM SIGCOMM workshop on Delay Tolerant Networking* (pp. 252-259).

Vahdat, A., & Becker, D. (2000). *Epidemic Routing for Partially Connected Ad Hoc Networks* (Tech. Rep. No. CS-2000-06). Durham, NC: Department of Computer Science, Duke University.

Yoneki, E., & Bacon, J. (2005). Content-Based Routing with On-Demand Multicast. In *ICDCSW'04, Proceedings of the 24ᵗʰ International Conference on Distributed Computing Systems Workshops*.

Zhang, Z. (2006). Routing in Intermittently Connected Mobile Ad Hoc Networks and Delay Tolerant Networks: Overview and Challenges. *IEEE Communications Surveys & Tutorials, 8*(1), 24–37. doi:10.1109/COMST.2006.323440

Zhao, W., Ammar, M., & Zegura, E. (2004). A Message Ferrying Approach for Data Delivery in Sparse Mobile Ad Hoc Networks. In *MobiHoc'04, Proceedings of the 5ᵗʰ ACM International Symposium on Mobile ad hoc networking and computing* (pp. 187-198).

Zhou, D., Schwan, K., Eisenhauer, G., & Chan, Y. (2001). JECho – Interactive High Performance Computing with Java Event Channels. In *IPDPS'01, Proceedings of the International Parallel and Distributed Processing Symposium*.

Chapter 19
Panel:
Current State and Future of Event-Based Systems

Annika Hinze
University of Waikato, New Zealand

Mani Chandi
California Institute of Technology, USA

Jean Bacon
University of Cambridge, UK

Avigdor Gal
Israel Institute of Technology, Israel

Alejandro Buchmann
Technische Universität Darmstadt, Germany

Dieter Gawlick
Oracle Incorporated, USA

Sharma Chakravarthy
The University of Texas at Arlington, USA

Richard Tibbetts
StreamBase Systems, USA

ABSTRACT

This chapter is a panel discussion in writing. The field of event-based systems finds researchers from a number of different backgrounds: distributed systems, streaming data, databases, middleware, and sensor networks. One of the consequences is that everyone comes to the field with a slightly different mindset and different expectations and goals. In this chapter, we try to capture some of the voices that are influential in our field. Seven panellists from academia and industry were invited to answer and discuss questions about event-based systems. The questions were distributed via email, to which each participant replied their initial set of answers. In a second round every panelist was given the opportunity to expand their statement and discuss the contributions of the other panellists. The questions asked can be grouped into two types. Questions in the first group refer to each participant's understanding of the basic concepts of event-based systems (EBS), the pros and cons of EBS, typical assumptions of the field and how they understood EBS to fit into the overall landscape of software architectures. The second group of questions pointed to the future of EBS, possible killer applications and the challenges that EBS researchers in academia and industry need to address in the medium and long term. The next section gives each panellist's initial statements as well as their comments to other participants' contributions. Each participant's section starts with a short introduction of the panellist and their work. In the final section, we compare and reflect on the statements and discussions that are presented by the seven panellists.

DOI: 10.4018/978-1-60566-697-6.ch019

PANELLISTS' CONTRIBUTIONS

Jean Bacon

Jean Bacon is Professor of Distributed Systems at the University of Cambridge Computer Laboratory, UK. She leads the Opera research group, with focus on middleware for large-scale, widely distributed or geographically concentrated ubiquitous systems. The Opera group pioneered event-based systems from the early 1990s, with applications including healthcare, pollution and transport monitoring, typically comprising multiple administration domains. Opera has carried out substantial work on access control, security and privacy for these and other applications, and the specification and enforcement of policy.

Jean is a Fellow of the IEEE and BCS and was an IEEE Fellowship Committee member in 2008. She was a member of the Governing Body of IEEE-CS, 2001-2007; and founding EIC of Distributed Systems Online 2000-2008, IEEE's first online-only magazine. She is currently on the Editorial Board of Computer, IEEE-CS's flagship magazine.

Annika: Where do you see the difference between publish/subscribe, event-based systems, Complex event processing (pub/sub, EBS, CEP)? How do you see message filtering technologies fitting into the picture? [Q1]

Jean: I suggest starting the discussion by giving some basic definitions:

Primitive event: Some define a primitive event as representing a state change (being notified asynchronously to interested parties). Others may accept the asynchronous transmission of a value e.g. a static sensor reading, as an event, even if there has not been a change of state in the process being monitored by the sensor.

We might extend the definition to include an event's manifestation as a message e.g. with topic/type, attributes and their values, plus a system-generated source timestamp that might be a point timestamp with some tolerance or an uncertainty interval to allow for the impossibility of representing time exactly.

One could regard the difference in time, if not in attribute values, as a state change for consistency with the definition, but this would be against the intuition for ``tell me only when some specified attribute changes value (significantly - by some specified amount).''

If mobile publishers are supported then a location stamp as well as a timestamp may be included in a message, or location might be deemed an attribute of an event.

Composite event: An expression combining primitive and composite events using a number of defined operators. The representation of a composite event raises many issues, as does engineering detection in a distributed system.

"Event-based system" is a general term that includes many different system styles. Common to all is asynchronous communication between decoupled communicating entities. EBSs include GUIs and server architectures (but these are sufficiently well-known, centralised and homogeneous not to be covered in this book), sensor-based systems in ubiquitous computing, active database triggers, stream processing, complex event processing, publish/subscribe applications such as news/stockquote notifications.

In order to address the questions below we need to distinguish client and system components' functionalities and interfaces and colocation. For example, do we assume a dedicated network of event brokers to implement communication or might there be a peer-to-peer model with each node in a system containing both application and system components?

I see EBS as the umbrella term for all event-driven systems.

The differences lie in the characteristics and requirements of the supported applications such as security, reliability of communication, whether a transactional model is needed by the client level, the scale required, whether real-time delivery is required, whether communicating entities are mobile or stationary.

An engineering difference is whether time in a distributed system is taken into account or whether the timestamp associated with an event is its arrival at a database or a stream. In the latter case we have abstracted above one of the realities of distributed systems and cannot carry out an analysis of physical causality, for example, ``did one explosion/fire set off others or did they start independently?'' ``did an increase in pressure cause a rupture in a container or did a rupture cause pressure to fall and be increased automatically in the monitored system?''

I see publish/subscribe both as an application interface and as a many-to-many communication implementation (one-to-many for any given publication). A pub/sub interface need not necessarily be supported by pub/sub communication.

Message filtering capability determines the expressiveness of subscriptions (and the complexity of the subscription language) and the complexity of algorithms for routing messages from publishers to subscribers, assuming communication is to be optimised. Monitored objects might be mobile and subscriptions might be able to include values of place and time, which increases the complexity of the matching process.

Annika: What are the pros and cons of publish/ subscribe / event-based systems? [Q2]

Jean: EBS of all kinds have the potential to support immediate response to occurrences. System engineering dictates whether any guaranteed real-time response is possible. Advantages and disadvantages are those of asynchronous systems (loose coupling, heterogeneity, potentially large scale) but their behaviour is harder to reason about than synchronous systems.

Publish/subscribe allows interest in events to be expressed and satisfied in terms of data names and values. The location and identity of publishers and subscribers is, or can be, hidden from the client level i.e. we have mutual anonymity of publishers and subscribers, and network independence of applications.

This makes it simple to build applications but has security implications in that publishers have to trust that subscribers are authorised to receive the data and that system components such as event-brokers are trusted to handle the data. For example, in content-based routing, brokers are trusted to decrypt message content for routing purposes. Yet the broker network might be dynamically reconfigurable for reliability reasons and/or to allow brokers to join and leave the network.

The source-anonymity of pub/sub makes it hard for a composite event recogniser to be engineered in a pub/sub-based distributed system. A source-aware CE recogniser can use a heartbeat protocol with each source to detect communication delay or failure and therefore whether an event might have been missed.

Without this detailed knowledge it is impossible to reliably implement some composition operators and consumption policies, and use in monitoring a safety-critical system would be impossible.

At an implementation level, a system design would be optimized very differently for many publishers and few subscribers (sensor systems reporting to servers) compared with few publishers and many subscribers (stock quotes/newsfeeds/ sports-result feeds).

Annika: What are typical assumptions made in the different pub/sub, EBS, CEP fields? [Q3]

Jean: I'm taking EBS as the umbrella term. I'm taking CEP systems to be in the usiness/banking world (not general, distributed event composition). I believe CEP requires reliable communication and a transaction model and abstracts above the time at distributed event sources. Composite event detection in a ubiquitous computing system would have very different assumptions, as mentioned for the previous question.

Again, with pub/sub it depends how you are defining the system. If you mean news/stockquote feeds then I suppose we assume a dedicated network of trusted brokers, probably statically configured.

In general, pub/sub assumes that mutual anonymity of publishers and subscribers is acceptable.

Annika: How do event-based systems fit into the overall landscape of software architectures (e.g., SOA and others)? [Q4]

Jean: SOA and EDA are already integrated in enterprise architectures. EDA changes the role of services into (active) producers and/or (reactive) consumers of messages. In this integration, a number of SOA concepts (e.g. endpoints, message types, and contracts) are adopted by EDA to specify the interaction between autonomous components in a more formal way. Other concepts, general to distributed systems (e.g. transactions) involve very different semantics in SOA and EDA.

Annika: Is pub/sub the way of the future? To what extent will small or large-scale pub/sub, EBS, CEP systems find application? [Q5]

Jean: Asynchronous communication has many properties that make it more widely applicable than synchronous communication, such as loose coupling, scalability, composition of heterogeneous components. Pub/sub adds anonymous interaction in terms of data. This is not a panacea – many issues remain to be resolved. There are pros and cons as mentioned above. Pub/sub can be a useful interface for active database triggers and can replace continuous queries. Integration of database and pub/sub communication has vast scope.

Pub/sub is good for highly dynamic systems (MANETs, DTNs – opportunistic computing, some WSNs) allowing mobile entities to communicate, and their leaving and joining to be masked. However, there is likely to be increased internet connectivity of mobile devices, making the research that shows how to do without it less widely applicable.

Annika: What are killer applications to pub/sub, EBS, CEP? [Q6]

Jean: Supporting immediate notification of changes and the ability to integrate databases and services has wide application. Healthcare, police, social services, transport monitoring, environmental monitoring in general e.g. pollution, disaster response (man-made and natural) as well as current financial and business applications.

Systems that support mobile entities communicating wirelessly are probably best based on pub/sub, at least until internet connectivity becomes more widespread. As mentioned in the discussion about assumptions made, pub/sub comes with security issues and the need for full trust of implementations by clients when data is sensitive.

Annika: What are the issues to be addressed in the medium term, long term? [Q7]

Jean: As discussed for the previous question - security and trust: some data is sensitive for human lifetimes and longer. There is a risk that encryption may be broken on a long timescale. Key spaces currently thought of as vast may come to be enumerable.

Transmitting encrypted data may not provide long-term security. Content-based routing, proposed for efficient communication (with path sharing) in large-scale systems, isn't an option for such data. There is unlikely to be sufficient basis for trust of all the brokers that might see it.

Reliability: for reliable delivery persistent memory has to be used. Integrating databases with pub/sub has many advantages for applications and in implementations - see the discussion for the previous question.

Mobility: many researchers are specialising in mobile computing, opportunistic/social networking and delay-tolerant networks. Attempts to use the mobile phone infrastructure have yet to demonstrate that it is capable of supporting continuous monitoring rather than the random, discrete calls for which it was designed. An economic model for continuous monitoring by mobile phones is needed. And software development is currently hampered by the need to build on legacy software, developed for resource-constrained systems.

In general, event composition is for deriving ``high level'' knowledge from lower-level events. To support this, we need to establish the event patterns of interest, by data mining or other techniques. We also need to place event aggregation and fusion and statistical analysis of events within a system.

The research community can decide on conceptual issues. If and when they are deemed understood and solved there is then the issue of engineering and maintaining large-scale systems and making them usable by developers and end-users.

Reflections and Comments

Richard: I would just like to react to Jean's comments about anonymity in pub/sub architectures. Whether pub/sub provides anonymity is a design decision, not a fundamental issue. In my experience pub/sub systems often pass along metadata about sources or routes. This additional information can be very helpful in certain application domains.

Alex Buchmann

Alejandro Buchmann is Professor in the Department of Computer Science of Technische Universität Darmstadt since 1991 and is responsible for the area of Databases and Distributed Systems. He studied chemical engineering at the Universidad Nacional Autónoma de Mexico and received his PhD from the University of Texas, Austin, in 1980. He was an Assistant/Associate Professor from 1980 to 1986 at IIMAS/UNAM and held positions as a senior researcher at Computer Corporation of America, Cambridge Mass. (86-89) and GTE Laboratories, Waltham Mass. (89-91) before joining TUD. He was responsible for the graduate research program in enabling technologies for e-commerce at TUD (1998-2007) and initiated a research program in Cooperative, Adaptive and Responsive Monitoring in Mixed Mode Environments (2006-). Alejandro's current research interests are in the areas of event-based and reactive systems, heterogeneous distributed systems, middleware, performance evaluation, peer-to-peer systems and new paradigms for data management and information processing in Cyberphysical systems.

Alejandro's involvement with event based systems dates back to the late 1980's in the HiPAC project on active databases. Subsequent work at TU Darmstadt included distributed event based systems, temporal uncertainty in distributed event systems, quality of service and transactional properties, content based publish/subscribe and concept based pub/sub for heterogeneous event systems, as well as performance modeling and benchmarking of event based systems.

Alex has been involved as general and/or program chair in ICDE 01 and ICDE08, VLDB 96, SIGMOD 98, DEBS 08 and AmI 07, and is on the editorial board of several journals.

He has held visiting positions at UT Austin, ICSI Berkeley, HP Labs Palo Alto, University of Virginia, Ecole Politechnique Federal de Lausanne (EPFL), IIT Bombay, and Caltech.

Annika: Where do you see the difference between publish/subscribe, event-based systems, Complex event processing (pub/sub, EBS, CEP)? How do you see message filtering technologies fitting into the picture? [Q1]

Alex: This is not a straight-forward comparison, since it involves very general concepts, such as event-based systems, and particular notification mechanisms, such as publish/subscribe.

To me an event is a meaningful change of state. What is meaningful is determined by the application. Since I include time in the description of state, two observations of a given property at different time are two events, even if the observed property did not change. For a sensor signal to be considered as an event, it must be discretized and identified as an event with its own timestamp. In a distributed system timestamps may be imprecise. Events have a representation, usually as a tuple of attribute values. Events are reified as notifications for dissemination. Events can be composed using the operators of an event algebra and can be enriched with context information to derive more abstract events.

Event-based systems are reactive systems that observe events, put them into the proper context and react to them. The event notification mechanism determines the degree of coupling between event source and event consumer. While there are some event based systems in which the event source delivers the events directly to the consumer, in the more general case we expect loose coupling and asynchronous communication between event source and consumer. Publish/subscribe in its different manifestations (channel-, subject-, topic-, content-, or concept-based) has become widely accepted as an adequate mechanism for the asynchronous delivery of event notifications decoupling event sources and consumers.

Complex Event Processing refers to the processing of event streams (often centralized), the composition of events, their enrichment with context information and the derivation of complex, application level events that trigger an application-level reaction. The difference between Distributed Event-Based Systems and Complex Event Processing is simply one of perception and emphasis in the respective communities. CEP has concentrated more on the aspects of deriving application level events and their semantics regardless of the notifiction mechanism. DEBS has concentrated more on event dissemination (often pub/sub), the efficiency and correctness of event filtering, composition and delivery, and quality of service at the technical level.

Event-based Systems is therefore the generic term that encompasses both TComplex Event Processing and Distributed Event-Based Systems, and Publish/Subscribe is just a very convenient delivery mechanism for event notifications. Publish/Subscribe can be centralized, as in most current JMS implementations or distributed as proposed by most research systems.

Annika: What are the pros and cons of publish/subscribe / event-based systems? [Q2]

Alex: The biggest advantage of event-based paradigm is that it reflects very well the requirements of many monitoring applications in which events and their describing data are pushed to the application. It is quite inefficient to implement such applications based on a request-reply interaction mode through application level polling cycles. For a push-based system, publish/subscribe offers several advantages: it decouples producers and consumers of events, it can be a very effective dissemination mechanism that supports asynchronous processing, and it allos us to filter and compose events on the broker nodes, thereby reducing the load on the event consumers.

The biggest disadvantage of event-based systems is the lack of control and the complexity derived from it. Since the event source typically is not aware of who subscribes to its events, quality of service must be defined and to some extent enforced by the consumer in cooperation with the notification middleware. To provide transactional guarantees this implies including the notification middleware in the transaction semantics and enforcing these semantics on consumer side. Rollback and recovery mechanisms must be adapted.

Quality of service guarantees must be based on the expectation of the event consumer.

Annika: *What are typical assumptions made in the different pub/sub, EBS, CEP fields? [Q3]*

Alex: The assumptions are application-domain dependent. Financial applications and fraud detection, for example, are high volume, run on stream processing engines on big iron, have stable high bandwidth communication and no energy constraints, and often depend on additional data sources for event enrichment. At the other extreme of the spectrum applications on wireless sensor networks assume resource constrained nodes, unstable communication, energy constraints, and stress routing and in-network data processing. Then you have systems in the health domain where the proper assumptions about false positives and false negatives are critical, as well as privacy concerns, and extreme heterogeneity between body area sensors and in the future body area networks and the large backends at hospitals and mobile equipment for doctors and care providers.

The application also determines the assumptions that are reasonable with respect to expressiveness of the event algebra and event composition, the treatment of time, the degree of distribution, the need for real-time constraints, security, persistance, auditability, mobility, delivery guarantees, event consumption, event replication and disposal to name just some of the most important ones.

Annika: *How do event-based systems fit into the overall landscape of software architectures (e.g., SOA and others)? [Q4]*

Alex: Event-based systems are complementary to other software architectures. They are particularly well suited for environments that must deal with large event streams. As sensors and cyberphysical systems become more widely used, event-based systems will be gaining importance. They will not and should not replace client/server architectures where stability is required and a pull-based interaction using request/reply is adequate. In conjunction with service oriented architectures event-based mechanisms can be used to trigger service invocation but also in the form of maintenace processes that alert service-based applications of changes in the underlying services or the availability of new services with expanded functionality.

Annika: *Is pub/sub the way of the future? To what extent will small or large-scale pub/sub, EBS, CEP systems find application? [Q5]*

Alex: Event-based systems will be used for many reactive applications, both large and small. To accelerate their use we must standardize the platforms and solve a variety of problems that today are, at best, solved in an ad hoc manner for individual applications. To call pub/sub the way of the future appears overblown to me. It is a powerful mechanism that blends very well with the event-based paradigm and as such it will also gain importance with the adoption of event-based systems.

Annika: *What are killer applications to pub/sub, EBS, CEP? [Q6]*

Alex: I see a variety of applications that have the potential to be killer applications. The most demanding, because it encompasses many other applications and is extremely rich is a smart city. Within a smart city we must provide the necessary infrastructure for navigation, context-based services, traffic monitoring and management, use-based billing, health-monitoring, emergency response, etc. Smart cities will have a huge density of mobile and stationary, heterogeneous sensor and actors that may interfere, must be self-organizing and self-configuring, the infrastructure must be secure and respect the privacy of individuals.

Another application that partially overlaps but has some additional aspects are integrated health systems, in particular as these expand into the domain of continuous health monitoring, remote response to medical conditions through automatic actuation of devices, and the integration with ambient assisted living.

A third application that has the potential of being a killer application are huge mobile sensor networks, for example, a million sensors mounted on the railway cars of a large national railroad system.

Annika: What are the issues to be addressed in the medium term, long term? [Q7]

Alex: I would like to group the issue I see thematically:

a.   **Event algebra and correlation** encompasses the need for well defined event algebras, that are rich enough but remain manageable and can, eventually, be standardized. The other main aspect in this category is event correlation. Since this often occurs through timestamps, the correct handling of time, particularly in distributed event-based systems remains an issue.

b.   **Event lifecycle issues** refer to the validity of events, their consumption and their purging from the system either after consumption or through garbage collection, the replication of event notifications for multiple consumers, long-lived events and the notification of new subscribers.

c.   **Event enrichment and event semantics** refers to the combination of events with external data sources, the management of event metadata, and the visualization of events. I would also group in this category the various event languages, extensions to SQL and XML, etc., although these are closely related to the event algebra.

d.   **Mobility, forwarding and staging of events** is critical to support many applica-

tion domains, yet there is a huge deficit in reliable mechanisms for event delivery in mobile distributed systems, particularly with respect to completeness, ordering, delivery after reconnection, and quality of service in general and timeliness in particular.

e.   **Security and privacy** are key issues for acceptance of event-based systems in many application domains, yet they are often just an afterthought and not an integral part of the architecture of event-based systems. Issues range from security and privacy policies, situation-aware security, to structuring of wireless sensor systems to limit visibility and secure publish/subscribe notification mechanisms.

f.   **Heterogeneity** will be a major issue in the killer applications, since the different parts will grow independently, have different life cycles and many disciplines and industries contribute, each with its own solutions and standards. In addition to the interoperability issues there will be scalability and interference problems. Perhaps the biggest challenge will be to provide the elements for large systems to be self-configuring and self-healing.

g.   **Software engineering** of event based systems will require new approaches to defining the requirements and expectation of applications, new testing methods that can account or anticipate future subscription patterns, very long deployment cycles typical in cyberphysical systems, and new maintenance and governace policies.

h.   Once functionality issues are solved **performance** becomes a distinguishing feature. Robust performance models for event-based systems are needed and their application to capacity planning. Finally, reliable **benchmarks** must be defined that allow an objectve comparison of event-based platforms.

Reflections and Comments

Richard: I'll echo what Alex said about the importance of development paradigms. Event processing is a new experience for most developers, and their familiar tools and techniques often do not apply. Our experience at StreamBase is that guiding developers towards best practices, and supporting those best practices with sophisticated and user-friendly developer tools is very helpful when training up new developers.

Sharma Chakravarthy

Sharma Chakravarthy is Professor of Computer Science and Engineering Department at The University of Texas at Arlington, Texas. He established the Information Technology Laboratory at UT Arlington in Jan 2000 and currently heads it. Sharma Chakravarthy has also established the NSF-funded, Distributed and Parallel Computing Cluster (DPCC@UTA) at UT Arlington in 2003. He is the recipient of the university-level "Creative Outstanding Researcher" award for 2003 and the department level senior outstanding researcher award in 2002. He is an associate editor of IEEE TKDE (Transactions of Knowledge and Data Engineering). He has served as financial chair for ICDE 2008, corporate sponsors chair for SIGMOD 2000, Publications chair for VLDB 2000, and as PC co-chair for ODBASE 2009.

He is well known for his work on semantic query optimization, multiple query optimization, active databases (HiPAC project at CCA and Sentinel project at the University of Florida, Gainesville), and more recently scalability issues in graph mining and its applications. His group at UTA has developed DBSubdue and DB-FSG – scalable versions of corresponding approaches for graph mining, and InfoSift – a classification system for text, email, and web that uses graph mining techniques.

His current research includes information retrieval, all aspects of database research, web technologies and ranking, stream data processing, complex event processing, mining and knowledge discovery – association, graph and text, push/pull technologies, web content monitoring, and information integration. He has published over 150 papers in refereed international journals and conference proceedings. He has given tutorial on a number of database topics, such as graph mining, database mining, active, real-time, distributed, object-oriented, and heterogeneous databases in North America, Europe, and Asia. He is listed in Who's Who Among South Asian Americans and Who's Who Among America's Teachers. He has published a book on Stream Data Processing, published by Springer in April 2009. He has been a consultant and is on the advisory board of companies and educations institutions.

Prior to joining UTA, he was with the University of Florida, Gainesville. Prior to that, he worked as a Computer Scientist at the Computer Corporation of America (CCA) and as a Member, Technical Staff at Xerox Advanced Information Technology, Cambridge, MA.

Sharma Chakrvarthy received the B.E. degree in Electrical Engineering from the Indian Institute of Science, Bangalore and M.Tech from IIT Bombay, India. He worked at TIFR (Tata Institute of Fundamental Research), Bombay, India for a few years. He received M.S. and Ph.D degrees from the University of Maryland in College park in 1981 and 1985, respectively.

Annika: Where do you see the difference between publish/subscribe, event-based systems, Complex event processing (pub/sub, EBS, CEP)? How do you see message filtering technologies fitting into the picture? [Q1]

Sharma: The terminology/nomenclature is mainly (in my opinion) due to the way different disciplines have arrived at the need for handling the same or similar situation at different points in time and with respect to different application domains. As I see it, event-based systems came out of simulation area, and (database) researchers realized the need for situation monitoring in several applications and tried to make that part of a DBMS initially. Later, its utility was recognized for a broader, general-purpose application and proceeded in that direction. CEP is a more recent phenomenon which is a re-engineering of event-based systems to meet the needs of large, continuous, (mostly) real-time applications. Some of the same players who worked on earlier event-based systems are also in this space. Also, stream processing has influenced CEP to a large extent as in some cases the difference between stream processing, CEP, and their combination is hard to differentiate. Pub/sub came from the distributed systems side mainly as a mechanism to decouple and/or accommodate the architectural advancements. Pub/sub, to some extent, can be viewed as a asynchronous (communication/exchange) mechanism to support either event-based or other (non-event based) systems.

I see message filtering as orthogonal to all of the above (pub/sub, EBS, and CEP) as filtering is useful and can be applied in any of these contexts and at different levels. It can be applied at the source of message generation or can be applied after the message is generated using different techniques. In fact, reduction of large quantities of raw data to obtain actionable knowledge assumes some form of filtering and aggregation, in general.

Annika: What are the pros and cons of publish/ subscribe / event-based systems? [Q2]

Sharma: As I have indicated earlier, I can visualize event-based systems with and without pub/sub systems. There are significant implications of these. Making it a pub/sub system is certainly better in terms of its utility and coupling with other systems. Otherwise, its utility is different and may be limited. As a pub/sub system, there are implications on latency issues and hence each may have different types of advantages and utility based on application needs.

Annika: What are typical assumptions made in the different pub/sub, EBS, CEP fields? [Q3]

Sharma: The assumptions for each of these (as explained in question 1) come from the disciplines from which these concepts/systems have emerged. For example, the initial event-based systems assumed that an underlying system (whether DBMS or something else) generated events that would be of interest to the applications that were running on the system. Of course, these assumptions were relaxed and generalized. Current CEP systems assume that you are getting a stream of events (source does not matter) and you are interested in looking for complex (event) patterns. Pub/sub systems were driven by the decoupling of consumers and producers as this would allow greater flexibility in only knowing what to consume and not where it came from.

As has happened over the years, the generalizations have created significant overlaps and that needs to be recognized to leverage and exploit the commonality among these approaches.

Annika: How do event-based systems fit into the overall landscape of software architectures (e.g., SOA and others)? [Q4]

Sharma: SOA provides a method for system development and integration through flexible exchange of data in a loosely coupled, distributed environment. Pub/sub can be supported (and are already) as services. Similarly, CEP can also be supported as a service. Some CEP systems are already using XML and other formats for the representation of events and data to bring it closer to the systems out there.

I think most of the current event-based systems are stand-alone systems. SOA support is needed for them to become components of larger systems. Again, this will require some attention to the services that needs to be supported by CEP/EBS systems. Perhaps we are still addressing the stand-alone mode and have not gotten to thinking in the SOA mode yet. But pub/sub has certainly happened with respect to CEP systems in general.

Annika: Is pub/sub the way of the future? To what extent will small or large-scale pub/sub, EBS, CEP systems find application? [Q5]

Sharma: It is hard to say. Anything is the way of the future until the next thing comes along -- as we have seen so many times ☺. There is always refinement, generalization, and search for better ways/abstractions for accomplishing the same things more effectively and with less capital/time. This is going to be the case here also.

I do believe these systems are being used and are useful in a number of applications. The question to me is "whether these technologies will become mainstream and be embraced by the community-at-large?" This is hard to answer as there are barriers (mostly business related) for this. For example, many financial institutions use in-house software for program trading and CEP applications. Bringing them around to use a commercial CEP system is a daunting task.

Annika: What are killer applications to pub/sub, EBS, CEP? [Q6]

Sharma: I think all of the above are still searching for a killer application to establish a permanent foothold. I do not see different applications becoming killer applications for the above. The same killer application is likely to serve most or all of them well. Pervasive applications based on sensors or large-scale monitoring applications (with real-time needs) are likely to be the killer applications which will include both CEP and PS. I see EBS becoming subsumed by the larger (in scope) of CEP systems.

Annika: What are the issues to be addressed in the medium term, long term? [Q7]

Sharma: I believe that there is a lot of confusion among these systems; the developers and end users are confused about the roles of each of the above. This is further exacerbated by the vendors claiming that their system is all of the above. One thing that can be done is to clarify the roles (if we can get consensus) resulting in the concentration of effort with a better focus. As I have indicated earlier, most of the systems currently claim to be all of the above (and stream processing) without clearly identifying the components and specifying their usage.

Currently, the expressiveness of systems varies considerably based on the domain they have focused on and the operators supported. The medium term should focus on consolidating the minimal set of features (including operators) that need to be supported in the above systems. This may be happening to some extent through streamSQL. Without this, the myriads of systems available will be difficult to compare for a given application as can be witnessed by the discussions on the CEP web site.

This brings us to the long-term issues to be addressed. Currently, most of the systems support a set of operators/computations based on the applications they have looked at. As can be inferred from the discussions on the CEP website, each application seems to have a different/unique need and we have not been able to cull out an acceptable set of minimal features.

This is clearly indicative of the lack of understanding of the expressiveness of the language/operations needed to support a large class of applications being targeted by CEP/PS/stream processing. The easiest approach is to provide a Turing-complete language. Even with that, it is easy to express some of the examples that we have been seeing on the CEP discussion site. Currently, there is no design methodology for developing applications or guidelines for decomposing the requirements and making it easier to develop applications that can be understood and maintained. The burden of coming up with a non-procedural way to write these applications is a long-term challenge. Currently, the burden of writing these applications in the chosen system is on the user and the user seems to have a lot of difficulty.

Reflections and Comments

Sharma: In summary, I see the same old wine presented in newer bottles with minor changes to the label. This, in itself, may not be bad as long as the end-user can see the subtle differences and can use them effectively for the applications. The proof of the pudding (apart from research potential and pushing the boundaries) is in the utility of these general purpose systems for applications where customized systems are being used. This will require performance compatibility as well as software engineering aspects.

Richard: Sharma's insight about the medium term issue is spot on. Researchers and implementers need to converge on the feature set required in an event processing platform, so that architects and developers know what to expect when designing systems. Over the last few years the CEP vendors in capital markets have made this convergence, and it has made user adoption much easier. Now the broader community has to follow along, expanding the footprint of event processing as appropriate but in the end building consensus about what users can expect.

Mani Chandy

Mani Chandy received his bachelors from IIT Madras, a Master's from the Polytechnic University of New York and a PhD from MIT. He has worked for Honeywell and IBM; was a co-founder of iSpheres, an early company in the event-processing space (1998); and has consulted for several companies. He was a professor at the University of Texas at Austin from 1970 to 1987, and has been a professor at the California Institute of Technology since then. He is a member of the U.S. National Academy of Engineering and has received several awards. He has a blog on sense and respond systems at http://senseandrespond. blogspot. com and he has coauthored a book in 2009 called "Event Processing: Designing IT Systems for Agile Companies."

Annika: Where do you see the difference between publish/subscribe, event-based systems, Complex event processing (pub/sub, EBS, CEP)? How do you see message filtering technologies fitting into the picture? [Q1]

Mani: All these technologies fall into the class of asynchronous matching technologies. The technologies are asynchronous because event-detection conditions (also called persistent queries, event patterns, and subscriptions) are specified asynchronously with respect to results that match the conditions. In particular, an event-condition posted at the current time may have results from the past, from the present, and from the future. These technologies deal with matching because they deal primarily with detecting an event.

Message filtering is one of the technologies that falls within the general rubric of asynchronous matching.

Sense and respond systems have three major components: sensing or acquiring raw data about the state, detecting events from the raw data, and then responding to them. The reaction of the environment to the response is sensed by the system, thus creating a feedback loop: sense-detect-respond-sense-... Asynchronous matching is one of the event-detection technologies used in sense and respond systems.

The differences between publish/subscribe, event-based systems, and complex event processing are about the strategies by which asynchronous matching is accomplished. Engineers working on each technology claim, quite correctly, that their technology solves the asynchronous matching problem. But, each of these technologies serves different aspects of the problem.

Event processing, particularly the complex variety, emphasizes the detection of events by aggregating and analyzing data from multiple sources. The aggregation may include analysis of analogue data such as radar signals from multiple stations and data from hundreds of seismometers. Pub/sub systems emphasize filtering more than complex aggregation. Some pub/sub systems limit subscriptions to topics but focus on delivering low latency. Event-based systems emphasize loosely-coupled application integration engendered by asynchrony. Tuple spaces, database systems using triggers, and stream-based systems provide powerful asynchronous matching technologies with variations of emphasis on high performance and reliability.

Students will benefit by having researchers analyze the broad space of asynchronous matching with a view to partitioning the space into problem domains that share characteristics. These characteristics include different requirements for semantics such as: transactional semantics, delivery of all event objects in the strict order in which they were created, delivery of events by each publisher in order (but possibly interleaving events from different publishers out of time sequence), ensuring at least once delivery, guaranteeing exactly once delivery, and so on. Characteristics include types of event-detection conditions such as: SQL-like query conditions, probabilistic conditions, geospatial conditions, fuzzy matches, and complex time sequences. Equally important characteristics are requirements for latency, throughput and reliability.

Annika: What are the pros and cons of publish/ subscribe / event-based systems? [Q2]

Mani: They all do the same thing: asynchronous matching. Publish/subscribe has been used since the first messaging systems. IBM's MQ Series was one of the earlier widely-used commercial systems. Many databases include mechanisms for queuing. JMS and enterprise service buses have

pub/sub capability. Listing pros and cons of pub/ sub systems is difficult because they vary so much.

Event-based systems may or may not use pub/ sub. For example, event-based systems may use a database. A comparison of pros and cons between event-based systems and pub/sub requires a granular definition of each term, and doing so at this time is premature.

Annika: What are typical assumptions made in the different pub/sub, EBS, CEP fields? [Q3]

Mani: Typical assumptions made by designers of publish/subscribe systems include the following. (Note the emphasis on "typical" because different designs are based on different assumptions.)

- There are many more subscribers than publishers.
- Publishers generally have a common ontology. For example, the word "stock" means roughly the same thing to all publishers; so subscribers can use these words knowing that they will get meaningful matches from most publishers.
- Subscribers care more about receiving data and less about its source. If a subscriber only wanted news stories from two or three sources – say, only the International Herald Tribune and the Economist then designers are likely to choose different architectures.
- There are huge numbers (hundreds of thousands or millions) of subscribers. If there were only smaller numbers (say tens of thousands) of subscribers then conventional technologies (including conventional pub/sub such as JMS, and filtering only at the periphery of a message network) are likely to be adequate.

Typical assumptions made in event-based systems are that loose-coupling in application integration is essential. Assumptions made by CEP systems are that aggregation of information from the "event cloud" is important in detecting complex events. Typical assumptions made by designers of stream-based systems are that data arrives, at high rates, in streams of values of the same type, and that very rapid detection of events is critical.

Annika: How do event-based systems fit into the overall landscape of software architectures (e.g., SOA and others)? [Q4]

Mani: Many (and perhaps most) systems will be hybrid systems; they won't be pure plays of event processing or pub/sub or request-reply. Service-oriented architecture, in its emphasis on modularity, will be used in event-driven systems as well. In this sense event-based systems will be SOA for asynchronous matching

Annika: Is pub/sub the way of the future? To what extent will small or large-scale pub/sub, EBS, CEP systems find application? [Q5]

Mani: Asynchronous matching has a major role to play in the future. It will not, however, replace synchronous queries. Many systems will be hybrid systems using different models for different parts of an application.

Annika: What are killer applications to pub/sub, EBS, CEP? [Q6]

Mani: There are many applications that are based on asynchronous matching including applications that deal with loosely-coupled enterprise application integration, business process management, processes based on sensor networks, citizen science projects, community-based sense and response applications, finance, defence and smart systems.

Annika: What are the issues to be addressed in the medium term, long term? [Q7]

Mani: The work by EPTS and other groups on identifying use cases is a superb step. A next step is to evaluate different technologies, in an unbiased way, to identify technologies that are best suitable for different types of problems. In many cases, conventional technologies may be adequate. There are, however, many problem domains for which different forms of asynchronous matching offer value superior to that of competing approaches. These domains need to be identified and characterized.

Avigdor Gal

Avigdor Gal is an Associate Professor at the Technion – Israel Institute of Technology, specializing in data management. His research on event-based systems concentrate on temporal events and rules, cooperative construction of event-based systems, event and rule uncertainty, and efficient pull-mechanisms of event streams.

Avigdor is a member of CoopIS (Cooperative Information Systems) Advisory Board, a member of IFIP WG 2.6, and a recipient of the IBM Faculty Award for 2002-2004. He is a member of the ACM and a senior member of IEEE. Avigdor served as a Program co-Chair and General Chair of CoopIS, and in various roles in ER and CIKM. He served as a program committee member in DEBS, SIGMOD, VLDB, ICDE and others. Avigdor is an Area Editor of the Encyclopedia of Database Systems.

Annika: Where do you see the difference between publish/subscribe, event-based systems, Complex event processing (pub/sub, EBS, CEP)? How do you see message filtering technologies fitting into the picture? [Q1]

Avigdor: While it is hard to put clear borders (and I would also argue that it has little benefit from a research point of view) I would like to make the following characteristics of the different disciplines. Pub/sub focuses on the matching problem of events (between publishers and subscribers). Event-based systems are more concerned with efficiency aspects of processing events (with traditionally less semantics than pub/sub). Complex event processing requires complex events around. Therefore, there are pub/sub systems with only simple events and there are event-based systems that process simple events only (in streaming systems for example). Complex events are generated from other events and therefore require a mechanism to handle the transformation itself. Pub/sub can fit in when considering distributed processing of complex events by doing matching on basic events. Event-based systems can use the extra semantics in knowing the structure of a complex event to compute it more efficiently.

Annika: What are the pros and cons of publish/subscribe / event-based systems? [Q2]

Avigdor: My response is a general one for event processing systems. On the pros side I would say that event processing is a natural paradigm to handle certain kinds of data, such as data present in reactive systems. On the cons side I would argue that sometimes the mechanism is not sufficiently lightweight and may generate a lot of overhead. In particular, one should worry about large rule sets that may bring a system to a halt.

Annika: What are typical assumptions made in the different pub/sub, EBS, CEP fields? [Q3]

Avigdor: There are many assumptions that are being taken. What is important, I believe, are those assumptions we find to be detrimental. One such common assumption is that events are deterministic. If they were reported to occur than this is the case. In my view, adding uncertainty management to event processing is an important aspect that when dealt with can elevate the usability of event-based systems significantly.

Annika: How do event-based systems fit into the overall landscape of software architectures (e.g., SOA and others)? [Q4]

Avigdor: SOA poses nice problems to event-based processing:

1. It is hard to get timely response
2. Event semantics are hard to figure out
3. It increases the performance time in general.

Annika: Is pub/sub the way of the future? To what extent will small or large-scale pub/sub, EBS, CEP systems find application? [Q5]

Avigdor: pub/sub is a good paradigm but not one-size-fits-all. In some cases, the publishers can be modelled in a very simplistic way. In other cases, it is true for the subscribers. Sometimes, the matching is trivial. In all of these cases, applying pub/sub may be an overkill.

Annika: What are killer applications to pub/sub, EBS, CEP? [Q6]

Avigdor: Real time applications that involve low bandwidth and many interruptions and require a lot of semantics (such as context, etc.) which prevent offline calculations is a killer application. Concrete examples: disaster mgmt, traffic control, minimizing the time to find parking in a busy mall...

Annika: What are the issues to be addressed in the medium term, long term? [Q7]

Avigdor: I'm not sure about medium vs. long term. I would say we need to handle:

1. *Semantics:* good tools for understanding context and interpreting data accordingly.

2. *Handling uncertainty:* producing quality results and ranking results based on perceived quality.

3. *Performance:* making the overhead of complex event processing lighter

Dieter Gawlick

Dieter Gawlick joined IBM in 1968. As member of the IMS development team he proposed, architected, and implemented products that enabled high-end transaction technology. Core database/transaction technologies such as 2phase commit, group-commit, partitioning, data replication, online utilities, escrow technology, and hot standby are among the achievements of this effort. At Amdahl, Dieter worked on I/O related problems. He developed methods for the usage of electronic data storage; at Digital, Dieter developed the first workflow manager with full integration into database and transaction technology.

Dieter joined Oracle in 1994. He architected Oracle/AQ (Advanced Queuing), an integral part of the Oracle database, and was a key contributor to Oracle's integration and sensor technologies. Dieter's current focus is leveraging and evolving database technologies to accelerate the evolution of event processing.

Annika: Where do you see the difference between publish/subscribe, event-based systems, Complex event processing (pub/sub, EBS, CEP)? How do you see message filtering technologies fitting into the picture? [Q1]

Dieter: Event processing can be characterized by three levels:

- Continuous/timely information (and knowledge) discovery
- Information dissemination
- Interrupt handling

These three levels reflect the historical evolution of this field. Each level depends on the level below and all three levels continue to evolve. Currently, the most attention is paid to the topmost level.

Interrupt Handling: Interrupt handling allows computer systems to coordinate independent threads of computing. Major use cases are process scheduling, I/O handling, and reactions to user input. Interrupt handling must provide a very high level of performance and scalability.

Information Dissemination: Information dissemination allows publishers to send messages to consumers asynchronously and without regard to their location. Major use cases are e-mail delivery, queuing (e.g. JMS), instant messaging, RSS, and enterprise service buses (ESB). Subscriptions are the preferred way to specify which messages should be disseminated to whom. Subscriptions are specified by subscribers; publishers as well as consumers can act as subscribers, subscription can be transient and persistent. In addition to high performance and scalability, information dissemination has to provide high availability and reliability, security, non-repudiation, and auditing and tracking.

Continuous/Timely Information (and Knowledge) Discovery: Continuous and timely information discovery allows users (consumers) to be aware of important information as soon as it becomes available. Instead of creating predetermined messages, publishers provide access to a set of data sources. These data sources are evaluated based on the directives of users (subscribers). Examples of data sources (and directives) include:

- Programs providing a callback – RulesML, ERP business events, database triggers
- Streams providing a continuous query language – CQL
- Databases supporting registered queries – a persistent query with a call back or dissemination instructions, scoring of models

- Systems supporting data mining for knowledge discovery – the resulting models will be (continuously) scored

Subscribers define under which condition what message has to be created; i.e., messages are different from events. Information Dissemination systems are used to distribute these messages (information); dissemination directive are typically part of the directives for information discovery. Information Discovery systems require a rich and extensible type system combined with a rich and extensible language. They also need to provide all of the operational characteristics expected from Information Dissemination systems; the number of directive is potentially very large (100K+ not counting customization to individual users)

In the upcoming questions I will equate Information Dissemination with pub/sub and Information Discovery systems with event based systems and CEP systems; I am aware that this is only a rough equivalence.

Annika: What are the pros and cons of publish/subscribe / event-based systems? [Q2]

Dieter: See my answer to your first question. Publish/subscribe and event-based systems deal with different issues. Information Discovery systems represent event based systems of this question as well as CEP systems in future questions.

Annika: What are typical assumptions made in the different pub/sub, EBS, CEP fields? [Q3]

Dieter: There are too many simplifying assumptions; e.g., the consumer is the subscriber in ps; events are confused with messages (and alerts) in CEP; data sources are limited to event streams (or event clouds). Making these simplifying assumptions limits the applicability of the technologies and also hurts their evolution.

Annika: How do event-based systems fit into the overall landscape of software architectures (e.g., SOA and others)? [Q4]

Dieter: Event based systems can be used as infra-structure for existing technologies, e.g., work flow. Additionally event based technology can extend the functionality of existing systems; e.g., registered queries can be used for continuous and timely information discovery in data bases.

Annika: Is pub/sub the way of the future? To what extent will small or large-scale pub/sub, EBS, CEP systems find application? [Q5]

Dieter: Pub/sub, EBS, CEP will become an important part of the future especially when used in cooperation with other technologies; there are many "killer applications" that require event processing.

Annika: What are killer applications to pub/sub, EBS, CEP? [Q6]

Dieter: The timely extraction of meaningful information will be the killer application, assuming it can be used and controlled by domain experts without the help of IT experts. Health care, financial services, utilities, software, and supply chains are just a few examples of domains with a high demand for this type of application. The SICU prototype developed by UUHSC, the University of Coimbra, and Oracle is a very good example of these types of applications

Annika: What are the issues to be addressed in the medium term, long term? [Q7]

Dieter: The short and medium term challenge is to promote the use of event technology as infra-structure and as well as the improvement of existing technologies through the use of event technology. This will force the event community to interact much more with the communities of existing and emerging technologies. I'd like to forgo a long term guess.

Richard Tibbets

Richard Tibbetts is Chief Technology Officer and co-founder of StreamBase Systems, a leading CEP vendor. Richard provides technical leadership for the company and leads architecture design for StreamBase's Event Processing Platform. Richard is also responsible for furthering new StreamBase capabilities such as StreamBase's 'white-box' application frameworks. As CTO, Richard directs the next-generation of StreamSQL, the event programming language developed by Richard and the StreamBase team, which applies the benefits of SQL for stored data to real-time transitory data.

Richard earned both his BS in Computer Science and Engineering, and Masters of Engineering Degrees at MIT. His graduate work included the Aurora project, which developed into StreamBase, and the Linear Road Benchmark for performance of event processing systems.

Annika: Where do you see the difference between publish/subscribe, event-based systems, Complex event processing (pub/sub, EBS, CEP)? How do you see message filtering technologies fitting into the picture? [Q1]

Richard: Publish/Subscribe systems focus on the distribution of data, making data accessible to subscribers, reducing load on publishers, and assuring the whole system functions efficiently and scalability. Of course, many Pub/Sub systems push the edge of this envelope, examining how the messaging layer can be adapted to the problem domain. For example, there are pub/sub systems tailored for distribution of real time financial market data.

Complex Event Processing instead focuses on the analysis of events. Early CEP systems analyzed events by temporal correlation, but today CEP systems use a variety of techniques, including rules, relational queries, and imperative programming. The goal is to empower developers and business analysis to build systems which react to events, using tools which are specific to the event processing domain.

"Event-Based Systems" is currently used in the community it is a catchall term that includes both data distribution and application logic. At StreamBase we use the term "Event Processing." As many systems span the area, some term will likely see broad adoption in the coming years. Message filtering is certainly an aspect of any event-based system, but not the primary technology. Application development, and tools which make the development of large event processing applications both possible and easy, are the real enabling technology.

Annika: What are the pros and cons of publish/subscribe / event-based systems? [Q2]

Richard: The biggest benefit of event processing is in the construction of decoupled systems. Using event-based semantics in application development can avoid tightly coupling applications to their data sources or sinks. Traditional application development methodologies make it difficult to build systems spanning trust and administrative domains. IT administrators are reluctant to let new systems take advantage of data that already exists in old systems, and require expensive safety strategies such as data warehouses and data marts. These intermediaries reduce the agility of business and increase the cost of new systems. With event driven architectures, data can safely be made available in real-time. And with event driven tools, workers can make use of this data in new systems quickly, easily, and profitably.

A secondary benefit comes in the form of performance and scalability. Systems architected according to event processing principles are generally more scalable and higher performing than traditional RPC systems. With multi-core systems becoming the norm, and data volumes continuing to increase, many organizations are forced to adopt an event driven architecture.

The biggest challenge is managing complexity for developers. Decoupling is a good thing, but too much decoupling makes applications difficult to understand, debug, and maintain. Developers who go too far in the direction of pub/sub find themselves with millions of events of unknown provenance, intention, or semantic meaning. It is important to have tools that can manage this complexity, and to use more constrained distribution models than full pub/sub when appropriate.

Annika: What are typical assumptions made in the different pub/sub, EBS, CEP fields? [Q3]

Richard: The fields of Pub/Sub and Complex Event Processing are duals of one another. Pub/Sub presumes that data distribution is the challenging problem, while CEP focuses on the data analysis and application development side of the problem. Many Pub/Sub researchers assume the barrier to broad adoption is a lack of sophistication in their message filtering, and seek to make more powerful and complex messaging layers. CEP researchers, on the other hand, generally focus on the syntax and semantics of languages, and the performance of executors, without addressing the complexities of data access.

Annika: How do event-based systems fit into the overall landscape of software architectures (e.g., SOA and others)? [Q4]

Richard: Event-based (or message-based) systems cover a large fraction of the distributed software architecture landscape, with the rest dominated by remote-procedure-call (RPC) systems. RPC systems are more natural for developers experienced with single-thread imperative programming languages. However, as multi-threaded programming becomes the norm, all developers

will become comfortable with notions of event dispatch and event based architecture patterns, for both single-machine and distributed systems.

To the extent SOA has become synonymous with web services, it is a largely RPC discipline. However, the core concept of service-orientation, and service descriptions, is orthogonal to the interaction model used between services. So Event Driven Architecture (EDA) isn't a competitor to SOA, or even a complementary system on the side. Event Driven Architecture represents another way to implement a truly Service Oriented Architecture.

Annika: Is pub/sub the way of the future? To what extent will small or large-scale pub/sub, EBS, CEP systems find application? [Q5]

Richard: Many applications can benefit from event processing, and will in the future. Decoupling producers and consumers via pub/sub is only one aspect of event processing.

Enterprise scale Event Processing will be important for the future of business intelligence applications that empower analysts to combine and analyze data from across the enterprise. This enterprise-wide intelligence will lead to improved real time decision-making systems that can better execute business processes. This will grant competitive advantage to firms who can stitch together disparate systems using event processing, and will eventually become a key business practice.

Annika: What are killer applications to pub/sub, EBS, CEP? [Q6]

Richard: The first killer applications for these architectures are systems where the volume of data and response time requirements are overwhelming for humans, and where systems require regular redesign to cope with changing business requirements. One canonical example of this is algorithmic trading systems in the financial space. Others include monitoring systems at Internet businesses, and data collection systems at intelligence agencies.

The second wave of applications will benefit from the decoupling effect of the technology. These will replace the current crop of narrowly focused Business Process Management, Business Activity Monitoring, and Business Intelligence applications. Event driven architecture will make more data available, and modern tools will make that data easier to understand and more efficient to process. These applications will not only increase visibility into the activities of the business, but they will control and manage those activities. By applying event-processing intelligence to millions of small decisions that affect the business every day, organizations will become more responsive and more profitable.

Annika:What are the issues to be addressed in the medium term, long term? [Q7]

Richard: In the medium term, the principle issue remains developer education and adoption. The best way to make this new paradigm accessible to developers is through rich tools that manage the complexity for them. Visual programming environments are seeing wide adoption, but there is much more that can be done. Deployments, data distribution, message formats, semantics, and syntax all tax the abilities of developers to understand the systems they maintain. Better tools and standardization will help to rein in this complexity.

In the long term, the issues will be around documenting and managing event semantics. When an institution has tens of thousands of different event types, distributed across dozens of event management systems, there will be considerable duplication and confusion. Systems to manage this complexity will be critical, and they will manage it not just statically but also dynamically, as the event producers and consumers evolve over time.

REFLECTION AND SUMMARY OF CONTRIBUTIONS

This section reflects on the statements and discussions provided by the panellists.

Understanding of the Central Terms of the Field

The variations in defining the central terms of the area reflect the different backgrounds of the panellists and illustrate the variety of opinion within the community. At a first glance, all panellists seem to agree on the general meaning of terms. On closer inspection, however, each is giving emphasis to different aspects. To the list of terms that were mentioned in the question (pub/sub, EBS, CEP, filtering), further terminology was introduced: streaming, event processing, information discovery and dissemination; sense and respond systems.

Four of the panellists (Sharma, Alex, Jean, Richard) see *event-based systems* as the most general approach, or even an umbrella term for all event-driven systems. On the other hand, Mani is using 'asynchronous matching' as the umbrella concept, with event-based systems emphasizing loosely coupled application integration. Dieter distinguishes between information discovery and information dissemination, with the latter referring to publish/subscribe and the former to event-based systems and CEP. For Avigdor, event-based systems are more concerned with efficiency of event distribution. Alex explicitly identifies distributed event-based systems as a sub-group of EBS focussed on event filtering and quality of service at a technical level.

Publish/subscribe is seen as an influence from the distributed systems side to decouple components; it can be viewed as a communication mechanism to support either event-based or other (non-event based) systems in a distributed system. (Sharma, Richard, Jean, Dieter). However, Alex stresses that pub/sub may be implemented as either centralised or distributed delivery mechanism. Mani and Avigdor emphasize the focus on filtering or matching in pub/sub systems (instead of complex aggregations). Jean additionally suggests pub/sub application interfaces are not necessarily supported by pub/sub communication.

Sharma identifies *complex-event processing* (CEP) as a more recent phenomenon that uses event-based concepts to meet the needs of large, continuous, (mostly) real-time applications. He notes that in some cases the difference between stream processing and CEP is hard to identify. Alex, Mani, Avigdor and Richard observe CEP's a strong focus on aggregation, transformation, and the analysis of data from multiple sources. Dieter expresses the same point aspect in a more data-focused manner. Jean points to a difference in the engineering of the systems: for example, in CEP the realities of the distributed system (such as timestamps) may not be available for the analysis of physical causality. Richard on the other hand emphasizes the multitude of techniques available to analyze the data for patterns beyond temporal correlation.

Filter technologies are seen by Richard as being part of EBS but not as a core technology. For Jean, however, message filtering capabilities are at the centre as they determine the expressiveness of subscriptions and the complexity of the matching and routing algorithms.

The panellists identified a number of characteristics that may be used to group applications into problem domains that share characteristics:

- Semantics requirements: transactional semantics, ordering of messages, at-least-once delivery, exactly-once delivery, real-time delivery (Mani, Jean), and future reference (Dieter).
- Event-detection conditions: SQL-like query conditions, probabilistic conditions, geospatial conditions, fuzzy matches, and mobile clients, complex time sequences. (Mani, Jean)

- Requirements for latency, throughput and reliability, security (Mani, Jean)

Assumptions, Pros and Cons

One assumption that has been identified for all types of systems is that events are deterministic. Avigdor argues that it may be necessary to question this assumption and that adding uncertainty management to event processing may elevate the usability of event-based systems significantly.

Further assumptions are that the event source does not matter (Sharma, Mani) and that the mutual anonymity is acceptable (Jean) and desired (Mani).

As assumptions for publish/subscribe systems, Mani names that

- there are many more subscribers than publishers.
- Publishers and subscribers follow a common terminology so that subscribers can rely on names trusting that they will receive meaningful matches from publishers.
- Subscribers care more about receiving data and less about its source. If it were otherwise, a different architecture without source anonymity would need to be employed.

Jean argues that CEP abstracts above the time at distributed event sources and that composite event detection in a ubiquitous computing system would have very different assumptions. Alex argues even more strongly that all assumptions made are application-domain dependent: each application field brings its own requirements and constraints to be taken into account (e.g., available bandwidth, real-time constraints, and quality of service expectations).

Other panellists focus more on the difference between pub/sub and EBS+CEP: that aggregation of information from events is important in detecting complex events (Mani, Sharma), and that p/s is driven by decoupling (Sharma) based on a trusted network of brokers (Jean).

The majority of the panellists name decoupling and source anonymity as advantages of event-based systems. The effect of decoupling are highlighted by Richard: "IT administrators are reluctant to let new systems take advantage of data that already exists in old systems, and require expensive safety strategies such as data warehouses and data marts. These intermediaries reduce the agility of business and increase the cost of new systems. With event driven architectures, data can safely be made available in real-time. And with event driven tools, workers can make use of this data in new systems quickly, easily, and profitably." Jean stresses that due to the hidden location and identity of publishers and subscribers, mutual anonymity of publishers and subscribers, and network independence of applications is achieved. Alex points to the reduced load on event consumers due to filtering and composition in broker nodes.

However, decoupling and anonymity area also named as posing challenges for the developer: Richard here refers to the complexity in debugging and maintaining, as well as in the event semantics of decoupled event-based systems. Alex makes the same point when referring to a "lack of control": quality of service considerations are eft to the consumer and the middleware. Jean stresses the security implication that the parties have to implicitly trust each other, and the limitations to reliable event composition.

Many panellists stress again that event-based systems may be implemented without a publish/subscribe mechanism and that one also needs to keep in mind that depending on the application and the design, each system may have different utility and advantages.

EBS and other Software Architectures

Event-based systems can be used as infrastructure for existing technologies (such as workflow) or as extensions of existing systems (continuous queries in databases). Most systems will be a hybrid of event-based and request-reply communication. Remote procedure call and event-based techniques are identified by Richard as the two communication techniques for distributed systems. He expresses the hope that event-based architectures will become more common as developers become more familiar with multi-threaded programming. Alex sees event-based systems gaining importance as cyberphysical systems become more widely used.

The link to SOA was approached from two angles: how can EBS be integrated using SOA (Sharma) and can EBS be used to implement SOA (Mani, Jean, Avigdor, Alex). Both Mani and Jean observe how EBS provide an alternative SOA concept. Finally, Avigdor looked at the technical issues of combining SOA and event-based processing.

Future of EBS: Applications and Challenges

All panellists confirm that there are and will be applications that require event processing.

Each of the participants elaborated on their suggestions; here we just give a quick list of applications that were suggested:

- Finance & trading system
- Business processes
  - process management, intelligence applications, activity monitoring
- Change notification and data integration
  - Healthcare, assisted living, police, social services
  - Transport and cargo monitoring, mobile sensor networks
  - Internet-based monitoring
  - Environmental monitoring, disaster management
  - Smart City system as collaboration of numerous sub-systems
  - Citizen science projects
- Defence systems

Large-scale distribution as well as sensor-based, wire-less and mobile systems were named as current areas of interest. However, it is also pointed out that event-based systems will not replace synchronous querying. Mani predicts the use of many hybrid systems using different models for different parts of the application. Avi reminds us that there is no single solution and that publish/subscribe may not be appropriate for many event-based applications.

Sharma identifies business-related barriers for CEP technologies to become mainstream and be embraced by a larger business community. One of the issues he identifies is the use of in-house software that may prevent the uptake of commercial CEP systems. Richard, on the other hand, predicts that "enterprise-wide intelligence will lead to improved real time decision-making systems that can better execute business processes. This will grant competitive advantage to firms who can stitch together disparate systems using event processing, and will eventually become a key business practice."

The question for future challenges that need to be addressed by EBS researchers is the one where the panellists' answers were most diverse.

Sharma and Mani argue for a clarification in the comparison between systems and the matches between systems and applications. Sharma observes that "developers and end users are confused about the roles of [the different system, which is] further exacerbated by the vendors claiming that their system is all of the above." He calls for clarification, identification of components and specification of usage. A first step in this direction is the use-case initiative by the EPTS. Mani sees the next step then in an identification of the technologies that are best suited for given problem types. As medium term goals for vendors of EBS, Sharma identifies the consolidation of a minimal set of features (including operators) that need to be supported in each of the system types. As the long-term issue Sharma identifies the "lack of understanding of the expressiveness of the language/operations needed to support a large class of [event-based] applications".

Alex named eight groups of issues to be addressed, the first three of which refer to the handling and understanding of events: the need for event algebras and the issue of event correlation (including the issue of time in distributed environments as systems also mentioned by Jean), event lifecycle, event enrichment and semantics. The issue of semantics is also stressed by Richard and Avigdor: context and semantics of event data need to be understood, interpreted, stored and communicated appropriately.

Alex and Jean both stress that security, trust, reliability and mobility are issues to be addressed. Jean points out that for data that is sensitive for human lifetimes and longer, content-based routing in large-scale systems may not be an option. She also observes that it still needs to be demonstrated that the mobile phone infrastructure is capable of supporting continuous monitoring.

Dieter sees as immediate and medium term challenges the promotion of event technology as infrastructure as well as the improvement of existing solutions by the use of event-based technology. One of the advantages will be the increased interaction between developers of existing and emergent technologies. A similar point is made by Richard: he sets the focus on developer education and adoption that should be aided by tools and standardisation.

Alex predicts that heterogeneity will become a major issue as different parts of the applications will grow independently. He calls for new methods for software engineering appropriate for event-based systems, including testing strategies and performance benchmarks.

ACKNOWLEDGMENT

Annika Hinze likes to thank all panellists for participating in this experiment.

Compilation of References

Abadi, D. J., Ahmad, Y., Balazinska, M., Cetintemel, U., Cherniack, M., Hwang, J., & Zdonik, S. (2005). The design of the Borealis stream processing engine. In *Proceedings of CIDR*, (pp. 277–289).

Abadi, D. J., Carney, D., Cetintemel, U., Cherniack, M., Convey, C., Lee, S., & Zdonik, S. (2003). Aurora: A new model and architecture for data stream management. *The VLDB Journal, 12*(2), 120–139. doi:10.1007/s00778-003-0095-z

Abdurazik, A., & Offutt, J. (2000, October). Using UML collaboration diagrams for static checking and test generation. In A. Evans, S. Kent, & B. Selic (Eds.), *Proceedings of UML 2000 - the unifed modeling language. advancing the standard. third international conference*. York, UK, (Vol. 1939, pp. 383-395). London: Springer.

Adaikkalavan, R., & Chakravarthy, S. (2006). SnoopIB: Interval-based event specification and detection for active databases. *Data & Knowledge Engineering, 59*(1), 139–165. doi:10.1016/j.datak.2005.07.009

Adam, D., Thiemo, V., & Juan, A. (2004). Making TCP/IP Viable for Wireless Sensor Networks. *European Workshop on Wireless Sensor Networks* (p. 4). Berlin, Germany: Springer.

Adi, A., & Etzion, O. (2004). Amit - the situation manager. *The VLDB Journal, 13*(2), 177–203. doi:10.1007/s00778-003-0108-y

Adi, A., Biger, A., Botzer, D., Etzion, O., & Sommer, Z. (2003). *Context Awareness in Amit* (pp. 160–167). Actuve Middleware Services.

Adler, M., Ge, Z., Kurose, J., Towsley, D., & Zabele, S. (2001). Channelization Problem in Large Scale Data Dissemination. In *Proceedings of the Ninth International Conference on Network Protocols* (pp.100-110). Washington, DC: IEEE Computer Society.

Aekaterinidis, I., & Triantafillou, P. (2006). *PastryStrings: A comprehensive content based publish/subscribe DHT network*. New York: Springer.

Aekaterinidis, I., & Triantafillou, P. (2007). Publish-Subscribe Information Delivery with Substring Predicates. *IEEE Internet Computing, 11*(4), 16–23. doi:10.1109/MIC.2007.90

Agha, G. (1986). *Actors: a Model of Concurrent Computation in Distributed Systems*. Cambridge, MA: MIT Press.

Aguilera, M. K., Strom, R., E., Sturman, D., C., Astley, M., & Chandra, T., D. (1999). Matching Events in a Content-Based Subscription System. In *PODC '99: Proceedings of the eighteenth annual ACM symposium on Principles of distributed computing* (pp. 53-61). Atlanta, GA: ACM Press.

Allaoui, F., Yehdih, A., & Donsez, D. (2005, August 20). *Open-source Java-based DDS (Data Distribution Service) Implementation*. Retrieved August 25, 2008, from Open-source Java-based OMG DDS Implementation: http://www-adele.imag.fr/users/Didier.Donsez/dev/dds/readme.html

Allen, J. (1983). Maintaining knowledge about temporal intervals. *CACM, 26*(1), 832–843.

Andrews, T. (2003). *Business Process Execution Language for Web Services.* Retrieved Oct. 31 2006, from http://www.ibm.com/developerworks/library/specification/ws-bpel/

Android. (n.d.). *An Open Handset Alliance Project.* Retrieved from http://code.google.com/android/

Anurag, D., & Anu, P. K. (2008). The Chronon Based Model for Temporal Databases. (DASFAA) (pp. 461-469)

Apple Developer Connection. (n.d.). *iPhone SDK.* Retrieved from http://developer.apple.com/iphone/program/download.html

Arasu, A., Babcock, B., Babu, S., Cieslewicz, J., Datar, M., & Ito, K. (2008). STREAM: The Stanford Data Stream Management System. In Garofalakis, M., Gehrke, J., & Rastogi, R. (Eds.), *Data Stream Management: Processing High-Speed Data Streams.* Berlin, Germany: Springer.

Arasu, A., Babcock, B., Babu, S., Datar, M., Ito, K., & Nishizawa, I. Widom, J. (2003). Stream: The Stanford stream data manager. In *Proceedings of ACM SIGMOD*, (pp. 665).

Arasu, A., Babu, S., & Widom, J. (2002). *An abstract semantics and concrete language for continuous queries over streams and relations.* (Technical Report 2002-57). Palo Alto, CA: Stanford University.

Ashayer, G., Leung, H., & Jacobsen, H.-A. (2002). Predicate Matching and Subscription Matching in publish/subscribe Systems. In *Proceedings of the 22nd International Conference on Distributed Computing Systems* (pp. 539 - 548). Washington, DC: IEEE Computer Society.

Aspnes, J., Kirsch, J., & Krishnamurthy, A. (2004, July). Load balancing and locality in range-queriable data structures. In *PODC '04: Proceedings of the twenty-third annual ACM symposium on principles of distributed computing* (pp. 115–124). New York: ACM Press.

Atul, A., Jon, H., Marvin, T., William, J. B., & John, R. D. (2002). Cooperative Task Management Without Manual Stack Management. In *Proceedings of the General Track of the annual conference on USENIX Annual Technical Conference*, (pp. 289-302).

Avvenuti, M., Vecchio, A., & Turi, G. (2005). A cross-layer approach for publish/subscribe in mobile ad hoc networks. In MATA'05, Proceedings of Mobility Aware Technologies and Applications: Second International Workshop (pp. 203-214). Monteral, Canada.

Babcock, B., Babu, S., Datar, M., Motwani, R., & Widom, J. (2002). *Models and Issues in Data Stream Systems* (pp. 1–16). PODS.

Babich, F., & Deotto, L. (2002). Formal methods for specification and analysis of communication protocols. *IEEE Communication Survey and Tutorials, 4*(1), 2–20. doi:10.1109/COMST.2002.5341329

Bacon, J., Bates, J., Hayton, R., & Moody, K. (1995). Using events to build distributed applications. In *Proc. IEEE SDNE*, (pp.148–155).

Bacon, J., Beresford, A., Evans, D., Ingram, D., Trigoni, N., Guitton, A., et al. (2008, January). Time: An Open Platform for Capturing, Processing and Delivering Transport-Related Data. In *Proceedings of the fifth IEEE consumer communications and networking conference (CCNC)* (pp. 687–691). Las Vegas, NV: IEEE Press. (Session on Sensor Networks in Intelligent Transportation Systems)

Bacon, J., Eyers, D. M., Moody, K., & Pesonen, L. I. W. (2005, November). Securing publish/subscribe for multi-domain systems. In G. Alonso (Ed.), Middleware (Vol. 3790, pp. 1–20). Grenoble, France: Springer Verlag. doi:10.1007/11587552_1

Bacon, J., Eyers, D. M., Peter, R., & an Pietzuch, J. S. (2008, July). Access control in publish/subscribe systems. In *Proceedings of the 2nd International Conference on Distributed Event-based Systems (DEBS'08)* (pp. 23-34). Rome: ACM Press. (Chair Roberto Baldoni)

Bacon, J., Eyers, D. M., Singh, J., & Pietzuch, P. R. (2008). Access control in publish/subscribe systems. In Baldoni, R. (Ed.), *DEBS* (Vol. 332, pp. 23–34). New York: ACM.

Bacon, J., Moody, K., & Yao, W. (2001, November). Access control and trust in the use of widely distributed services. In Middleware '01, IFIP/ACM international conference on distributed systems platforms (Vol. 2218, pp. 295–310). Heidelberg, Germany: Springer Verlag.

Bacon, J., Moody, K., & Yao, W. (2002, November). A model of OASIS role-based access control and its support for active security. [TISSEC]. *ACM Transactions on Information and System Security, 5*(4), 492–540. doi:10.1145/581271.581276

Bacon, J., Moody, K., Bates, J., Hayton, R., Ma, C., & McNeil, A. (2000). Generic support for distributed applications. *Computer, 33*(3), 68–76.

Baduel, L., Baude, F., Caromel, D., Contes, A., Huet, F., & Morel, M. (2006). Programming, Deploying, Composing, for the Grid. In Cunha, J., & Rana, O. (Eds.), *Grid Computing: Software Environments and Tools*. Berlin, Germany: Springer-Verlag. doi:10.1007/1-84628-339-6_9

Baehni, S., Chabra, C. S., & Guerraoui, R. (2005). Frugal Event Dissemination in a Mobile Environment. In *Proceedings of the 6th ACM International Middleware Conference*. Grenoble, France.

Balazinska, M., Balakrishnan, H., & Karger, D. (2002, August). INS/Twine: A scalable peer-to-peer architecture for intentional resource discovery. In *Pervasive '02: Proceedings of the first international conference on pervasive computing* (pp. 195–210). London: Springer-Verlag.

Balazinska, M., Khoussainova, N., & Suciu, D. (2008). PEEX: Extracting probabilistic events from rd data. In *Proceedings of ICDE*.

Baldoni, R., Beraldi, R., Cugola, G., Migliavacca, M., & Querzoni, L. (2005). Structure-less content-based routing in mobile ad hoc networks. In *Proceedings of the IEEE International Conference on Pervasive Services*, (pp. 37-46).

Baldoni, R., Beraldi, R., Quema, V., Querzoni, L., & Tucci-Piergiovanni, S. (2007). TERA: topic-based event routing for peer-to-peer architectures. In Proceedings of Distributed event-based systems, (DEBS2007) New York.

Baldoni, R., Querzoni, L., & Virgillito, A. (2006). Distributed Event Routing in Publish/Subscribe Communication Systems: a Survey (Technical report – MIDLAB 1/2006). Rome: Dipartimento di Informatica e Sistemistica "A.Ruberti", Università di Roma la Sapienza.

Bamba, B., Liu, L., Pesti, P., & Wang, T. (2008, April). Supporting Anonymous Location Queries in Mobile Environments with PrivacyGrid. In *Proceedings of 17th International World Wide Web Conference* (WWW). Beijing, China

Bamba, B., Liu, L., Yu, P. S., & Iyengar, A. (2009). Distributed Processing of Spatial Alarms: A Safe Region-based Approach. In *Proceedings of IEEE Int. Conf. on Distributed Computing*, June 22-26, Montreal, Quebec, Canada.

Bamba, B., Liu, L., Yu, P. S., Zhang, G., & Doo, M. (2008, December). Scalable Processing of Spatial Alarms. In *Proceedings of 15th International Conference on High Performance Computing*. Bangalore, India.

Banavar, G., Candra, T. D., Strom, R. E., & Sturman, D. C. (1999). A Case for Message Oriented Middleware. *International Symposium on Distributed Computing* (pp. 1-18). Bratislava, Slovak Republic: Springer.

Banavar, G., Kaplan, M., Shaw, K., Strom, R. E., Sturman, D. C., & Tao, W. (1999). Information flow based event distribution middleware. In *Electronic commerce and web-based applications/middleware workshop at the international conference on distributed computing systems 1999*. Austin, TX: IEEE.

Barbier, F., & Belloir, N. (2003, April). Component behavior prediction and monitoring through built-in test. In *Proceedings of ECBS '03: The 10th IEEE International Conference and Workshop on the Engineering of Computer-based Systems* (pp. 17-22). Huntsville, AL: IEEE Computer Society.

Baresi, L., Ghezzi, C., & Mottola, L. (2007). On accurate automatic verification of publish-subscribe architectures. In *ICSE '07: Proceedings of the 29th international conference on software engineering* (pp. 199–208). Washington, DC: IEEE Computer Society.

Bass, L., Clements, P., & Kazman, R. (2003). *Software Architecture in Practice*. Reading, MA: Addison-Wesley.

Baughmann, N. E., & Levinne, B. N. (2001). Cheat-proof playout for centralized and distributed online games, In *Proceedings of IEEE INFOCOMM*.

Bazinette, V., Cohen, N., Ebling, M., Hunt, G., Lei, H., Purakayastha, A., Stewart, G., Wong, L., & Yeh, D. (2001). *An Intelligent Notification System*. IBM Research Report RC 22089 (99042).

Beaver, J., Chrysanthis, P. K., Pruhs, K., & Liberatore, V. (2006). To Broadcast Push or Not and What? In *Proceedings of the 7th International Conference on Mobile Data Management* (MDM 2006).

Bellavista, P., & Corradi, A. (Eds.). (2006). *The Handbook of Mobile Middleware*. Boca Raton, FL: Auerback Publications. doi:10.1201/9781420013153

Belokosztolszki, A., Eyers, D. M., Pietzuch, P. R., Bacon, J., & Moody, K. (2003, June). Role-based access control for publish/subscribe middleware architectures. In Proceedings of the *2nd International Workshop on Distributed Event-based Systems (DEBS'03)*. San Diego, CA: ACM Press. (Program Chair Hans-Arno Jacobsen)

Bharambe, A. R., Agrawal, M., & Seshan, S. (2004, October). Mercury: supporting scalable multi-attribute range queries. *SIGCOMM Computer Communication Review, 34*(4), 353–366. doi:10.1145/1030194.1015507

Bharambe, A. R., Rao, S., & Seshan, S. (2002). Mercury: a scalable publish-subscribe system for internet games. In *Proceedings of the 1st Workshop on Network and System Support for Games,* (pp. 3-9).

Bhola, S., Strom, R. E., Bagchi, S., Zhao, Y., & Auerbach, J. S. (2002). Exactly-once delivery in a content-based publish-subscribe system. In *Proceedings of DSN '02: The 2002 International Conference on Dependable Systems and Networks* (pp. 7-16). Washington, DC: IEEE Computer Society.

Birman, K. P., Hayden, M., Ozkasap, O., Xiao, Z., Budiu, M., & Minsky, Y. (1999). Bimodal multicast. *ACM Transactions on Computer Systems, 17*(2), 41–88. doi:10.1145/312203.312207

Bittner, S., & Hinze, A. (2007). The arbitrary Boolean publish/subscribe model: making the case. In *Proceedings of the 2007 inaugural international conference on Distributed event-based systems,* (pp 226 - 237). New York: ACM.

Bloom, B. (1970). Space/Time Trade-offs in Hash Coding with Allowable Errors. *Communications of the ACM, 13*(7), 422–426. doi:10.1145/362686.362692

Bolchini, C., Curino, C., Quintarelli, E., Schreiber, F. A., & Tanca, L. (2007). A data-oriented survey of context models. *SIGMOD Record, 36*(4), 19–26.

Boonma, P., & Suzuki, J. (2008). Exploring Self-star Properties in Cognitive Sensor Networking. *International Symposium on Performance Evaluation of Computer and Telecommunication Systems.* Edinburgh, Scottland: IEEE/SCS.

Boonma, P., & Suzuki, J. (2008). Middleware Support for Pluggable Non-Functional Properties in Wireless Sensor Networks. *International Workshop on Methodologies for Non-functional Properties in Services Computing* (pp. 360-367). Honolulu, HI: IEEE.

Boonma, P., & Suzuki, J. (2009). Self-Configurable Publish/Subscribe Middleware for Wireless Sensor Networks. *International Workshop on Personalized Networks.* Las Vegas, NV: IEEE.

Boonma, P., & Suzuki, J. (2009). Toward Interoperable Publish/Subscribe Communication between Wireless Sensor Networks and Access Networks. *International Workshop on Information Retrieval in Sensor Networks.* Las Vegas, NV: IEEE.

Borgida, A., & Patel-Schneider, P. (1994). A semantics and complete algorithm for subsumption in the classic description logic. *Journal of Artificial Intelligence Research*, 277–308.

Borsje, J., Levering, L., & Frasincar, F. (2008, March 16-20). Hermes: a Semantic Web Based News Decision Support System. In *Proceedings of The 23rd Annual ACM Symposium on Applied Computing* Fortaleza, Ceará, Brazil.

Bose, P., Morin, P., Stojmenović, I., & Urrutia, J. (2001, November). Routing with guaranteed delivery in ad hoc wireless networks. *ACM Wireless Networks, 7*(6), 609–616. doi:10.1023/A:1012319418150

Boudec, J.-Y. L., & Vojnovic, M. (2005). Perfect simulation and stationarity of a class of mobility models. *Proceedings - IEEE INFOCOM, 2005*, 2743–2754.

Boukerche, A., Dash, T., & Pinotti, M. C. (2004). Performance analysis of a hybrid push-pull algorithm with QoS adaptations in wireless networks. In *Proceedings of the IEEE Symposium on Computers and Communications* (ISCC).

Braginsky, D., & Estrin, D. (2002, October). Rumor routing algorithm for sensor networks. In *Proceedings of the first workshop on sensor networks and applications (WSNA '02), Atlanta, GA* (pp. 22–31). New York: ACM Press.

Brenner, D., Atkinson, C., Malaka, R., Merdes, M., Paech, B., & Suliman, D. (2007). Reducing verification effort in component-based software engineering through built-in testing. *Information Systems Frontiers, 9*(2-3), 151–162. doi:10.1007/s10796-007-9029-4

Brito, A., Fetzer, C., Sturzrehm, H., & Felber, P. (2008). Speculative out-of-order event processing with software transaction memory. In R. Baldoni (Ed.), *Proceedings of the Second International Conference on Distributed Event-Based Systems* (pp. 265-275). New York: ACM Press.

Brodén, B., Hammar, M., & Nilsson, B. J. (2004). Online and Offline Algorithms for the Time-Dependent TSP with Time Zones. *Algorithmica, 39*(4), 299–319. doi:10.1007/s00453-004-1088-z

Brose, G. (1997). JacORB: Implementation and Design of a Java ORB. *International Working Conference on Distributed Aplications and Interoperable Systems* (pp. 143-154). Cottbus, Germany: Chapman & Hall.

Bry, F., & Eckert, M. (2008). *On static determination of temporal relevance for incremental evaluation of complex event queries* (pp. 289–300). DEBS.

Buchmann, A., Bornhövd, C., Cilia, M., Fiege, L., Gärtner, F., & Liebig, C. (2004). *DREAM: Distributed Reliable Event-based Application Management*. Berlin, Germany: Springer.

Bultan, T., Fu, X., Hull, R., & Su, J. (2003). Conversation specification: a new approach to design and analysis of e-service composition. In *Proceedings of WWW '03: The 12th International Conference on World Wide Web* (pp. 403-410). New York: ACM Press.

Burcea, I., Petrovic, M., & Jacobsen, H.-A. (2003). I know what you mean: Semantic Issues in Internet-scale Publish/Subscribe Systems In *Proceedings of the International Workshop on Semantic Web and Databases* (SWDB03), Berlin, Germany.

Buschnmann, F., Meunier, R., Rohnert, H., Sommerland, P., & Stal, M. (1996). *Pattern-Oriented Software Architecture - A: System of Patterns*. New York: Wiley and Sons.

Busetta, P., Merzi, M., Rossi, S., & Zancanaro, M. (2003, October). Group communication for real-time role coordination and ambient intelligence. In A. Kruger & R. Malaka (Eds.), *Proceedings of workshop on AI in mobile systems (AIMS 2003), 5th international conference on ubiquitous computing (UbiComp 2003)* (pp. 9–16). Seattle, WA: Springer-Verlag.

Cabri, G., Leonardi, L., & Zambonelli, F. (2000). MARS: A Programmable Coordination Architecture for Mobile Agents. *IEEE Internet Computing, 4*(4), 26–35. doi:10.1109/4236.865084

Cai, L. R., Bradbury, J. S., & Dingel, J. (2007). Verifying distributed, event-based middleware applications using domain-specific software model checking. In Bonsangue, M. M., & Johnsen, E. B. (Eds.), *FMOODS (Vol. 4468*, pp. 44–58). Berlin, Germany: Springer-Verlag.

Cai, M., & Frank, M. (2004, May). RDF Peers: a scalable distributed RDF repository based on a structured peer-to-peer network. In *Proceedings of the 13th international conference on world wide web (WWW 2004), New York, NY, USA* (pp. 650–657). New York: ACM Press.

Cai, M., Chervenak, A., & Frank, M. (2004, November). A peer-to-peer replica location service based on a distributed hash table. In *Proceedings of the ACM/IEEE SC2004 conference on high performance networking and computing, Pittsburgh, PA* (pp. 56–56). Washington, DC: IEEE Computer Society.

Cai, Y., & Hua, K. (2002). An Adaptive Query Management Technique for Efficient Real-Time Monitoring of Spatial Regions in Mobile Database Systems. In *Proceedings of the IEEE IPCCC*, (pages 259–266).

Callsen, C. J., & Agha, G. (1994). Open Heterogeneous Computing in ActorSpace. *Journal of Parallel and Distributed Computing, 21*(3), 289–300. doi:10.1006/jpdc.1994.1060

Campanile, F., Coppolino, L., Giordano, S., & Romano, L. (2008). A business process monitor for a mobile phone recharging system. *The EUROMICRO Journal of Systems Architecture, 54*(9), 843–848. doi:10.1016/j.sysarc.2008.02.005

Cao, F., & Singh, J. P. (2004). Efficient event routing in content-based publish-subscribe service networks. *Proceedings - IEEE INFOCOM, 2004*, 929–940.

Cao, F., & Singh, J. P. (2005). MEDYM: Match-Early with Dynamic Multicast for Content-Based Publish-Subscribe Networks. In *Proceedings of Middleware 2005* (pp. 292–313). Berlin/Heidelberg, Germany: Springer Verlag. doi:10.1007/11587552_15

Cao, X., & Shen, C.-C. (2007). Subscription-aware publish/subscribe tree construction in mobile ad hoc networks. In *Proceedings of the 13th IEEE International Conference on Parallel and Distributed Systems*, (pp. 1-9).

Caporuscio, C., Carzaniga, A., & Wolf, A. L. (2003). Design and Evaluation of a Support Service for Mobile, Wireless Publish/Subscribe Applications. *IEEE Transactions on Software Engineering, 29*(12). doi:10.1109/TSE.2003.1265521

Carey, M. J., & DeWitt, D. J. (1996). Of Objects and Databases: A Decade of Turmoil. VLDB'96, In *Proceedings of 22th International Conference on Very Large Data Bases*, (pp. 3-14).

Carlson, J., & Lisper, B. (2003). An interval-based algebra for restricted event detection. In *Proceedings of FORMATS*, (pp. 121–133).

Carlson, J., & Lisper, B. (2004). An event detection algebra for reactive systems. In *Proceedings of EMSOFT '04: The 4th ACM International Conference on Embedded Software* (pp. 147-154). New York: ACM Press.

Carriero, N., & Gelernter, D. (1989). Linda in context. *Communications of the ACM, 32*(4), 444–458. doi:10.1145/63334.63337

Carroll, J., Dickinson, I., & Dollin, C. (2004, May 17-22). Jena: Implementing the Semantic Web Recommendations. In *Proceedings of World Wide Web Conference 2004*, New York. Retrieved from http://jena.sourceforge.net/

Carvalho, N., Araujo, F., & Rodrigues, L. (2006). Reducing Latency in Rendezvous-Based Publish-Subscribe Systems for Wireless Ad Hoc Networks. In *Proceedings of the 26th IEEE International Conference Workshops on Distributed Computing Systems*, (p. 28).

Carzaniga, A. & Hall., C. P. (2006). Content-based communication: a research agenda. In *Proceedings of the International Workshop on Software Engineering and Middleware*, (pp. 2-8).

Carzaniga, A. (1998). *Architectures for an Event Notification Service Scalable to Wide area Networks*. (PhD thesis), Politecnico di Milano.

Carzaniga, A. (2008). *Siena – Software*. Retrieved from http://www.inf.unisi.ch/carzaniga/siena/software/index.html

Carzaniga, A., & Wolf, A. (2003). Forwarding in a Content-Based Network. In *Proceedings of ACM SIGCOMM 2003* (pp. 163-174). New York: ACM Press.

Carzaniga, A., Rosenblum, D. S., & Wolf, A. L. (2001). Design and evaluation of a wide-area event notification service. *ACM Transactions on Computer Systems*, *19*(3), 332–383.

Casalicchio, E., & Morabito, F. (2007). Distributed subscriptions clustering with limited knowledge sharing for content-based publish/subscribe systems. In Proceedings of Network Computing and Applications, (pp 105-112). Cambridge, MA.

Castellote, G.-P., & Bolton, P. (2002, February). Distributed real-time applications now have a data distribution protocol. *RTC Magazine*. Retrieved August 23, 2008, from http://www.rti.com/docs/RTC_Feb02.pdf

Castro, M., Druschel, P., Kermarrec, A. M., & Rowstron, A. (2002). SCRIBE: A large-scale and decentralized application-level multicast infrastructure. *IEEE Journal on Selected Areas in Communications*, *20*(8), 1489–1499. doi:10.1109/JSAC.2002.803069

Cayuga Project. (2005). *Cayuga technical report*, (Technical Report). Ithica, NY: Cornell, University. Retrieved from http://www.cs.cornell.edu/mshong/cayuga-techreport.pdf

Chakraborty, D., Joshi, A., Yesha, Y., & Finin, T. W. (2006). Toward Distributed Service Discovery in Pervasive Computing Environments. *IEEE Transactions on Mobile Computing*, *5*(2), 97–112. doi:10.1109/TMC.2006.26

Chakravarthy, S. (1994). Snoop: an expressive event specification language for active databases. *Data & Knowledge Engineering*, *14*(1), 1–26. doi:10.1016/0169-023X(94)90006-X

Chakravarthy, S., & Mishra, D. (1996). Snoop: An expressive event specification language for active databases. *Data & Knowledge Engineering*, *14*(1), 1–26. doi:10.1016/0169-023X(94)90006-X

Chambel, T., Moreno, C., Guimaraes, N., & Antunes, P. (1994). *Concepts and architecture for loosely coupled integration of hyperbases*. In (Vol. 3, 4). Broadcast Secretariat, Department of Computing Science, University of Newcastle-upon-Tyne, UK.

Chan, C. Y., & Ni, Y. (2006). Content-based Dissemination of Fragmented XML Data. In *Proceedings of International Conference on Distributed Computing Systems*, (ICDCS 2006).

Chandler, R., Lam, C. P., & Li, H. (2005). Ad2us: An automated approach to generating usage scenarios from uml activity diagrams. In *Apsec '05: Proceedings of the 12th Asia-Pacific software engineering conference* (apsec'05) (pp. 9-16). Washington, DC: IEEE Computer Society.

Chandrasekaran, S., Cooper, O., Deshpande, A., Franklin, M., Hellerstein, J., Hong, W., & Shah, M. (2003). Continuous dataflow processing for an uncertain world. In *Proceedings of Innovative Data System Research* (pp. 269–280). TelegraphCQ.

Chau, T., Muthusamy, V., Jacobsen, H. A., Litani, E., Chan, A., & Coulthard, P. (2008). Automating SLA modeling. In *Proceedings of the 2008 conference of the Centre for Advanced Studies on Collaborative Research*, Richmond Hill, Canada.

Chawathe, Y. D. (2000). *Scattercast: An architecture for Internet broadcast distribution as an infrastructure service*. Unpublished doctoral dissertation, University of California, Berkeley.

Chen, Y., & Schwan, K. (2005). Opportunistic Brokers: Efficient Content Delivery in Mobile Ad-Hoc Networks. In *Proceedings of the 6th ACM/IFIP/USENIX International Middleware Conference, Volume 3790* (pp. 354-374, Lecture Notes in Computer Science), Berlin, Germany: Springer.

Cherniack, M., Balakrishnan, H., Balazinska, M., Carney, D., Çetintemel, U., Xing, Y., & Zsonik, Y. (2003). Scalable distributed stream processing. In *Proceedings of the 2003 Conference on Innovative Data Systems Research*. Asilomar, CA.

Cheung, A., & Jacobsen, H.-A. (2006). Dynamic Load Balancing in Distributed Content-based Publish/Subscribe. In *Proceedings of the 7th ACM/IFIP/USENIX International Middleware Conference* (pp 249-269). New York: Springer.

Cheung, A., & Jacobsen, H.-A. (2008). *Efficient Load Distribution in Publish/Subscribe (Technical report).* Toronto, Canada: Middleware Systems Research Group, University of Toronto.

Chirita, P. A., Damian, A., Nejdl, W., & Siberski, W. (2005, November). Search strategies for scientific collaboration networks. In *P2PIR '05: Proceedings of the 2005 ACM workshop on information retrieval in peer-to-peer networks* (pp. 33–40). New York: ACM Press.

Chirita, P.-A., Idreos, S., Koubarakis, M., & Nejdl, W. (2004, May). Publish/subscribe for RDF-based P2P networks. In *Proceedings of the European semantic web symposium.* Heraklion, Greece.

Chirita, P.-A., Idreos, S., Koubarakis, M., & Nejdl, W. (2006, May 10-12). Publish/Subscribe for RDF-Based P2P Networks. In *Proceedings of the 1st European Semantic Web Symposium* (ESWS 2004), Heraklion, Greece.

Choi, Y., & Park, D. (2005, June). Mirinae: A peer-to-peer overlay network for large-scale content-based publish/subscribe systems. In *NOSSDAV '05: Proceedings of the international workshop on network and operating systems support for digital audio and video* (pp. 105–110). New York: ACM Press.

Christian, H.-W., & James, H. (2007). Toward expressive syndication on the web. In *Proceedings of the 16th international conference on World Wide Web.* Alberta, Canada: ACM.

Chu, Y. hua, Rao, S. G., & Zhang, H. (2000, June). A case for end system multicast (keynote address). In *Sigmetrics '00: Proceedings of the 2000 ACM SIGMETRICS international conference on measurement and modeling of computer systems* (pp. 1–12). New York: ACM Press.

Cilia, M., Antollini, M., Bornhovd, C., & Buchmann, A. (2004). Dealing with Heterogeneous Data in Pub/Sub Systems: The Concept-Based Approach. In *Proceedings of DEBS'04, International Workshop on Distributed Event-Based Systems.*

Cilia, M., Bornhövd, C., & Buchmann, A. P. (2003). An Infrastructure for Distributed, Heterogeneous Event-Based Applications. In *Proceedings of CoopIS 2003.* Catania, Italy: CREAM.

Cilia, M., Haupt, M., Mezini, M., & Buchmann, A. (2003). The convergence of AOP and active databases: towards reactive middleware. In *Proceedings of GPCE '03: The Second International Conference on Generative Programming and Component Engineering* (pp. 169-188). New York: Springer.

Clark, J., & DeRose, S. (1999). *XML path language (xpath).* Retrieved from http://www.w3.org/TR/xpath

comScore. (2008, December). *comScore Media Metrix.* Retrieved December, 2008 from http://comscore.com/metrix

CORBA-ES. (2004). *Corba event service, version 1.2.* Retrieved from http://www.omg.org/technology/documents/formal /event service.htm

Corporation, I. B. M. (2002, May). *WebSphere MQ Event Broker.* Retrieved from http://www.ibm.com/software/integration/mqfamily/eventbroker/

Corson, S., & Macker, J. (1999). Routing protocol performance issues and evaluation considerations (RFC 2501). *Network Working Group.*

Costa, P., & Picco, G. P. (2005). Publish-Subscribe on Sensor Networks: A Semi-Probabilistic Approach. *International Conference on Mobile Adhoc and Sensor Systems* (p. 332). Washington, DC: IEEE Press.

Costa, P., Coulson, G., Mascolo, C., Mottola, L., Picco, G. P., & Zachariadis, S. (2007). A Reconfigurable Component-based Middleware for Networked Embedded Systems. *Springer Journal of Wireless Information Networks, 14*(2), 149–162. doi:10.1007/s10776-007-0057-2

Costa, P., Gavidia, D., Koldehofe, B., Miranda, H., Musolesi, M., & Riva, O. (2008). When cars start gossiping. In *Proceedings of the 6th Workshop on Middleware for Network Eccentric and Mobile Applications*, (pp. 1-4).

Costa, P., Mascolo, C., Musolesi, M., & Picco, G. P. (2008). Socially-aware Routing for Publish-Subscribe in Delay-tolerant Mobile Ad Hoc Networks. *IEEE Journal on Selected Areas in Communications, 26*(5), 748–760. doi:10.1109/JSAC.2008.080602

Costa, P., Migliavacca, M., Picco, G. P., & Cugola, G. (2003). Introducing Reliability in Content-Based Publish-Subscribe through Epidemic Algorithms. In *Proceedings of DEBS'03, International Workshop on Distributed EventBased Systems*.

Costa, P., Migliavacca, M., Picco, G. P., & Cugola, G. (2004). Epidemic algorithms for reliable content-based publish-subscribe: an evaluation. In *Proceedings of the 24th International Conference on Distributed Computing Systems*, (pp. 552-561).

Coulouris, G., Dollimore, J., & Kindberg, T. (2001). *Distributed Systems, Concepts and Design* (3rd ed.). Reading, MA: Addision Wesley.

Cox, L. P., Murray, C. D., & Noble, B. D. (2002, Winter). Pastiche: making backup cheap and easy. *SIGOPS Operating Systems Review, 36*(SI), 285–298.

Crespo, A., & Garcia-Molina, H. (2002, July). Routing indices for peer-to-peer systems. In *Proceedings of the 22nd international conference on distributed computing systems (ICDCS'02)* (pp. 23–32). Washington, DC: IEEE Computer Society.

CSRG. (1986). Unix programmer's reference manual (4.3 BSD ed.) [Computer software manual]. Berkley, CA: Computer Systems Research Group, University of California.

Cugola, G., & di Nitto, E. (2008). On adopting Content-Based Routing in service-oriented architectures. *Information and Software Technology, 50*(1-2), 22–35. doi:10.1016/j.infsof.2007.10.004

Cugola, G., & Picco, G. P. (2006). REDS: a reconfigurable dispatching system. In *Proceedings of the 6th International Workshop on Software Engineering and Middleware (SEM'06)* (S. 9-16). Portland, OR: ACM Press.

Cugola, G., Nitto, E. D., & Fugetta, A. (2001). The JEDI Event-Based Infrastrucutre and its Application to the Development of the OPSS WFMS. [TSE]. *IEEE Transactions on Software Engineering, 27*(9), 827–850. doi:10.1109/32.950318

Cugola, G., Nitto, E. D., & Fuggetta, A. (1998). Exploiting an event-based infrastructure to develop complex distributed systems. In *Proceedings of ICSE '98: The 20th International Conference on Software Engineering* (pp. 261-270). Washington, DC: IEEE Computer Society.

Curbera, F., Doganata, Y. N., Martens, A., Mukhi, N., & Slominski, A. (2008). Business Provenance - A Technology to Increase Traceability of End-to-End Operations. In *Proceedings of the 16th International Conference on Cooperative Information Systems (CoopIS'08)* (S. 100-119). Monterrey, Mexico: Springer.

Cutting, D. (2007). *Balancing implicit group messaging over peer-to-peer networks*. Unpublished doctoral dissertation, School of Information Technologies, University of Sydney.

Cutting, D., Quigley, A., & Landfeldt, B. (2007). *Special Interest Messaging: A Comparison of IGM Approaches.* The Computer Journal.

Cutting, D., Quigley, A., & Landfeldt, B. (2008). SPICE: Scalable P2P implicit group messaging. *Computer Communications, 31*(3), 437–451. doi:10.1016/j.comcom.2007.08.026

Dabek, F., Cox, R., Kaashoek, F., & Morris, R. (2004, August). Vivaldi: A decentralized network coordinate system. In *Proceedings of the ACM SIGCOMM '04 conference.* Portland, Oregon, USA (pp. 15–26). New York: ACM Press.

Daily, G. P. S. (2008, February 6). *NXP Fuels Rise of Mobile Location-Based Services.* Retrieved from http://www.gpsdaily.com/reports/NXP_Fuels_Rise_Of_Mobile_Location_Based_Services_999.html

Datta, A., Quarteroni, S., & Aberer, K. (2004). Autonomous gossiping: A self-organizing epidemic algorithm for selective information dissemination in wireless mobile ad-hoc networks. In *Proceedings of 1st International IFIP Conference on Semantics of a Networked world*, (pp. 126-143).

Daws, C., Olivero, A., Tripakis, S., & Yovine, S. (1996). *The tool kronos in Hybrid Systems III.* (LNCS 1066), (pp. 208-219). New York: Springer.

Dayal, U. (1988). The HiPAC project: Combining active databases and timing constraints. *SIGMOD Record, 17*(1), 51–70. doi:10.1145/44203.44208

de Alfaro, L., & Henzinger, T. A. (2001). Interface automata. *ACM SIGSOFT Software Engeneering Notes, 26*(5), 109–120.

Dean, J. (2006). Experiences with MapReduce, an Abstraction for Large-Scale Computation. In Proceedings of Keynote PACT06.

Dean, J., & Ghemawat, S. (2008). MapReduce: simplified data processing on large clusters. *Communications of the ACM, 51*(1), 107–113. doi:10.1145/1327452.1327492

Dedecker, J., Van Cutsem, T., Mostinckx, S., D'Hondt, T., & De Meuter, W. (2006). Ambient-oriented Programming in AmbientTalk. In *Proceedings of the 20th European Conference on Object-oriented Programming (ECOOP), 4067*, (pp. 230-254).

Deering, S. (1989, August). *Host extensions for IP multicasting.* Stanford University: Network Working Group, Internet Engineering Task Force. Retrieved March, 2007 from http://www.ietf.org/rfc/rfc3170.txt

Deering, S., & Cheriton, D. R. (1990). Multicast routing in datagram internetworks and extended LANs. *ACM Transactions on Computer Systems, 8*(2), 85–111. doi:10.1145/78952.78953

Deveaux, D., & Collet, P. (2006). *Specification of a contract based built-in test framework for fractal.*

Dey, A. K. (2001). Understanding and using context. *Personal and Ubiquitous Computing, 5*(1), 4–7.

Dey, A. K., & Abowd, G. D. (2000). The Context Toolkit: Aiding the Development of Context-Aware Applications. In *Proceedings of the Workshop on Software Engineering for Wearable and Pervasive Computing.* Limerick, Ireland.

Dey, A. K., & Abowd, G. D. (2000). Towards a Better Understanding of Context and Context-Awareness. In: CHI 2000 Workshop on the What, Who, Where, When, and How of Context-Awareness, Dyreson, C. E., & Snodgrass, R.T. (1993). Timestamp semantics and representation. *Inf. Syst. (IS) 18*(3), 143-166)

Dey, A., & Abowd, G. (2000). CybreMinder: A Context-Aware System for Supporting Reminders. In *Proceedings of the Second International Symposium on Handheld and Ubiquitous Computing*, (pages 172–186).

Diao, Y., Altinel, M., Franklin, M. J., Zhang, H., & Fischer, P. (2003). Path sharing and predicate evaluation for high-performance XML filtering. [TODS]. *ACM Transactions on Database Systems, 28*(4), 467–516. doi:10.1145/958942.958947

Dias, M. S., & Vieira, M. E. R. (2000). Software architecture analysis based on statechart semantics. In *Proceedings of IWSSD '00: The 10th International Workshop on Software Specification and Design* (p. 133). Washington, DC: IEEE Computer Society.

Dierks, T., & Allen, C. (1999, January). The TLS protocol version 1.0. *RFC 2246.*

Ding, C., Chi, C.-H., Deng, J., & Dong, C.-L. (1999, October). Centralized content-based web filtering and blocking: How far can it go? In IEEE international conference on systems, man and cybernetics (pp. 115–119). SMC.

Dingel, J., Garlan, D., Jha, S., & Notkin, D. (1998). Reasoning about implicit invocation. In *Proceedings of SIGSOFT '98/FSE6: The 6th ACM SIGSOFT International Symposium on Foundations of Software Engineering* (pp. 209-221). New York: ACM Press.

Douence, R., & Südholt, M. (2002). *A model and a tool for event-based aspect-oriented programming (EAOP)* (Tech. Rep. No. 02/11/INFO). Ecole des Mines de Nantes.

Drugan, O. V., Plagemann, T., & Munthe-Kaas, E. (2005). Building resource aware middleware services over MANET for rescue and emergency applications. In *Proceedings of the 16th Annual IEEE International Symposium on Personal Indoor and Mobile Radio Communications, International Congress Center (ICC)*.

Eaton, P. R. (2002, November). *Caching the web with OceanStore* (Tech. Rep. No. UCB/CSD-02-1212). UC Berkeley, Computer Science Division.

el Hindi, K., & Lings, B. (1994). Using Truth Maintenance Systems to Solve the Data Consistency Problem. In Proceedings of CoopIS (pp. 192-201).

Elkin, M., Emek, Y., Spielman, D. A., & Teng, S. (2008). Lower-Stretch Spanning Trees. *SIAM Journal on Computing, 38*(2), 608–628. doi:10.1137/050641661

Embedded Systems Institute. (2007). *The Poseidon project*. Retreived from http://www.esi.nl/poseidon

Ericson, C. (2005). *Real-Time Collision Detection*. San Francisco: Morgan Kaufmann.

EsperTech. (2008). *Esper Reference Documentation*. Retrieved August 25, 2008, from http://esper.codehaus.org/

Estrin, D., Govindan, R., Heidemann, J., & Kumar, S. (1999). Next Century Challenges: Scalable Coordination in Sensor Networks. *International Conference on Mobile Computing and Networks* (pp. 263-270). Seattle, WA: ACM.

Etzion, O. (2005). Towards an Event-Driven Architecture: An Infrastructure for Event Processing Position Paper. In. *Proceedings of RuleML, 2005*, 1–7.

Etzion, O., Gal, A., & Segev, A. (1994). Retroactive and Proactive Database processing. In *Proceedings of RIDE-ADS*, (pp. 126-131)

EU Commission. (2007, October). *An integrated maritime policy for the European Union*. Brussels, Belgium: European Commission, Maritime Affairs.

Eugster, P. (2007). Type-Based Publish/Subscribe: Concepts and Experiences. *ACM Transactions on Programming Languages and Systems, 29*(1).

Eugster, P. T., Felber, P. A., Guerraoui, R., & Kermarrec, A. M. (2003). The many faces of publish/subscribe. *ACM Computing Surveys, 35*(2), 114–131. doi:10.1145/857076.857078

Eugster, P. T., Guerraoui, R., & Damm, C. H. (2001). On objects and events. In *OOPSLA '01: Proceedings of the 16th ACM SIGPLAN conference on Object oriented programming, systems, languages, and applications*, (pp. 254-269).

Eugster, P. T., Guerraoui, R., & Sventek, J. (2000). Distributed Asynchronous Collections: Abstractions for Publish/Subscribe Interaction. In *ECOOP '00: Proceedings of the 14th European Conference on Object-Oriented Programming*, (pp. 252-276).

Eugster, P., Garbinato, B., & Holzer, A. (2005). Location-based Publish/Subscribe. In *Proceedings of the Fourth IEEE International Symposium on Network Computing and Applications*, (pp. 279-282).

Extensions, X. M. P. P. (n.d.). *XMPP Extensions*. Retrieved from http://www.xmpp.org/extensions/

Fabret, F., Jacobsen, H. A., Llirbat, F., Pereira, J., Ross, K. A., & Shasha, D. (2001). Filtering algorithms and implementation for very fast publish/subscribe systems. In *SIGMOD '01: Proceedings of the 2001 ACM SIGMOD international conference on Management of data* (pp. 115-126). New York: ACM Press.

Facebook. (2008, December). *Facebook*. Retrieved December, 2008 from http://facebook.com

Fakas, G. J., & Karakostas, B. (2004). A peer to peer architecture for dynamic workflow management using web services. *Information and Software Technology Journal, 46*(6), 423–431. doi:10.1016/j.infsof.2003.09.015

Fan, L., Cao, P., Almeida, J., Broder, A.Z (2000). Summary cache: a scalable wide-area web cache sharing protocol. *IEEE/ACM Transactions on Networking, 8*(3), 281-293.

Fawcett, T., & Provost, F. (n.d.). Activity monitoring: Noticing interesting changes in behavior. In *Proceedings of the fifth ACM SIGKDD international conference on Knowledge discovery and data mining* (pp 53-62), New York: ACM.

Fayad, M. E., Schmidt, D. C., & Johnson, R. E. (1999). *Implementing application frameworks: Object-oriented frameworks at work.* New York: John Wiley & Sons, Inc.

Fenkam, P., Jazayeri, M., & Reif, G. (2004, May). On methodologies for constructing correct event-based applications. In A. Carzaniga & P. Fenkam (Eds.), *3rd International Workshop on Distributed Event-based Systems (DEBS'04)* (pp. 38-43). Edinburgh, UK: IEEE Computer Society.

Ferraiolo, D., & Kuhn, R. (1992, October). Role-based access controls. In *15th NIST-NCSC National Computer Security Conference* (pp. 554-563). Baltimore, MD.

Fidler, E., Jacobsen, H. A., Li, G., & Mankovski, S. (2005). The PADRES Distributed Publish/Subscribe System. In Proceedings of Feature Interactions in Telecommunications and Software Systems (pp. 12-30).

Fiege, L., Gartner, F., Kasten, O., & Zeidler, A. (2003). Supporting mobility in content-based publish/subscribe middlewares. In *Proceedings of ACM/IFIP/USENIX International Middleware Conference 2003*, 103-122.

Fiege, L., Mezini, M., Mühl, G., & Buchmann, A. P. (2002). Engineering Event-Based Systems with Scopes. In *Proceedings of the 16th European Conference on Object-Oriented Programming* (pp 309-333), Berlin, Germany: Springer.

Fischer, G. (2001, August). Communities of interest: Learning through the interaction of multiple knowledge systems. In A. M. S. Bjornestad R. Moe & A. Opdahl (Eds.), *24th annual information systems research seminar in Scandinavia (IRIS'24),* Ulvik, Hardanger Fjord, Norway (pp. 1–14).

Flautner, K., & Mudge, T. (2002, December). Vertigo: Automatic Performance-Setting for Linux. *Operating Systems Review, 36*(5S), 105–116. doi:10.1145/844128.844139

Flinn, J., & Satyanarayanan, M. (1999). *SOSP* (pp. 48–63). Energy-Aware Adaptation for Mobile Applications. In Proceedings of.

Foote, B., & Johnson, R. (1989). Reflective Facilities in Smalltalk-80. In *Proceedings of the 4th International Conference on Object-Oriented Programming Systems, Languages and Applications (OOPSLA 89),* (pp. 327-335).

Forgy, C. L. (1982). Rete: A Fast Algorithm for the Many Pattern/Many Object Pattern Match Problem. *Artificial Intelligence, 19*(1), 17–37. doi:10.1016/0004-3702(82)90020-0

Fowler, M. (2004). *Inversion of Control Containers and the Dependency Injection Pattern.* Retrieved January 23, 2004, from http://martinfowler.com/articles/injection.html

Francis, P. (2000, April). *Yoid: Extending the Internet multicast architecture.* Retrieved March, 2007 from http://www.icir.org/yoid/docs/index.html

Freeman, E., Arnold, K., & Hupfer, S. (1999). *JavaSpaces Principles, Patterns, and Practice.* Essex, UK: Addison-Wesley Longman Ltd.

Friedman-Hill, E. J. (2003). *Jess, The Rule Engine for the Java Platform.* Retrieved from http://herzberg.ca.sandia.gov/jess/

Gal, A. (1999). Semantic Interoperability in Information Services: Experiencing with CoopWARE. *SIGMOD Record, 28*(1), 68–75. doi:10.1145/309844.310061

Gal, A. (2007). Why is Schema Matching Tough and What Can We Do About It? *SIGMOD Record, 35*(4), 2–5. doi:10.1145/1228268.1228269

Galton, A., & Augusto, J. C. (2002). Two approaches to event definition. In *Proceedings of DEXA,* (pp. 547–556).

Gamma, E., Helm, R., Johnson, R., & Vlissides, J. (1995). *Design patterns: elements of reusable object oriented software.* Reading, MA: Addison-Wesley Professional.

Ganesan, P., Yang, B., & Garcia-Molina, H. (2004, June). One torus to rule them all: multi-dimensional queries in P2P systems. In *WEBDB '04: Proceedings of the 7th international workshop on the web and databases, Paris, France* (pp. 19–24). New York: ACM Press.

Gao, J. (2004). *A distributed and scalable peer-to-peer content discovery system supporting complex queries.* Unpublished doctoral dissertation, Computer Science Department, Carnegie Mellon University. (CMU-CS-04-170)

Gao, J. Z., Tsao, H. J., & Wu, Y. (2003). *Testing and quality assurance for component-based software.* Norwood, MA: Artech House.

Garlan, D., & Scott, C. (1993). Adding implicit invocation to traditional programming languages. *In Proceedings of ICSE '93: The 15th International Conference on Software Engineering* (pp. 447-455). Los Alamitos, CA: IEEE Computer Society Press.

Garlan, D., & Shaw, M. (1994, January). *An introduction to software architecture* (Tech. Rep. No. CMUCS-94-166). Pittsburgh, PA: School of Computer Science, Carnegie Mellon University.

Garlan, D., Khersonsky, S., & Kim, J. S. (2003). Model checking publish-subscribe systems. In *Proceedings of the 10th International SPIN Workshop on Model Checking of Software (SPIN 03)*. Portland, OR.

Gatziu, S., & Dittrichothers, K. R. (1994). Detecting composite events in active database systems using Petri Nets. In *Proceedings of RIDE-AIDS*, (pp. 2–9).

Gay, D., Levis, P., Behren, R. v., Welsh, M., Brewer, E., & Culler, D. (2003). *The nesC Language: A Holistic Approach to Networked Embedded Systems. Programming Language Design and Implementation* (pp. 1–11). San Diego, CA: ACM.

Gedik, B., & Liu, L. (2008, January). Protecting Location Privacy with Personalized k-Anonymity: Architecture and Algorithms. In. *Proceedings of IEEE Transactions on Mobile Computing, 7*(1), 1–18. doi:10.1109/TMC.2007.1062

Gehani, N. H., Jagadish, H. V., & Shmueli, O. (1992). Composite event specification in active databases: Model and implementation. In *Proceedings of VLDB*, (pp. 327–338).

Geihs, K. (2001). Middleware challenges ahead. *Computer, 34*(6), 24–31.

Gelernter, D. (1985). Generative communication in Linda. *ACM Transactions on Programming Languages and Systems, 7*(1), 80–112.

Geominder (n.d.). *Geominder.* Retrieved from http://ludimate.com/products/geominder/

Ghodsi, A., Alima, L. O., & Haridi, S. (2005, January). Low-bandwidth topology maintenance for robustness in structured overlay networks. In *Proceedings of the 38th annual Hawaii international conference on system sciences (HICSS'05), Big Island, HI, USA.* Washington, DC: IEEE Computer Society.

Ghodsi, A., Alima, L. O., & Haridi, S. (2005). *Symmetric replication for structured peer-to-peer systems. In 3rd intl. workshop on databases, information systems and peer-to-peer computing* (pp. 74–85).

Girod, L., Elson, J., Cerpa, A., Stathopoulos, T., Ramanathan, N., & Estrin, D. (2004). Emstar: A Software Environment for Developing and Deploying Wireless Sensor Network. *USENIX Technical Conference* (pp. 24-38). Boston, MA: USENIX.

Gnawali, O. (2002). *A keyword set search system for peer-to-peer networks.* Unpublished master's thesis. Cambridge, MA: Massachusetts Institute of Technology (MIT).

Golab, L., & Tamer Özsu, M. (2003). Issues in Data Stream Management. *SIGMOD Record, 32*(2), 5–14. doi:10.1145/776985.776986

Goldman, R. (2008, June 1). *Using the SPOT Emulator in Solarium.* Retrieved December 24, 2008, from SunSPOTWorld: http://www.sunspotworld.com/docs/Blue/SunSPOT-Emulator.pdf

Gomez, R., & Augusto, J. C. (2004). Durative events in active databases. In *Proceedings of ICEIS*, (pp. 511–516).

González, A., Piel, É., & Gross, H.-G. (2008, September). Architecture support for runtime integration and verification of component-based systems of systems. In *Proceedings of the 1st international workshop on automated engineering of autonomous and run-time evolving systems* (ARAMIS 2008). L'Aquila, Italy: IEEE Computer Society.

Google Inc. (2008, October). *Blogger*. Retrieved October, 2008 from http://blogger.com

Google Inc. (2008, October). *Google*. Retrieved October, 2008 from http://google.com

Gopalakrishnan, V., Silaghi, B., Bhattacharjee, B., & Keleher, P. (2004, March). Adaptive replication in peer-to-peer systems. In *Proceedings of the 24th international conference on distributed computing systems (ICDCS'04), Hachioji, Tokyo, Japan* (pp. 360–369). Washington, DC: IEEE Computer Society.

Gottschalk, K., Graham, S., Kreger, H., & Snell, J. (2002). Introduction to Web services architecture. *IBM Systems Journal, 41*(2), 170–177. doi:10.1147/sj.412.0170

Graham, S., Niblett, P., Chappell, D., Lewis, A., Nagaratnam, N., Parikh, J., et al. (2004). *Publish-Subscribe Notification for Web services*. Retrieved August 23, 2008, from http://www.ibm.com/developerworks/webservices/library/specification/ws-pubsub/

Grimm, R., Davis, J., Lemar, E., Macbeth, A., Swanson, S., & Anderson, T. (2004). System support for pervasive applications. *ACM Transactions on Computer Systems, 22*(4), 421–486. doi:10.1145/1035582.1035584

Gross, H.-G. (2005). *Component-based software testing with UML*. Heidelberg, Germany: Springer.

Gross, H.-G., & Mayer, N. (2004). Built-in contract testing in component integration testing. *Electronic Notes in Theoretical Computer Science, 82*(6), 22–32. doi:10.1016/S1571-0661(04)81022-3

Grossnickle, J., Board, T., Pickens, B., & Bellmont, M. (2005, October). RSS - Crossing Into the Mainstream. In Proceedings of Yahoo! IPSOS Insight.

Gruber, B., Krishnamurthy, B., & Panagos, E. (1999). The architecture of the READY event notification service. In *Proceedings of ICDCS Workshop on Electronic Commerce and Web-Based Applications*, (pp. 1–8).

Guo, S., Keeney, J., O'Sullivan, D., & Lewis, D. (2007, November 27-29). Adaptive Semantic Interoperability Strategies for Knowledge Based Networking. In *Proceedings of the International Workshop on Scalable Semantic Web Knowledge Base Systems (SSWS '07) at OTM 2007*. Vilamoura, Portugal.

Guo, S., Keeney, J., O'Sullivan, D., & Lewis, D. (2008, April 7-11). Coping with Diverse Semantic Models when Routing Ubiquitous Computing Information. In *Proceedings of the Workshop on Managing Ubiquitous Communications and Services (MUCS2008) at NOMS 2008*. Bahia, Brazil.

Guo, Y., & Heflin, J. (2005). LUBM: A Benchmark for OWL Knowledge Base Systems. *Journal of Web Semantics, 3*(2). doi:10.1016/j.websem.2005.06.005

Guo, Y., Heflin, J., & Pan, Z. (2004). An Evaluation of Knowledge Base Systems for Large OWL Datasets (Technical Report), Bethlehem, PA: CSE department, Lehigh University.

Gupta, A., & Suciu, D. (2003). Stream processing of xpath queries with predicates. In *Proceedings of 2003 ACM SIGMOD Intl conference on Management of data* (pages 419–430).

Gupta, A., Sahin, O. D., Agrawal, D., & Abbadi, A. E. (2004). Meghdoot: Content-Based publish/subscribe over P2P Networks. In *Proceedings of the 5th ACM/IFIP/USENIX International Middleware Conference* (pp 254-273), New York: Springer.

Guttman, A. (1984). R-trees: A dynamic index structure for spatial searching. In *Proceedings of ACM SIGMOD*, (pp. 47–57).

Haarslev, V., & Moller, R. (2001). RACER System Description. In *Proceedings of IJCAR 2001, volume 2083 of LNAI*, (701–706). Siena, Italy: Springer.

Hadar, E., & Silberman, G. (2008, October 19-23). Agile Architecture Methodology: Long Term Strategy Interleaved with Short Term Tactics. In *Proceedings of the International Conference on Object Oriented Programming, Systems, Languages and Applications*, (OOPSLA 2008), Nashville, TN.

Haddow, G. D., & Bullock, J. A. (2004). *Introduction to Emergency Management*. Amsterdam: Butterworth-Heinemann.

Hadim, S., & Mohamed, N. (2006). Middleware Challenges and Approaches for Wireless Sensor Networks. *IEEE Distributed Systems Online*, 7(3), 1. doi:10.1109/MDSO.2006.19

Hall, D. L., & Llina, J. (2001). *Handbook of Multisensor Data Fusion*. Boca Raton, FL: CRC Press.

Haller, P., & Odersky, M. (2006). Event-Based Programming without Inversion of Control. In *Proc. Joint Modular Languages Conference, 4228*, (pp. 4-22).

Hanak, D., Szijarto, G., & Takacs, B. (2007). A Mobile Approach to Ambient Assisted Living. In *Proceedings of IADIS Wireless Applications and Computing*. Lisbon, Portugal.

Hao-Ping Hung & Ming-Syan Chen. (2005). A general model of hybrid data dissemination. In *Proceedings of the 6th international conference on Mobile data management*, (pp. 220–228), Ayia Napa, Cyprus.

Hapner, M. (2002). *Java Message Service Specification (Version 1.1)*. Redwood Shores, CA: Sun Microsystems, Inc.

Harel, D. (1987, June). Statecharts: a visual formalism for complex systems. *Science of Computer Programming*, 8(3), 231–274.

Harter, A., Hopper, A., Stegglesand, P., Ward, A., & Webster, P. (2002). The anatomy of a context-aware application. *Springer Wireless Networks*, 8(2-3), 187–197. doi:10.1023/A:1013767926256

Hartmann, J., Vieira, M., Foster, H., & Ruder, A. (2005). A UML-based approach to system testing. *Innovations in Systems and Software Engineering*, 1(1), 12–24. doi:10.1007/s11334-005-0006-0

Hauswirth, M. (1999). *Internet-Scale Push Systems for Information Distribution - Architecture, Components, and Communication*. (Dissertation). Wien, Austria: Technischen Universitet Wien.

Hayton, R. (1996). *OASIS: An Open architecture for Secure Inter-working Services*. (PhD thesis), Cambridge, UK: University of Cambridge.

Hedetniemi, S. M., Hedetniemi, S. T., & Liestman, A. L. (1988). A survey of gossiping and broadcasting in communication networks. *Networks*, 18(4), 319–349. doi:10.1002/net.3230180406

Henricksen, K., & Robinson, R. (2006). A Survey of Middleware for Sensor Networks: State-of-the-art and Future Directions. *International Workshop on Middleware for Sensor Networks* (pp. 60-65). Melbourne, Australia: ACM.

Herlihy, M., Eliot, J., & Moss, B. (1993). Transactional memory: Architectural support for lock-free data structures. In *Proceedings of the 20th Annual International Symposium on Computer Architecture, 1993* (pp. 289-300). New York: ACM Press.

Herwono, I., Sachs, J., & Keller, R. (2005). Provisioning and performance of mobility-aware personalized push services in wireless broadband hotspots. *Computer Networks*, 49(3), 364–384. doi:10.1016/j.comnet.2005.05.010

Hinze, A., & Bittner, S. (2002). Efficient distribution-based event filtering. In Proceedings of Distributed Event Based Systems, (pp. 525–532).

Hoffmann-Wellenhof, B. H., Lichtenegger, H., & Collins, J. (1994). *GPS: Theory and Practice*. New York: Springer.

Hoffmeister, C., Nord, R., & Soni, D. (2000). *Applied Software Architecture*. Reading, MA: Addison-Wesley.

Holzmann, G. J., & Smith, M. H. (1999). A practical method for verifying event-driven software. In *ICSE '99: Proceedings of the 21st international conference on software engineering* (pp. 597-607). Los Alamitos, CA: IEEE Computer Society Press.

Howe, A. E., & Somlo, G. (1997). Modelling Discrete Event Sequences as State Transition Diagrams. In *Proceedings of IDA* (pp. 573-584)

Hu, H., Xu, J., & Lee, D. (2005). A Generic Framework for Monitoring Continuous Spatial Queries over Moving Objects. In *Proceedings of ACM SIGMOD*, 2005.

Hu, S., Muthusamy, V., Li, G., & Jacobsen, H.-A. (2008). Distributed Automatic Service Composition in Large-Scale Systems. *Proceedings of the 2nd International Conference on Distributed Event-Based Systems (DEBS'08)* (S. 233-244). Rome: ACM Press.

Huang, Y., & Garcia-Molina, H. (2003). Publish/subscribe tree construction in wireless ad-hoc networks. In *Proceedings of the 4th International Conference on Mobile Data Management*, (pp. 122-140).

Hunkeler, U., Truong, H. L., & Stanford-Clark, A. (2008). *MQTT-S — A Publish/Subscribe Protocol for Wireless Sensor Networks. Communication Systems Software and Middleware and Workshops* (pp. 791–798). Dublin, Ireland: ICST.

Hwang, J.-H., Çetintemel, U., & Zdonik, S. Fast and Highly-Available Stream Processing over Wide Area Networks. In *Proceedings of the 24th International Conference on Data Engineering* (804-813) Washington, DC: IEEE Computer Society.

IBM. (2003). *Web Service Level Agreements (WSLA) Project*. Retrieved July 12th, 2007, from http://www.research.ibm.com/wsla/

IEEE. (n.d.). *IEEE 802.11*. Retrieved from http://grouper.ieee.org/groups/802/11/index.html

IETF Network Working Group. (2004). RFC 3920 Extensible Messaging and Presence Protocol (XMPP): Core. *IETF*. Retrieved from http://www.ietf.org/rfc/rfc3920.txt

Intanagonwiwat, C., Govindan, R., & Estrin, D. (2000, August). Directed diffusion: a scalable and robust communication paradigm for sensor networks. In *Proceedings of the 6th annual international conference on mobile computing and networking, Boston* (pp. 56–67). New York: ACM Press.

International Telecommunication Union. (2001). *Recommendation ITU-R M.1371-1*. Geneva, Switzerland: ITU.

iPhone. (2008, December). *Apple iPhone*. Retrieved December, 2008 from http://apple.com/iphone

Iwasaki, Y., Farquhar, A., Saraswat, V. A., Bobrow, D. G., & Gupta, V. (1995). Modeling Time in Hybrid Systems: How Fast Is "Instantaneous"? In *Proceedings of the IJCAI* (pp. 1773-1781).

Jacobsen, H.-A. (2004). *PADRES User Guide*. Retrieved July 19, 2006, from http://research.msrg.utoronto.ca/Padres/UserGuide

Jacobsen, H.-A. (2006). *eQoSystem*. http://research.msrg.utoronto.ca/Eqosystem

Jacobsen, H.-A. (2007). *Enterprise Application Integration*. http://research.msrg.utoronto.ca/EAI/

Jacobsen, H.-A., Mühl, G., & Jaeger, M. A. (Eds.). (2007, June). *Proceedings of the inaugural conference on distributed event-based systems (DEBS'07)*. New York: ACM Press. Retrieved from http://debs.msrg.utoronto.ca

Jennings, B., van der Meer, S., Balasubramaniam, S., Botvich, D., O'Foghlu, M., Donnelly, W., & Strassner, J. (2007, October). Towards Autonomic Management of Communications Networks. *IEEE Communications Magazine, 45*(10). doi:10.1109/MCOM.2007.4342833

Jennings, C., & Mahy, R. (2008). *Managing Client Initiated Connections in the Session Initiation Protocol (SIP)*. *IETF*. Internet Draft.

Jensen, C. Lin, D., & Ooi. B. (2004). Query and Update Efficient B+-Tree based Indexing of Moving Objects. In *Proceedings of VLDB*, (pages 768–779).

Jensen, C. S., & Snodgrass, R. T. (1999). Temporal Data Management. [TKDE]. *IEEE Transactions on Knowledge and Data Engineering*, *11*(1), 36–44. doi:10.1109/69.755613

Jerzak, Z., & Fetzer, C. (2007). Prefix Forwarding for Publish/Subscribe. In *DEBS '07: Proceedings of the 2007 Inaugural International Conference on Distributed Event-Based Systems* (pp. 238-249). Toronto, Canada: ACM Press.

Jerzak, Z., & Fetzer, C. (2008). Bloom Filter Based Routing for Content-Based Publish/Subscribe. In *DEBS '08: Proceedings of the second international conference on Distributed event-based systems* (pp. 71-81). Rome, Italy: ACM Press.

Jin, Y., & Strom, R. (2003). Relational Subscription Middleware for Internet-Scale Publish-Subscribe. In *DEBS '03, Proceedings of the 2nd international workshop on Distributed event-based systems.*

Jobst, D., & Preissler, G. (2006). Mapping clouds of SOA- and business-related events for an enterprise cockpit in a Java-based environment. In *Proceedings of the 4th International Symposium on Principles and Practice of Programming in Java (PPPJ'06)* (S. 230-236). Mannheim, Germany: ACM Press.

Jordan, M., & Stewart, C. (2005). Adaptive Middleware for Dynamic Component-level Deployment. In *Proceedings of the 4th workshop on Reflective and Adaptive Middleware Solutions*. Grenoble, France: ACM.

Juang, H. Oki, Wang, Y., Martonosi, M., Peh, L., & Rubens, D. (2002). Energy-Efficient Computing for Wildlife Tracking: Design Tradeoffs and Early Experiences with Zebra Net. In ASPLOS'02, Proceedings of Architectural Support for Programming Languages and Operating Systems.

Jung, D., & Hinze, A. (2004). A Meta-Service for Event Notification. *Lecture Notes in Computer Science*, *3290*, 283–300.

Junginger, M., & Lee, Y. (2004). A self-organizing publish/subscribe middleware for dynamic peer-to-peer networks. *IEEE Network*, *18*(1), 38–43. doi:10.1109/MNET.2004.1265832

Kaminsky, A., & Bischof, H.-P. (2002). Many-to-Many Invocation: a new object oriented paradigm for ad hoc collaborative systems. In OOPSLA '02: Companion of the 17th annual ACM SIGPLAN conference on Object-oriented programming, systems, languages, and applications, (pp. 72-73).

Keeney, J., Jones, D., Roblek, D., Lewis, D., & O'Sullivan, D. (2008). Knowledge-based Semantic Clustering. In *Proceedings of ACM Symposium on Applied Computing*, Fortaleza, Brazil.

Keeney, J., Lewis, D., & O'Sullivan, D. (2006, July 19-21). Benchmarking Knowledge-based Context Delivery Systems. In Proceedings of ICAS06, Silicon Valley, CA.

Keeney, J., Lewis, D., & O'Sullivan, D. (2007, March). Ontological Semantics for Distributing Contextual Knowledge in Highly Distributed Autonomic Systems. *Journal of Network and Systems Management*, 15.

Keeney, J., Lewis, D., O'Sullivan, D., Roelens, A., Wade, V., Boran, A., & Richardson, R. (2006, April). Runtime Semantic Interoperability for Gathering Ontology-based Network Context. In *Proceedings of Network Operations and Management Symposium* (NOMS 2006), Toronto, Canada.

Keeney, J., Roblek, D., & Jones, D. Lewis, & O'Sullivan, D., (2008). Extending Siena to support more expressive and flexible subscriptions. In *Proceedings of the 2nd International Conference on Distributed Event-Based Systems* (DEBS 2008), Rome, Italy.

Keoh, S. L., Dulay, N., Lupu, E., Twidle, K., Schaeffer-Filho, A. E., Sloman, M., et al. (2007). Self-Managed Cell: A Middleware for Managing Body-Sensor Networks. *International Conference on Mobile and Ubiquitous Systems* (pp. 1-5). Philadelphia, PA: IEEE.

Kephart, J. O., & Chess, D. M. (2003). The Vision of Autonomic Computing. *Computer*, *36*(1), 41–50. doi:10.1109/MC.2003.1160055

Khambatti, M. (2003). *Peer-to-peer communities: architecture, information and trust management.* Unpublished doctoral dissertation, Arizona State University.

Khelil, A., Marrón, P. J., Becker, C., & Rothermel, K. (2007). Hypergossiping: A Generalized broadcast strategy for mobile ad hoc networks. *Ad Hoc Networks*, 5(5), 531–546. doi:10.1016/j.adhoc.2006.03.001

Kiczales, G., Lamping, J., Mendhekar, A., Maeda, C., Lopes, C. V., Loingtier, J.-M., et al. (1997, June). Aspect-oriented programming. In *Proceedings of the 11th European Conference on Object-oriented Programming (ECOOP)*. Berlin, Germany: Springer-Verlag.

Kim, S., Kim, M., Park, S., Jin, Y., & Choi, W. (2004). Gate Reminder: A Design Case of a Smart Reminder. In *Proceedings of the Conference on Designing Interactive Systems*, (pages 81–90).

Kleinberger, T., Becker, M., Ras, E., Holzinger, A., & Müller, P. (2007). Ambient Intelligence in Assisted Living: Enable Elderly People to Handle Future Interfaces. In Proceedings of Universal Access in Human-Computer Interaction. Ambient Interaction (LNCS Vol. 4555/2007, pp. 103-112,), Berlin, Germany, Springer.

Koenig, I. (2007). Event Processing as a Core Capability of Your Content Distribution Fabric. In *Proceedings of the Gartner Event Processing Summit,* Orlando, FL.

Konana, P., Liu, G., Lee, C.-G., & Woo, H. (2004). Specifying timing constraints and composite events: An application in the design of electronic brokerages. *IEEE Transactions on Software Engineering, 30*(12), 841–858.

Koparanova, M., & Risch, T. (2004). High-performance grid stream database manager for scientific data. In F. Fernández Rivera, Marian Bubak, A. Gómez Tato & Ramon Doallo (Eds.), *European Across Grids Conference* (pp. 86-92). Berlin/Heidelberg, Germany: Springer Verlag.

Krishnamurthy, B., & Rosenblum, D. S. (1995). Yeast: A general purpose event-action system. *IEEE Transactions on Software Engineering, 21*(10), 845–857.

Kshemkalyani, A. D. (2007). Temporal Predicate Detection Using Synchronized Clocks. [TC]. *IEEE Transactions on Computers, 56*(11), 1578–1584. doi:10.1109/TC.2007.70749

Kulkarni, S. (2006, September). Video streaming on the Internet using split and merge multicast. In *P2P '06: Proceedings of the sixth IEEE international conference on peer-to-peer computing* (pp. 221–222). Washington, DC: IEEE Computer Society.

Kumar, R., Novak, J., Raghavan, P., & Tomkins, A. (2003, May). On the bursty evolution of blogspace. In *WWW '03: Proceedings of the 12th international conference on world wide web* (pp. 568–576). New York: ACM Press.

Kumar, V., & Cai, Z. (2006). Implementing Diverse Messaging Models with Self-Managing Properties using IFLOW. *IEEE International Conference on Autonomic Computing* (pp 243-252). IEEE Computer Society. Washington, DC.

Lamport, L. (1978). Time, Clocks, and the Ordering of Events in a Distributed System. [CACM]. *Communications of the ACM, 21*(7), 558–565. doi:10.1145/359545.359563

Landers, M., Zhang, H., & Tan, K.-L. (2004, August). Peerstore: Better performance by relaxing in peer-to-peer backup. In 4th international conference on peer-to-peer computing (P2P 2004), Zurich, Switzerland (p. 72-79). IEEE Computer Society.

Lave, J., & Wenger, E. (1991). *Situated learning: Legitimate peripheral participation (learning in doing: Social, cognitive & computational perspectives).* Cambridge, UK: Cambridge University Press.

Lazaridis, I., & Mehrotra, S. (2003). Capturing sensor-generated time series with quality guarantees. In *Proceedings of ICDE*, (pp. 429–440).

Le, A., Jaitner, T., & Litz, L. (2007). Sensor-based Training Optimization of a Cyclist Group. In *Proceedings of the 7th International Conference on Hybrid Intelligent Systems* (pp. 265-270). Kaiserslautern, Germany.

Lee, C., Nordstedt, D., & Helal, S. (2003). Enabling Smart Spaces with OSGi. *Pervasive Computing, 2*(3), 89–94. doi:10.1109/MPRV.2003.1228530

Lee, J. (2003, February). An end-user perspective on file-sharing systems. *Communications of the ACM, 46*(2), 49–53. doi:10.1145/606272.606300

Lee, S., Su, W., Hsu, J., Gerla, M., & Bagrodia, R. (2000). A performance comparison study of ad hoc wireless multicast protocols. In *Proceedings of 9th Annual Joint Conference of the IEEE Computer and Communications Societies* (pp 565-574), IEEE Computer Society. Washington, DC.

Lehman, T.J., Cozzi, A., Xiong, Y., Gottschalk, J., Vasudevan, V., & Landis, S. (2001). Hitting the distributed computing sweet spot with TSpaces. *Computer Networks, 35*(4), 457–472. doi:10.1016/S1389-1286(00)00178-X

Leitner, P., Michlmayr, A., Rosenberg, F., & Dustdar, S. (2008). End-to-End Versioning Support for Web Services. In *Proceedings of the International Conference on Services Computing (SCC'08)*. Honolulu, HI: IEEE Computer Society.

Leitner, P., Rosenberg, F., & Dustdar, S. (2009). Daios - Efficient Dynamic Web Service Invocation. *IEEE Internet Computing, 13*(3), 72–80. doi:10.1109/MIC.2009.57

Lekova, A., Skjelsvik, K. S., Plagemann, T., & Goebel, V. (n.d.). Fuzzy Logic-Based Event Notification in Sparse MANETs. In *AINA'07, Proceedings of the 21st International Conference on Advanced information Networking and Applications Workshop* (Vol. 2, pp. 296-301), Niagara Falls, Canada.

Leslie, M., Davies, J., & Huffman, T. (2006, April). Replication strategies for reliable decentralised storage. *The First International Conference on Availability, Reliability and Security (ARES 2006)*.

Leveson, N. G., Heimdahl, M. P. E., Hildreth, H., & Reese, J. D. (1994). Requirements specification for process-control systems. *IEEE Transactions on Software Engineering, 20*(9), 684–707.

Levis, P., Madden, S., Polastre, J., Szewczyk, R., Whitehouse, K., Woo, A., et al. (2005). TinyOS: An Operating System for Sensor Networks. In W. Weber, J. Rabaey, & E. Aarts, Ambient Intelligence (pp. 115-148). Berlin, Germany: Springer.

Lewis, D., Keeney, J., O'Sullivan, D., & Guo, S. (2006, October 23-25). Towards a Managed Extensible Control Plane for Knowledge-Based Networking. In *Proceedings of Distributed Systems: Operations and Management Large Scale Management, (DSOM 2006)*, at Manweek 2006, Dublin, Ireland.

Lewis, D., O'Sullivan, D., Power, R., & Keeney, J. (2005, October). Semantic Interoperability for an Autonomic Knowledge Delivery Service. In *Proceedings of Workshop on Autonomic Communication* (WAC 2005), Athens, Greece.

Li, G., & Jacobsen, H.-A. (2005). Composite Subscriptions in Content-Based publish/subscribe Systems. In *Proceedings of the 6th ACM/IFIP/USENIX International Middleware Conference* (pp 249-269), Berlin, Germany: Springer.

Li, G., Cheung, A., Hou, S., Hu, S., Muthusamy, V., Sherafat, R., et al. (2007). Historic data access in publish/subscribe. In *Proceedings of the 2007 inaugural international conference on Distributed event-based systems* (pp 80-84), Toronto, Canada.

Li, G., Hou, S., & Jacobsen, H.-A. (2005). A Unified Approach to Routing, Covering and Merging in Publish/Subscribe Systems based on Modified Binary Decision Diagrams. In *Proceedings of the 25th IEEE International Conference on Distributed Computing Systems* (pp 447-457), Columbus, OH.

Li, G., Hou, S., & Jacobsen, H.-A. (2008). Routing of XML and XPath Queries in Data Dissemination Networks. In *Proceedings of the 28th IEEE International Conference on Distributed Computing Systems* (pp 627-638), Beijing, China.

Li, G., Muthusamy, V., & Jacobsen, H.-A. (2007). *NI-ÑOS: A Distributed Service Oriented Architecture for Business Process Execution. (Technical report)*. Toronto, Canada: Middleware Systems Research Group, University of Toronto.

Li, G., Muthusamy, V., & Jacobsen, H.-A. (2008). Adpative content-based routing in general overlay topologies. In *Proceedings of the 9th ACM/IFIP/USENIX International Middleware Conference* (pp 249-269), Berlin, Germany: Springer.

Li, G., Muthusamy, V., & Jacobsen, H.-A. (2008). *Subscribing to the past in content-based publish/subscribe. (Technical report)*. Toronto, Canada: Middleware Systems Research Group.

Li, H., & Jiang, G. (2004). Semantic Message Oriented Middleware for Publish/Subscribe Networks. *Proceedings of the Society for Photo-Instrumentation Engineers, 5403*, 124–133.

Li, S., Lin, Y., Son, S. H., Stankovic, J. A., & Wei, Y. (2004). Event Detection Services Using Data Service Middleware in Distributed Sensor Networks. *Springer Telecommunication Systems, 26*(2-4), 351–368. doi:10.1023/B:TELS.0000029046.79337.8f

Liebig, C., Cilia, M., & Buchmann, A. (1999). Event composition in time-dependent distributed systems. In *Proceedings of IFIP CoopIS*, (pp. 70–78).

Liebig, C., Cilia, M., & Buchmann, A. P. (1999). Event Composition in Time-dependent Distributed Systems. In Proceedings of CoopIS (pp. 70-78).

Lindahl, C., & Blount, E. (2003, November). Weblogs: simplifying web publishing. *Computer, 36*(11), 114–116. doi:10.1109/MC.2003.1244542

Lindgren, A., Doria, A., & Schelén, O. (2003). Probabilistic Routing in Intermittently Connected Networks. In *MobiHoc'03, Proceedings of the Fourth ACM International Symposium on Mobile Ad Hoc Networking and Computing*.

LinkedIn. (2008, December). *LinkedIn*. Retrieved December, 2008 from http://linkedin.com

Liu, G., Mok, A. K., & Konana, P. (1998). A unified approach for specifying timing constraints and composite events in active real-time database systems. In *Proceedings of the Real-Time Technology and Applications Symposium*, (pp. 2–9).

Liu, H., & Jacobsen, H. (2004). Modeling uncertainties in publish/subscribe system. In *Proceedings of ICDE*, (pp. 510–521).

Liu, H., & Jacobsen, H. A. (2002). A-ToPSS: A Publish/Subscribe System Supporting Approximate Matching. In *Proceedings of 28th International Conference on Very Large Data Bases* (pp 1107-1110), Hong Kong, China.

Liu, H., & Jacobsen, H. A. (2004). Modeling uncertainties in publish/subscribe systems. In *Proceedings of the 20th International conference on Data Engineering* (pp 510-522), Boston, MA.

Liu, H., Muthusamy, V., & Jacobsen, H. A. (2009). *Predictive Publish/Subscribe Matching. (Technical report)*. Toronto, Canada: Middleware Systems Research Group, University of Toronto.

Liu, H., Ramasubramanian, V., & Sirer, E. G. (2005). Client behavior and feed characteristics of RSS, a publish-subscribe system for web micronews. In *Proceedings of the 5th ACM SIGCOMM conference on Internet Measurement* (pp 3-3), Berkeley, CA: USENIX Association.

Liu, L. Pu, C., & Tang, W. (2000). WebCQ - Detecting and Delivering Information Changes on the Web. In *Proceedings of CIKM*, (pages 512–519).

Liu, L., Ryu, K. D., & Lee, K.-W. (2004, April). *Keyword fusion to support efficient keyword-based search in peer-to-peer file sharing*. Presented at the Fourth IEEE International Symposium on Cluster Computing and the Grid (CCGrid'04) Chicago, IL.

Lockheed Martin. (2008). *Maritime safety, security & surveillance integrated systems for monitoring ports, waterways and coastlines*. Retrieved from http://www.lockheedmartin.com

Loser, A., Naumann, F., Siberski, W., Nejdl, W., & Thaden, U. (2003). Semantic overlay clusters within super-peer networks. In *Proceedings of Workshop on Databases, Information Systems and Peer-to-Peer Computing in Conjunction with the (VLDB '03).*

Lost in Klaus. (n.d.). *Lost in Kluas.* Retrieved from http://www.cc.gatech.edu/~ulee3/LostInKlaus/

Lu, T., Sinha, S., & Sudan, A. (2002). Panaché: A scalable distributed index for keyword search (Tech. Rep.). Cambridge, MA: Massachusetts Institute of Technology (MIT).

Luckham, D. (2002). *The Power of Events: An Introduction to Complex Event Processing in Distributed Enterprise Systems.* Reading, MA: Addison-Wesley.

Luckham, D. C., & Vera, J. (1995). An Event-Based Architecture Definition Language. *IEEE Transactions on Software Engineering, 21*(9), 717–734. doi:10.1109/32.464548

Ludford, P., Frankowski, D., Reily, K., Wilms, K., & Terveen, L. (2006). Because I Carry My Cell Phone Anyway: Functional Location-Based Reminder Applications. In *SIGCHI Conference on Human Factors in Computing Systems,* (pages 889–898)

Lupu, E., Dulay, N., Sloman, M., Sventek, J., Heeps, S., & Strowes, S. (2008). Amuse: autonomic management of ubiquitous e-health systems. *Concurrency and Computation, 20*(3), 277–295. doi:10.1002/cpe.1194

Lv, Q., Cao, P., Cohen, E., Li, K., & Shenker, S. (2002). Search and replication in unstructured peer-to-peer networks. In *ICS '02: Proceedings of the 16th international conference on supercomputing* (pp. 84–95). New York: ACM.

Lynch, D., Keeney, J., Lewis, D., & O'Sullivan, D. (2006, May). A Proactive Approach to Semantically Oriented Service Discovery. In *Proceedings of Innovations in Web Infrastructure (IWI 2006). at World-Wide Web Conf.,* Edinburgh, Scotland.

Ma, C., & Bacon, J. (1998). COBEA: A corba-based event architecture. In *Proceedings of COOTS,* (pp. 117–132).

Madden, S., Franklin, M. J., Hellerstein, J., & Hong, W. (2002). TAG: a tiny aggregation tree for ad-hoc sensor networks. In *Proceedings of USENIX Symposium on Operating Systems Design and Implementation,* (pp. 131–146).

Madhyastha, H. V., Anderson, T., Krishnamurthy, A., Spring, N., & Venkataramani, A. (2006, October). A structural approach to latency prediction. In *IMC '06: Proceedings of the 6th ACM SIGCOMM on Internet measurement* (pp. 99–104). New York: ACM Press.

Mahambre, S. P., Kumar, M. S., & Bellur, U. (2007). A Taxonomy of QoS-Aware, Adaptive Event-Dissemination Middleware. *IEEE Internet Computing, 11*(4), 35–44. doi:10.1109/MIC.2007.77

Mahemoff, M. (2006, March). Comet: A New Approach to Ajax Applications. *Ajaxian.* Retrieved from http://ajaxian.com/archives/comet-a-new-approach-to-ajax-applications

Mamei, M., & Zambonelli, F. (2004). Programming Pervasive and Mobile Computing Applications with the TOTA Middleware. In *PERCOM '04: Proceedings of the Second IEEE International Conference on Pervasive Computing and Communications,* (pp. 263-276).

Mansouri-Samani, M., & Sloman, M. (1997). Gem: A generalized event monitoring language for distributed systems. *IEE/IOP/BCS Distributed systems Engineering Journal, 4*(2), 96–108.

Marchiori, A., & Han, Q. (2008). A Foundation for Interoperable Sensor Networks with Internet Bridging. *Workshop on Embedded Networked Sensor.* Charlottesville, VA: ACM.

Mariani, L., Papagiannakis, S., & Pezze, M. (2007). Compatibility and regression testing of COTS-Component-Based software. In *ICSE '07: Proceedings of the 29th international conference on software engineering* (pp. 85-95). Washington, DC: IEEE Computer Society.

Marmasse, N., & Schmandt, C. (2000). Location-Aware Information Delivery with Com-Motion. In *Proceedings of HUC,* (pages 157–171).

Marques, E. R. B., Goncalves, G. M., & Sousa, J. B. (2006). The use of real-time publish-subscribe middleware in networked vehicle systems. In *Proceedings of the 1st IFAC Workshop on Multivehicle Systems*.

Marrón, P. J., Lachenmann, A., Minder, D., Gauger, M., Saukh, O., & Rothermel, K. (2005). Management and Configuration Issues for Sensor Networks. *Wiley International Journal of Network Management*, 15(4), 235–253. doi:10.1002/nem.571

Martin, D., Burstein, M., Hobbs, J., Lassila, O., McDermott, D., McIlraith, S., Narayanan, S., Paolucci, M., Parsia, B., Payne, T., Sirin, E., Srinivasan, N., & Sycara, K. (2004, November 22). OWL-S: Semantic Markup for Web Services. *W3C Member Submission*.

Mascolo, C., Capra, L., & Emmerich, W. (2002). Mobile Computing Middleware. In *Proceedings of Advanced lectures on networking* (pp. 20–58). New York: Springer-Verlag. doi:10.1007/3-540-36162-6_2

Matsuoka, S., & Kawai, S. (1988). Using tuple space communication in distributed object-oriented languages. *SIGPLAN Not. Special issue: 'OOPSLA 88 Conference Proceedings, 23*(11), 276-284.

Maymounkov, P., & Mazieres, D. (2002, March). Kademlia: A peer-to-peer information system based on the XOR metric. In Druschel, P., Kaashoek, M. F., & Rowstron, A. I. T. (Eds.), *First international workshop on peer-to-peer systems at IPTPS 2002, Cambridge, MA, USA* (pp. 53–65). New York: Springer.

McGrew, D., & Sherman, A. (1998, May). *Key establishment in large dynamic groups using one-way function trees* (Tech. Rep.). Glenwood, MD: TIS Labs at Network Associates, Inc.

Meier, R., & Cahill, V. (2002). STEAM: Event-Based Middleware for Wireless Ad Hoc Networks. In *DEBS'02, Proceedings of the 1ˢᵗ International Workshop on Distributed Event-Based Systems*, Vienna, Austria.

Meier, R., & Cahill, V. (2003). Exploiting Proximity in Event-Based Middleware for Collaborative Mobile Applications. In *Proceedings of the 4th IFIP International Conference on Distributed Applications and Interoperable Systems (DAIS'03)*, (pp, 285-296).

Meier, R., & Cahill, V. (2005). Taxonomy of Distributed Event-Based Programming Systems. *The Computer Journal*, 48(5), 602–626. doi:10.1093/comjnl/bxh120

Memon, A., Banerjee, I., & Nagarajan, A. (2003, November). GUI ripping: reverse engineering of graphical user interfaces for testing. In *Proceedings of the 10th working conference on reverse engineering* (pp. 260-269). Piscataway, NJ.

Meyer, B. (1993). Systematic concurrent object-oriented programming. *Communications of the ACM*, 36(9), 56–80. doi:10.1145/162685.162705

mGraffiti (n.d.). *mGraffiti*. Retrieved from http://www.cc.gatech.edu/projects/disl/mGraffiti/

Michelson, B. M. (2006). *Event-Driven Architecture Overview*. Boston, MA: Patricia Seybold Group.

Michlmayr, A. (2010). Event Processing in QoS-Aware Service Runtime Environments, PhD Thesis, Vienna University of Technology.

Michlmayr, A., Fenkam, P., & Dustdar, S. (2006, September). Architecting a testing framework for publish/subscribe applications. In *Proceedings of the 30th annual international computer software and applications conference* (compsac'06), (Vol. 1, pp. 467-474).

Michlmayr, A., Fenkam, P., & Dustdar, S. (2006, July). Specification based unit testing of publish/subscribe applications. In *Proceedings of the 26th IEEE international conference on distributed computing systems workshops* (ICDCSW '06), (p. 34).

Michlmayr, A., Rosenberg, F., Leitner, P., & Dustdar, S. (2008). Advanced Event Processing and Notifications in Service Runtime Environments. In *Proceedings of the 2nd International Conference on Distributed Event-Based Systems (DEBS'08)* (pp. 115-125). Rome: ACM Press.

Michlmayr, A., Rosenberg, F., Leitner, P., & Dustdar, S. (2009). Service Provenance in QoS-Aware Web Service Runtimes. In *Proceedings of the 7th International Conference on Web Services (ICWS'09)*, (S. 115-122). Los Angeles: IEEE Computer Society.

Michlmayr, A., Rosenberg, F., Platzer, C., Treiber, M., & Dustdar, S. (2007). Towards Recovering the Broken SOA Triangle - A Software Engineering Perspective. In *Proceedings of the Second International Workshop on Service Oriented Software Engineering (IW-SOSWE'07)* (pp. 22-28). Dubrovnik, Croatia: ACM Press.

Miller, M. (2006). *Robust Composition: Towards a Unified Approach to Access Control and Concurrency Control*. Baltimore, MD: John Hopkins University.

Miller, M., Tribble, E. D., & Shapiro, J. (2005). Concurrency among strangers: Programming in E as plan coordination. In. *Proceedings of the Symposium on Trustworthy Global Computing*, *3705*, 195–229. doi:10.1007/11580850_12

Miniwatts Marketing Group. (2008, October). *Internet usage statistics. Internet World Stats*. Retrieved October, 2008 from http://www.internetworldstats.com/stats.htm

Mislove, A., Post, A., Reis, C., Willmann, P., Druschel, P., & Wallach, D. S. (2003, May). POST: A secure, resilient, cooperative messaging system. In M. B. Jones (Ed.), *Proceedings of HOTOS'03: 9th workshop on hot topics in operating systems, Lihue (Kauai), Hawaii, USA* (pp. 61–66). USENIX.

Mitre, J., & Navarro-Moldes, L. (2004, June). P2P architecture for scientific collaboration. In *WETICE '04: Proceedings of the 13th ieee international workshops on enabling technologies: Infrastructure for collaborative enterprises (WETICE '04)* (pp. 95–100). Washington, DC: IEEE Computer Society.

Mokbel, M., Xiong, X., & Aref, W. (2004). Scalable Incremental Processing of Continuous Queries in Spatio-Temporal Databases. In *ACM SIGMOD* (pp. 623–634). SINA.

Moody, K. (2000, August). Coordinating policy for federated applications. In 14th IFIP WG3 working conference on databases and application security (pp. 127–134). Schoorl, The Netherlands: Kluwer.

Morton, G. M. (1966). *A computer oriented geodetic data base; and a new technique in file sequencing (Tech. Rep.)*. Ottawa, Canada: IBM Canada Ltd.

Motik, B., & Sattler, U. (2006). Practical DL Reasoning over Large A Boxes with KAON2. Retrieved from http://kaon2.semanticweb.org/

Mottola, L., Cugola, G., & Picco, G. P. (2005). *A Self-Repairing Tree Overlay Enabling Content-Based Routing on Mobile Ad Hoc Networks (Tech. Rep.)*. Politecnico di Milano.

Mottola, L., Cugola, G., & Picco, G. P. (2008). A self repairing tree topology enabling content-based routing in mobile ad hoc networks. *IEEE Transactions on Mobile Computing*, *7*(8), 946–960. doi:10.1109/TMC.2007.70789

Motwani, R., Widom, J., Arasu, A., Babcock, B., Babu, S., Datar, M., & Varma, R. (2003). Query processing, approximation, and resource management in a data stream management system. In Proceedings of Innovative Data Systems Research, (pp. 245–256).

Mowshowitz, A., & Kawaguchi, A. (2002, September). Bias on the web. *Communications of the ACM*, *45*(9), 56–60. doi:10.1145/567498.567527

mTrigger (n.d.). *mTrigger Project*. Retrieved from http://www.cc.gatech.edu/projects.disl/mTriggers/

Muccini, H., Dias, M., & Richardson, D. J. (2005). Reasoning about software architecture-based regression testing through a case study. In *Proceedings of the 29th annual international computer software and applications conference* (Vol. 2, pp. 189-195). Washington, DC: IEEE Computer Society.

Mudge, T. (2001). Power: A First-Class Architectural Design Constraint. *Computer*, *34*(4), 52–58. doi:10.1109/2.917539

Mühl, G. (2002). *Large-scale content-based publish / subscribe systems*. Unpublished doctoral dissertation, Technische Universität Darmstadt, Darmstadt, Germany.

Mühl, G., Fiege, F., & Pietzuch, P. (2006). *Distributed Event-Based Systems*. Berlin, Germany: Springer-Verlag.

Mühl, G., Ulbrich, A., Herrmann, K., & Weis, T. (2004, May). Disseminating information to mobile clients using publish/subscribe. *IEEE Internet Computing, 8*(3), 46–53. doi:10.1109/MIC.2004.1297273

Muhl, G., Fiege, L., & Buchmann, A. (2002). Filter similarities in content-based publish/subscribe systems. In *Proceedings of ARCS*, (pp. 224–238).

Muhl, G., Fiege, L., & Pietzuch, P. (2006). *Distributed event-based systems*. Berlin, Germany: Springer.

Mukherjee, B., Heberlein, L. T., & Levitt, K. N. (1994). Network intrusion detection. *IEEE Network, 8*(3), 26–41. doi:10.1109/65.283931

Munthe-Kaas, E., Drugan, O., Goebel, V., Plagemann, T., Pužar, M., Sanderson, N., & Skjelsvik, K. S. (2006). Mobile Middleware for Rescue and Emergency Scenarios. In Bellavista, P., & Corradi, A. (Eds.), *Mobile Middleware*. Boca Raton, FL: CRC Press. doi:10.1201/9781420013153.ch48

Murphy, A., Picco, G., & Roman, G.-C. (2001). LIME: A Middleware for Physical and Logical Mobility. In *Proceedings of the The 21st International Conference on Distributed Computing Systems*, (pp. 524-536).

Murugappan, A., & Liu, L. (2008, June 30-July 2). An Energy Efficient Approach to Processing Spatial Alarms on Mobile Clients. In *Proceedings of the ISCA 17th International Conference on Software Engineering and Data Engineering* (SEDE-2008).

Musolesi, M., Hailes, S., & Mascolo, C. (2005). Adaptive Routing for Intermittently Connected Mobile Ad Hoc Network. In *WoWMom'05, Proceedings of the IEEE 6th International Symposium on a World of Wireless, Mobile, and Multimedia Networks*, Taormina, Italy.

Musolesi, M., Mascolo, C., & Hailes, S. (2005). EMMA: Epidemic Messaging Middleware for Ad hoc networks. *Personal and Ubiquitous Computing, 10*(1), 28–36. doi:10.1007/s00779-005-0037-4

Muthitacharoen, A., Morris, R., Gil, T. M., & Chen, B. (2002, Winter). Ivy: a read/write peer-to-peer file system. *SIGOPS Operating Systems Review, 36*(SI), 31–44.

Muthusamy, V., & Jacobsen, H.-A. (2005). Small-scale Peer-to-peer Publish/Subscribe. *Proceedings of the MobiQuitous Conference* (pp 109-119), New York: ACM.

Muthusamy, V., & Jacobsen, H.-A. (2007). *Infrastructure-less Content-Based Pub. (Technical report)*. Toronto, Canada: Middleware Systems Research Group, University of Toronto.

Muthusamy, V., & Jacobsen, H.-A. (2008). SLA-driven distributed application development. In *Proceedings of the 3rd Workshop on Middleare for Service Oriented Computing* (pp 31-36), Leuven, Belgium.

Muthusamy, V., Jacobsen, H.-A., Coulthard, P., Chan, A., Waterhouse, J., & Litani, E. (2007). SLA-Driven Business Process Management in SOA. In *Proceedings of the 2007 conference of the center for advanced studies on Collaborative research* (pp 264-267), Ontario, Canada.

Muthusamy, V., Petrovic, M., & Jacobsen, H. (2005). Effects of routing computations in content-based routing networks with mobile data sources. In *Proceedings of the 11th Annual International Conference on Mobile Computing and Networking 2005*, (pp. 103-116).

Naggie 2.0 (n.d.). *Naggie 2.0: Revolutionize Reminders with Location!* Retrieved from http://www.naggie.com/

Nagpal, R. (2002). Programmable self-assembly using biologically inspired multiagent control. In *Proceedings of AAMAS*, (pp. 418–425).

Nam, C.-S., Jeong, H.-J., & Shin, D.-R. (2008). Design and Implementation of the Publish/Subscribe Middleware for Wireless Sensor Networks. *International Conference Networked Computing and Advanced Information Management* (pp. 270-273). Gyeongju, South Korea: IEEE.

NASA Software Technology Branch. L. B. J. S. C. (1995, February). *C language integrated production system.* Retrieved from http://www.cs.cmu.edu/afs/cs/project/ai-repository/ai/areas/expert /systems/clips/0.html

Nayate, A., Dahlin, M., & Iyengar, A. (2004). Transparent information dissemination. In *Proceedings of the 5th ACM/IFIP/USENIX International Middleware Conference* (pp 212 - 231), Berling, Germany: Springer.

Ni, S. Y., Tseng, Y. C., Chen, Y. S., & Sheu, J. P. (1999). The broadcast storm problem in a mobile ad hoc network. *In MobiCom'99, Proceedings of the 5ᵗʰ Annual ACM/IEEE International Conference on Mobile Computing and Networking* (pp. 151-162), Seattle, WA.

Niebuhr, D., Klus, H., Anastasopoulos, M., Koch, J., Weiß, O., & Rausch, A. (2007). *DAiSI – Dynamic Adaptive System Infrastructure* (IESE-Report No. 051.07/E). Kaiserslautern, Germany: Fraunhofer Institut Experimentelles Software Engineering.

Notkin, D., Garlan, D., Griswold, W. G., & Sullivan, K. J. (1993). Adding implicit invocation to languages: Three approaches. In *Proceedings of the First JSSST International Symposium on Object Technologies for Advanced Software* (pp. 489-510). London: Springer-Verlag.

OASIS International Standards Consortium. (2005). *ebXML Registry Services and Protocols.* Retrieved August 22, 2008, from http://oasis-open.org/committees/regrep

OASIS International Standards Consortium. (2005). *Universal Description, Discovery and Integration (UDDI).* Retrieved August 22, 2008, from http://oasis-open.org/committees/uddi-spec/

OASIS International Standards Consortium. (2006). *Web Services Notification (WS-Notification).* Retrieved August 22, 2008, from http://www.oasis-open.org/committees/wsn

OASIS. (2004). *Oasis web services reliable messaging (WSRM) TC.* Retrieved from http://www.oasis-open.org/committees/wsrm/

OASIS. (2007). *Web Services Business Process Execution Language.* OASIS Standard Version 2.0 from http://docs.oasis-open.org/wsbpel/2.0/wsbpel-v2.0.pdf

Object Management Group (2002, December) *The Common Object Request Broker Architecture: Core Specification, Revision 3.0.* Needham, MA: OMG.

Object Management Group. (2007). *Common Object Request Broker Architecture (CORBA) Specification, Version 3.1; Part 2: CORBA Interoperability.* Needham, MA: OMG.

Object Management Group. (2007). *Data Distribution Service (DDS) for real-time systems, v1.2.* Needham, MA: OMG.

OhmyNews. (2008, October). *OhmyNews International.* Retrieved October, 2008 from http://english.ohmynews.com

Oki, B., Pfluegl, M., Siegel, A., & Skeen, D. (1993). The information bus: an architecture for extensible distributed systems. In *Proceedings of SOSP '93: The Fourteenth ACM Symposium on Operating Systems Principles* (pp. 58-68). New York: ACM Press.

OMG. O. M. G. (2007, February). *UML 2.1.1 superstructure specification.* Retrieved from http://www.omg.org/technology/documents/formal/uml.htm

Opyrchal, L., Astley, M., Auerbach, J., Banavar, G., Strom, R., & Sturman, D. (2000). Exploiting IP multicast in content-based publish-subscribe systems. *IFIP/ACM International Conference on Distributed systems platforms,* (pp 185-207), New York: Springer.

Opyrchal, L., Prakash, A., & Agrawal, A. (2007). Supporting privacy policies in a publish-subscribe substrate for pervasive environments. *JNW, 2*(1), 17–26.doi:10.4304/jnw.2.1.17-26

Oracle. (2005, June). *Oracle® streams advanced queuing user's guide and reference 10g Release 2 (10.2), Part number b14257-01.* Retrieved from http://download-east.oracle.com/docs/cd/B1930601/server.102/b14257/toc.htm

Oreizy, P., Medvidovic, N., & Taylor, R. N. (1998). Architecture-based runtime software evolution. In *ICSE '98: Proceedings of the 20th international conference on software engineering* (pp. 177-186). Washington, DC: IEEE Computer Society.

Organisation for Economic Co-operation and Development (OECD). (2008, October). *OECD Broadband Portal.* OECD Retrieved October, 2008 from http://www.oecd.org/sti/ict/broadband

Orso, A., Do, H., Rothermel, G., Harrold, M. J., & Rosenblum, D. S. (2007). Using component metadata to regression test component-based software. *Software Testing. Verification and Reliability, 17*(2), 61–94. doi:10.1002/stvr.344

OSGi. (2003). *OSGi Service Platform: The OSGi Alliance.* Amsterdam: IOS Press.

Osigma, T. (2007). *Evaluation of MIDAS Middleware in a Mobile Healthcare Scenario.* (Master thesis), Enschede, Netherlands: University of Twente.

Ostrowski, K., & Birman, K. (2006). Extensible Web Services Architecture for Notification in Large-Scale Systems. *Proceedings of the IEEE International Conference on Web Services* (pp 383-392), Washington, DC: IEEE Computer Society.

Ousterhout, J. (1996). Why Threads Are A Bad Idea (for most purposes), In *Proceedings of the Presentation given at the 1996 Usenix Annual Technical Conference.*

OWL. (2004). *W3C Recommendation: OWL Web Ontology Language Overview.* Retrieved June, 2008 from http://www.w3.org/TR/owl-features/

Özsu, M. T., & Valduriez, P. (1999). *Principles of Distributed Database Systems.* Upper Saddle River, NJ: Prentice Hall.

Pallickara, S., & Fox, G. (2003). Naradabrokering: A distributed middleware framework and architecture for enabling durable peer-to-peer grids. In *Proceedings of ACM/IFIP/USENIX International Middleware Conference Middleware-2003* (Vol. 2672, pp. 41-61). Berlin, Germany: Springer-Verlag.

Pan, Z. (2005). Benchmarking DL Reasoners Using Realistic Ontologies. In Proceedings of Intl workshop on OWL: Experience and Directions (OWL-ED2005). Galway, Ireland.

Papazoglou, M. P., Traverso, P., Dustdar, S., & Leymann, F. (2007). Service-Oriented Computing: State of the Art and Research Challenges. *IEEE Computer, 40*(11), 38–45.

Parsia, B., & Sirin, E. (2004). Pellet: An OWL-DL Reasoner. Poster at ISWC 2004, Hiroshima, Japan.

Patroumpas, K., & Sellis, T. K. (2006). Window Specification over Data Streams. In *Proceedings of the EDBT Workshops* (pp. 445-464).

Pei, G., Ravindran, B., & Jensen, E. D. (2008). On A Self-organizing MANET Event Routing Architecture with Causal Dependency Awareness. In *Proceedings of the 2nd IEEE International Conference on Self-Adaptive and Self-Organizing Systems*, (pp. 339-348).

Peleg, D. (2002). Low stretch spanning trees. In *Proceedings of 27th International Symposium of Mathematical Foundations of Computer Science*, (LNCS, pp. 68-80).

Pellet Performance. (2003). *Pellet Performance.* Retrieved from http://www.mindswap.org/2003/pellet/performance.shtml

Pereira, J., Fabret, F., Llirbat, F., & Shasha, D. (2000). Efficient Matching for Web-Based Publish/Subscribe Systems. In *CoopIS '02: Proceedings of the 7th International Conference on Cooperative Information Systems* (pp. 162-173). London: Springer.

Pesonen, L. I. W., & Bacon, J. (2005, September). Secure Event Types in Content-Based, Multi-domain Publish/Subscribe Systems. In *SEM '05: Proceedings of the 5th international workshop on Software Engineering and Middleware* (pp. 98-105). New York: ACM Press.

Pesonen, L. I. W., & Eyers, D. M. (2007, June). Encryption-Enforced Access Control in Dynamic Multi-Domain Publish/Subscribe Networks. In H.-A. Jacobsen, G. Mühl, & M. A. Jaeger (Eds.), *Proceedings of the inaugural conference on distributed event-based systems* (pp. 104–115). New York: ACM Press. Available from http://debs.msrg.utoronto.ca

Pesonen, L. I. W., Eyers, D. M., & Bacon, J. (2006, January). A capabilities-based access control architecture for multi-domain publish/subscribe systems. In *Proceedings of the symposium on applications and the internet (SAINT 2006)* (pp. 222–228). Phoenix, AZ: IEEE.

Petrovic, M. Burcea, I. & Jacobsen, H. A. (2003). S-ToPSS: Semantic Toronto Publish/subscribe Systems. In *Proceedings of CLDB* (pp. 1101-1104).

Petrovic, M., Burcea, I., & Jacobsen, H. A. (2003). S-ToPSS: semantic Toronto publish/subscribe system. In *Proceedings of the 29th international conference on Very large data bases* (VLDB03), Berlin, Germany.

Petrovic, M., Liu, H., & Jacobsen, H.-A. (2005, September). CMS-ToPSS: Efficient Dissemination of RSS Documents. In *Proceedings of 31st International Conference on Very Large Data Bases* (VLDB).

Petrovic, M., Muthusamy, V., & Jacobsen, H.-A. (2005). Content-based routing in mobile ad hoc networks. In *Proceedings of the Second Annual International Conference on Mobile and Ubiquitous Systems: Networking and Services* (pp 45-55), San Diego, CA.

Pew Internet & American Life Project. (2004, February). *Content creation online: 44% of U.S. Internet users have contributed their thoughts and their files to the online world.* Pew Internet. Retrieved from March, 2007 from http://www.pewinternet.org/PPF/r/113/report_display.asp

Picco, G. P., Cugola, G., & Murphy, A. L. (2003). Efficient content-based event dispatching in the presence of topological reconfiguration. In *Proceedings of the 23th International Conference on Distributed Computing Systems*, (pp. 234–243).

Picco, G. P., Cugola, G., & Murphy, A. L. (2003). Efficient content-based event dispatching in the presence of topological reconfiguration. In ICDCS 03. In *Proceedings of the 23rd International Conference on Distributed Computing Systems* (pp. 234-243).

Pietzuch, P. (2004). *Hermes: A Scalable Event-Based Middleware.* (PhD Thesis) Cambridge, UK: Computer Laboratory, Queens' College, University of Cambridge.

Pietzuch, P. R., & Bacon, J. (2002, July). Hermes: A distributed event-based middleware architecture. In *Proceedings of the 22nd international conference on distributed computing systems workshops (ICDCSW '02), Vienna, Austria* (pp. 611–618). Washington, DC: IEEE Computer Society.

Pietzuch, P. R., & Bacon, J. M. (2003, June). Peer-to-peer overlay broker networks in an event-based middleware. In *Proceedings of the 2nd international workshop on distributed event-based systems (DEBS'03).* New York: ACM SIGMOD.

Pietzuch, P. R., & Bhola, S. (2003, June). Congestion Control in a Reliable Scalable Message-Oriented Middleware. In M. Endler & D. Schmidt (Eds.), In *Proceedings of the 4th int. conf. on middleware (Middleware '03)* (pp. 202–221). Rio de Janeiro, Brazil: Springer.

Pietzuch, P. R., Shand, B., & Bacon, J. (2004). *Composite Event Detection as a Generic Middleware Extension. IEEE Network Magazine, Special Issue on Middleware Technologies for Future Communication Networks* (pp. 44–55). Washington, DC: IEEE Computer Society.

Pietzuch, P., & Bacon, J. (2003, June). Peer-to-Peer Overlay Broker Networks in an Event-Based Middleware. Distributed Event-Based Systems (DEBS'03). In *Proceedings of the ACM SIGMOD/PODS Conference*, San Diego, CA.

Pietzuch, P., Eyers, D., Kounev, S., & Shand, B. (2007, June). Towards a Common API for Publish/Subscribe. In H.-A. Jacobsen, G. Mühl, & M. A. Jaeger (Eds.), *Proceedings of the inaugural conference on distributed event-based systems* (pp. 152–157). New York: ACM Press. Retrieved from http://debs.msrg.utoronto.ca

Pietzuch, P., Ledlie, J., Shneidman, P., Roussopoulos, M., Welsh, M., & Seltzer, M. (2004). Network-Aware Operator Placement for Stream-Processing Systems. *International Conference on Data Engineering* (p. 49). Atlanta, GA: IEEE.

Pietzuch, P., Shand, B., & Bacon, J. (2004). Composite event detection as a generic middleware extension. *IEEE Network, 18*(1), 44–55. doi:10.1109/MNET.2004.1265833

Pitoura, T., Ntarmos, N., & Triantafillou, P. (2006, March). Replication, load balancing and efficient range query processing in DHTs. In Y. E. Ioannidis et al. (Eds.), *Proceedings of 10th international conference on extending database technology (EDBT06)* (pp. 131–148). Berlin, Germany: Springer.

Plagemann, T., Andersson, J., Drugan, O., Goebel, V., Griwodz, C., Halvorsen, P., et al. (2006). Middleware Services for Information Sharing in Mobile Ad-hoc Networks – Challenges and Approach. In *Proceedings of the Workshop on Challenges of Mobility, IFIP TC6 World Computer Congress*, Toulouse, France.

PlanetLab. (2006). *PlanetLab*. Retrieved from http://www.planet-lab.org/

Pleisch, S., & Birman, K. (2006). Senstrac: Scalable querying of sensor networks from mobile platforms using tracking-style queries. In *Proceedings of the 3rd IEEE International Conference on Mobile Ad-hoc and Sensor Systems*, (pp. 306-315).

Podnar, I., Hauswirth, M., & Jazayeri, M. (2002). Mobile Push: Delivering Content to Mobile Users. In *Proceedings of the International Workshop on Distributed Event-Based Systems* (ICDCS/DEBS'02).

Povinelli, R. J., & Feng, X. (2003). A New Temporal Pattern Identification Method for Characterization and Prediction of Complex Time Series Events. [TKDE]. *IEEE Transactions on Knowledge and Data Engineering, 15*(2), 339–352. doi:10.1109/TKDE.2003.1185838

Powell, D. (1996). Group communication. [New York: ACM.]. *Communications of the ACM, 39*(4), 50–53. doi:10.1145/227210.227225

Pyarali, I., Harrison, T., Schmidt, D. C., & Jordan, T. D. (1997). Proactor -- An Object Behavioral Pattern for Demultiplexing and Dispatching Handlers for Asynchronous Events. *Pattern Languages of Programming Conference*. Monticello, IL: Washington University.

Pyun, Y. J., & Reeves, D. S. (2004, August). Constructing a balanced, (log(N)/loglog(N))-diameter super-peer topology for scalable P2P systems. In 4th international conference on peer-to-peer computing (P2P 2004), Zurich, Switzerland (p. 210-218). IEEE Computer Society.

Rafaeli, S., & Hutchison, D. (2003). A survey of key management for group communication. *ACM Computing Surveys, 35*(3), 309–329. doi:10.1145/937503.937506

Raiciu, C., Rosenblum, D. S., & Handley, M. (2006). Revisiting Content-Based Publish/subscribe. In *ICDCSW'06, Proceedings of the 26th IEEE International Conference on Distributed Computing Systems Workshops*.

Ramasubramanian, V., & Sirer, E. G. (2004, March). Beehive: O(1) lookup performance for power-law query distributions in peer-to-peer overlays. In *Nsdi'04: Proceedings of the 1st conference on symposium on networked systems design and implementation* (pp. 99–112). Berkeley, CA: USENIX Association.

Ratnasamy, S. P. (2002). *A scalable content-addressable network*. Unpublished doctoral dissertation, Berkely, CA: University of California at Berkeley.

Ratnasamy, S., Francis, P., Handley, M., Karp, R., & Shenker, S. (2001, August). A scalable content-addressable network. In *Sigcomm '01: Proceedings of the 2001 conference on applications, technologies, architectures, and protocols for computer communications* (pp. 161–172). New York: ACM Press.

Ratnasamy, S., Francis, P., Handley, M., Karp, R., & Shenker, S. (2001). A scalable content-addressable network. In *Proceedings of the ACM SIGCOMM*, (pp. 161-172).

Ratnasamy, S., Handley, M., Karp, R. M., & Shenker, S. (2001, November). Application-level multicast using content-addressable networks. In *NGC '01: Proceedings of the third international COST264 workshop on networked group communication* (pp. 14–29). London: Springer-Verlag.

Real-Time Innovations, Inc. (2007). *RTI Data Distribution Service*. Retrieved August 23, 2008, from http://www.rti.com/products/data_distribution/RTIDDS.html

Research, I. B. M. (2001). *Gryphon: Publish/Subscribe over public networks*. Retrieved from http://researchweb.watson.ibm.com/grypohn/Gryphon/gryphon.html

Reynolds, P., & Vahdat, A. (2003, June). Efficient peer-to-peer keyword searching. In *Proceedings of acm/ifip/usenix international middleware conference (middleware 2003)* (pp. 21–40). Rio de Janeiro, Brazil: Springer.

Rhea, S. C., & Kubiatowicz, J. (2002, June). Probabilistic location and routing. In *Proceedings of INFOCOM 2002: Twenty-first annual joint conference of the IEEE computer and communications societies, New York, USA* (pp. 1248–1257).

Rheingold, H. (2000). *The virtual community: Homesteading on the electronic frontier* (Revised ed.). Cambridge, MA: MIT Press.

Rheingold, H. (2002). *Smart mobs: The next social revolution*. New York: Perseus Books Group.

Riabov, A., Liu, Z., Wolf, J. L., Yu, P. S., & Zhang, L. (2002). Clustering algorithms for content-based publication-subscription systems. In *Proceedings of the 22nd International Conference on Distributed Computing Systems* (pp 133-142), Washington, DC: IEEE Computer Society.

Riabov, A., Liu, Z., Wolf, J. L., Yu, P. S., & Zhang, L. (2003). New Algorithms for Content-Based Publication-Subscription Systems. In *Proceedings of the 23nd International Conference on Distributed Computing Systems* (pp 678-686), Washington, DC: IEEE Computer Society.

Roblek, D. (2006). Decentralized Discovery and Execution for Composite Semantic Web Services. (M.Sc. Thesis, Computer Science), Dublin Ireland: Trinity College.

Roitman, H., Carmel, D., & Yom-Tov, E. (2008). Maintaining dynamic channel profiles on the web. In *Proceedings of the 34th Conference on Very Large Data Bases* (VLDB 2008), Auckland, New-Zealand.

Romer, K., Kasten, O., & Mattern, F. (2002). Middleware Challenges for Wireless Sensor Networks. *ACM Mobile Computing and Communications Review, 6*(4), 59–61. doi:10.1145/643550.643556

Rose, I., Murty, R., Pietzuch, P., Ledlie, J., Roussopoulos, M., & Welsh, M. (2007). Cobra: Content-based Filtering and Aggregation of Blogs and RSS Feeds. In *Proceedings of the 4th USENIX Symposium on Networked Systems Design & Implementation* (pp 231-245), Cambridge, MA.

Rosenberg, F., Celikovic, P., Michlmayr, A., Leitner, P., & Dustdar, S. (2009). An End-to-End Approach for QoS-Aware Service Composition. In *Proceedings of the 13th IEEE International Enterprise Computing Conference (EDOC'09)*. Auckland, New Zealand: IEEE Computer Society.

Rosenberg, F., Leitner, P., Michlmayr, A., & Dustdar, S. (2008). Integrated Metadata Support for Web Service Runtimes. In *Proceedings of the Middleware for Web Services Workshop (MWS'08)* (S. 24-31). Munich, Germany: IEEE Computer Society.

Rosenberg, F., Platzer, C., & Dustdar, S. (2006). Bootstrapping Performance and Dependability Attributes of Web Services. In *Proceedings of the IEEE International Conference on Web Services (ICWS'06)*, (pp. 205-212). Chicago: IEEE Computer Society.

Rosenblum, D., & Wolf, A. (1997). A design framework for Internet-scale event observation and notification. *SIGSOFT Software Engineering Notes, 22*(6), 344–360. doi:10.1145/267896.267920

Rowstron, A., & Druschel, P. (2001, November). Pastry: Scalable, decentralized object location and routing for large-scale peer-to-peer systems. In Middleware '01, IFIP/ACM international conference on distributed systems platforms (pp. 329–350). Berlin / Heidelberg, Germany: Springer Verlag.

Rowstron, A., Kermarrec, A., Castro, M., & Druschel, P. (2001). SCRIBE: The design of a large-scale event notification infrastructure. In *Proceedings of the Networked Group Communication: Third International COST264 Workshop* (pp. 30-43, NGC 2001), London.

Royer, E. M., & Perkins, C. E. (2000). *Multicast Ad hoc On-Demand Distance Vector (MAODV) Routing (INTER-NET DRAFT)*. Mobile Ad Hoc Network Working Group.

Rozsnyai, S., Schiefer, J., & Schatten, A. (2007, June). Concepts and models for typing events for event based systems. In H.-A. Jacobsen, G. Mühl, & M. A. Jaeger (Eds.). In *Proceedings of the Inaugural Conference on Distributed Event-based Systems*. New York: ACM Press. Retrieved from http://debs.msrg.utoronto.ca

Rozsnyai, S., Vecera, R., Schiefer, J., & Schatten, A. (2007). Event Cloud - Searching for Correlated Business Events. In *Proceedings of the 9th IEEE International Conference on E-Commerce Technology and The 4th IEEE International Conference on Enterprise Computing, E-Commerce and E-Services (CEC-EEE'07)* (S. 409-420). Tokyo, Japan: IEEE Computer Society.

Ryu, M., Kim, J., & Maeng, J. C. (2006, June). Reentrant statecharts for concurrent real-time systems. In H. R. Arabnia (Ed.). In *Proceedings of the International Conference on Parallel and Distributed Processing Techniques and Applications & Conference on Real-time Computing Systems and Applications* (pp. 1007- 1013). Las Vegas, NV: CSREA Press.

Saif, U., & Greaves, D. J. (2001). Communication primitives for ubiquitous systems or RPC considered harmful. In *Proceedings of the International Conference on Distributed Computing Systems*, (pp. 240-245).

Samet, H. (1984, June). The quadtree and related hierarchical data structures. *ACM Computing Surveys, 16*(2), 187–260. doi:10.1145/356924.356930

Samet, H. (1990). *The Design and Analysis of Spatial Data Structures*. Reading, MA: Addison-Wesley.

Sanderson, N. C., Skjelsvik, K. S., Drugan, O., Pužar, M., Goebel, V., Munthe-Kaas, E., & Plagemann, T. (2007). *Developing Mobile Middleware – An Analysis of Rescue and Emergency Operations* (Tech. Rep. No. 358. ISBN 82-7368-317-8, ISSN 0806-3036). Oslo, Norway: Department of Informatics, Unversity of Oslo.

Sanderson, N., Goebel, V., & Munthe-Kaas, E. (2005). Metadata Management for Ad-Hoc InfoWare – A Rescue and Emergency Use Case for Mobile Ad-Hoc Scenarios. In R. Meersman and Z. Tari (Eds.), *ODBASE'05: International Conference on Ontologies, Databases and Applications of Semantics*. (LNCS 3761, pp. 1365-1380).

Sandhu, R. S., Ferraiolo, D. F., & Kuhn, R. (2000). The NIST model for role-based access control: towards a unified standard. In *Rbac '00: Proceedings of the fifth ACM workshop on role-based access control* (pp. 47–63). New York: ACM Press.

Sandhu, R., Coyne, E., Feinstein, H. L., & Youman, C. E. (1996). Role-based access control models. *IEEE Computer, 29*(2), 38–47.

Sarkar, S. K., Basavaraju, T., & Puttamadappa, C. (2007). *Ad hoc mobile wireless networks: Principles, protocols and applications*. Boca Raton, FL: Auerbach Publications. doi:10.1201/9781420062229

Sarshar, N., Boykin, P. O., & Roychowdhury, V. P. (2004, August). Percolation search in power law networks: Making unstructured peer-to-peer networks scalable. In *4th international conference on peer-to-peer computing* (p. 2P). Zurich, Switzerland: IEEE Computer Society.

Sayre, R. (2005). Atom: The Standard in Syndication. *IEEE Internet Computing, 9*(4), 71–78. doi:10.1109/MIC.2005.74

Schieferdecker, I. (2007, February). *The Testing and Test Control Notation version 3*; Core language (Tech. Rep). Sophia-Antipolis Cedex, France: European Telecommunications Standards Institute.

Schlosser, M., Sintek, M., Decker, S., & Nejdl, W. (2002). HyperCuP — Hypercubes, Ontologies and Efficient Search on P2P Networks. In *Proceedings on the International Workshop on Agents and Peer-to-Peer-Systems*. Bologna, Italy: Springer.

Schmidt, D. C., Stal, M., & Rohnert, H. (2000). *Pattern-Oriented Software Architecture - Patterns for Concurrent and Networked Objects* (*Vol. 2*). Reading, MA: Addison-Wesley.

Schneider, M. (1993, March). Self-stabilization. [CSUR]. *ACM Computing Surveys, 25*(1), 45–67. doi:10.1145/151254.151256

Schönherr, J. H., Parzyjegla, H., & Mühl, G. (2008). Clustered Publish/Subscribe in Wireless Actuator and Sensor Networks. *International Workshop on Middleware for Pervasive and Ad-Hoc Computing* (pp. 60-65). Leuven, Belgium: ACM.

Schott, W., Gluhak, A., Presser, M., Hunkeler, U., & Tafazolli, R. (2007). e-SENSE Protocol Stack Architecture for Wireless Sensor Networks. Mobile and Wireless Communications Summit (pp. 1-5). Budapest, Hungary: IST.

Schuler, C., Schuldt, H., & Schek, H. J. (2001). Supporting Reliable Transactional Business Processes by publish/subscribe Techniques. In *Proceedings of the Second International Workshop on Technologies for E-Services* (pp 118-131), London, UK: Springer-Verlag.

Schulzrinne, H., & Wedlund, E. (2000). Application-layer mobility using SIP. *SIGMOBILE Mob. Comput. Commun. Rev., 4*(3), 47–57. doi:10.1145/372346.372369

Schwiderski, S. (1996). *Monitoring the Behavior of Distributed Systems*. (PhD thesis), Cambridge, UK: University of Cambridge.

Segall, B., & Arnold, D. (1997, September). Elvin has left the building: A publish/subscribe notification service with quenching. In *Proceedings of Australian UNIX and open systems user group conference (AUUG 97), Brisbane, Australia.*

Segall, B., & Arnold, D. (1998). Elvin has left the building: A publish/subscribe notification service with quenching. In *Proceedings of AUUG.* Retrieved from http://www.dtsc.edu.au/

Segall, B., Arnold, D., Boot, J., Henderson, M., & Phelps, T. (2000). Content-Based Routing in Elvin4. In Proceedings of AUUG2K, Canberra, Australia.

Shah, R. Roy, S., Jain, S., & Brunette, W. (2003). Data MULEs: Modeling a Three-tier Architcture for Sparse Sensor Networks. In *IEEE SNPA'03, Proceedings of the 1ˢᵗ IEEE International Workshop on Sensor Network Protocols and Applications* (pp. 30-41).

Sharon, G., & Etzion, O. (2008). Event Processing Networks – model and implementation. *IBM Systems Journal, 47*(2), 321–334. doi:10.1147/sj.472.0321

Shavit, N., & Touitou, D. (1995). Software Transactional Memory. In *Proceedings of the Symposium on Principles of Distributed Computing,* (pp. 204-213).

Shaw, R., & Larson, R. R. (2008). Event Representation in Temporal and Geographic Context. In *Proceedings of ECDL* (pp. 415-418).

Sherafat Kazemzadeh, R., & Jacobsen, H.-A. (2007). *Fault-Tolerant Publish/Subscribe systems. (Technical report).* Toronto, Canada: Middleware Systems Research Group, University of Toronto.

Sherafat Kazemzadeh, R., & Jacobsen, H.-A. (2008). *Highly Available Distributed Publish/Subscribe Systems. (Technical report).* Toronto, Canada: Middleware Systems Research Group, University of Toronto.

Shevade, B., Sundaram, H., & Xie, L. (2007). Exploiting Personal and Social Network Context for Event Annotation. In *Proceedings of ICME* (pp. 835-838)

Shmueli, G., & Fienberg, S. (2006). Current and potential statistical methods for monitoring multiple data streams for biosurveillance. In *Statistical Methods in Counterterrorism* (pp. 109–140). Berlin, Germany: Springer Verlag. doi:10.1007/0-387-35209-0_8

Shnayder, V., Hempstead, M., Chen, B.-r., Allen, G., & Welsh, M. (2004). Simulating the Power Consumption of Large-Scale Sensor Network Applications. *International Conference on Embedded Networked Sensor Systems* (pp. 188-200). Baltimore, MD: ACM.

Shu, L., Wang, J., Xu, H., Jinsung, C., & Sungyoung, L. (2006). Connecting Sensor Networks with TCP/IP Network. *International Workshop on Sensor Networks* (pp. 330-334). Harbin, China: Springer.

Simon, D., & Cifuentes, C. (2005). The Squawk Virtual Machine: Java™ on the Bare Metal. *Conference on Object Oriented Programming Systems Languages and Applications* (pp. 150-151). San Diego, CA: ACM.

Singh, J., Bacon, J., & Moody, K. (2007, April). Dynamic trust domains for secure, private, technology-assisted living. In *Proceedings of the the second international conference on availability, reliability and security (ARES'07)* (pp. 27-34). Vienna, Austria: IEEE Computer Society.

Singh, J., Eyers, D. M., & Bacon, J. (2008). Decemberin press). Credential management in event-driven healthcare systems. In *Middleware*. Leuven, Belgium: Springer Verlag.

Singh, J., Vargas, L., & Bacon, J. (2008, January). A Model for Controlling Data Flow in Distributed Healthcare Environments. In Pervasive Health 2008: Second international conference on pervasive computing technologies for healthcare. Tampere, Finland: IEEE Press.

Singh, J., Vargas, L., Bacon, J., & Moody, K. (2008, June). Policy-based information sharing in publish/subscribe middleware. In *IEEE workshop on policies for distributed systems and networks (Policy 2008). IBM Palisades*. New York: IEEE Press.

Sivaharan, T., Blair, G., & Coulson, G. (2005). *GREEN: A Configurable and Re-configurable Publish-Subscribe Middleware for Pervasive Computing. On the Move to Meaningful Internet Systems* (pp. 732–749). Agia Napa, Cyprus: Springer.

Skjelsvik, K. S. (2008). *A Distributed Event Notification Service for Sparse Mobile Ad-hoc Networks*. (Ph.d thesis), Oslo, Norway: University of Oslo.

Skjelsvik, K. S., Lekova, A., Goebel, V., Munthe-Kaas, E., Plagemann, T., & Sanderson, N. (2006). Supporting Multiple Subscriptions Languages by a single Event Notification Overlay in Sparse MANETs. In *Proceedings of ACM MobiDE*.

Skjelsvik, K. S., Søberg, J., Goebel, V., & Plagemann, T. (2007). Using Continuous Queries for Event Filtering and Routing in Sparse MANETs. In *FTDCS'07, Proceedings of the 11th IEEE International Workshop on Future Trends of Distributed Computing Systems* (pp. 138-148).

Skovronski, J., & Chiu, K. (2006, December 4-6). Ontology Based Publish Subscribe Framework. In *Proceedings of International Conference on Information Integration and Web-based Applications Services*. Yogyakarta, Indonesia.

Snoeren, A. C., Conley, K., & Gifford, D. K. (2001). Mesh-based content routing using XML. *ACM SIGOPS Operating Systems Review.*, *35*(5), 160–173. doi:10.1145/502059.502050

Snoeren, A., Conley, K., & Gifford, D. (2001). Mesh based content routing using XML. In *Proc. ACM SOSP*, (pp. 160–173).

Sohn, T., Li, K., Lee, G., Smith, I., Scott, J., & Griswold, W. (2005). A Study of Location-Based Reminders on Mobile Phones. In *Proceedings of UbiComp*. Place-Its.

Songlin, H., Muthusamy, V., Li, G., & Jacobsen, H.-A. (2008). Distributed Automatic Service Composition in Large-Scale Systems. In *Proceedings of the 2nd International Conference on Distributed Event-Based Systems* (DEBS 2008), Rome, Italy.

Souto, E., Guimarães, G., Vasconcelos, G., Vieira, M., Nelson, R., & Ferraz, C. (2005). Mires: A Publish/Subscribe Middleware for Sensor Networks. *Springer Personal Ubiquitous Computing*, *10*(1), 37–44. doi:10.1007/s00779-005-0038-3

Spatial Alarms Project. (n.d.). *Spatial Alarms Project*. Retrieved from http://www.cc.gatech.edu/projects/disl/SpatialAlarms/

Spatial Data. (n.d.). *Spatial Data Transfer Format*. Retrieved from http://www.mcmcweb.er.usgs.gov/sdts/

Spiess, P., Vogt, H., & Jütting, J. (2006). *Integrating Sensor Networks With Business Processes. Real-World Sensor Networks Workshop*. Uppsala, Sweden: ACM.

Spring, J. H., Privat, J., Guerraoui, R., & Vitek, J. (2007). Streamflex: high-throughput stream programming in java. In *OOPSLA '07: Proceedings of the 22nd annual ACM SIGPLAN conference on Object oriented programming systems and applications* (pp. 211-228). New York: ACM Press.

Spyropoulos, T., Psounis, K., & Raghavendra, C. S. (2005). Spray and Wait: An Efficient Routing Scheme for Intermittently Connected Mobile Networks. In *SIGCOMM'05, Proceedings of ACM SIGCOMM workshop on Delay Tolerant Networking* (pp. 252-259).

Sterling, T., Becker, D. J., Savarese, D., Dorband, J. E., Ranawake, U. A., & Packer, C. V. (1995). BEOWULF: A parallel workstation for scientific computation. In D. P. Agrawal (Ed.), *Proceedings of the 24th International Conference on Parallel Processing.* London: CRC Press.

Stoica, I., Morris, R., Karger, D., Kaashoek, M. F., & Balakrishnan, H. (2001, August). Chord: A scalable peer-to-peer lookup service for Internet applications. In *Proceedings of the 2001 conference on applications, technologies, architectures, and protocols for computer communications* (pp. 149–160). New York: ACM Press.

Stoica, I., Morris, R., Liben-Nowell, D., Karger, D.R., Kaashoek, M.F., Dabek, F., & Balakrishnan, H. (2003). Chord: a scalable peer-to-peer lookup protocol for internet applications. *IEEE/ACM Transactions on Networking, 11*(1), 17-32.

Stonebraker, M., Cetintemel, U., & Zdonik, S. (2005). The 8 requirements of real-time stream processing. *SIGMOD Record, 34*(4), 42–47. doi:10.1145/1107499.1107504

Stuckenholz, A., & Zwintzscher, O. (2004). Compatible component upgrades through smart component swapping. In Reussner, R. H., Stafford, J. A., & Szyperski, C. A. (Eds.), *Architecting systems with trustworthy components* (*Vol. 3938*, pp. 216–226). Berlin, Germany: Springer. doi:10.1007/11786160_12

Suliman, D., Paech, B., Borner, L., Atkinson, C., Brenner, D., Merdes, M., et al. (2006, September). The MORABIT approach to runtime component testing. In *Proceedings of the 30th annual international computer software and applications conference* (Vol. 2, pp. 171-176).

Sun Microsystems. (1999). *Dynamic Proxy Classes.* Retrieved August 25, 2008, from http://java.sun.com/j2se/1.3/docs/guide/reflection/proxy.html

Sun Microsystems. (2008, February 14th 2008). *Concurrency in Swing.* Retrieved August 12th 2008, from http://java.sun.com/docs/books/tutorial/uiswing/concurrency

SUN-JINI. (2003). Jini's distributed events specification, version 1.0. Retrieved from http://java.sun.com/products/jini/2.1/doc/specs/html/event-spec.html

SUN-JMS. (2002). Java message service (JMS) specification, version 1.1. Retrieved from http://java.sun.com/products/jms/docs.html

Szyperski, C. (1998). *Component software: Beyond object-oriented programming.* New York: ACM Press.

Tai, W. (2007, December). Fault Management System using Semantic Publish/Subscribe approach. (M.Sc. Thesis, Computer Science). Dublin Ireland: Trinity College.

Tai, W., O'Sullivan, D., & Keeney, J. (2008, April). Distributed Fault Correlation Scheme using a Semantic Publish/Subscribe system. In *Proceedings of Network Operations and Management Symposium* (NOMS 2008), Salvador, Brazil.

Tanin, E., Harwood, A., & Samet, H. (2005, April). A distributed quadtree index for peer-to-peer settings. In *Proceedings of the 21st international conference on data engineering (ICDE'05), Tokyo, Japan* (pp. 254–255). Washington, DC: IEEE Computer Society.

Tarkoma, S. (2008). Dynamic filter merging and mergeability detection for publish/subscribe. *Pervasive and Mobile Computing, 4*(5), 681–696. doi:10.1016/j.pmcj.2008.04.007

Tarkoma, S., & Kangasharju, J. (2006). On the cost and safety of handoffs in content-based routing systems. *Computer Networks, 51*(6), 1459–1482. doi:10.1016/j.comnet.2006.07.016

Tarkoma, S., Heikkinen, J., & Pohja, M. (2007). Secure push for mobile airline services. *Telecommunication Systems, 35*(3-4), 177–187. doi:10.1007/s11235-007-9048-y

Technorati Inc. (2008, October). *State of the blogosphere / 2008.* Technorati. Retrieved August, 2008 from http://www.technorati.com/blogging/state-of-the-blogosphere/

Tempich, C., Staab, S., & Wranik, A. (2004). REMIN-DIN': semantic query routing in peer-to-peer networks based on social metaphors. In *Proceedings of the International World Wide Web Conference* (WWW), New York.

Thales Group. (2007). *Maritime safety and security.* Retrieved from http://shield.thalesgroup.com/offering/port_maritime.php

Tock, Y., Naaman, N., Harpaz, A., & Gershinsky, G. (2005). Hierarchical Clustering of Message Flows in a Multicast Data Dissemination System. In *Proceedings of the* 17th *IASTED International Conference on Parallel and Distributed Computing and Systems* (pp 320-326), Calgary, Alberta: ACTA Press.

Tomasic, A., Garrod, C., & Popendorf, K. (2006). *Symmetric publish/subscribe via constraint publication* (Tech. Rep. No. CMU-CS-06-129R). Pittsburgh, PA: Carnegie Mellon University.

Tosi, D. (2003). An Advanced Architecture for Push Services. In *Proceedings of the Fourth International Conference on Web Information Systems Engineering Workshops*, (pp 193–200).

Treiber, M., & Dustdar, S. (2007). Active Web Service Registries. *IEEE Internet Computing, 11*(5), 66–71. doi:10.1109/MIC.2007.99

Triantafillou, P., & Economides, A. Subscription Summarization: A New Paradigm for Efficient publish/subscribe Systems. In *Proceedings of the 24nd International Conference on Distributed Computing Systems,* (pp 562-571), Washington, DC: IEEE Computer Society.

Tryfonopoulos, C., Zimmer, C., Weikum, G., & Koubarakis, M. (2007). Architectural Alternatives for Information Filtering in Structured Overlays. *IEEE Internet Computing, 11*(4), 24–34. doi:10.1109/MIC.2007.79

Tsai, W.-C., & Chen, A.-P. (2008). Service oriented architecture for financial customer relationship management. In *Proceedings of the 2nd international conference on Distributed event-based systems*, (pp. 301-304).

Twitter. (2008, December). *Twitter.* Retrieved December, 2008 from http://twitter.com

USGS. (n.d.). *U.S. Geological Survey.* Retrieved from http://www.usgs.gov

Vahdat, A., & Becker, D. (2000). *Epidemic Routing for Partially Connected Ad Hoc Networks* (Tech. Rep. No. CS-2000-06). Durham, NC: Department of Computer Science, Duke University.

Van Cutsem, T. (2008). *Ambient References: Object Designation in Mobile Ad Hoc Networks.* Brussels, Belgium: Programming Technology Lab, Faculty of Sciences, Vrije Universiteit Brussel.

Van Cutsem, T., Mostinckx, S., Gonzalez Boix, E., Dedecker, J., & De Meuter, W. (2007). AmbientTalk: object-oriented event-driven programming in Mobile Ad hoc Networks. In *Proceedings of the XXVI International Conference of the Chilean Computer Science Society (SCCC 2007)*, (pp. 3-12).

Vargas, L., Bacon, J., & Moody, K. (2008). Event-Driven Database Information Sharing. In *British national conference on databases (BNCOD)* (*Vol. 5071*, pp. 113–125). Cardiff, UK: Springer.

Veanes, M., Campbell, C., Schulte, W., & Kohli, P. (2005, January). *On-the-fly testing of reactive systems* (Tech. Rep. No. MSR-TR-2005-05). Microsoft Research.

Vincent, J., King, G., Lay, P., & Kinghorn, J. (2002). Principles of Built-In-Test for Run-Time-Testability in component-based software systems. *Software Quality Journal, 10*(2), 115–133. doi:10.1023/A:1020571806877

Vinoski, S. (1997). CORBA: Integrating Diverse Applications within Distributed Heterogeneous Environments. *Communications Magazine*, 46-55.

Vinoski, S. (2003). Service Discovery 101. *Internet Computing*, 69-71.

Völgyesi, P., Maróti, M., Dóra, S., & Osses, E., & Lédeczi Ákos. (2005). Software composition and verification for sensor networks. *Science of Computer Programming, 56*(1-2), 191–210.

von der Beeck, M. (1994). A comparison of statecharts variants. In *Proceedings of PROCOS: The Third International Symposium, organized jointly with the Working Group Provably Correct Systems on Formal Techniques in Real-time and Fault-tolerant Systems* (pp. 128-148). London, UK: Springer-Verlag.

W3C. (2004). *W3C xml schema*. Retrieved from http://www.w3.org/XML/Schema

W3C: The Wine Ontology (2003). *The Wine Ontology*. Retrieved from http://www.w3.org/TR/owl-guide/wine.rdf

Walzer, K., Breddin, T., & Groch, M. (2008). Relative temporal constraints in the Rete algorithm for complex event detection. In *Proceedings of DEBS* (pp. 147-155).

Wang, C., Alqaralleh, B. A., Zhou, B. B., Brites, F., & Zomaya, A. Y. (2006, September). Self-organizing content distribution in a data indexed DHT network. In *P2P '06: Proceedings of the sixth IEEE international conference on peer-to-peer computing* (pp. 241–248). Washington, DC: IEEE Computer Society.

Wang, J., Jin, B., & Li, J. (2004). An ontology-based publish/subscribe system. In *Proceedings of ACM/IFIP/USENIX International Conference on Middleware*.

Wang, M.-M., Cao, J.-N., Li, J., & Dasi, S. K. (2008). Middleware for Wireless Sensor Networks: A Survey. *Springer Journal of Computer Science, 23*(3), 305–326.

Wang, T., & Liu, L. (2009, September). Privacy-Aware Mobile Services over Road Networks. In *Proceedings of the 35th International Conference on Very Large Data Bases*. PVLDB 2(1): 1042-1053 (2009).

Wasserkrug, S., Gal, A., Etzion, O., & Turchin, Y. (2008). Complex event processing over uncertain data. In *Proceedings of DEBS* (pp. 253-264)

Waterhouse, S. (2001, May). *JXTA search: Distributed search for distributed networks* (Tech. Rep.). Palo Alto, CA: Sun Microsystems, Inc. Retrieved January, 2008 from http://gnunet.org/papers/JXTAsearch.pdf

Weerawarana, S., Curbera, F., Leymann, F., Storey, T., & Ferguson, D. F. (2005). *Web Services Platform Architecture: SOAP, WSDL, WS-Policy, WS-Addressing WS-BPEL, WS-Reliable Messaging, and More*. Upper Saddle River, NJ: Prentice Hall.

Weiser, M. (1991). *The Computer for the 21*[st] *Century*. Scientific American.

Weiser, M. (1991). The computer for the twenty-first century. *Scientific American*, 94–104. doi:10.1038/scientificamerican0991-94

Weiser, M. (1993). Some computer science issues in ubiquitous computing. *Communications of the ACM, 36*(7), 75–84.

Weisstein, E. W. (2002). *Multiset. MathWorld – Wolfram Resource*. Retrieved from http://mathworld.wolfram.com/Multiset.html

Weitzner, D. J., Abelson, H., Berners-Lee, T., Feigenbaum, J., Hendler, J., & Sussman, G. J. (2008). Information accountability. *Communications of the ACM, 51*(6), 82–87. doi:10.1145/1349026.1349043

Welsh, M., Culler, D. E., & Brewer, E. A. (2001). Seda: An architecture for well-conditioned, scalable internet services. In *Proceedings of the 18th ACM Symposium on Operating Systems Principles (SOSP '01)*, (pp. 230-243). New York: ACM Press.

Welsh, M., Culler, D., & Brewer, E. (2001). SEDA: An Architecture for Well-Conditioned, Scalable Internet Services. *Symposium on Operating Systems Principles* (pp. 230-243). Banff, Canada: ACM.

Widom, J., & Ceri, S. (1996). *Active Database Systems: Triggers and Rules For Advanced Database Processing*. San Francisco: Morgan Kaufmann Publishers.

Wieringa, R. J. (2002). *Design methods for software systems: Yourdon, Statemate and UML*. San Francisco: Morgan Kaufmann Publishers.

Wong, T., Katz, R. H., & McCanne, S. (2000). An evaluation of preference clustering in large-scale multicast applications. In *Proceedings of the conference on computer communications* (pp 451-460), Washington, DC: IEEE Computer Society.

Workman, S., Parr, G., Morrow, P. J., & Charles, D. (2005). Relevance-Based Adaptive Event Communication for Mobile Environments with Variable QoS Capabilities. In *Proceedings of MMNS* (pp. 59-70).

World Wide Web Consortium. (2006). *Web Services Eventing (WS-Eventing).* Retrieved August 22, 2008, from http://www.w3.org/Submission/WS-Eventing/

Wun, A., & Jacobsen, H.-A. (2007). A policy management framework for content-based publish/subscribe. In *Middleware '07* (pp. 368–388). Newport Beach, CA: Springer.

Xue, T., Feng, B., & Zhang, Z. (2004, October). P2PENS: Content-based publish-subscribe over peer-to-peer network. In H. Jin, Y. Pan, N. Xiao, & J. Sun (Eds.), *Proceedings of third international conference on grid and cooperative computing (GCC 2004), Wuhan, China* (pp. 583–590). Berlin, Germany: Springer-Verlag.

Yahoo. Inc. (2008, October). *Flickr.* Retrieved October, 2008 from http://flickr.com

Yan, W., Hu, S., Muthusamy, V., Jacobsen, H.-A., & Zha, L. (2009). Efficient event-based resource discovery. In *Proceedings of the 2009 inaugural international conference on Distributed event-based systems,* Nashville, TN.

Yang, B., & Garcia-Molina, H. (2002, July). Efficient search in peer-to-peer networks. In *Proceedings of the 22nd international conference on distributed computing systems.* Vienna, Austria.

Yoneki, E. (2006). *ECCO: Data Centric Asynchronous Communication.* (PhD thesis), (Technical Report UCAM-CL-TR677), Cambridge, UK: University of Cambridge.

Yoneki, E., & Bacon, J. (2005). Content-Based Routing with On-Demand Multicast. In *ICDCSW'04, Proceedings of the 24th International Conference on Distributed Computing Systems Workshops.*

Yoneki, E., & Bacon, J. (2005). Unified Semantics for Event Correlation over Time and Space in Hybrid Network Environments. *International Conference on Cooperative Information Systems* (pp. 366-384). Agia Napa, Cyprus: IFIP.

Yoneki, E., & Bacon, J. (2007). ecube: Hypercube event for efficient filtering in content-based routing. In *Proceedings of the International Conference on Grid computing, High-performance and Distributed Applications,* (LNCS 4804, pp. 1244–1263).

Yonezawa, A., Briot, J.-P., & Shibayama, E. (1986). Object-oriented concurrent programming in ABCL/1. In *Conference Proceedings on Object-oriented programming systems, languages and applications,* (pp. 258-268).

Yu, X., Pu, K., & Koudas, N. (2005). Monitoring k-Nearest Neighbor Queries over Moving Objects. In *Proceedings of ICDE,* (pages 631–642)

Yuan, X., & Memon, A. M. (2008, August). Alternating GUI test generation and execution. In Testing: Academic and industry conference - practice and research techniques (taic part'08) (pp. 23{32). Windsor, UK: IEEE Computer Society.

Zeng, L., Benatallah, B., Ngu, A. H., Dumas, M., Kalagnanam, J., & Chang, H. (2004). QoS-Aware Middleware for Web Services Composition. *IEEE Transactions on Software Engineering, 30*(5), 311–327. doi:10.1109/TSE.2004.11

Zhang, H., Bradbury, J. S., Cordy, J. R., & Dingel, J. (2006). Using source transformation to test and model check implicit-invocation systems. Science of Computer Programming, *62*(3), 209 - 227. (Special issue on Source code analysis and manipulation (SCAM 2005))

Zhang, J., Zhang, G., & Liu, L. (2007, September). GeoGrid: A Scalable Location Service Network. In *Proceedings of the 27th IEEE International Conference on Distributed Computing Systems.*

Zhang, Y., & Weiss, M. (2003, Fall). Virtual communities and team formation. *Crossroads, 10*(1), 5. doi:10.1145/973381.973386

Zhang, Z. (2006). Routing in Intermittently Connected Mobile Ad Hoc Networks and Delay Tolerant Networks: Overview and Challenges. *IEEE Communications Surveys & Tutorials*, 8(1), 24–37. doi:10.1109/COMST.2006.323440

Zhao, B. Y., Huang, L., Stribling, J., Rhea, S. C., Joseph, A. D., & Kubiatowicz, J. (2004, January). Tapestry: A resilient global-scale overlay for service deployment. *IEEE Journal on Selected Areas in Communications*, 22(1), 41–53. doi:10.1109/JSAC.2003.818784

Zhao, W., Ammar, M., & Zegura, E. (2004). A Message Ferrying Approach for Data Delivery in Sparse Mobile Ad Hoc Networks. In *MobiHoc'04, Proceedings of the 5th ACM International Symposium on Mobile ad hoc networking and computing* (pp. 187-198).

Zhao, Y., Sturman, D., & Bhola, S. (2004). Subscription propagation in highly available publish/subscribe middleware. In *Middleware '04: Proceedings of the 5th ACM/IFIP/USENIX International Conference on Middleware* (pp. 274-293). New York: Springer-Verlag New York, Inc.

Zhong, M., Moore, J., Shen, K., & Murphy, A. L. (2005, June). An evaluation and comparison of current peer-to-peer full-text keyword search techniques. In *Proceedings of the 8th International Workshop on the Web & Databases* (WebDB2005), ACM SIGMOD/PODS 2005 Conference. Baltimore, MD.

Zhou, D., Schwan, K., Eisenhauer, G., & Chan, Y. (2001). JECho – Interactive High Performance Computing with Java Event Channels. In *IPDPS'01, Proceedings of the International Parallel and Distributed Processing Symposium.*

Zhu, Y., & Hu, Y. (2007). Ferry: A P2P-Based Architecture for Content-Based publish/subscribe Services. *IEEE Transactions on Parallel and Distributed Systems*, 18(5), 672–685. doi:10.1109/TPDS.2007.1012

Zhuang, S. Q., Zhao, B. Y., Joseph, A. D., Katz, R. H., & Kubiatowicz, J. D. (2001, June). Bayeux: An architecture for scalable and fault-tolerant wide-area data dissemination. In *NOSSDAV '01: Proceedings of the 11th international workshop on network and operating systems support for digital audio and video* (pp. 11–20). New York: ACM Press.

Zimmer, D., & Unland, R. (1999). On the semantics of complex events in active database management systems. In *Proceedings of ICDE*, (pp. 392-399).

Zou, Y., Finin, T., & Chen, H. (2004, April). F-OWL: an Inference Engine for the Semantic Web. In *Proceedings of Workshop on Formal Approaches to Agent Based Systems*, (LNCS 3228).

Zuehlke, D. (2008). SmartFactory – from Vision to Reality in Factory Technologies. In *Proceedings of the 17th World Congress, The International Federation of Automatic Control*. Seoul, Korea.

About the Contributors

Annika Hinze is a Senior Lecturer (Associate Professor) at the University of Waikato, New Zealand, where she heads the research group on Information Systems and Databases. She received her Master's from the Technical University Berlin in the area of Technical Mathematics. Annika received her Ph.D. degree in Computer Science from the Freie Universität Berlin in 2003. In her PhD thesis, she proposed an adaptive event-based system. Annika's current research focuses on context-aware systems, event-based systems, interaction design and mobility. She combines methods from Information Systems with Formal Methods and Human Computer Interaction. During 2009, while this book was prepared, Annika was a visiting Professor at the Humboldt University Berlin for the area of context-aware systems. Annika was a co-chair for ODBASE and APCCM, and served in various roles for the event community's central workshop/conference DEBS.

Alejandro Buchmann is Professor in the Department of Computer Science of Technische Universität Darmstadt since 1991 and is responsible for the area of Databases and Distributed Systems. He studied chemical engineering at the Universidad Nacional Autónoma de Mexico and received his PhD from the University of Texas, Austin, in 1980. He was an Assistant/Associate Professor from 1980 to 1986 at IIMAS/UNAM and held positions as a senior researcher at Computer Corporation of America, Cambridge Mass. (86-89) and GTE Laboratories, Waltham Mass. (89-91) before joining TUD. He was responsible for the graduate research program in enabling technologies for e-commerce at TUD (1998-2007) and initiated a research program in Cooperative, Adaptive and Responsive Monitoring in Mixed Mode Environments (2006-). Alejandro's current research interests are in the areas of event-based and reactive systems, heterogeneous distributed systems, middleware, performance evaluation, peer-to-peer systems and new paradigms for data management and information processing in Cyberphysical systems. Alejandro's involvement with event based systems dates back to the late 1980's in the HiPAC project on active databases. Subsequent work at TU Darmstadt included distributed event based systems, temporal uncertainty in distributed event systems, quality of service and transactional properties, content based publish/subscribe and concept based pub/sub for heterogeneous event systems, as well as performance modeling and benchmarking of event based systems. Alex has been involved as general and/or program chair in ICDE 01 and ICDE08, VLDB 96, SIGMOD 98, DEBS 08 and AmI 07, and is on the editorial board of several journals. He has held visiting positions at UT Austin, ICSI Berkeley, HP Labs Palo Alto, University of Virginia, Ecole Politechnique Federal de Lausanne (EPFL), IIT Bombay, and Caltech.

* * *

Paulo Alencar is currently a research associate professor with the Computer Systems Group at the University of Waterloo. His research, teaching, and consulting activities have been directed to software engineering in general and his current research interests specifically include software design, architecture, evolution, software processes, and formal methods. He has been the principal or co-principal investigator in projects supported by NSERC (individual and strategic), IBM, CSER, Bell, CITO, Precarn, Sybase, Siebel, i-Anywhere Solutions, and funding agencies in Germany, Argentina, and Brazil. He has been a visiting professor at the University of Karlsruhe, at the Imperial College of Science and Technology (London, UK), and at the University of Waterloo. He has also held faculty positions at the University of Brasilia, Brazil. He has published more than 120 technical papers. He is a member of the IEEE, the IEEE Computer Society, and the ACM.

Andé Appel holds a Diploma in computer science and is a PhD student at at the Software Systems Engineering chair at Clausthal University of Technology. His main research interests are means to provide an application state transfer among components in dynamic adaptive systems. He worked together with H. Klus and D. Niebuhr in several research projects with industrial partners.

Jean Bacon is Professor of Distributed Systems at the University of Cambridge Computer Laboratory, UK. She leads the Opera research group, with focus on middleware for large-scale, widely distributed or geographically concentrated ubiquitous systems. The Opera group pioneered event-based systems from the early 1990s, with applications including healthcare, pollution and transport monitoring, typically comprising multiple administration domains. Opera has carried out substantial work on access control, security and privacy for these and other applications, and the specification and enforcement of policy. Jean is a Fellow of the IEEE and BCS and was an IEEE Fellowship Committee member in 2008. She was a member of the Governing Body of IEEE-CS, 2001-2007; and founding EIC of Distributed Systems Online 2000-2008, IEEE's first online-only magazine. She is currently on the Editorial Board of Computer, IEEE-CS's flagship magazine.

Bhuvan Bamba is a Ph.D. student in the School of Computer Science, College of Computing at Georgia Institute of Technology. He received a Bachelor of Technology degree from Indian Institute of Technology, Madras, India in 2003 and Masters of Sciences in Computer Science from Georgia Institute of Technology in 2009. His current research is focused on location-based services and applications, distributed systems and privacy. His research experience in the industry with IBM and NTT covered areas in data management, information retrieval and information integration. His list of honors includes the Dean's Fellowship from Georgia Institute of Technology in 2006 and a best paper award from the ACM Conference on Information and Knowledge Management in 2005.

Rolando Blanco is a PhD candidate with the School of Computer Science, University of Waterloo. His current research focuses on the engineering of Distributed Event-based Systems, in particular, the modeling and verification of DEBSs and DEBS applications. Prior to joining the University of Waterloo, he worked as technical lead in several database and application integration projects. Before industry, he was research assistant with the Ocean Pollution Research Center (OPRC) at the University of Miami, FL. While at OPRC, and funded by NOAA and the USCG, he worked developing systems for managing atmospheric and oceanographic data. He was also a graduate research assistant with the Database Systems Group at Universidad de Los Andes, Bogota, Colombia, where, funded by Oracle Corp, his research focus was in the area of data distribution and replication.

Pruet Boonma is a second year Ph.D. student at the University of Massachusetts, Boston. His research interests include adaptive distributed systems and biologically-inspired wireless sensor networks. Mr. Boonma received his B.Eng. in Computer Engineering from Chiangmai University, Thailand, and M.IT. in Computer Science from Monash University, Australia. He worked for the Department of Computer Engineering, Chaingmai University, as a faculty member since 1997. In 2001, Mr. Boonma received Australian Government Scholarship (AusAID) for his master program in Australia. In 2005, he received Thai Government Scholarship for his PhD. work in the United States.

Andrey Brito received his diploma in Electronic Engineering in 2002 from the Federal University of Paraiba, Brazil. In 2004 he received his diploma in Computer Science and his Master Degree in Informatics from the Federal University of Campina Grande (at the Distributed Systems Lab), his Master thesis was focused on the project and implementation of failure detectors in distributed systems. Since April 2006 he has been working at the Systems Engineering Group at the Dresden University of Technology. His Ph.D. focuses on the usage of speculation to reduce processing latency in distributed event stream processing systems. In 2008 he received the OpenSPARC Community Innovation Award for his work with speculation in stream processing operators in multicore machines.

Sharma Chakravarthy is Professor of Computer Science and Engineering Department at The University of Texas at Arlington, Texas. He established the Information Technology Laboratory at UT Arlington in Jan 2000 and currently heads it. Sharma Chakravarthy has also established the NSF-funded, Distributed and Parallel Computing Cluster (DPCC@UTA) at UT Arlington in 2003. He is the recipient of the university-level "Creative Outstanding Researcher" award for 2003 and the department level senior outstanding researcher award in 2002. He is an associate editor of IEEE TKDE (Transactions of Knowledge and Data Engineering). He has served as financial chair for ICDE 2008, corporate sponsors chair for SIGMOD 2000, Publications chair for VLDB 2000, and as PC co-chair for ODBASE 2009. He is well known for his work on semantic query optimization, multiple query optimization, active databases (HiPAC project at CCA and Sentinel project at the University of Florida, Gainesville), and more recently scalability issues in graph mining and its applications. His group at UTA has developed DBSubdue and DB-FSG – scalable versions of corresponding approaches for graph mining, and InfoSift – a classification system for text, email, and web that uses graph mining techniques. His current research includes information retrieval, all aspects of database research, web technologies and ranking, stream data processing, complex event processing, mining and knowledge discovery – association, graph and text, push/pull technologies, web content monitoring, and information integration. He has published over 150 papers in refereed international journals and conference proceedings. He has given tutorial on a number of database topics, such as graph mining, database mining, active, real-time, distributed, object-oriented, and heterogeneous databases in North America, Europe, and Asia. He is listed in Who's Who Among South Asian Americans and Who's Who Among America's Teachers. He has published a book on Stream Data Processing, published by Springer in April 2009. He has been a consultant and is on the advisory board of companies and educations institutions. Prior to joining UTA, he was with the University of Florida, Gainesville. Prior to that, he worked as a Computer Scientist at the Computer Corporation of America (CCA) and as a Member, Technical Staff at Xerox Advanced Information Technology, Cambridge, MA. Sharma Chakrvarthy received the B.E. degree in Electrical Engineering from the Indian Institute of Science, Bangalore and M.Tech from IIT Bombay, India. He worked at TIFR (Tata Institute of Fundamental Research), Bombay, India for a few years. He received M.S. and Ph.D degrees from the University of Maryland in College park in 1981 and 1985, respectively.

Mani Chandy received his bachelors from IIT Madras, a Master's from the Polytechnic University of New York and a PhD from MIT. He has worked for Honeywell and IBM; was a co-founder of iSpheres, an early company in the event-processing space (1998); and has consulted for several companies. He was a professor at the University of Texas at Austin from 1970 to 1987, and has been a professor at the California Institute of Technology since then. He is a member of the U.S. National Academy of Engineering and has received several awards. He has a blog on sense and respond systems at http://sense-andrespond.blogspot.com and he has coauthored a book in 2009 called "Event Processing: Designing IT Systems for Agile Companies."

Alex Cheung's research interests center around overlay routing, load balancing, load optimization, and security in distributed publish/subscribe systems. His M.A.Sc research involved building a dynamic load balancing algorithm for heterogeneous content-based publish/subscribe messaging systems. In his spare time he enjoys playing computer games and learning Japanese and Korean languages.

Daniel Cutting's research interests include operating systems, distributed data structures and P2P networks. He is particularly excited by the potential of ubiquitous mobile devices that can extend the fringes of massively distributed systems. He earned his Ph.D. in 2007 at the University of Sydney under the supervision of Drs. Aaron Quigley and Björn Landfeldt. He has published several articles in leading journals, conferences and workshops and was nominated for the Australian Computer Society Distinguished Dissertation award. Daniel has worked in both Sydney and London in a variety of industry roles. Since 2007, he has been the discrete messaging architect for Livestation, a live P2P television distribution platform.

Wolfgang De Meuter is a Professor at the Vrije Universiteit Brussel and also heads the ambient-oriented programming group of the University's Programming Technology Laboratory (PROG). He has been active in the world of object-orientation since the early nineties. His research interests include programming languages and their evaluators, aspect-oriented programming, meta-programming and more recently also programming language constructs for ambient-oriented systems. He has organized numerous successful workshops at previous ECOOP and OOPSLA conferences. Recently, in 2008, he won the prestigious AITO Dahl-Nygaard Junior Prize awarded to researchers who have made significant technical contributions to the field of object-orientation. His coordinates can be found at http://prog.vub.ac.be/~wdmeuter/WolfHome.

Myungcheol Doo is a Ph.D. student at College of Computing, Georgia Institute of Technology and also a member of Distributed Data Intensive Systems Lab. He received the B.S. degree in Computer Science Education from Korea University, and the M.S. degree in C.S. from Georgia Institute of Technology. His research interests include location-based systems, indexing algorithms for location objects, spatial database, and mobility model. You can contact him by mcdoo@cc.gatech.edu.

Schahram Dustdar is Full Professor of Computer Science with a focus on Internet Technologies heading the Distributed Systems Group, Institute of Information Systems, Vienna University of Technology (TU Wien) where he is director of the Vita Lab. He is also Honorary Professor of Information Systems at the Department of Computing Science at the University of Groningen (RuG), The Netherlands. He is Chair of the IFIP Working Group 6.4 on Internet Applications Engineering and a founding member of the Scientific Academy of Service Technology. More information can be found at http://www.infosys. tuwien.ac.at/Staff/sd.

Opher Etzion is IBM Senior Technical Staff Member, and Event Processing Scientific Leader in IBM Haifa Research Lab, Previously he has been lead architect of event processing technology in IBM Websphere, and a Senior Manager in IBM Research division, managed a department that has performed one of the pioneering projects that shaped the area of "complex event processing". He is also the chair of EPTS (Event Processing Technical Society). In parallel he is also an adjunct professor at the Technion – Israel Institute of Technology. He has authored or co-authored around 70 papers in refereed journals and conferences, on topics related to: active databases, temporal databases, rule-base systems, complex event processing and autonomic computing, and co-authored the book "Temporal Database – Research and Practice", Springer-Verlag, 1998. Prior to joining IBM in 1997, he has been a faculty member and Founding Head of the Information Systems Engineering department at the Technion, and held professional and managerial positions in industry and in the Israel Air-Force.

David Eyers is a post-doctoral research associate at the University of Cambridge Computer Laboratory, UK. His PhD, completed at the University of Cambridge, explores the use of distributed event-based middleware to effect, and to support, dynamic access control and workflow systems. As a member of the Opera research group, he has ongoing research interests in the application of event-based middleware to wide-area, multi-domain, distributed systems. The unique security requirements of healthcare systems have been a running theme in his research work. His previous education includes a B.E. / B.Sc. (Hons) from the University of New South Wales, in Sydney, Australia.

Robert Fach has a diploma degree in Computer Science from the Dresden University of Technology. He continued as a Ph.D. student and is working as a researcher at the Systems Engineering group in the Department of Computer Science. His research interests include real-time and large-scale distributed systems with current focus on large-scale fault-tolerant stream processing systems.

Christof Fetzer received his diploma in computer science from the University of Kaiserlautern, Germany and his Ph.D. from UC San Diego. As a student he received a two-year scholarship from the DAAD, won two best student paper awards (SRDS and DSN) and was a finalist of the 1998 Council of Graduate Schools/UMI distinguished dissertation award. He also won an IEE mather premium in 1999. Dr. Fetzer joined AT&T Labs-Research in August 1999 and since April 2004 he has an endowed chair (Heinz-Nixdorf endowment) in Systems Engineering in the Computer Science Faculty at the Dresden University of Technology. Prof. Dr. Fetzer has published over 70 research papers.

Avigdor Gal is an Associate Professor at the Faculty of Industrial Engineering & Management at the Technion. He received his D.Sc. degree from the Technion in 1995 in the area of temporal active databases. He has published more than 80 papers in journals (e.g. Journal of the ACM (JACM), ACM Transactions on Database Systems (TODS), IEEE Transactions on Knowledge and Data Engineering (TKDE), ACM Transactions on Internet Technology (TOIT), and the VLDB Journal), books (Temporal Databases: Research and Practice) and conferences (ICDE, ER, CoopIS, BPM) on the topics of data integration, temporal databases, information systems architectures, and active databases. Avigdor is a steering committee member of IFCIS, a member of IFIP WG 2.6, and a recepient of the IBM Faculty Award for 2002-2004. He is a member of the ACM and a senior member of IEEE.

Dieter Gawlick joined IBM in 1968. As member of the IMS development team he proposed, architected, and implemented products that enabled high-end transaction technology. Core database/transaction technologies such as 2phase commit, group-commit, partitioning, data replication, online utilities, escrow technology, and hot standby are among the achievements of this effort. At Amdahl, Dieter worked on I/O related problems. He developed methods for the usage of electronic data storage; at Digital, Dieter developed the first workflow manager with full integration into database and transaction technology. Dieter joined Oracle in 1994. He architected Oracle/AQ (Advanced Queuing), an integral part of the Oracle database, and was a key contributor to Oracle's integration and sensor technologies. Dieter's current focus is leveraging and evolving database technologies to accelerate the evolution of event processing.

Hans-Gerhard Gross received an MSc in Computer Science (1996) from the University of Applied Sciences, Berlin, Germany, and a PhD in Software Engineering (2000) from the University of Glamorgan, Wales, UK. Following his PhD, Dr. Gross joined the Fraunhofer Institute for Experimental Software Engineering in Kaiserslautern, Germany, where he was responsible for a number of public research projects, devising software testing strategies, and for consulting projects with major German software organizations. Since 2005, Dr. Gross is employed as Assistant Professor at Delft University of Technology, The Netherlands. His research interests encompass all phases of software development, in general, and software testing, in particular.

Vera Goebel is professor in the Distributed Multimedia Systems group at the Department of Informatics of the University of Oslo, Norway. She obtained a PhD degree from the University of Zurich, Switzerland in 1994 and an MSc from the University of Erlangen-Nuremberg, Germany in 1989. Her research interests are Distributed Systems, Database Systems, Middleware, Operating Systems, and the Future Internet.

Alberto González born in Valladolid, Spain, is a Computer Science PhD Student under the supervision of Prof. Dr. Hans-Gerhard Gross, in the Software Technology Department of the Delft University of Technology, The Netherlands. His main research interest is Runtime Testing and Testability of dynamic component-based architectures. He received his BSc in Computer Engineering in 2005, and his MSc in Computer Science in 2007, from the University of Valladolid, Spain. During his Master's Thesis he worked on automatic error detection on embedded software based on dynamic invariants. This work was performed in the Embedded Systems Laboratory of the Delft University of Technology as an Erasmus exchange student. He has also worked in industry in the IT department of the parliament of Castilla y Leon, Spain.

Song Guo holds a First Class B.Eng degree in Automation & electronics at the Liaoning Shihua University in China. Song gained an Advanced MSc as student on the Telecommunications and Internet Systems course in the department of Computing & Information Engineering at the University of Ulster in the UK. After completing his M.Sc. in Sept 2004 Song moved to Trinity College Dublin as a Ph.D. candidate, researching how flexible a semantic interoperability service can be achieved dynamically in Knowledge-based Networks. Song currently has six publications in the area of mobile wireless networking, autonomic network and ubiquitous computing. Song submitted his Ph.D. thesis entitled "Using Semantic Mapping for Semantic based Publish/Subscribe Systems" in June 2009.

Ethan Hadar is a VP for Research at CA Labs, CA Inc. His responsibilities include leading strategic research in architecture, design and modeling, in collaboration with R&D groups and academia. Ethan's recent research interests include software engineering, architecture and design, SOA; Design Tools and Techniques, Architecture Centric Evolution (ACE); Object Oriented Design (OOD); Knowledge Management Architecture; Multidisciplinary Applications in ITIL; Management of Computing and Information Systems; and "Time to Value" projects that accelerate the assimilation of tools within CA customers' sites. Prior to joining CA, Ethan was the principal architect at Mercury-HP Software, where he developed new methodologies in software engineering, service oriented architectures, and object oriented technology. Ethan has over 18 years of experience in consulting and mentoring R&D teams on topics related to software architecture and product design. He holds a Ph.D. from the Department of System Analysis and Operations Research at the Technion.

Jani Heikkinen received his M.Sc. degree in Computer Science from the Helsinki University of Technology (TKK), Finland. He has participated in WeSAHMI and Trustworthy Internet research projects funded by the Finnish National Technology Agency Tekes and industry. His research interests include distributed software systems and mobile communications, with an emphasis on information systems security, privacy, and identity management.

Hans-Arno Jacobsen holds the Bell University Laboratories Chair in Software, and he is a faculty member in the Department of Electrical and Computer Engineering and the Department of Computer Science at the University of Toronto, where he leads the Middleware Systems Research Group. His principal areas of research include the design and the development of middleware systems, distributed systems, and information systems. Arno's current research focus lies on publish/subscribe, content-based routing, event processing, and aspect-orientation.

Zbigniew Jerzak received his M.Sc. degree in Computer Science from the Silesian University of Technology, Department of Computer Science. He is currently finishing his Ph.D. studies in the Systems Engineering group, led by Prof. Christof Fetzer, Ph.D., at the Dresden University of Technology. His research interests include real-time and large-scale distributed systems. Zbigniew Jerzak is the author of the XSiena family of publish/subscribe systems. His work has been supported by the Polish Ministry of Science and Higher Education grant number N N516 375034.

Dominic Jones is a Ph.D. candidate in the Knowledge and Data Engineering Group, Trinity College, Dublin. Dominic received a B.Sc. in Computer and Network Technology, with First Class honours, from Manchester Metropolitan University in June 2005. He went on to receive a M.Sc. in Computer

Science, from Trinity College Dublin, with a focus on Networks and Distributed Systems in October 2006. He is currently in the final stages of a Ph.D. at Trinity where he is researching the management of semantic clustering in Knowledge-based Networks. Dominic currently has nine publications in the area of Distributed Event-based Systems, Policy-based Network Management and Semantic clustering.

Reza Sherafat Kazemzadeh received his BSc degree from Sharif University of Technology in 2004. He finished his MASc degree in Software Engineering at McMaster University in 2006. Since then, he has been working towards his PhD in the Electrical and Computer Engineering department at University of Toronto. The focus of his dissertation is fault-tolerance and recovery of large distributed Publish/Subscribe messaging middleware. His other research interests include content-based routing, reliable messaging platforms, and reliability and load balancing of distributed systems and services. Since 2006, he is a member of the Middleware Systems Research Group (MSRG) at University of Toronto.

John Keeney is a Research Fellow with the Knowledge and Data Engineering Group (KDEG) in the Department of Computer Science of Trinity College, Dublin (TCD). His research focuses on the use of semantics in the management of autonomic adaptable systems, particularly networking and telecoms systems. John graduated from Trinity College, Dublin in 1999 with an undergraduate degree in Computer Engineering (B.A.I, B.A). His Ph.D. thesis, entitled "Completely Unanticipated Dynamic Adaptation of Software", was completed in 2004. Dr. Keeney has published in excess of 28 papers in significant journals, conferences and workshops.

Holger Klus holds a Diploma in computer sciences and is a PhD student at the Software Systems Engineering chair at Clausthal University of Technology. His main research interests are mobile and context-aware applications. He gained experiences in this area by developing middleware solutions and applications in several research projects and together with industrial partners. Currently he works in the OPEN project (http://www.ict-open.eu/), a project which is funded by the European Union. It aims to develop an environment, which provides people with the ability to continue to perform their tasks when they move about and change their interaction device.

Philipp Leitner has a BSc and MSc in business informatics from Vienna University of Technology. He is currently a PhD candidate and university assistant at the Distributed Systems Group at the same university. Philipp's research is focused on middleware for distributed systems, especially for SOAP-based and RESTful Web services. Additionally, he has done work in the area of P2P computing, network management and security of distributed systems. More information can be found at http://www.infosys.tuwien.ac.at/Staff/leitner.

David Lewis has 16 years R&D experience in academia and industry, with over 100 publications. He gained his PhD from University College London, where he worked for 12 years before moving to TCD in 2002. He has an international research reputation in the engineering of open distributed systems for network and service management and for the knowledge-driven engineering of Autonomic Pervasive Computing and Communication systems. He is on the editorial board of Elsevier's Journal of Network and System Management, as well as serving on the programme committees of numerous conferences and reviewing for journals in the management systems and autonomic systems areas (ICAC, Autonomic Networking, EASE, MACE). He has worked as a senior consultant to a Danish SME developing network management platforms and has contributed to international standards bodies such as the ITU-T Study Group 12 and the TeleManagement Forum in the area of management system modelling.

Guoli Li is a Ph.d candidate at Computer Science department in University of Toronto. Her research focus on "historic data access in Publish/Subscribe". Guoli received her master degree in computer science from University of Toronto in 2005. She also received her Master degree in electronic engineering from Xi'an jiaotong university in 2002. Guoli finished her bachelor's degree in Inforamtion Techniques from Xi'an Jiaotong University in 1999.

Ling Liu is a full Professor in the School of Computer Science, College of Computing, at Georgia Institute of Technology. There she directs the research programs in Distributed Data Intensive Systems Lab (DiSL), examining various aspects of data intensive systems with the focus on performance, availability, security, privacy, and energy efficiency. Dr. Liu has published over 250 International journal and conference articles in the areas of databases, data engineering, and distributed computing systems. She is a recipient of the best paper award of ICDCS 2003, WWW 2004, the 2005 Pat Goldberg Memorial Best Paper Award, and the best data engineering paper award of Int. conf. on Software Engineering and Data Engineering 2008. Dr. Liu served on the editorial board of IEEE Transactions on Knowledge and Data Engineering (TKDE) and International Journal of Very Large Databases (VLDBJ) from 2004 – 2008. She is currently on the editorial board of several international journals, including Distributed and Parallel Databases (DAPD, Springer), IEEE Transactions on Service Computing (TSC), and Wireless Network (WINET, Springer). Dr. Liu's current research is primarily sponsored by NSF, IBM, and Intel.

Balasubramaneyam (Maniy) Maniymaran is a research assistant in the middleware system research group (MSRG) at University of Toronto. Maniy was born in Sri Lanka and moved to Canada in 2001 to pursue graduate studies. He designed and implemented an optimized resource allocation algorithm for utility computing in his Masters research and his Ph.D. work presented a scalable, distributed resource discovery scheme for peer-to-peer networks, that can discover resources based on resource attributes, network location, and bandwidth. He is currently working on designing a universal language and matching engine for complex event processing. He is also a member of the event processing technical society (EP-TS.) His personal interests vary from bicycling to photography.

Anton Michlmayr received the MSc degree in computer science from Vienna University of Technology in April 2005. He is currently a PhD candidate and university assistant in the Distributed Systems Group at Vienna University of Technology. His research interests include software architectures and middleware for distributed systems with an emphasis on service-oriented architectures and distributed event-based systems. More information can be found at http://www.infosys.tuwien.ac.at/Staff/michlmayr.

Vinod Muthusamy is a PhD candidate in the Middleware Systems Research Group in the Department of Electrical and Computer Engineering at the University of Toronto. His research interests include publish/subscribe systems and distributed workflow processing. Vinod holds a BASc degree from the University of Waterloo and an MASc degree from the University of Toronto.

Dirk Niebuhr holds a Diploma in computer sciences and is a PhD student at the Software Systems Engineering chair at Clausthal University of Technology. His main research interests are means to provide dependable dynamic adaptive systems. A middleware designed especially for these systems - the Dynamic Adaptive System Infrastructure DAiSI - has been developed under his lead. He carried out the technical lead of the middleware workpackage within the BelAmI project (http://www.belami-project.org/). Furthermore he has been responsible for the tailoring aspects during the development of the V-Modell XT (http://www.v-modell-xt.de/).

Declan O'Sullivan is the Director of the Knowledge and Data Engineering Group in TCD and has over 20 years research and development experience in both industry and academia. He holds a Phd, MSc and BA (Mod) in Computer Science from TCD. His particular research interest is in knowledge driven approaches to achieving semantic interoperability, especially applied to the network and service management of distributed networks. During his time in industry, he was involved in industry fora such as the TeleManagement Forum and the Object Management Group (OMG). He has over 70 publications, and has contributed to several organising and programme committees (such as DSOM, LCN, MACE).

Peter Pesti is a PhD student in the College of Computing at the Georgia Institute of Technology. He received his MS in Computer Science from Georgia Tech, and his MSc in Technical Informatics from the Budapest University of Technology and Economics, both in 2006. His current research interests include scalability issues of location-based, mobile and distributed systems, and all things geospatial. In the past, he has worked on biometric identification using blink patterns, and corpus-based speech synthesis for the Hungarian language.

Guanhong Pei is a Ph.D. student in Electrical and Computer Engineering (ECE) at Virginia Tech, Blacksburg, Virginia, USA, from Aug, 2006. Before that, he received his B.S. in ECE from Shanghai Jiao Tong University (SJTU), Shanghai, China in June 2006. He also received his second-major bachelor's degree in Business Administration from Shanghai Jiao Tong University in 2006. Guanhong's interest includes cross-layer design and algorithms, distributed networks system design, wireless spectrum allocation, and stochastic network optimization.

Éric Piel born in France, is currently post-doc in Software Technology Department of the Delft University of Technology (The Netherlands) working in the Poseidon project, in partnership with Thales Nederland. The subject of the research is to ease and improve the integration of large-scale component-based systems. Previously, he received his PhD in 2007 at INRIA Lille (France) on the the subject of embedded system specification and model transformations, and his engineer diploma in Computer Science at the University of Technology of Compiègne (France) in 2003. He has also worked in the industry in the R&D department of Bull on the subject of mixing Real-Time capabilities and parallel processing.

Thomas Plagemann is Professor at the University of Oslo since 1996. Currently, he leads the research group in Distributed Multimedia Systems at the Department of Informatics. He has successfully managed national and international projects, like INSTANCE, Ad-Hoc InfoWare and Midas; and he leads the Research WP in the Network-of-Excellence CONTENT. He has a Dr.SC degree from Swiss Federal Institute of Technology (ETH) in 1994 and received in 1995 the Medal of the ETH Zurich for his excellent Dr.Scient thesis. From 1994 to 1996 he was a researcher at the Center of Technology at Kjeller in Norway and at Telenor R&D. His research interests include protocol architectures and middleware solutions for multimedia communication and mobile systems, and the integration of communication protocols operating systems, and data management mechanisms. He has published over 100 papers in peer reviewed journals, conferences and workshops in his field. He serves as Associate Editor for ACM Transactions of Multimedia Computing, Communications and Applications and as Editor-in-Chief for the Springer Multimedia Systems Journal.

Aaron Quigley is a college lecturer in the School of Computer Science & Informatics, University College Dublin, a co-Principal Investigator of the CLIQUE SFI funded SRC, an IBM CAS visiting scientist, UCD director of ODCSSS, coordinator of the EU FP7 CAPSIL project, a researcher in Lero the Irish Software Engineering Research Centre and a collaborator in CLARITY, The Centre for Sensor Web Technologies. His research interests include pervasive computing, software engineering, information visualisation, human computer interaction, graph drawing, location and context awareness, peer-to-peer computing, surface interaction and network analysis. In 2009 he is the program co-chair for the 4th International Symposium on Location- and Context-Awareness, (LoCA 2009) in Tokyo Japan and in 2010 the workshop co-chair for Pervasive 2010 in Helsinki. He has published over 80 internationally peer-reviewed publications including edited volumes, journal papers, book chapters, conference and workshop papers along with 3 patents.

Andreas Rausch is the head of the chair for Software Systems Engineering at the Clausthal University of Technology. Until early 2007 he was head of the chair for Software Architecture at the University of Kaiserslautern. In 2001 he obtained his doctorate at the University of Munich under Prof. Dr. Manfred Broy. His research in the field of software engineering focuses on software architecture, model-based software engineering and process models, with more than 70 publications worldwide. Prof. Dr. Andreas Rausch is project leader for the development of the new V-Modell XT, the standard system development process model for IT systems of the Federal Republic of Germany. In addition to his research activities he participated in various commercial software projects developing large distributed systems. He is one of the four founders and shareholders of the software and consulting company 4Soft GmbH, Munich.

Binoy Ravindran is an Associate Professor in the Department of Electrical and Computer Engineering at Virginia Tech, Blacksburg, Virginia, USA. His research interests include real-time and embedded systems, operating systems, distributed systems, and network algorithms and protocols. He has published over 150 papers on these topics. His research has also resulted in experimental software systems (e.g., real-time middleware, timing analysis toolkits), many of which have also been transitioned to US DoD programs. Dr. Ravindran's 2006 ACM Design, Automation, and Test in Europe (DATE) conference paper was selected as one of The Most Influential Papers of 10 Years of DATE. He is an Associate Editor of ACM Transactions on Embedded Computing Systems and was a Co-Guest-Editor for IEEE Transactions on Computers. Dr. Ravindran currently serves as an ACM Distinguished Speaker and as an IEEE Distinguished Visitor.

Florian Rosenberg is currently a university assistant in the Distributed Systems Group at the Vienna University of Technology. He received his PhD in June 2009 with a thesis on "QoS-Aware Composition of Adaptive Service-Oriented Systems". His general research interests include service-oriented computing and software engineering. He is particularly interested in all aspects related to QoS-aware service composition and adaptation. More information can be found at http://www.infosys.tuwien.ac.at/Staff/rosenberg.

Jatinder Singh is a research student at the University of Cambridge, currently completing his Ph.D. thesis entitled "Event-Based Data Control in Healthcare". His research interests surround data privacy, particularly concerning infrastructure for the wide-scale dissemination of personal information. He has industry experience in the design and development of systems supporting health and judicial services. He holds a B.Sc.(Hons) from the University of Western Australia, where he has also studied Law.

Katrine Stemland Skjelsvik is post.doc in the Distributed Multimedia Systems group at the Deparment of Informatics of the University of Oslo, Norway. She obtained a PhD degree from the same department in April 2008 and an MSc in 2002. Her reasearch interest are mobile middleware, event notification services and pervasive systems. Currently she is working on a project related to middleware services for pervasive applications.

Junichi Suzuki received a Ph.D. in computer science from Keio University, Japan, in 2001. He joined the University of Massachusetts, Boston in 2004, where he is currently an assistant professor of computer science. From 2001 to 2004, he was with the School of Information and Computer Science, the University of California, Irvine (UCI), as a postdoctoral research fellow. Before joining UCI, he was with Object Management Group Japan, Inc., as Technical Director. His research interests include wireless sensor networks, autonomous adaptive distributed systems, biologically-inspired software designs and model-driven software/performance engineering. In these areas, he has authored two books and published over 95 refereed papers including five award papers. He has chaired or co-chaired four conferences including ICSOC'09, and served on program committees for over 55 conferences including AINA, ICCCN, SECON, SASO, CEC, BIOSIGNALS and BIONETICS. He is an active participant and contributor in ISO SC7/WG19 and Object Management Group, Super Distributed Objects SIG. He is a member of IEEE and ACM.

Sasu Tarkoma received his M.Sc. and Ph.D. degrees in Computer Science from the University of Helsinki, Department of Computer Science. He has been recently appointed as full professor at University of Helsinki, Department of Computer Science; he is also currently professor at Helsinki University of Technology, Department of Computer Science and Engineering. He has managed and participated in national and international research projects at the University of Helsinki, Helsinki University of Technology, and Helsinki Institute for Information Technology (HIIT). He has worked in the IT industry as a consultant and chief system architect, and he is principal member of research staff at Nokia Research Center. He has over 100 publications, and has also authored two recent books.

Richard Tibbetts is Chief Technology Officer and co-founder of StreamBase Systems, a leading CEP vendor. Richard provides technical leadership for the company and leads architecture design for StreamBase's Event Processing Platform. Richard is also responsible for furthering new StreamBase capabilities such as StreamBase's 'white-box' application frameworks. As CTO, Richard directs the next-generation of StreamSQL, the event programming language developed by Richard and the StreamBase team, which applies the benefits of SQL for stored data to real-time transitory data. Richard earned both his BS in Computer Science and Engineering, and Masters of Engineering Degrees at MIT. His graduate work included the Aurora project, which developed into StreamBase, and the Linear Road Benchmark for performance of event processing systems.

Tom Van Cutsem is a researcher at the Programming Technology Laboratory (PROG) of the Vrije Universiteit Brussel, Belgium. He recently finished his PhD dissertation on object designation in mobile ad hoc networks. His PhD research focussed on the topics described in this chapter, such as Ambient-Talk and ambient references. His broad research interests include programming language design and implementation, distributed programming and reflective architectures. He is a post-doctoral fellow of the Research Foundation, Flanders (FWO). His coordinates can be found at http://prog.vub.ac.be/~tvcutsem.

Jilles van Gurp received a licentiate degree from the Blekinge Institute of Technology (2001) in Sweden and a Ph. D. Degree from the University of Groningen in the Netherlands (2003). Since 2005 he has worked in Nokia first as a industrial researcher in the Nokia Research Center in Helsinki Finland and currently as a software architect in Nokia Gate5 in Berlin. During his research career he has published over 30 peer reviewed articles in journals, conference & workshop proceedings on topics including object oriented frameworks, software architecture, software product lines and pervasive computing.

Matt Weber is currently a PhD student in the College of Computing at the Georgia Institute of Technology. He received his BS in Computer Science from Georgia Tech in 2007. His current research interests include map matching, secure localization and indoor positioning. Matt also enjoys mobile and Location Based Service development.

Eiko Yoneki is a Post Doctoral Research Associate in the University of Cambridge Computer Laboratory, Systems Research Group, working on the Haggle Project (EU-FP6) and the SOCIALNETs project (EU-FP7). Haggle is a new autonomic networking architecture designed to enable communication in the presence of intermittent network connectivity, which exploits autonomic opportunistic communications. Her research interests are distributed systems over mobile/wireless networks including delay tolerant networks, bio/socio-inspired networks, wireless sensor networks, event-driven systems, event correlation, and data synchronization. She has received her Ph.D. degree from the University of Cambridge in December, 2006 (ECCO: Data Centric Asynchronous Communication) and a Postgraduate Diploma in Computer Science from the University of Cambridge in 2002. Previously, she has spent several years with IBM (US, Japan, Italy and UK) working on various networking products. She is a member of ACM and IEEE Computer Society. Contact her at eiko.yoneki@cl.cam.ac.uk.

Index